Homology, Genes, and Evolutionary Innovation

Homology, Genes, and Evolutionary Innovation

Günter P. Wagner

PRINCETON UNIVERSITY PRESS

Princeton and Oxford

press.princeton.edu

The epigraph at the beginning of chapter 8 is reprinted by permission from
Macmillan Publishers Ltd: Nature Reviews Genetics, "The Evolution of Cell
Types in Animals: Emerging Principles from Molecular Studies,"
by Detlev Arendt, vol. 9, number 11, copyright 2008.

Cover design by Pamela Lewis Schnitter

First paperback printing, 2018
Paper ISBN: 978-0-691-18067-0
The Library of Congress has cataloged the cloth edition as follows:

Wagner, Günter P., 1954–
Homology, genes, and evolutionary innovation / Günter P. Wagner.
pages cm
Includes bibliographical references and index.
ISBN 978-0-691-15646-0 (hardcover : alk. paper) 1. Developmental genetics.
2. Evolution (Biology) 3. Genetic regulation. I. Title.
QH430.W335 2014
571.8'5—dc23
2013025386

British Library Cataloging-in-Publication Data is available

This book has been composed in Minion Pro and Myraid Pro

Printed on acid-free paper ∞

Printed in the United States of America

To my parents and all who,

through the years,

have sustained me

by their love and kindness,

their mentorship and encouragement.

CONTENTS

PREFACE

This book is the product of an intellectual journey that began in the spring term of 1972 when I first heard a talk given by my future dissertation mentor, Rupert Riedl. I was then a junior in a chemical engineering program, although my brother had encouraged me to attend a weekly seminar series on evolution given at the University of Vienna. Riedl was one of the speakers and had just returned from a five-year stint at the University of North Carolina. He had brought with him a manuscript of his major work on evolution, *Order in Living Organisms* (German: *Die Ordnung des Lebendigen*, 1975; English edition 1978). In that book as well as the seminar I attended, Riedl outlined a theory of evolutionary change that addressed many of the questions that, at the time, had been neglected in mainstream evolutionary theory. He laid out a vision of evolutionary biology that relied heavily on developmental considerations and anticipated the agenda of what later would become known as evolutionary developmental biology or, perhaps better, devo-evo. He also argued that the key to understanding macroevolutionary patterns was the origin and persistence of homologous body parts. Homology, the correspondence of body parts in sometimes distantly related organisms, is the key to understanding what the evolution of complex body plans is about. I immediately realized that this was an intellectual challenge that, at the very least, would keep me entertained for a lifetime.

This early fascination with questions of morphological evolution should have made me become a comparative anatomist, although that is not what I became. Even though I have a handful of strictly anatomical papers to my credit, the bulk of my work is in areas that are seemingly unrelated to morphology, including population genetic theory, molecular evolution, developmental evolution and, more recently, the molecular biology of gene regulation. This came about because, as I was taught by Riedl, the questions raised by the patterns of morphological evolution are broader and deeper than the morphological patterns themselves. They raise questions regarding developmental constraints and evolvability, the evolution of gene regulatory networks, and many others. To take these problems seriously, the researcher is drawn to various fields that seem to be far afield from traditional morphology. And so is the narrative offered in this book.

It may seem strange to some to find chapters with a conceptual focus and even a philosophical focus side-by-side with chapters on gene regulation and transposable elements or transcription factor biochemistry. However, I could not see a way of talking about these issues by leaving out one or the other of these directions: the conceptual or the molecular and mechanistic. Biology is the science of a holistic phenomenon (life) and even any reasonably complex aspect of biology tends to require knowledge in several fields. Any other way would shortchange the subject of study.

There is no broad consensus, nor even a narrow consensus on the subject of homology and its mechanistic foundations. For this reason and by necessity, the perspective developed in this book is a very personal one derived from decades of thinking and research on this topic and topics related to it in various ways. I draw mostly on papers that influenced me at certain stages of my intellectual development, and the research done in my lab and with collaborators. Thus, the references I have cited are highly biased toward these sources. It would have been impossible for me to do justice to the sprawling literature that underpins the many areas of research on homology, developmental evolution, the evolution and molecular biology of gene regulation, population genetics of modularity, and many more. I tried to be fair to differing views, although I am certain that I failed in many instances and I apologize for any oversights and misrepresentations. I can only hope that vigorous debate will ultimately correct these flaws in my presentation.

Every biologist will agree that homology is a confused and confusing subject. To forge a view on this subject that reduced the confusion, at least to the satisfaction of this author if nobody else's, was a feat that depended on the contributions of and the interactions with a great many people, beginning with my doctoral and postdoctoral mentors Rupert Riedl (Vienna), Peter Schuster (Vienna), Joachim Wolff (Göttingen), and Alfred Gierer (Tübingen). This also includes colleagues and students throughout the last decades with whom I had the privilege of working in biological research and teaching. To be comprehensive would necessitate a very long list. However, I must mention a few who, lately, have been wonderful companions and sources of inspiration. I must acknowledge my colleagues at Yale: Frank Ruddle, Michael Donoghue, Richard Prum, Leo Buss, Steve Stearns, Jacques Gauthier, and Paul Turner. Colleagues and friends at other institutions with whom I shared part of the journey include James Cheverud, Ron Amundson, Roberto Romero, Reinhard Bürger, Benedikt Hallgrimmsson, Gerd Müller, Peter Stadler, Chris Amemiya, Walter Fontana, Joe Thornton, Lee Altenberg, Mihaela Pavličev, Ingo Brigandt, Alan Love, and many more. Of course, the closest associates are always former and current members of one's academic family, my students and postdocs, including Bernhard Misof, Andreas Wagner, Manfred Laubichler, Thomas Hansen, Hans Larsson, Joachim Hermisson, Karen Crow, Alex Vargas, Chi-hua Chu, Vincent (Vinny) Lynch, Mauris

Nnamani, Deena Emera, Jake Musser, Koryu Kin, and many others. I want to specifically acknowledge my colleagues who helped me by providing comments on the draft of this manuscript: Mihaela Pavličev, Thomas Hansen, Alessando Minelli, Peter Godfrey-Smith, Ingo Brigandt, and Koryu Kin. Their critical eyes and incisive questions, more often than not, helped me avoid premature convergence. My gratitude goes to all of those people, both mentioned and unmentioned by name, for their incredible generosity. Finally, I thank my editor at Princeton University Press, Alison Kalett, for her patience and guidance through the publication process of this project.

Günter P. Wagner
Cheshire, CT
March 2013

Homology, Genes, and Evolutionary Innovation

Introduction: What This Book Aims
to Do and What It Is Not

Homology, the correspondence of characters from different species or even within the same organism, is a fundamental concept in evolutionary biology and biology in general (Wake 1999). It is broadly recognized that homology is explained by derivation from a common ancestor that had the same character or trait. This explanation applies at least to characters from different species. Accordingly, this concept has applications in many fields of biology by referring to morphological characters, behaviors (Lorenz 1981; Prum 1990; Griffiths 1997; Scholes-III 2008), proteins and genes, as well as to gene regulatory networks (Abouheif 1999) and developmental mechanisms and processes (Bolker and Raff 1996; Gilbert and Bolker 2001). Each of these fields has its own conceptual and technical challenges. However, beyond the most general statement that homology is a hypothesis of descent from a common ancestor, little of substance can be said regarding all of these notions of homology. Hence, it would not be particularly useful to write a book that pertains to all these forms of homology. The issues are far too heterogeneous to allow a synthetic treatment, at least at this point in history. Consequently, this book is not meant to cover homology in all its breadth. Rather, it focuses on one class of homology relationships: that between morphological characters. I made this choice for the following reasons.

Not all of the various notions of homology face the same conceptual and biological challenges. For example, the notion of gene homology may, at times, be technically challenging to demonstrate; yet at the conceptual level, it is relatively straightforward. DNA is directly inherited by cells, by individuals, and by different generations. DNA is directly copied in that the copying process itself duplicates the relevant information-carrying structure[1] (i.e., the nucleotide sequence). There is even material overlap between the parental DNA and daughter DNA strands; that is, in the "semi-conservative replication" of DNA, one strand of the double helix of daughter DNA is physically one of the strands of parental DNA.

Ironically, one of the most difficult classes of homology relationships to explain is the one from which the homology concept originated: the homology of morphological characters (Owen 1848). There are two reasons why

---

[1] An exception are those species in which genes do not contain all of the information necessary for mRNA (i.e., species in which extensive RNA editing occurs, as in some ciliates).

the homology of morphological characters is harder to understand than that of genes. First is the obvious fact that morphological characters are not passed on from generation to generation by direct copying. In each generation, all morphological characters need to develop de novo from a fertilized or unfertilized egg cell (Wagner 1989). An obvious way forward would be to say that what is inherited are the genes necessary to engender the development of the character (Van Valen 1982). In other words, a morphological character is transmitted indirectly through the transmission of the genes that control the development of the character. This is certainly true, at least in the short term, that is, over several generations and likely within the lifetime of a single species. Over longer periods of evolutionary history, however, the correspondence between genes and character identity decreases.

There is mounting evidence that homologous characters from distantly related organisms, like grasshoppers and fruit flies, often use quite different genes for the development of clearly homologous characters, like insect body segments (see chapter 3). Hence, the identity of morphological characters cannot be explained by the *identity* of the set of genes that directs their development (Spemann 1915; DeBeer 1971; Roth 1984; Roth 1988; Wagner and Misof 1993; Hall 1994; Wagner 1994; Wray and Abouheif 1998; Hall 2003).

Because of these heterogeneous problems that are attached to the various uses of the homology concept, this book does not aim to cover all notions of homology. There will be no discussion of molecular homology, nor will there be a discussion on the homology of behavioral patterns or of physiological and developmental functions per se. Rather, the goal of this book is quite specific—namely, to propose a solution to the homology problem for morphological characters as sketched out in the previous paragraph. While homology of morphological characters can be viewed as being quite "retro," going back to the comparative anatomy of the eighteenth and nineteenth centuries, the contemporary context of these problems is all but retro. Homology of morphological characters became a major theme in evolutionary developmental biology during the last two decades, mostly due to progress made in evolutionary developmental biology (see references cited above).

The developmental genetic revolution, which began with the discovery of the homeobox genes, has deepened our understanding of how the development of morphological characters is controlled by genes. It has also deepened our understanding of how changes in the expression and activity of developmental genes underlie the evolution of developmental pathways and, thus, the evolution of morphological structures (Carroll, Grenier et al. 2001; Wilkins 2002). Given that we can now understand the evolution of morphological characters as a consequence of the modification of gene regulatory networks, it is time to revisit character identity and homology of morphological characters.

This book, although ostensibly about homology, is really a book on developmental evolutionary biology. The claim can be made that a mechanistic understanding of homology can become a unifying theme for evolutionary developmental biology, as well as any other branches of science that are concerned with the structure and development of organisms. I will make the argument that one of the main benefits of a deeper understanding of homology is that it enables an empirical research program on the major transitions in evolution, in particular the origin of evolutionary novelties. I argue that the origin of novel characters and novel body plans is one of the most important but least researched questions in evolutionary biology.

Novelty and major transitions underlie the broad patterns of biological diversity by explaining the differences between major groups of animals, plants, fungi, and the vast array of protozoans and microbes. Yet very limited progress has been made in this area compared to the rich knowledge we have acquired on speciation, behavioral evolution, and adaptation. However, during the last ten to twenty years, the situation has changed, and this book aims to synthesize this new knowledge into an account of the evolution of character identity and character origination.

The argument that understanding homology is essential for making progress in understanding evolutionary novelties is based on an analogy. Ideas regarding the nature of species are essential for setting the research agenda on the mechanisms of species origination. Whatever one may think of the biological species concept of Ernst Mayr and his contemporaries (Mayr 1942), it remains a historical fact that it enabled a rich, productive study of speciation. By conceptualizing species as collections of interbreeding populations that are reproductively isolated from other populations, it is clear that the objective of a research program on the origin of species has to be the question: how do barriers to sexual reproduction and, thus, to gene flow arise? Prior to that it was not clear what species are and, thus, it was not clear how they originated, or even what the question was. This was a failure of conceptual clarity that prevented progress in empirical research. Similarly, without an understanding of what characters are, we cannot determine how they originate.

Here I will argue that the key to understanding the nature of character identity is to ask what individuates body parts both developmentally and genetically. I will propose that the key is the existence of gene regulatory networks that *enable the expression of different developmental programs* in different parts of the body,[2] so-called Character Identity Networks (Wagner 2007). It turns out that these networks are evolutionarily more conserved

[2] Note that here I am talking about gene regulatory networks that enable differential gene expression, and not about the totality of the gene regulatory network that underlies a particular character. This helps to resolve the problem of conserved and variable parts of the development of homologous characters (see chapter 3).

than other parts of the developmental program, and are often rigidly associ-ated with the identity of the character that they enable. From this perspective, the evolutionary origin of characters and body plans is the origin of those gene regulatory networks that underlie character identity.

At its core, this is the argument made in the remainder of this book. It will be shown that this move requires a number of conceptual adjustments, including recognizing the difference between character identity and char-acter states (chapter 2). In addition, because certain conceptualizations of homology, as for example those made in the cladistic tradition of taxonomy, are incompatible with this program, certain adjustments in the metaphysical assumptions underlying evolutionary biology will have to be made (chapter 7). The complexity of this undertaking necessitated a book format, as this task could not have been achieved even in a long article. It is my hope that this book comes close to achieving this goal.

For really impatient readers, a one-page summary of the core conclusions of this book is provided in chapter 13 in the section entitled, "What Are the Core Claims of This Model of Homology?"

PART I

· · · · · · · ·

Concepts and Mechanisms

This book is divided into two parts that assume different perspectives on the problem of homology and character identity. The first part begins with the conceptual issues that surround the idea of homology and attempts to sort out these complications so that mechanistic facts can be productively connected to the homology concept. Thus, the first part is a mix of highly conceptual chapters along with chapters on developmental, genetic, and evolutionary mechanisms in an attempt to build a bridge between these two areas. The integrity of this bridge will come under scrutiny in the second part of this book where specific systems are discussed from the perspective of a mechanistic explanation of character identity and its origin.

Chapter 1 opens the discussion by reintroducing the so-called, and much maligned, typological thinking into evolutionary biology. Taking homology seriously inevitably leads one to a mode of thinking that was out of favor during most of the twentieth century (Mayr 1982). Although typological thinking originally emanated mostly from morphology and paleontology, it was soundly rejected during the heydays of the evolutionary synthesis. However, toward the end of the twentieth century, typological thinking was introduced, somewhat surprisingly, by one of the most progressive fields in biology at the time: developmental biology and, in particular, developmental genetics (Amundson 2005). Thus, this chapter will argue that typology naturally emerged from the facts of evolutionary developmental biology and it would be seriously problematic to try to avoid it.

Chapter 2 opens the debate regarding the conceptual status of homology. No lasting progress can be made in explaining body plan evolution without a thorough housecleaning. The problem is that many contradictory positions on homology made sense within the research programs in which they were introduced. The position to be developed in this chapter will aim at a mechanistic developmental and evolutionary explanation of body plan evolution and the origin of character identities. Clearly, this is not the only legitimate research program and the proposal put forward here is not one that claims to be the ultimate truth about homology.

What I am aiming at, however, is to develop the most productive notion of homology for the purposes of developmental evolution. Slightly differing uses of the term can easily be accommodated by minimal terminological adjustments. At the core of the current proposal is the distinction between character identity and character states, which at least in systematics is controversial. The next chapter will provide a developmental justification for this distinction.

Chapter 3 addresses the most challenging problem when attempting to explain character identity; namely, unquestionable homologies (i.e., character identities across species) are often associated with extensive variations in the developmental pathways and mechanisms that produce these characters. This problem has prevented establishing a mechanistic science of character identity since the beginning of experimental developmental biology with Spemann (Spemann 1915). In this chapter I propose to resolve this problem by observing that genes responsible for character identity determination vary much less between species than those responsible for positional information. This approach leads to the concept of character identity networks. Historical continuity of character identity networks (ChINs) is hypothesized to be the genetic basis of character identity and, thus, of homology.

The notions of character identity and of ChINs that are introduced in chapters 2 and 3 imply a re-definition of the problem of evolutionary novelties. A novelty is understood to be the evolutionary origin of a novel character identity and, thus, the evolution of a novel ChIN. Chapters 4, 5, and 6 will explore what is known regarding the developmental, genetic, and evolutionary mechanisms of character origination.

The last chapter in this section, chapter 7, returns to purely conceptual issues, even more abstract than those addressed in chapter 2. Reintroducing homology as a scientifically credible concept has implications for the metaphysics of evolutionary biology, that is, the question of whether such words as "characters" can refer to real things. This chapter will explore how the notions of class, individuals, and natural kinds relate to the conceptual proposal in this book. It will be argued that characters can be understood as natural kinds, if the latter notion is appropriately modified.

1

The Intellectual Challenge of

Morphological Evolution: A Case for

Variational Structuralism

Throughout the history of evolutionary biology, as well as many other sciences, there has been a conflict between two styles of thinking (Mayr 1982; Hughes and Lambert 1984; Ghiselin 1997; Amundson 2005). One is conventionally called functionalism, although in evolutionary biology the term "adaptationism" is more frequently used today because a trait's "functional fit for its office" is produced through adaptation by natural selection (i.e., function is explained by adaptation through natural selection). The functionalist stance is one that explains organismal traits through their functional and adaptive values.

The alternative style of thinking does not have a generic name in biology, although in other areas of study it is called "structuralist."[1] In evolutionary biology, the most influential structuralist manifesto is "The Spandrels of San Marco and the Panglossian Paradigm" by Gould and Lewontin (Gould and Lewontin 1978). In their essay, Gould and Lewontin attacked the adaptationist assumption that each feature of an organism had to have an adaptive explanation. The "spandrels"[2] of the *Basilica di San Marco* were not built to accept an angel picture; rather, they were a geometric consequence of a structure that was built with arches. Similarly, the protruding "chin" of the human lower jaw was a consequence of the progressive reduction of the tooth row relative to the body of the mandible rather than a directly selected character.

The differences in perspective between these two styles of thinking was best expressed by Rudy Raff in his quip, "They [the population geneticists]

[1] There is a fringe movement in evolutionary biology that identifies itself as structuralists (e.g., Goodwin 2009); however, their stance derives its claim to structuralism from a denial of natural selection, which is a position that I do not consider productive.

[2] A spandrel is the space between two arches or between an arch and a rectangular enclosure.

are interested in species while we [devo-evo researchers] are interested in bodies."[3] I hasten to add that this contrast is most often exemplified by the different perspectives of population *geneticists* and developmental biologists; however, there is no scientific or conceptual reason for a substantial conflict between *population genetics* and a structuralist view.

The tension between functionalist and structuralist views of nature is as pervasive as the tension between explanations based on nature or nurture, and probably as unnecessary and unproductive in the long term. With respect to the nature/nurture debate, any serious biologist or psychologist understands that organisms and their traits (including behaviors) are the product of the interactions between genes (nature) and the environment (nurture). Hence, the contrast "nature *or* nurture" does not make any sense, because we are always faced with the combined effects of both "nature *and* nurture."[4]

In contrast, the conflict between functionalist and structuralist accounts of organisms has not been resolved, and will likely be more difficult to resolve than that between nature and nurture. The reason is that form (i.e., structure) and function (i.e., adaptation) are intertwined during the life and evolution of organisms in a complex manner such that each party, both structuralists and functionalists, can make a legitimate claim to a certain amount of truth. This book is about a project to overcome this conflict by addressing a specific biological phenomenon for which the conflict often crystallizes: the question of homology (i.e., the existence of the same body parts in different and often distantly related organisms—aka homologs). At its core, the question is whether homologs exist—that is, whether they are natural members of the "furniture of the world" or whether they are only transient traces of the phylogenetic past. In the latter case, they would have no biological, conceptual, or causal significance. In the former case, homologs would have to play a central role among the concepts of evolutionary theory.

To some this question may seem quaint and irrelevant, as the dominant school of evolutionary thinking during the twentieth century was thought to have disposed of this topic for good. This was ostensibly achieved by excluding issues of body organization, development, morphology, and, for a long time, even phylogeny from the research program of the New Synthesis (Mayr 1982). Yet this situation has changed due to the maturation of phylogenetic inference (Donoghue 1992), as well as that of developmental biology and the subsequent emergence of developmental evolution as a new branch of biology (Carroll, Grenier et al. 2001; Wilkins 2002). In developmental

---

[3] I heard Rudy say this at the Werkman conference on "Form and Function" at Florida State University in 2005, although I cite Amundson (2005) here.

[4] Of course it is meaningful to distinguish genetic and environmental influences and measure them separately. This is done for example using the methods of quantitative genetics by separating genetic variance components from non-genetic components; see Lynch, M. and B. Walsh (1998). *Genetics and Analysis of Quantitative Traits*. Sunderland, MA, Sinauer Associates, Inc.

evolution, questions regarding the nature of homology have resurfaced and have come into sharp focus (Hall 1994). Our ability to identify those genes responsible for the development of morphological characters opened up new avenues to address questions that were, for the most part, abandoned at the end of the nineteenth century (Abouheif 1997; Wray and Abouheif 1998; Abouheif 1999; Wang, Young et al. 2011). In the nineteenth century, however, the program of evolutionary morphology collapsed under the weight of these questions, which, at the time, where unmatched by scientific methodology (Nyhart 1995; Nyhart 2002).

After *Origin of Species* (1859) was published, the study of comparative anatomy experienced a revolution in which comparisons of anatomical structures were pursued in order to infer the phylogenetic relationships among species and higher taxa, and to explain the evolutionary origin of body plans and the characters that make up the bodies of animals and plants (Nyhart 1995; Laubichler and Maienschein 2003). However, at that time, there was neither a proper comparative methodology to infer phylogenetic relationships nor was there another source of phylogenetic information other than morphology, such as that provided now by molecular sequence data. Also, any understanding regarding the developmental mechanisms that create morphological characters during embryogenesis was virtually absent, and genes had not been (re)discovered. In retrospect, the program of evolutionary morphology was premature.

This situation, however, has changed dramatically during the last 20 to 30 years. Data at the molecular level have greatly improved the amount and quality of phylogenetic information. Conceptual progress and the availability of cheap computing power have vastly increased our ability to infer phylogenetic relationships. The discovery of conserved developmental genes made it possible to compare the development of non-model organisms with that of model organisms (Shubin, Tabin et al. 1997; Wilkins 2002), and high-throughput methods make high-quality molecular information available for a wide variety of species. The style and power of research are now fundamentally different from the situation 150 years ago. Thus, it is time to re-consider the questions that, at that time, were beyond the reach of any rational research program.

Contrasting Ontologies

I would like to briefly outline the conceptual incompatibilities that separate the functionalist and structuralist views of nature. Much deeper analyses were previosly published. Here I draw particularly on the work of pioneers like Ron Amundson (Amundson 2005) to remind my readers of the nature of this incompatibility.

Each field of science has its own ontology, which means a list of things that are considered as real and relevant to that part of reality pursued by that particular branch of science. For example, chemistry recognizes that there are atoms that have as their principal building blocks electrons, protons, and neutrons. Next, there are composite entities, like molecules, which consist of atoms that are bound together by what are called chemical bonds. In between are complexes and associations of atoms or molecules that are bound by forces weaker than covalent bonds. The list of naturally occurring atom types is comparatively short and consists of the 92 naturally occurring elements in the periodic table. The list of possible molecules is potentially unlimited, although they are formed by well-understood rules of combination among atoms. Then there are the attributes of systems of molecules and atoms, such as energy and entropy, which drive reaction dynamics, and the different kinds of bonds, such as covalent, ionic, and hydrogen bonds. This, essentially, is the ontology of classical chemistry. Nearly everything that can be claimed to be of concern for chemistry can be expressed in terms of these concepts.

The ontology of the New Synthesis of evolutionary biology (a.k.a. the Modern Synthesis) is also relatively straightforward. There are genes and their variants, called alleles; there are combinations of genes on chromosomes or haplotypes; and there are populations that, in the conceptions of population geneticists, are basically collections of genes and haplotypes that compete for their representation in the next generation. Among populations there are special collections that are labeled species, which are those populations that represent independent lines of evolutionary change. Organisms are sometimes mentioned, although only as vehicles for transmitting and replicating genes (Dawkins 1978; Hull 1980).

There are also quantitative parameters that explain evolutionary dynamics, defined as changes in gene frequency; these are fitness measures of competitive ability, which in mathematical models play a role similar to that of energy in chemistry (Fisher 1930). There is also effective population size, which is a measure of the influence of random effects on gene frequencies and, thus, is similar to the inverse of temperature in chemistry. Finally, there are the mutation rate and its cousin the mutational variance, V_m, and the recombination rate; these are all measures of the rates at which new genetic types are introduced into a population. That pretty much is what population geneticists care about, including me when I wear my hat as a population geneticist.

Just as atoms and molecules define what the realm of chemistry is, so genes and populations define what the realm of population genetics is. This is not a problem and I certainly agree with Michael Lynch who wrote, "nothing in evolution makes sense except in the light of population genetics" (Lynch 2007). However, *I disagree that only those ideas that can be expressed in the language of population genetics make sense.* This is the case because the ontology and language of population genetics are far too limited.

Life is so much richer than the arid, abbreviated language of population genetics can represent. I say this as someone who spent considerable amounts of my time working in theoretical population genetics. The one glaring omission in the ontology of population genetics is its lack of a concept for "organism" (Wagner and Laubichler 2000), except as a collection of genes and as the producer of fitness differences ("vehicle"; Dawkins 1978).

This is fine for what population genetics wishes to achieve: understanding changes in gene and genotype frequencies. However, it is a problem if it is seen as the only legitimate perspective on evolution, because important facts about life and evolution are screened from view. Specifically, any aspects of the structure and development of organisms are off the list of relevant phenomena because, in population genetics, there is simply no way to talk about these things (Reeve and Sherman 1993). For example, the question of how complex body plans arise is not within the reach of population genetics neither are questions on how complex organisms can arise from random mutations and selection. The latter is perhaps the biggest problem, given that Darwinism aims to explain all of evolution.

For many people, the origin of complex organisms arising from random genetic changes is simply not credible, or at least not intuitive, and thus we, as the experts in this area, better have an answer to this question. Nevertheless, questions of evolvability were neglected for most of the twentieth century, despite repeated attempts to make it an issue to be dealt with (Dawkins 1989; Wagner and Altenberg 1996; Wagner 2005; Brigandt 2007; Hendrikse, Parsons et al. 2007; Wagner 2010).

The ontology of developmental evolution is not as crisp as those of chemistry and population genetics because it is not yet as well worked out. Regardless, the list of relevant entities already includes gene regulatory networks, signaling cascades, cell types, and what Amundson calls "developmental types," such as the tetrapod limb (Shubin and Alberch 1986; Shubin 1994), the feather (Prum 1999; Prum and Brush 2002), and the nymphalid ground plan of wing pigmentation and the eyespots in the wings of butterflies (Nijhout 1991). All of these terms refer to parts and subsystems of the developmental and structural machinery of organisms. It is probably premature to say precisely what the ontology of developmental evolution will end up becoming, although it tends to fill in what is left out from biology by the ontology of population genetics.

Given the complementary ontologies of population genetics and devo-evo, their explanatory schemes are also difficult to reconcile, as convincingly argued by Ron Amundson (2005). The argument is that in population genetics, the random assembly of genetic variation gets sorted into predictable outcomes by natural selection. That is to say, the main causal force in the worldview of adaptationist (= functionalist) biology is natural selection. Natural selection is a causal process that acts within populations. Speciation is important because it breaks the reach of natural selection between populations that no longer interbreed.[5] In contrast, in the structuralist worldview, things like the nymphalid ground plan (figure 1.1) are entities that exert their influence over large numbers of species and, thus, they claim causal relevance where natural selection cannot have any.

Of course, we know that the nymphlid ground plan of wing pigmentation arose in a common ancestor of the nymphalid butterflies; however, this does not make things much better. To a population geneticist, this view sounds as though the nymphalid ancestor "is reaching from its grave to clutch the throats of its descendents" (Reeve and Sherman 1993). No explanation or model based on natural selection can do that. To make this contrast between explanations based on natural selection and those based on what Amundson calls "developmental types" more real, I want to interrupt this conceptual argument and take a little detour into biological facts.

What Do Developmental Types Mean?

A key set of studies from the pre-molecular era of developmental evolution were two papers by Pere Alberch and Emily Gale published in 1983 and in 1985 (Alberch and Gale 1983; Alberch and Gale 1985). These studies were important because they represented the first experimental evidence for the causal relevance of developmental types for the pattern of morphological evolution. The patterns that Alberch and Gale investigated were the contrasting regularities of digit reduction in urodeles and all other tetrapods (a.k.a. "eu-tetrapods"; i.e., amniotes and anurans, although this term does not signify a clade). In eu-tetrapods, digits are lost in a pattern known as "Morse's law"[6] (Morse 1872). This "law" states that in eu-tetrapods, digits are lost in a stereotypical sequence starting with digit 1, followed by digits 5, 2, and 3, and finally digit 4 (figure 1.2). In contrast, the sequence of digit loss in urodeles begins at the posterior with digit 5, and then proceeds anteriorly with digits 4 and 3 (figure 1.3). Hence, there is a pattern of digit loss that is specific to the urodele clade, whereas all other tetrapods follow a different pattern. The question then is, what explains this group specific pattern of digit loss?

A functionalist explanation of these contrasting patterns of digit loss in urodeles and other tetrapods would propose that there are differences between the manner in which urodeles and other tetrapods use their limbs and, thus, the idea is that contrasting patterns of natural selection explain that urodeles

---

[5] Interbreeding is just one population process that causes competition among genes within a population. More generally, species boundaries are defined by the limits of demographic replacability, as proposed by Alan Templeton in the cohesion concept of species (Templeton 1989).

[6] This "law" is not always followed, in particular if there are functional reasons for the retention of digit 1, as in the case of theropod dinosaur hands and, thus, bird wings (Wagner and Gauthier 1999), or in the case of the three-digited skinks of the genus Chalcides (Young et al. 2009).

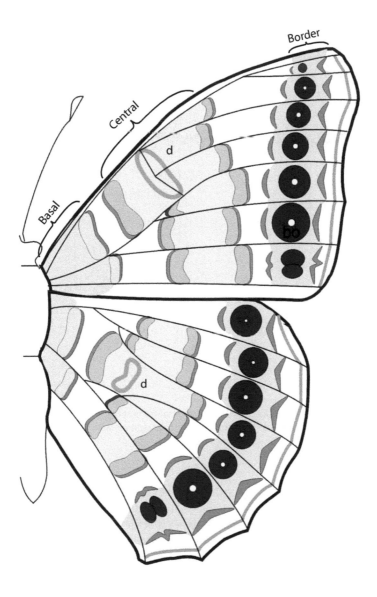

FIGURE 1.1: The nymphalid ground plan of butterfly wing patterns (Nijhout 1991). All nymphalid wing pattern variety can be understood as variations of three so-called symmetry systems: basal, central, and border symmetry systems. This was one of the first typological concepts reintroduced to mainstream evolutionary biology at the end of the twentieth century based on developmental considerations and the analysis of diversity patterns. (Source: http://biology.duke.edu/nijhout/patternspage2.htm.) Reproduced by permission of Fred Nijhout.

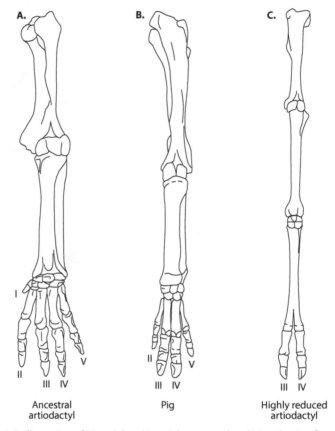

FIGURE 1.2: Illustration of Morse's law. Morse's law states that digit reduction first affects the anterior and posterior digits, as shown here for artiodactyl mammals (from Sears et al., 2011, *Evol. Dev.* 13:6, 533-541).

lose digits in a different sequence than that of other tetrapods. The structuralist explains these contrasting patterns of digit reduction by differences in the developmental makeup of urodele limbs versus the limbs of eu-tetrapods. In other words, the structuralist position proposes that "the urodele limb" and the "eu-tetrapod limb" are two different developmental types, which differ in their natural tendencies for change.[7] Pere Alberch, then on the faculty of Harvard's EOB department, and Emily Gale set out to test this idea.

Alberch's idea was to experimentally induce digit loss and see whether the same experimental treatment resulted in different patterns of digit loss in salamanders and frogs. Based on the work of Bretscher (Bretscher 1949) on

---

[7] For now I leave aside the question of why urodele limbs and eu-tetrapod limbs are different. I will come back to this question in the chapter on fins and limbs.

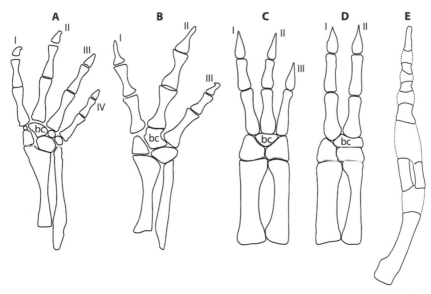

FIGURE 1.3: Sequence of digit reduction in urodele forelimbs. The ancestral forelimb had four digits. A) *Siren lacertina*, B) *Pseudobranchus striatus*, C) *Amphiuma tridactylum*, D) *Amphiuma means*, E) *Amphiuma pholeter*. Note that in urodeles, digit reduction starts from posterior (digit IV) and proceeds to anterior (compare to Figure 1.2 for the pattern in mammals). (Stopper and Wagner, 2007, *Dev. Dynamics* 236:321-331).

Xenopus limb buds, it was known that inhibiting cell proliferation at a critical stage of limb bud development would result in phalange and digit loss. Alberch and Gale used this approach for *Ambystoma* and *Xenopus* larvae by treating early limb buds with a reversible mitosis inhibitor (Alberch and Gale 1983; Alberch and Gale 1985). They found that the pattern of phalange and digit loss induced experimentally was different between *Ambystoma* and *Xenopus* and that this difference paralleled the differences in evolutionary digit loss between urodeles and anurans. Hence, the difference in evolutionary digit loss between urodeles and anurans was not necessarily caused by functional factors; rather it can be explained by factors intrinsic to the developmental makeup of the urodele and anuran limbs.

In fact, it has been known for a long time that there are radical differences in the development sequences of urodele and anuran limbs (Shubin and Alberch 1986). In anurans, as in all other eu-tetrapods, the sequence of digit development is 4 → 3 → 2 → 5 → 1, with some variations in the sequences during the emergence of the last two digits. In contrast, the digits in the urodele limb emerge in the sequence 2 and then 1 → 3 → 4 → 5.[8] The sequence

---

[8] This description does not take into account that amphibian forelimbs have at most four digits. Hence, the sequence is for hindlimbs with a full digit complement.

of digit development is the inverse of the sequence of digit loss during evolution. Hence, the taxon specific patterns of evolutionary digit loss are likely caused by taxon specific mechanisms of digit development.

In support of this conclusion, other studies have shown parallelism between population variation and the patterns of evolutionary divergence (Shubin and Wake 1996). This approach was also pioneered by Alberch (Alberch 1983) when he investigated the fusions of carpal and tarsal elements in plethodontid salamanders. Interestingly, the pattern of population variation is even stable within species hybrids like those between *Triturus alpestris* and *T. taenata* (Rienesl and Wagner 1992). Again, these variational patterns can be explained by the sequences and modes of skeletal development (Blanco and Alberch 1992). The weakness of this type of evidence, however, is that with the variational data from wild populations alone, one cannot exclude the possibility that the particular pattern of population variation is influenced or caused by natural selection.

When describing the studies of Alberch and Gale, it becomes easy to talk about the limbs of urodeles and anurans as causal agents—that is, to talk about "the urodele limb" and "the anuran limb" as distinct and meaningful entities. In fact, this is what these studies demonstrated. There are phylogenetically stable differences in the developmental make-ups of urodele and anuran limbs and these differences have consequences for their evolutionary fates (in this case, the pattern of digit loss).

Of course, there is nothing mystical about these observations because the manner in which limbs develop is different between salamanders and frogs and these developmental differences result in different variational tendencies. Because natural selection is downstream of the production of phenotypic variation during development, any stable difference in variational tendencies will have evolutionary consequences that are independent of the actions of natural selection. Naturally, a second question is why would these developmental and variational differences be evolutionarily stable (i.e., why are they conserved attributes of larger clades?)

Structuralist Concepts Arising from Macromolecular Studies

A radically novel type of evidence that supported a structuralist account of evolution derived from the work of Joe Thornton and his collaborators regarding the evolution of protein structure and function. Thornton has been investigating the evolution of novel glucocorticoid receptors' ligand specificities in vertebrates. In his most important study, he identified the amino acid substitutions that resulted in the novel ligand specificity of the ancestral metalo-corticoid receptor (Bridgham, Ortlund et al. 2009). However, reversing the amino acid substitutions in the derived protein did not restore the ancestral function. The reason was that the original forward amino acid substitutions were functionally adaptive only in the presence of a number of amino acids, which were in the ancestral protein, but are absent in the derived protein. These enabled these substitutions to have their adaptive effects.

After the changes that led to the new ligand specificity had occurred, however, the enabling amino acids became neutral and mutated away, which essentially made a return to the ancestral function impossible by reversing the amino acid substitutions that brought about the derived function in the first place. It is important to note that the "enabling" amino acids are themselves neutral and, thus, natural selection could not restore them should there be selection for restoring the original function. Hence, by losing "enabling" amino acid residues, the protein had burned the bridge that brought it to the derived function. The protein can not return to the ancestral function along the same path that had been used to arrive at the derived function. The point here is that natural selection cannot create such an opportunity again because, in the derived protein, the enabling amino acids are neutral and, thus, are not seen by natural selection. This detailed study shows that the power of natural selection is limited by factors intrinsic to the complexity of the protein under investigation. Naturally, selection is not able to explain the irreversibility of the transition from metalo-corticoid to glucocorticoid receptor specificity, but rather the intrinsice interdependencies among the amino acids in the protein are responsible for this pattern.

Another important conceptual advance resulted from computational studies of RNA secondary structures. The secondary structure of RNA is ideally suited for computational studies because the energy of folded RNA is dominated by base pairing, which is more easily treated computationally than the larger number of moderate energy contributions that play a role in protein folding (Schuster, Fontana et al. 1994). The secondary structure of RNA is a simple form of a phenotype and the space of RNA sequences is a genotype space. One can think of the folding process of a particular RNA as a simple form of "development" that influences the evolutionary dynamics of RNA phenotypes.

Computational studies of RNA secondary structures led to the first theory of the consequences of developmental factors on evolution, because this computational approach was the first for which the structure of a genotype-phenotype map could be explored in detail (Fontana 2002). Comprehensive computational explorations of RNA secondary structures were performed in the laboratory of Peter Schuster at the University of Vienna. These studies led to important conceptual advances that are particularly helpful for understanding the "structuralist" perspective of evolution. Schuster, his students, and postdocs showed that a fundamental property of a genotype-phenotype map is that there are many orders of magnitude more RNA sequences of a particular length than there are secondary structure phenotypes (Schuster,

Fontana et al. 1994). A consequence of this many-to-few mapping is the existence of large, mostly connected "neutral networks."

A neutral network is a set of sequences that are connected by single nucleotide substitutions and in which each sequence represents the same minimal energy secondary structure. One can think of neutral networks as being a large tangle of differently colored ropes in which each rope represents the neutral network corresponding to a different phenotype (as indicated by the color of the rope). Evolutionary dynamics at the phenotype level is then determined by the neighborhood relationships among these "ropes." The chance of mutating from one phenotype to another is determined by the fraction of sequences in a neutral network that is close to another particular neutral network. This conceptual image makes it easy to explain, or at least understand many otherwise counterintuitive variational properties of phenotypes, such as developmental constraints and historicity (Stadler, Stadler et al. 2001).

For example, if two phenotypes are represented by neutral networks of very different sizes, say A has a larger neutral network than B, then the mutational transitions from B → A are more likely than those from A → B. This is because, in a smaller network, a larger fraction of sequences will be near the transition points to A than in the larger network. Hence, variational biases and constraints among phenotypic states can be understood as consequences of the topology of neutral networks and the transition points between them.

In the studies of Alberch, Schuster, and Thornton, we have three causal paradigms that explain the structuralist intuition that complex systems (limbs, RNA, and proteins) play a causal role in determining their evolutionary fates. These variational tendencies and, thus, the evolutionary possibilities are strongly influenced by the patterns of development (the differences between urodele and anuran limbs; Alberch and Gale 1983; 1985), the folding dynamics of RNA, and by the interdependencies (interaction, or epistasis) among adaptive and neutral variations (irreversibility of protein evolution; Bridgham, Ortlund et al. 2009).

Back to Concepts and Ontologies

Returning to the problem of what is considered causally relevant in evolutionary biology by different schools of thinking, the adaptionist and the structuralist views, one can see that, since the late twentieth century and beginning with the work of Rupert Riedl (Riedl 1977; Riedl 1978), Stephen Gould (Gould 1977) and Richard Lewontin (Gould and Lewontin 1978), structuralist ideas are neither committed to non-material causes, as frequently was the case in the nineteenth and early twentieth centuries (e.g., the vitalistic ideas of Hans Driesch, 1847–1941, and Adolf Portmann, 1897–1982), nor are they necessarily tied to rejecting natural selection as a cause of evolutionary change. There are structuralist schools of thinking that achieve their rhetorical punch from rejecting natural selection. However, these radical positions have proven to be scientifically sterile (e.g., Ho and Saunders, Brian Goodwin, 1931–2009, and Jerry Fodor).

Rather, there is a broad understanding that the structuralist view reflects upon constraints on variation or variational accessibilities of phenotypes (e.g., Alberch 1983; Maynard-Smith, Burian et al. 1985). In my view, the realization that complex organisms/systems have unique and historically contingent variational constraints and biases paves the way for a seamless unification of functionalist and structuralist agendas, at least in principle. One can call this view "*variational structuralism*" to differentiate it from other forms of structuralist thinking. As far as I can see, however, variational structuralism requires a specific correction in the ontology of neo-Darwinian science.

The New Synthesis theory of evolution incorporates an ontological commitment that prevents building a bridge to variational structuralism—namely, the idea that the most fundamental level of biological reality is variation, at least for evolutionary biology (e.g., Ghiselin 1997). Variation here stands for the realized different forms of genes and phenotypes. This move was explicitly aimed at excluding concepts like body plans and other ideas considered as "types." This means that developmental and variational commonalities among organisms, as reflected in terms like "urodele limb," cannot be relevant causal factors. According to the New Synthesis ontology, all that is real are the realized differences among organisms, and not their underlying variational tendencies.

At the time the New Synthesis was forged, that move was a good thing, as long as its purpose was to reject non-material or non-mechanistic models in biology. Unfortunately, it also excludes mechanistically meaningful concepts and is even inconsistent with the research practices of population genetics. To see this difficulty, one has to observe that variational structuralists talk about factors that influence the possibility and probability of phenotypic variation. Hence, for them, the important ontological commitment is to the existence and historical permanence of developmental mechanisms and molecular interactions that influence future phenotypic variations. However, if the ontology of neo-Darwinism stops at the level of realized variations, any discussion about factors that affect the production of future variations is "forbidden" due to the ontological commitment to variation as the most fundamental level of reality.

The radical commitment to variation as the most fundamental level of reality, however, is not even respected in the research practices of population genetics itself. In fact, the models of mathematical population genetics incorporate numerous parameters that are used to measure variational tendencies (i.e., the possibility and probability of future variations). The most obvious are the mutation rate, a mutational variance, V_m, and the mutational co-variance matrix, M. Each of these parameters is measurable and critical to the core business of population genetics: explaining changes in allele frequencies. For example, the amount of heritable phenotypic variation maintained under mutation, selection, and genetic drift depends not only on the strength of natural selection and the effective population size, but also on the rate at which new genetic variation is produced (i.e., the mutation rate and V_m; Bürger 2000). It is true that, in population genetics, these quantities are usually treated as parameters, and not as variables. Yet given that we know that, for example, mutation rates are influenced by genotypes, these quantities can be considered to be variables. Hence, it cannot be such a terrible thing to discuss variational tendencies, or even the evolution of variational tendencies.

To overcome this problem, it is useful to make a distinction between two terms that, until recently, were and sometimes still are used as synonyms: variation and variability (Wagner and Altenberg 1996). Variation refers to the actually realized differences between individuals and genotypes. For example, a nucleotide polymorphism (π), and additive genetic variance (V_g), are measures of *variation*. On the other hand, one can use the term "variability" to describe the "ability to vary" (i.e., the variational tendencies of a genotype or a phenotype). Mutation rate (μ and mutational variance (V_m) are measures of *variability*. Obviously, developmental type terms, like "urodele limb" and "nymphalid ground plan," are terms that refer to collections of entities that are characterized by their variability. The reason is that it is the pattern of variation that allows us to recognize the unity of nymphalid pigment patterns in contrast to, say, the wing patterns of moths (Nijhout 1991).

Now that this detour into recent scientific developments and their ontology is over, we can return to the intellectual history and context of evolutionary biology.

Facts and Ideas about Bodies

Darwin was part of an intellectual movement that began in the late eighteenth century with the goal to "discover the laws of life," which meant to discover the regularities of biological diversity and find an explanation for them (Rupke 1994; Amundson 2005; Thomson 2009). This movement was certainly inspired by parallel efforts in the inorganic sciences that led to the classification of minerals according to the symmetry properties of their crystals, to the periodic table of elements in chemistry, and the discovery of the stoichiometric regularities of molecular structures. At one time, Richard Owen's vision was that comparative anatomy might become as systematic a science as chemistry (Rupke 1994). He thought that it should be possible to describe the organization of an animal body in the form of a body "formula" just like the composition of a chemical can be written in the form of a chemical formula. For example, H_2SO_4 is just a configuration of "chemical elements" (H, S, and O) and their molecular proportions combined into a molecule (i.e., sulfuric acid).

For Owen, the analogues of chemical elements are the homologs in biology; that is, homologs are the anatomical "atoms" of the bodies that, in different combinations and configurations, make up the various specific bodies of actual animals. Even today there are some remnants of this project in biology, as for example the "tooth formula" that is still used today to describe the dentition of mammals. A tooth formula gives the numbers and kinds of tooth types (incisor, I, canine, C, premolar, P, and molar teeth, M) in the jaws of a mammal. For example, the tooth formula for a human is

$$\frac{I_2 C P_2 M_{2-3}}{I_2 C P_2 M_{2-3}}$$

Of course, this was and remains an acceptable mode of scientific conduct; namely, identifying quasi-invariant units and how they combine to create a composite whole. The fact that at that time the causes of evolutionary change or even the fact of evolutionary change was unknown does not invalidate this approach because, at the same time, the reasons for the stoichiometric regularities in chemistry were also unknown, as were the reasons for the symmetry properties of crystals.

Recent biographical research (Thomson 2009) has shown that Darwin was very much a child of his time in that he explicitly set out to do what all other ambitious naturalists of his time wanted to do—namely, develop an all-encompassing theory of life's diversity in all its forms, from the anatomical to the behavioral (i.e., explain why there are so many forms of life, how they came about, and what explains their features). Darwin competed with his contemporaries over the same intellectual territory as had at least half a century of naturalists who preceded him. What was that intellectual territory, what were the styles of thinking represented in this territory (which coincidentally are still with us today in some form), and what were the facts that supported the views on each side of that struggle?

The two styles of thinking introduced previously, functionalist and structuralist, correspond to two broad generalizations about life: (1) all organisms are exquisitely adapted for their stations in life; and (2) even organisms with very different lifestyles nevertheless can share "deep" similarities in their bodily structures, a phenomenon called "unity of type."

When our pet cat seeks my company and I have the leisure time to pet her, I cannot help but note the exquisite detailed adaptations that Nature (viz. natural selection) has endowed cats with for their life as (often) nocturnal hunters. Everyone knows about the whiskers on a cat's nose (as well as many other mammals) and that they function as part of the mechanoreceptors that provide the animal with spatial information in tight spaces in the dark. But, that is not all. There also are whiskers above the eyes, and on the cheeks, and there is longer hair that sticks up over the main bulk of the fur; obviously, these are more useful for mechanoreception than for keeping the cat warm.

These long hairs are placed at strategic places, like the shoulders and the elbows. It looks as though the cat can "feel" the space around her body just as a fish can feel the movement of water through its lateral line system and an electric fish can feel the shape of the space around it through the distortions in its electric field. We now explain these exquisitely detailed adaptations through the awesome power of natural selection, the relentless sorting of heritable phenotypic variations according to their contributions to survival and reproduction; in other words, we explain these features as adaptations.

When focusing on adaptation, a natural response is to think about organisms in terms of function; organisms are as they are because they have to meet those challenges that impinge upon their station in life (i.e., their ecological and social niches). This manner of thinking is not uniquely Darwinian, or even evolutionary. It was also a dominant style of pre-Darwinian thinking. The most prominent pre-Darwinian functionalist theory of life was that proposed by George Cuvier. He explained the differences between "types" of organisms (his *embranchments*) as different modes of functional organization. His law of correlation of parts, so he thought, was caused by the necessities of functional integration. During Darwin's time, the school of natural theology was also a functionalist mode of explanation, albeit a religious one.

To see the historical roots of structuralist thinking, one has to literally dig deeper than the fur on a cat. The classical and still persuasive argument for structuralism is Richard Owen's introduction to his "The Nature of Limbs" (Owen 1849). In this essay, Owen reflects upon those body parts that "chiefly are answering the call for locomotion": the paired appendages. He first notes (p. 4) that these body parts of vertebrates are quite different from those that perform the same function in invertebrates, such as in insects and crabs. In arthropods, the hard parts form the body cover and the limbs comprise different numbers and kinds of segments than do vertebrate limbs (figure 1.4). In contrast, the vertebrate limb has an internal skeleton and fewer segments.

Then, Owen calls upon vertebrate limbs (all mammalian) that are adapted to different forms of locomotion. His first witness is the paddle of the dugong, a marine mammal with a pectoral fin that is a "strong, stiff, short, broad, flat, and obtusely pointed paddle." Functionally, the dugong's pectoral fin has only one movable joint at that point where this paddle joins the body. Then, Owen moves on to the forelimb of a mole, which externally is similar to the dugong's paddle, except that its distal extremity, rather than being smooth and thin, is notched and "armed with a row of tooth like horny points, adapted for scraping and throwing back the soil." Next comes the bat wing, which he describes as a "vastly expanded sheet of membrane supported, like an umbrella, by slender rays" and moved through the air so as to lift the body above the ground. Next, he describes the adaptations of the hoofed animals and,

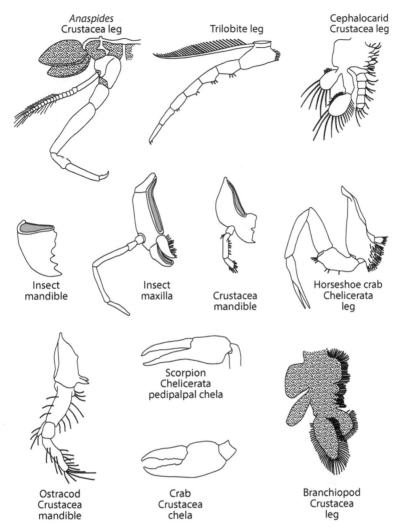

FIGURE 1.4: Diversity of arthropod appendages.

finally, arrives at the human hand, which chiefly is not an organ of locomotion, but adapted for grasping, touch, and "manipulation."

After Owen called his chief witnesses, he paused and considered the various contraptions devised by human ingenuity to effect the same purposes (swimming, flying, burrowing, and so forth) and noted that there is very little in common in terms of structure between "flying machines," cars, and tunneling machines. After this detour into human technology, Owen returns to his comparisons of mammalian limbs by showing at great length how the bat's wing, the dugong's paddle, and the mole's shovel, as well as the human hand, are all structured according to a common pattern. This pattern incorporates three main segments (a proximal segment with a single long bone, the humerus, then two bones in the middle segment, the radius and ulna), then a variable set of small carpal bones, followed by metacarpals, and the phalanges that form the fingers (figure 1.5). This is the classical case of "unity of type." That is, similarity of form and structure in spite of differences in function.

Clearly, these commonalities among mammalian limbs cannot be explained by functional necessities. Invertebrates accomplish similar feats using different limb structures and human inventions accomplish a variety of functions without having a common plan. What the explanation was for these commonalities among mammalian limbs Owen did not know. Much has been made of, and too much damage has been done to Owen's reputation, by allusions to Platonic "ideas" to explain these patterns. However, these philosophical diversions played no serious role in his scientific work, as he was happy to switch between explaining either these adaptations or the unity of type by Platonic ideas, depending on who complained about his lack of understanding of Platonic philosophy (Rupke 1994).

Clearly, he could not have cared less whether adaptations or the unity of type corresponded to Platonic ideas. Rather, Owen was more inspired by the possibility of being able to "infer the material (polarizing) forces" that shape the bodies of animals, much like those hypothesized to cause the structure of crystals (Rupke 1994; Amundson 2007). This is important

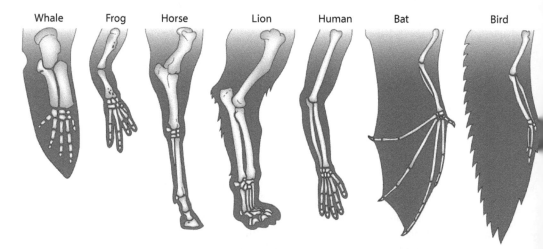

| Whale | Frog | Horse | Lion | Human | Bat | Bird |

FIGURE 1.5: Variations and unity of vertebrate limbs. Skeletal arrangements remain comparable despite different functions and shapes. This was one of the strongest arguments for Owen's homology concept and remains a textbook example of homology. (Wagner, 2007, *Nature Reviews Genetics* 8:473–479.)

to note because it shows that the structuralist stance never was necessarily linked to idealist philosophy or the assumption of non-mechanistic causes or forces (Amundson 2005).

Of course, Darwin made a lot of hay from Owen's structuralist argument. In Darwin's hands, the fact of non-adaptive similarities provided the strongest argument for descent from common ancestors. In other words, Owen's archetype became the ancestor of a group of species that was built according to some structural type. This move had two important consequences. First, it explained the nested distribution of homologs among animals (Wagner 1994), and secondly, it took away the ontological bite of the unity of type concept.

Owen's definition of homologous characters, "the same organ in different animals under every variety of form or function," does not explain or even address the fact that homologs tend to be distributed in nested sets. For example, the set of species with tetrapod limbs are a proper subset among all vertebrates that have jaws. Furthermore, those tetrapods with hair and those with feathers are both nested within the set of animals with limbs. Hierarchical schemes for arranging the diversity of species within a system have been used at least since Linné; however, whether these arrangements actually mean something in terms of biological reality was not clear. In fact, the most influential naturalists of the pre-Darwinian era were nominalists who thought that the hierarchy of the system of species was just an economic means to represent a large amount of information, although not reflecting an important fact about nature (Mayr 1982).

Darwin's idea of evolution by descent with modification and lineage splitting provided a realistic interpretation of the hierarchical nature of natural systems. It explained the nested distribution of homologous characters based on a bifurcating tree of descent. Animals with tetrapod limbs are nested within the set of animals with jaws because tetrapod vertebrates shared a more recent common ancestor than jawed vertebrates, and mammals (i.e., tetrapods with hair) shared an even more recent common ancestor. This was a powerful argument and explained a large number of facts regarding biological diversity.

As with any important scientific advance, this discovery could also be turned into a tool for further discovery. Given the Darwinian interpretation of homology as indicative of the prior existence of a common ancestor, identifying homologs became the principal type of evidence for inferring phylogenetic relationships. Morphological homologies remained the primary source of phylogenetic information until the arrival of molecular methods. Hence, Darwin's interpretation of homology gave rise to the first large-scale research program based on evolutionary theory: evolutionary morphology (Nyhart 1995). At the same time, however, homology and the unity of type became seen solely as a trace of history, and not as indicative of deeper causal mechanisms that underlay the conserved patterns of morphology.

Homology had lost its ontological bite and this paved the way to the eventual eclipse of structuralist views for most of the twentieth century. Case closed?

In fact, this case was closed with few dissenting voices until the discovery of conserved developmental genes, with homeobox genes as the first to be discovered (Gehring 1998). It is now well established and common place that all animals share a set of conserved genes[9] that are causally important for the development of body plan characters. This was a deeply surprising discovery because, in the tradition of neo-Darwinian evolutionary biology, the possibility of homologous genes among distantly related species was explicitly dismissed (Amundson 2005). Until the discovery of conserved developmental genes, the majority view was that the genome was a collection of hereditary factors that underwent constant, opportunistic turnover with no causally important structures remaining over longer periods of time.

To explain the similarity of characters or even the re-appearance of ancestral characters (atavisms) by the presence of homologous genes was not part of the picture. It did not matter that the empirical support for this view was based on the methodological limitations of transmission genetics. Transmission genetics works only among individuals of the same species, with occasional hybrids among closely related species. Hence, the neo-Darwinian stance on homologous genes was the result of a methodological blind spot, what Amundson called the Mendelian blind spots of the New Synthesis biology (Amundson 2005, p. 182).

Rejecting homologous genes also supported the view that organisms are "balls of wax in the hand of natural selection" with no relevant internal constraints or structure (Gould 2002). But all of this changed with the discovery of homeobox genes, which resulted in a resurgence of typological, structuralist ideas, a few of which will be sketched out below. Before we look at the structuralist models that were inspired by molecular developmental genetics, I want to briefly look back to late twentieth-century developments that foreshadowed the devo-evo revolution of the 1990s.

Re-focusing on the Role of Development

Despite the dominant position of neo-Darwinian theory duirng most of the twentieth century, dissenting voices were never fully silenced. Only a few attempts were made to formulate a credible alternative to the adaptationist research program. Here, I want to discuss two of these: Riedl's theory of the

---

[9] I use the term conserved here in a descriptive sense, meaning that they are recognizably homologous genes in very different animals. Conservation neither means that the sequence is absolutely unchanged nor that the function is truly the same (see Wagner and Lynch 2008).

"immitatory epigenotype" and Shubin and Alberch's developmental interpretation of the tetrapod limb.

Rupert Riedl (1925-2005) was trained as a comparative anatomist who made his name as a pioneer in marine ecology after World War II. When Riedl accepted a position at the University of North Carolina as the Kenan Professor of Marine Biology in 1968, he was confronted in his new workplace with the synthetic theory of evolution at its high point and began to question the generality of adaptive explanations. Upon returning to the University of Vienna, he summarized his ideas in his influential book, *Order in Living Organisms* (German 1975; English 1978), in which he made a connection between two important facts. First, organisms are highly integrated systems with many functional and developmental interdependencies. He pointed out that these interdependencies have to lead to limitations and biases involving the availability of potentially adaptive phenotypic variation and, thus, may influence the pattern and rate of evolution. In other words, he anticipated the notion of developmental constraints, even though he did not use this term.

The second fact was that it is difficult for random mutation to improve upon any system that has too many degrees of freedom (i.e., any system that is too complex faces what mathematicians call the "curse of dimensionality"). This latter point clearly was also a concern for the founding fathers of the synthetic theory, like Ronald A. Fisher, who developed his "geometric" model of adaptation to solve this problem to his satisfaction. Riedl's original idea was that these two phenomena, constraints through integration and complex adaptation through random mutations, could perhaps be linked.

If developmental integration preferentially eliminated those phenotypic degrees of freedom that were least likely to lead to improvements in fitness, then adaptation by random mutation would be facilitated and the curse of dimensionality would be lifted (for an informal summary of these arguments, see Wagner and Laubichler 2004). In other words, if epigenetic (i.e., developmental) integration "mimics" functional integration, then the problem of complex adaptations would be solved.

Of course, the remaining question was why developmental interdependencies should assume a shape in which adaptive variation was allowed and non-adaptive, unconditionally deleterious mutation effects were suppressed. To explain this, Riedl invoked the idea that natural selection should be able to directly improve evolvability, an idea that has gained considerable credence in recent years (Draghi and Wagner 2008; Draghi and Wagner 2008; Pavličev, Cheverud et al. 2011), although it remains controversial.

Riedl's theory was remarkable because it was the first structuralist theory that fully embraced the achievements of population genetics and the adaptationist research program. He did not invoke non-material forces, and development was considered an ally rather than an alternative or opponent to natural selection. Nevertheless, Riedl was uncompromising when it came to his conviction that many broad patterns of biological diversity ought to be explained by the constraints placed on natural variation through development. Thus, he clearly saw that developmental causation had to play a key role in explaining patterns of evolutionary diversification.

At about the same time, the Harvard paleontologist Stephen J. Gould (1941–2002) published a major book arguing for the importance of development in evolutionary biology. Gould's argument was both scholarly in its historical perspective as well as its focus on the role of heterochrony during evolutionary changes in development. His book *Ontogeny and Phylogeny* was published in 1977 (Gould 1977). This book had the effect of rehabilitating the idea that developmental factors may play a role in evolutionary processes. Gould liked to tell the story that, while he was writing this book, he would tell colleagues about his project and that more often than not the reaction was that they too were convinced of the importance of developmental factors in evolution; yet, they always acted as if they had admitted to a shameful sin.

The other significant pre-molecular attempt to establish the importance of developmental factors for evolution was the synthetic review of limb variation and development by Neil Shubin and Pere Alberch (1954–1998) published in *Evolutionary Biology* in 1986 (Shubin and Alberch 1986). At that time, Neil Shubin was a graduate student with Pere Alberch at Harvard University. They summarized what was known about limb development and diversity at that time. The main result of this undertaking was the discovery of a remarkable constancy in developmental sequences during limb development. Hence, the unity of type of vertebrate limbs demonstrated by Owen (1849) at the anatomical level was shown to be mirrored by an equally conservative pattern of development.

Shubin and Alberch went further and proposed that the processes of pattern formation in the mesenchyme, which precede the formation of the skeletal elements, constrain the possible shapes of tetrapod limb skeletons and, thus, explain the unity of anatomical types by the unity of developmental types. This model had a great heuristic influence and could also explain the patterns of digit reduction, which were briefly reviewed above with the empirical evidence for the explanatory value of structuralist ideas.

Between these two major synthetic achievements that called for a developmental extension of evolutionary biology, one in 1975 by Riedl and one by Shubin and Alberch in 1986, was the founding conference of the devo-evo movement; the Dahlem Conference on Evolution and Development in 1981 (published by Springer Publishers in 1982). This conference was the first attempt to draw together the various strands of developmental thinking in evolutionary biology. I mention these historical tidbits to show the fertile intellectual ground into which the discovery of the homeobox genes eventually fell in 1983.

Homeobox genes were discovered at about the same time by Walter Gehring and William McGinnis in Basel and Matthew P. Scott and Amy Weiner at Indiana University. Of course, the evolutionary importance of this discovery remained unclear until the discovery of homologous genes in mammals shortly thereafter. I argue that it was this confluence of developmental work in evolutionary biology (mostly in the United States by David Wake at the University of California Berkeley, who was Pere Alberch's mentor, and by Stephen J Gould at Harvard, and in Europe by Rupert Riedl at the University of Vienna, and the advances in molecular developmental biology made by Walter Gehring and Matthew Scott) that led to the emergence of the new field of evolutionary developmental biology and, thus, to the re-emergence of a credible structuralist research program.

The Emergence of Molecular Structuralism

I now wish to discuss a few recent examples of structuralist ideas that grew out of the emerging field of molecular developmental biology and its fusion with evolutionary biology during the 1990s. In all cases, these concepts arose from a synthesis of comparative anatomical and embryological facts and the new molecular biology of development.

The Phylotype

The idea that the unity of the anatomical type, as described by Owen and other biologists before him, is subscribed by conserved developmental mechanisms is old and has had a varied and at times tortured history. Similar to the problem with a developmental account of homology, which will occupy us in the remainder of this book, the problem is that conservation of development is not general and the patterns of developmental variations are a mosaic of conserved and variable elements. For example, it is not true that the earlier stages of development are more conserved in general, as suggested by the vulgarized version of Haeckel's theory of "recapitulation."

In fact, the very early stages of development are phylogenetically as plastic as the late stages, as they respond to differences in life history strategies, with variations in yolk mass and maternal provisioning (placentation) as the main drivers of early developmental variation (Raff and Kaufman 1983; Raff 1996). Late developmental stages are expected to be variable because they form the adult organism with its diverse functional adaptive needs. In addition, mid-developmental stages are not immune to radical modifications, as exemplified by metamorphosis and the adaptations of larvae if they encounter different ecological conditions than does the adult stage. For example, many body parts of an adult fly are derived from imaginal discs, which develop differently than the corresponding body parts of grasshoppers that have direct development.

The most enduring concept that accommodates many of these complications is the phylotypic stage of development. The phylotypic stage is defined as "the first stage that reveals the general characters shared by all members of the phylum" (Sander 1983). This idea was introduced by Seidel in 1960 and was called "Körpergrundgestalt," which is German for "basic body pattern" (Seidel 1960). The empirical basis of this concept derived from observations made during the early development of, for the most part, insects and vertebrate embryos.

The concept of a phylotypic stage recognizes the fact that different insect orders begin their development from quite different starting points, and then converge to a highly conserved stage, but subsequently diverge again; these observations led to the so-called hourglass model (figure 1.6). In insects, this developmental stage is the germ band that consists of head lobes, three gnathal segments with the buds for the mouthparts, and three thoracic and eight to eleven abdominal segments (figure 1.7). This arrangement is found in all insects at some stage of development and comprises the basic parts of the insect body plan.

Similarly, early vertebrate development, also called "primitive development" in classical comparative embryology, differs greatly depending on egg size and the presence or absence of extra-embryonic tissues and placentation. Nevertheless, all vertebrates converge to a stage that includes the chorda, the neural tube, a tubular heart, and somites. This stage is either called the pharyngula or tail bud stage. Whether other major groups also have a phylotypic stage remains a matter of debate. In any event, Seidel (1960) proposed phylotypic stages for hydrozoans, mollusks, and annelids and Williams (1994) proposed one for the majority of the crustacean taxa.

A conserved stage during mid-embryogenesis was recently confirmed by molecular evidence (Hazkani-Covo, Wool et al. 2005; Kalinka, Varga et al. 2010; Irie and Kuratani 2011). For example, Kalinka and collaborators compared gene expression divergence among Drosophila species across developmental stages and found that divergence was least at the morphologically defined phylotypic stage (figure 1.8). Similarly, Irie and collaborators (2011) compared the transcriptomes of mouse, chick, zebrafish, and *Xenopus* during different stages of development and found evidence of a mid-embryonic stage of maximal similarity among these vertebrates.

Before discussing the explanations and molecular extensions of the phylotypic stage concept, I will briefly discuss a number of minor objections to this concept (Slack 2003, p. 312). One easily avoidable problem is evoked by the modifier "phylo-" in the term "phylotypic," which refers to the Linnean category "phylum." The objection is that not all "phylotypic" stages characterize a phylum, as for example the insects that technically are a class of

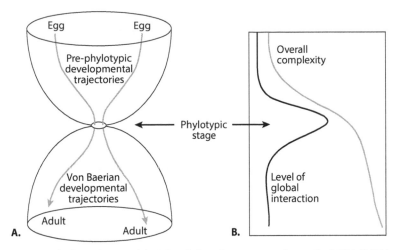

FIGURE 1.6: The hourglass model of the phylotypic stage according to Rudolf Raff. A) Variations in developmental pathways are minimal at intermediate stages of development, the so-called phylotypic stage. It is called this because at this stage the embryo has assumed phylum (or better clade) specific body plan characteristics. Before and after this stage, variation between species is greater. B) The explanation for this pattern is that, early during development, morphogenesis is dominated by flexible global pattern formation processes and, thus, divergence in the developmental pathway is readily achieved, as for example in response to changes in the amount of yolk or to the evolution of placentation. During the phylotypic stage, developmental pathways are more constrained because this is the stage at which the embryo subdivides into distinct cell populations, like the germ layers and organ anlagen. At this time, the interactions among cell populations are necessary to establish developmental identity and polarity. After the phylotypic stage, the developmental pathways diverge because development proceeds with quasi-modular parts that are easier to change than during the phylotypic stage when they are still strongly interdependent. (Source: http://scienceblogs.com/pharyngula/2010/12/27/the-molecular-foundation-of-th/.)

the phylum Arthropoda. This is correct, but irrelevant, since there is broad agreement that the supra-specific Linnean categories like phylum or class are biologically meaningless, other than that they signify some higher level clade. Hence, the distinction between a phylum and a class is biologically irrelevant.

In either case, one is referring to a highly conserved body plan that is characteristic of an old clade. To avoid this misunderstanding, a phylotypic stage could be called a "typogenetic" stage (i.e., that stage at which the typical body plan features arise that characterize a clade of animals). It should also be remembered that Seidel's original term was "Körpergrundgestalt" ("basic body shape"), a term that is silent with respect to Linnean categories.

The other somewhat hairsplitting objection is that the phylotypic stage "is not a stage at all." Rather, it is "a more-or-less elongated phase of early-middle development within which the most highly conserved features of

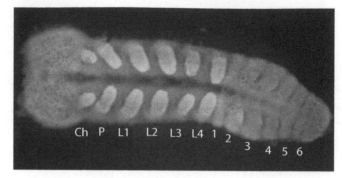

Ch P L1 L2 L3 L4 1 2 3 4 5 6

FIGURE 1.7: The phylotypic stage of insects and chelicerates is the extended germ band stage. At this stage, all segments are laid down and appendages are formed, although not yet differentiated according to species-specific patterns. As an example, the germ band stage of the spider *Cupiennius salei* is shown here. The embryo is removed from the egg and flattened out. Reproduced by permission from Damen et al., 1998, *PNAS* 95:10665–10670. Copyright 1998 National Academy of Sciences, USA.

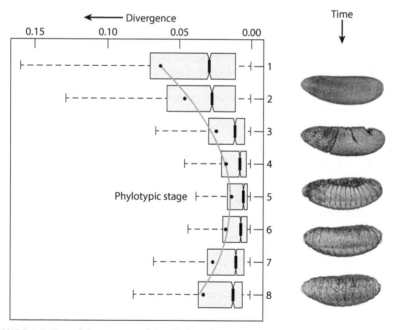

FIGURE 1.8: Test of the concept of the phylotypic stage using gene expression data from six Drosophila species. A comparison of transcriptomes shows that gene expression diverges least during the phylotypic stage of fly development. Reprinted by permission from Macmillan Publishers Ltd: *Nature*. Kalinka, Varga et al., 2010, *Nature* 468:811–814. Copyright 2010.

morphology are apparent" (Slack 2003, p. 312). Yet, it is not clear how this in any way affects the validity of this concept. Nothing depends on the assumption that the phylotypic stage is, in fact, a single time point in development.

What is important, however, is the question of whether the existence of a phylotypic stage, as defined by Sander (1983), is a biologically significant fact. One way to explain a phylotypic stage is to say that it is that stage that is least exposed to directional selection pressures because it is neither affected by an adaptation to different egg types and placentation, as are the early stages, nor is it affected by the adaptive, diversifying pressures of the larval and adult stages of development. To respond to this objection, one must observe that the stages of mid-development are not immune from adaptive modifications, as the differences between short and long germ development insects show as does the development of extra-embryonic tissues in vertebrates and insects. This also occurs at about the time of the phylotypic stage in most amniotes, with the exception of mammals in which extra-embryonic tissues form earlier.

For example, it is likely that the adaptive advantage of long germ development is a reduction in development time. Rather than going through a longer process of stepwise additions of segments, long germ insects simultaneously generate body segments in three short steps. Hence, it is not clear why the insects and other groups should retain a highly conserved stage of development. For example, Sander (1983) points out that the insect head and mouth parts are so modified and synorganized in many insect orders that it is surprising to see that they are all derived from three embryologically distinct segments.

What then is different during the phylotypic stage that makes it more conserved than other stages and aspects of development? The most satisfying idea, which also has some degree of empirical support, is Rudy Raff's model of different modes of developmental regulation (Raff 1996). According to this model, the very early stages of development are governed by global pattern formation processes, which easily accommodate different amounts of yolk. These early developmental stages also react to experimental perturbations as a unitary developmental field. During later developmental stages, the embryo is subdivided into autonomous developmental modules that allow for the diversification and divergence of different parts of the embryo toward the adult stage.

The phylotypic stage marks the transition from early global development to the modular mode later in development. During this transition, the developmental modules of the later stages arise and this is a process in which interdependencies among embryo parts are maximal and which requires a certain spatial arrangement of cell groups for the exchange of inductive signals. In other words, at this stage of development, the embryo's morphology is causally important for establishing basic body organization. This model implies that the early to middle developmental stages are highly vulnerable to perturbations and, thus, would explain the conservation of this stage by constraints resulting from a lack of modularity.

In an important paper published in 2001, Frietson Galis and Johan Metz at the University of Leiden in the Netherlands showed that, in the case of mammals, the phylotypic stage was a developmental period of highly enhanced vulnerability to external perturbations (Galis and Metz 2001). Galis and Metz analyzed the teratological literature and found that the phylotypic stage was, indeed, the most sensitive stage of development in terms of malformations and mortality due to toxins. In addition, they analyzed the kinds of malformations that arise during the phylotypic stage and concluded that malformations affected multiple organ systems simultaneously and, thus, were consistent with the idea of limited modularity and high interdependency during this stage of development. Hence, there is experimental evidence, at least for mammals, that the phylotypic stage is, indeed, different and more vulnerable to perturbations than are other developmental stages and, thus, may be conserved exactly because of this fact.

Note that this is the basic logic for structuralist explanations of animal form conservation. Because mutations that affect the phylotypic stage are unlikely to be adaptive, and because of the likelihood of unconditionally deleterious effects, the phylotypic stage is conserved and, thus, is inherited by descendent species. In each descendent species, these constraints still exert their influence, regardless of species boundaries. This logic is as elementary and persuasive as that of natural selection. It is also as equally un-mystical and mechanistic as natural selection.

To say that the phylotypic stage is conservative does not mean that it is totally invariant or completely immutable. For example, Terri Williams (Williams 1994) argued that the nauplius stage of crustacean development was their phylotypic stage, which is only replaced by a different mode of development in one group of crustaceans, the Malacostraca. It was also unnecessary and damaging to "tune" the images of vertebrate embryos, as Haeckel apparently did (Richardson, Hanken et al. 1997). Conservation means just that: conserved but not necessarily invariant given that even atoms and elementary particles are not immutable.

From the Phylotype to the Zootype

The idea that the diversity of life is, at various levels of generality, subscribed by a deep unity, that is the unity of type as first envisioned by Geoffroy Saint-Hilaire, Carus, Goethe and Owen, received a major boost with the discovery of conserved developmental genes duirng the 1980s. However, the difference this time was that these commonalities were not only at the level of a particular body plan, as defined by comparative anatomy and roughly corresponding to "phyla" of traditional taxonomy, but the unity of developmental type appeared to include all animals. The first conceptual expression of these ideas was the "zootype" proposed by Jonathan Slack, Peter Holland, and Chris Graham (Slack, Holland et al. 1993). In this influential article, the authors proposed a "morphological" definition of the term "animal" based on the shared expression and function of genes.

In a 2003 review of the zootype concept, Jonathan Slack included the following list of characters: an anterior-posterior body axis patterned by a cluster of homeobox genes (a.k.a. *Hox* genes); anterior patterning by *Otx* and *emx*; a photoreceptive organ based on *Pax6* activity; a heart dependent on *Nkx2.5*; and posterior structures that were dependent on the activities of *evx* and *cdx*. What was intriguing, and had already been pointed out in the original 1993 paper, was that these genes were all expressed in the phylotypic stage of the animals that were compared. These initial observations have been confirmed on a genome-wide basis in mouse in which it was found that, during the phylotypic stage of the mouse, the embryo will preferentially express widely shared developmental genes compared to other stages of development (Irie and Sehara-Fujisawa 2007) and the vertebrate phylotypic stage expresses more conserved genes than other stages of development (Hazkani-Covo, Wool et al. 2005).

The main challenges to the zootype concept arise from developmental studies of cnidarians and other diploblast animals. A phylogenetic analysis of cnidarian homeobox genes suggests that the Hox cluster is not a synapomorphy of all animals, but is probably a derived condition of bilaterians and, thus, the cnidarians do not have a body axis that is patterned in the same manner and by the same mechanisms as those of insects and mammals (Kamm, Schierwater et al. 2006). Rather, it appeared that the primary body axis was controlled by *Wnt* genes and that *Wnt* genes secondarily recruited homeobox genes into this function (Ryan, Mazza et al. 2007). Thus, it is likely that the zootype, as conceived by Slack and co-authors in 1993, is in fact a "bilaterian-type." I think that this is an important correction of the range of applicability of this concept, although it does not undermine the basic insight that there are deep commonalities shared among a wide array of animal lineages included in the major clade in the animal kingdom, bilaterians. Fortunately, there are metazoans that do not share this developmental type, which offer us a glimpse into the developmental evolution of early metazoan animals.

Developmental Character Types and Deep Homology

Soon after the zootype concept was introduced, progress in the detailed analysis of organ specific development also revealed deep developmental genetic commonalities among specific body parts. This view was forcefully argued by Neil Shubin, Cliff Tabin, and Sean Carroll with regard to limbs and body appendages in 1997 (Shubin, Tabin et al. 1997). Owen's argument was that vertebrate limbs share anatomical similarities that cannot be explained by functional needs, and his argument even runs contrary to functional explanations. Shubin, Tabin, and Carroll went two steps further.

First, they pointed out that tetrapod limbs and, in fact, fish fins as well, shared developmental gene expression patterns and patterning mechanisms. They all have a posterior signaling center based on the expression of the same gene, *Shh*. They also share many aspects of Hox gene expression and a distal signaling center, the Apical Ectodermal Ridge, which expresses *Fgf* signaling molecules. A similar developmental type can be formulated for insect and arthropod limbs depending on *Dll* for their outgrowth and so on. Hence, clearly homologous characters do share highly conserved developmental mechanisms.

Subsequently, Shubin, Tabin, and Carroll (STC) took a second step (Shubin, Tabin et al. 2009) by drawing attention to even broader similarities, even those between insect limbs and vertebrate limbs. STC did not argue that insect limbs and vertebrate limbs were homologous. Their argument is more radical—namely, homology of developmental programs that transcends the homologies of individual structures (i.e., what they called "deep homology"). Their idea was that even though vertebrate and insect limbs were not homologous as body parts, their genetic machinery might be homologous because it had evolved to pattern some kind of a body appendage in a common bilaterian ancestor.

This developmental machinery, dedicated to making things protrude from the body wall, has, according to their idea, been recruited to pattern independently derived appendages throughout the history of life. Good support for this view comes from the observation that clearly novel appendages, such as the horns on the pronotum of certain beetles, also use a similar set of genes (figure 1.9). Hence, it is possible that specific independently derived body appendages have evolved that share a similar molecular patterning mechanism. A dramatic demonstration of this principle was the recent discovery that the pronotum shield of treehoppers was likely homologous to insect wings (Prud'homme, Minervino et al. 2011).

Whatever the merits are of the detailed scenarios proposed by Slack, Holland, Graham, Shubin, Tabin, and Carroll, it is clear that the notion of conserved developmental types gained substantial credibility with the developmental biology revolution during the 1980s and 1990s. After all, this partial list of proponents of a new structuralism includes some of the best biologists of this generation. Thus, it is time to re-assess the role and conceptual makeup of structuralist ideas in evolutionary biology. This book was written to do exactly this by focusing on one of the most difficult and confusing concepts: homology.

A.

B.

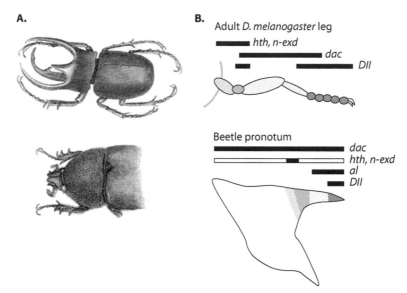

FIGURE 1.9: Similarity of horned beetle horn development and insect limb development. A) Pronotum of the horned beetle (top) compared to a non-horned beetle (below). B) Comparisons of gene expression along the insect leg and the beetle horn. Reprinted by permission from Macmillan Publishers Ltd: *Nature*, Shubin et al., 2009, *Nature* 457: 818, copyright 2009.

The Enigma of Developmental Variation

Before engaging in our project to rethink and re-evaluate the homology concept in the light of evolutionary developmental biology, we have to confront the main obstacle to a genetic and mechanistic understanding of developmental types. As impressive as the evidence is for conserved genetic mechanisms that underlie the development of distantly related animals, there are equally impressive amounts of data that show that aspects of development other than those summarized by the zootype and deep homology are variable. Variable development in itself is not surprising given the diversity of life. What is problematic, though, is the fact that clearly homologous characters can derive from different developmental mechanisms in different species (Butler and Saidel 2000). The best documented example is the development of the body axis and segmentation in insects in which homologous genes have different developmental roles in different insect orders. These and other examples will be discussed later in this book and do not need to be explained in any detail here.

The main point, however, is that the comparisons of developmental mechanisms across animals reveal a mix of conserved and variable components.

Only a theory regarding the role of "developmental types" that addresses this conundrum can be useful. Only when we can understand both the conservation and the variation of the development of homologous characters can we have a chance at successfully integrating developmental biology into evolutionary theory.

2

A Conceptual Roadmap to Homology

In chapter 1 it was argued that the core idea of variational structuralism is that organisms and their parts play causal roles in shaping the patterns of phenotypic evolution. It was stated that homology is a problem for which structuralist thinking has to prove its value. It was also noted that explaining homology, the origin and maintenance of character identity over phylogenetic time, faces considerable obstacles. In particular, this is because the developmental processes underlying homologous characters change over time. In this chapter I want to begin working toward the goal of proposing a mechanistic account of character identity, its origin, and its maintenance. To achieve this goal, we must first clear the ground at the conceptual level; there is a lot of ground to clear.

This is certainly not the first attempt to clarify the homology concept. Many of the previous attempts were successful in what they wanted to achieve (Remane 1952; Hennig 1966; Riedl 1978; Patterson 1982; VanValen 1982; Roth 1984; Sattler 1984; Wagner 1989; dePinna 1991; Donoghue 1992; Hall 1994; McKitrick 1994; Shubin 1994; Rieppel 1996; Müller and Newman 1999; Laubichler and Maienschein 2003; Ghiselin 2005; Brigandt 2007; Wiley 2007). For example, in molecular evolution it was necessary to differentiate the concept of homologous genes into two different notions: orthology (i.e., corresponding genes in the same genomic location), and paralogy (i.e., homologous genes created by gene duplication; Fitch 1970). In phylogenetic systematics it was necessary to distinguish between shared derived homologous characters (apomorphic) and shared ancestral characters (plesiomorphic characters).

What is distinct regarding those attempts to clarify homology and the conceptualization proposed here is the particular goal. My goals here are not those of a systematist, but are those of an evolutionary biologist. In other words, my purpose is not to facilitate the reconstruction of phylogeny and the description of biological diversity and disparity. Rather it is to support the research program of developmental evolution that seeks to explain the patterns of phenotypic (mostly morphological) diversity, as sketched out in chapter 1.

Two Observations: Sameness and Continuity

There are numerous examples of corresponding characters between species for which it is hard to escape the conclusion that organisms from different species are clearly composed of the same building blocks, such as heads, limbs, and brains. It is worth reflecting on what makes these examples so compelling. Take for example the human hand or foot and that of a chimpanzee (figure 2.1). It does not require any formal training to recognize that these are corresponding (i.e., "the same") body parts, in spite of their differences in size and shape. But why do we arrive at this conclusion? We need to refrain from the simple answer, namely "common sense," because that would distract from a more principled analysis of this problem. Of course, common sense often provides a reliable answer, although it does not help us to understand what causes Nature employs to create the patterns that we encounter.

The most obvious answer to the question of why the hands of a human and a chimpanzee are perceived as the same is similarity. The hands of humans and those of chimpanzees have the same sets of bones, although in different sizes and proportions, and the soft tissues are also similar, albeit not identical. Clearly, differences also exist in more interesting ways—in particular, the structures and functions of their thumbs. These are also more similar to each other (i.e., across species boundaries) than each of them is to any other part of the body of a human or chimpanzee (except for the hand on the other side of the body, which is a mirror image of the other).

The closest thing to a human hand in the chimpanzee body (other than the chimp hand) is the foot. However, the idea of identifying the human hand with the foot of a chimpanzee is quickly dismissed by the anatomical context. The hand is found attached to an extremity that is close to the head, whereas the foot is attached to the rear end of the body. These are the two classical criteria for homology: similarity of structure and similarity of position.

So far so good. But, let us reflect and ask what this teaches us about evolution? In cases where homology (i.e., correspondence or "sameness") is most obvious, a body part of one species is most similar in structure and position to a body part in another species than it is to any other part of the same organism. Thus, the inference is that the two homologous body parts in these two species have been inherited from the most recent common ancestor for these species. Hence, it is assumed that it is more likely mechanistically that the common ancestor had the same body part and that evolution caused only the small-to-moderate modifications like those that distinguish the human from the chimpanzee hand.

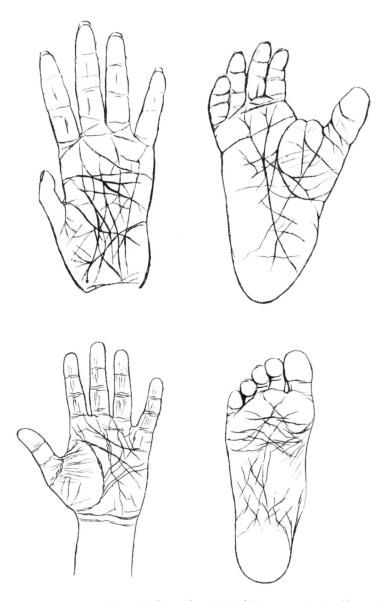

FIGURE 2.1: Comparison of hands (left) and feet (right) of chimpanzee (top) and human (bottom). Although they are not identical, their overall similarity leaves little doubt that these body parts are fundamentally the same kind of thing. This basic observation is the core of the homology concept. Homology becomes a problem with more distantly related organisms for which the correspondence of body parts is not that obvious and for which the units of comparison are debatable. (Source: http://www.kohts.ru/ladygina-kohts_n.n./ichc/html/apd.html.)

There are a number of conceivable alternative scenarios though. For example, it is possible that the most recent common ancestor of humans and chimpanzees had no such thing as a hand. In this case, one would have to assume that they evolved independently in the human and the chimpanzee lineage and only appeared to be similar because they served the same function (grasping). This is a far-fetched scenario, given that we know that all the other relatives of humans and chimps also have hands and all tetrapods probably originally had what technically is called an autopod. Yet it is useful to ask what this scenario would mean in terms of evolution.

Assuming that the human, the chimpanzee, and the human hand arose independently would mean that the origin of a complex anatomical structure was more likely than the modifications that distinguishes human from chimpanzee hands from that of their common ancestor. Of course, we know from other evidence that this is not true, because all mammals have hands (autopods) and we do observe a lot of variation in the sizes and proportions of digits and the loss of digits. Yet at least in mammals, no independent evolution of hands is likely based on studies of mammalian phylogeny and natural variation (McKitrick 1994).

There are even more (mechanistically) unlikely scenarios. Among them is the idea that the common ancestor of humans and chimps looked like a chimp, but, in the human lineage, the whole limb was lost and replaced by an outgrowth of the skin, say a mammary gland, which then morphed into something resembling an arm. Among German zoologists of the twentieth century, inventing such outlandish scenarios was a form of entertainment (Stümpke 1967). Again, the reason that we do not think this is likely is that the amount of genetic change that would be necessary to morph a skin gland into an arm and a hand is so much greater than the modifications that are necessary to account for the differences between human and chimp hands and any other mammalian autopod, regardless of how strongly modified it might be.

What all these scenarios lead to is that any hypothesis regarding homology (i.e., a statement regarding the correspondence between body parts of different species) is an implicit hypothesis regarding the mechanisms of evolution (Wagner 1995). Specifically, this assumes that the observed differences between two homologous body parts are more likely to occur by mutation and selection and drift than from (1) the independent origin of these structures and (2) the transformation of any other part of the body into these characters (say a hand). In other words, homology implies that inheritance with modification is more likely than losing the character and/or creating it, and it is more difficult to turn one character (say a nose) into another (say a limb).

Homology thus implies a pattern of evolution in which stability and variability are connected in a very specific way. Any character that can be homologized is assumed to have continuity in terms of its existence in a lineage of descent, as well as persistence of differences from other parts of the body (individuality). On the other hand, any character is variable in various ways that allows natural selection to modify it in response to particular requirements for survival and reproduction. These modifications are called adaptations. For example, limbs lose digits if an animal uses them for fast running because it decreases the mass that needs to be accelerated during each step. Most, if not all, of these differences among limbs are correlated with different lifestyles (running, swimming, flying, and so on) and habitat needs (running on solid ground versus on sand). On the other hand, the origin of a body part that forms lineages and thus gives rise to a set of homologous structures in descendent species is called a novelty.

Thus, we have two distinct phenomena to explain. We distinguish between the origin of homologs (i.e., novelties[1]) and their modification by natural selection (i.e., adaptation). The former is studied in developmental evolution, whereas the latter is studied in functional morphology, ecological genetics, life history theory, and other branches of evolutionary biology.

The Limits of Homology: Lack of Individuality

As every student of biology knows, not every circumscribable part of an organism has individually recognizable counterparts, even among close relatives. Examples of this include hair and red blood cells. Each cell and each hair is a physically distinct entity that is separated from other instances of the same kind of cell or hair either by a cell membrane, in the case of cells, or by having its own root and shaft, as in the case of hair. For these cases, the physical distinctness is not matched by any form of biological individuality that would allow us to recognize which specific cell in my bloodstream corresponds to which specific blood cell in the body of my brother.

Of course, we know why this is so. Each individual red blood cell is the product of the same genetic program that created all other red blood cells of the same type. It is not only impossible to tell which cell in my bloodstream corresponds to which cell in that of my brother, the question is biologically meaningless. Each of my red blood cells is an instance of the same thing as any other blood cell, as it expresses the same genetic information. This problem, of course, is not limited to blood cells and hairs, as it extends to many cases of repeated parts, like some of the segments in annelid worms and centipedes, the uniform row of legs in millipedes, multiple leaves on a plant, and so on.

---

[1] Instead of novelty the term innovation has been used. It has been found useful to distinguish between the origin of novel body parts and novel functions. The latter has been called innovation, while the former is called novelty. Müller, G. B. and G. P. Wagner (1991). Novelty in evolution: restructuring the concept. *Ann. Rev. Ecol. Syst.* 22: 229–256, Love, A. C. (2003). Evolutionary morphology, innovation and the synthesis of evolutionary and developmental biology. *Biol. Phil.* 18: 309–345.

No given hair is exactly the same as any other, at least not at the molecular level. Yet what these examples lack is a consistency of differences that distinguish between different instances of the same kind of hair.[2] What do we learn from this somewhat trivial example regarding evolution? We learn that not every physically distinct, recognizable part of an organism can be the subject of a meaningful homology statement (i.e., a statement that leads to a one-to-one identification of it in one species to its counterparts in another species). The difference between these examples, blood cells and hairs, and those in the previous section, hands, is that hands have a genetic/developmental machinery that is dedicated to making them different from all other parts of the body (including feet), whereas each hair on my head and each red blood cell in my bloodstream derives from the same genetic information.

The individuality of body parts, required for homology to make biological sense, requires specific genetic and developmental mechanisms to cause the distinctness of the body part during the life of an individual and continuity of distinctness in the course of evolution (Riedl 1978). Hence, homology is a reflection of the highly non-trivial manner in which developmental and genetic information is organized. Physical, descriptive distinctness is not sufficient to infer the kind of biological individuality that is necessary to recognize a part of the body as a homolog. Empirically, it is often difficult to recognize and decide upon whether two parts of the same body are individualized and, thus, should be treated as two different characters and merit distinct names.

A Detour into Genetics: Homologous Genes

The homology concept is an attempt to identify and name the units of phenotypic organization (i.e., those body parts that have historical continuity and that can often be found in many species derived from a common ancestor). The problem of identifying basic building blocks of reality, however, is not limited to comparative biology, but is an objective of every major branch of science, as discussed in chapter 1. Nowhere is this more evident than in chemistry with its clearly circumscribed inventory of objects: atoms and ions consisting of nucleus and electrons; molecules consisting of covalently bound atoms; and complexes consisting of loosely connected molecules.

For my present purpose, the most useful example, however, is genetics, with its distinctions between genetic loci and alleles, and between orthologous and paralogous genes. The formal structures of these distinctions are useful for sorting out the phenomenology of phenotypic variation, as well

---

[2] Of course there are different kinds of hairs that are individualized from other kinds, like whiskers and body hair and pubic hair versus head hair. These are examples for the origin of novel characters through differentiation (see chapter 4).

as evolution (i.e., to recast homology into a language that can support a research program on the evolution of organisms and their morphological organization).

Loosely speaking, a gene is a part of the genome that is largely unaffected during its transmission from generation to generation. Genes come in alternative forms known as alleles, which can take each other's place within the genome. Alleles are recognized because they cause differences in the phenotype, as for example different colors of flowers or shapes of peas. For this phenomenology to hold, it is important that one and only one copy of a gene and, hence, only one allele is transmitted from a parent to its offspring. An organism can have one or two different alleles, depending on whether the alleles of both the mother and the father are retained.

Under normal circumstances (i.e., in the absence of gene duplications or deletions), genes are transmitted from generation to generation and, thus, form a lineage of descent (figure 2.2A). In this respect, genes behave like species or phenotypic characters, or any other entity that is transmitted from generation to generation; for example, mitochondria and chloroplasts. If the species in which a gene resides forms another species, and if this sister species retains the gene, then the gene lineage also splits. Hence, the gene lineage will form a tree of descent that reflects the species tree.[3] This is the reason why comparisons of gene sequences can be used to reconstruct the phylogeny of a set of species. When I speak here of a gene lineage, I mean the tree of descent that reflects the various lines of descent following the species phylogeny.

If, however, a gene is duplicated (i.e., if there are two genomic locations with copies of a gene), then each gene begins to form its own lineage of descent (figure 2.2B). Once a gene is duplicated, it is formally a different gene, even though initially the two copies may have been identical at the molecular level. Duplicated genes are also homologous in the broad sense, as they are both derived from the same ancestral gene; but, they eventually will diverge and are then called paralogous genes. Paralogous genes are those that are related to each other through a gene duplication event and this relationship can also be visualized in a phylogenetic tree (figure 2.2B).

In contrast, genes that occupy the same genomic locus within different species are called orthologous genes. These are defined as genes that are related to each other only through speciation events. Examples of paralogous genes are red and green opsin genes in humans. These genes encode for the red- and green-light-sensitive pigments in our retina and allow us to discriminate between red and green light. In lower primates, there is only one

---

[3] A complication can arise if the population undergoing speciation contains an ancient polymorphism and only one of the alleles is assigned randomly to each descendent species (Felsenstein, J. 2003. *Inferring Phylogenies*, Sunderland, MA: Sinaure). But if during speciation the gene is non-polymorphic, the gene tree is congruent with the species tree.

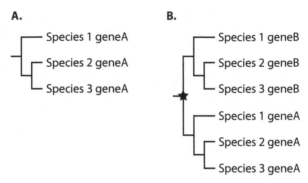

FIGURE 2.2: Relationship between gene and species genealogies and how gene duplications can be detected by phylogenetic analysis. A) Species tree and gene tree that are congruent, and all three genes are homologs. B) Gene tree indicating a gene duplication event. Genes B from different species form a clade that is separate from the A genes from the same set of species. Also, the branching order within the A and B gene clades is identical and reflects the species phylogeny. The gene duplication event is indicated by a star in (B). This is the most convincing evidence that genes A and B are paralogs even if they may be very similar at the sequence level.

red/green opsin gene, which encodes for an opsin with intermediate light sensitivity.

In the absence of gene duplication and along the line of descent of a gene, alleles will replace each other so that different forms of the gene represent the same gene at different times during phylogeny and in different species. Hence, a gene's identity is based on the continuity of descent and not primarily on the similarity of the gene itself in different species or at different times during evolution. The "same" gene in different species can be quite dissimilar because of the ongoing process of allele replacement due to mutation, selection, and drift during evolution. For example, the *bicoid* gene in Drosophila is a homolog of the *Hox3* gene in other arthropods (Stauber, Jackle et al. 1999), yet it is so modified that it is difficult to be recognized outside of higher flies.

Of course, similarity remains a valuable heuristic to find homologous genes in different species. For example, homologous transcription factor genes can be identified and cloned from different species because many of these have conserved sequence motifs, like the homeodomain. Homeodomain proteins are transcription factors that share a highly conserved 60 amino acid motif that is involved with, among other things, DNA binding. The homeodomain is coded in the genome by a corresponding sequence of 180 nucleotides called the homeobox (Gehring 1994).

Because this motif is so highly conserved, fragments of homologous genes can be found by using molecular probes that recognize this conserved sequence (Murtha, Lechman et al. 1991). However, these practical approaches should not be confused with the conceptual basis of gene identity, namely,

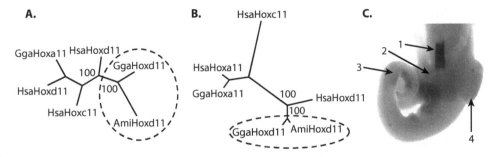

FIGURE 2.3: Example of how an ortholog gene can be identified in molecular evolution. A DNA fragment similar to mouse and chicken *HoxD11* genes was cloned from alligator (*Ami-Hoxd11*). This gene fragment was subjected to phylogenetic analysis to establish that it was a *HoxD11* gene. It was compared to chicken *HoxD11* (*GgaHoxd11*), human *HoxD11* (*HsaHoxd11*), as well as the paralogs chicken *HoxA11* (*GgaHoxa11*), human *HoxA11* (*HsaHoxa11*), and *HoxC11* (*HsaHoxc11*). A) Maximum Parsimony analysis, B) neighbor-joining analysis. The important point is that hypothesized alligator *HoxD11* forms a clade with other known *HoxD11* genes and that this tree reflects the structure of the known species tree to the exclusion of the paralogs, *HoxA11* and *HoxC11*. C) The determination that the alligator gene is in fact a *HoxD11* ortholog is also supported by its expression in alligator embryos in known *HoxD11* expression domains: hindgut (1), genital tubercle (2), tail (3), and limb (4). (Source: Vargas et al., 2008, *PLoS ONE* 3:e3325.)

the continuity of descent in a lineage of genes. Therefore, any gene cloned based on sequence similarity has to be investigated by phylogenetic analysis for whether it belongs to the same lineage as the gene one is attempting to clone. The purpose of phylogenetic analysis is to establish whether the gene isolated based on sequence similarity establishes a lineage with its orthologous genes in other species.

If two genes are "the same" (i.e., belong to the same lineage), then they need to be more closely related to each other than they are to paralogous genes. Figure 2.3 shows a simple example for which we identified a homeobox gene from alligator—the alligator ortholog of the mouse *HoxD11* gene (Vargas, Kohlsdorf et al. 2008). This gene was found in the alligator genome because a comparison of the mouse and chicken *HoxD11* genes identified two highly conserved stretches in the amino acid sequences of their corresponding proteins; one was the homeodomain and the other was a sequence at the N-terminal end of the protein.

We designed primers that recognized these two conserved sequences and allowed us to copy the intervening DNA sequence and clone it. To confirm that the gene fragment cloned was, in fact, the *HoxD11* gene from alligator rather than a paralog (for example *HoxA11*), we had to show that the sequence we isolated was more closely related to other *HoxD11* genes from other species than to any other paralogous genes in the genome.

In the mammalian genome there are 39 other genes that form a gene family that all share the homeodomain, called the Hox genes.[4] We tested the identity of our alligator gene using a phylogenetic analysis of our new alligator gene fragment, its putative homologs in mouse and chicken, as well as paralogous genes from the mouse/human genome, which are the most closely related genes to the targeted gene *HoxD11*; in this case, genes from the same "paralogous group": *HoxC11* and *HoxA11*. As seen in figure 2.3, our putative alligator *HoxD11* gene is more closely related to the chicken *HoxD11* gene than to the human *HoxD11* gene. This was expected because the crocodilians are more closely related to birds than they are to mammals.

In addition, all of the *HoxD11* genes included in our analysis formed a clade (i.e., more closely related to each other than to the other "group 11" genes, like mouse *HoxA11* and *HoxC11*). This showed that our putative *HoxD11* gene formed a lineage of descent with other *HoxD11* genes and was, thus, the "same" gene as the *HoxD11* gene in mouse. In contrast, the alligator *HoxA11* gene was more closely related to the mouse and chicken *HoxA11* genes than it was to the alligator *HoxD11* gene (Figure 2.3A and B). Hence, the *HoxA11* genes had formed another line of descent that ran parallel to that of the *HoxD11* genes.

The *HoxA11* and the *HoxD11* genes each individually reflected the phylogenetic relationships of the three species compared. That is, the alligator gene was always more closely related to the corresponding chicken gene than it was to either the mouse gene or the other homeobox genes in the alligator genome (alligator *HoxA11* and *HoxD11* genes). The interpretation of our newly sequenced gene from alligator was further corroborated by the similarity of mouse and alligator *HoxD11* expression domains—for example, in the genital tubercules and the hindgut (figure 2.3C). Hence, similarity is our guide to gene discovery, although gene identity is judged based on lines of descent that have to be tested with phylogenetic methods.

To summarize, genes are transmitted from generation to generation in a limited number of copies such that different forms of the gene (i.e., alleles) replace each other during the course of evolution. In this way, a gene establishes a lineage that most often follows the phylogenetic relationships among the species in which they are found. Gene attributes can change (e.g., their sequences or their intron-exon organizations and functions), yet they maintain their identity as long as they represent a lineage of descent in each species lineage in which they are present (unless the gene is lost from the genome). Deviations from this basic pattern include horizontal gene transfer[5] and gene duplication.

---

[4] The family of homeobox genes is actually much larger, but *HoxD-11* belongs to a subfamily of homeobox genes called Hox genes. These are defined as those genes that are homologous to the homeobox genes in the homeotic gene cluster in *Drosophila melanogaster*. This Hox gene family in humans has 39 members.

[5] Horizontal gene transfer is a process during which genes are transferred from one species to another or from one individual to another outside the parent offspring relationship. Examples are

Gene Lineages, Natural Selection, and Gene Function

As mentioned previously, in the case of gene duplication, a copy of a gene is produced that also resides within the same genome as the original copy. Gene duplication thus leads to independent lineages of descent and is, therefore, analogous to a speciation event. Even though both copies of a gene are homologous in the broad sense (i.e., both gene copies are related to each other through a gene duplication event), they play fundamentally different biological roles.

Orthologous genes are related to each other through allele substitutions and, thus, due to competition among their alleles. Competition results because only a fixed number of alleles (usually one) can be transmitted from each parent to the offspring. By compariosn, there is no theoretical upper limit for the number of paralogous genes within a genome. For example, the mouse genome has 26 paralogous copies of the gene that encodes for the prolactin hormone, whereas humans have only one prolactin gene. Hence, paralog gene copies do not compete with one another for their representation within a genome, but alleles do.

The mode of inheritance of a particular gene leads to competition among its alleles and, thus, is necessary for evolution by natural selection or genetic drift. For this reason, I think that the strict one-to-one mode of inheritance of only one allele per parent is a functional necessity for the efficacy of natural selection at the genetic level. Other forms of gene inheritance are possible, as for example the genes in the macronucleus of ciliates in which each gene is present in multiple copies and this "bag of genes" is simply randomly divided into two when the cell divides.

In contrast, the strict one-to-one mode of inheritance in most eukaryotic organisms is a highly contrived situation that requires a complicated machinery of chromosome sorting, or meiosis. In yeast, meiosis is made possible by at least 300 genes that encode for the machinery that guarantees the orderly transmission of only one gene copy and only one chromosome of each type to the next generation (Bresch, Müller et al. 1968; Thompson and Stahl 1999). Hence, the mode of inheritance that maintains gene identity by a strict lineage of descent from generation to generation is not a primitive consequence of physics and chemistry; rather, it is an emergent biological phenomenon that requires a complicated molecular machinery, presumably maintained by natural selection, and also necessary for evolution by natural selection.

Because the mode of inheritance that creates a single line of descent for each gene is also responsible for the competition among the alleles of that

---

fungal genes found in aphids necessary for carotene synthesis and the transfer of resistance factors between bacteria. Even though there are many documented examples, in particular in bacteria, horizontal gene transfer in higher organisms is a relatively rare event.

gene, gene identity by descent is also tied to the unique functional role that a gene plays. Hence, gene identity through descent with modification is not only a historical, genealogical fact but is deeply tied to the functional individuality of a gene. Each gene within a genome forms its own line of descent and experiences its own competitive dynamic among its alleles.

Each gene is a genetic unit of adaptive change. This adaptive dynamic at each gene locus molds each gene for its own functional role in the genome. Complementary to the competitive dynamics among alleles is that genes at different loci do not compete with each other. Thus, they can cooperate in supporting the organism. Competition is limited to alleles of the same gene, whereas genes at different genomic loci are insulated from the effects of competition and are, therefore, free to cooperate.

The fact that, under most circumstances, adaptation by natural selection depends on the competition among alleles at the same locus led to a way of thinking about evolution known as gene selectionism (Williams 1966; Brandon 1982; Sober 1987; Brandon 1999; Hull 2002). Gene selectionism asserts that genes are the fundamental units of biological organization, because it is the alleles at genetic loci that exhibit the competitive dynamics that lead to adaptation by natural selection. This way of thinking posits that any explanation of adaptive change has to be done at the level of the single gene. No higher levels of organization seem to play a causal role. For example, altruistic behavior at the organismal level is explained by "egoistic" genes that seek their selective advantage through "manipulating" organisms for their adaptive advantage (Dawkins 1976; Dawkins 1978).

While there is much to be discussed with respect to this view of evolution, gene selectionism nevertheless reflects a fundamental fact about biology: genes are set up to be a privileged locus of adaptive evolution because of their mode of inheritance. Their mode of inheritance leads to competition among alleles and no competition among genes at different loci.

The main point of this discussion is that the formal aspects of transmission genetics, the distinction between genes, alleles, and paralogous genes, is not just a random curiosity of life; rather, it is deeply connected with the mechanism that makes life possible: natural selection. Natural selection, in turn, shapes genes for their life-sustaining physiological functions. Notably, morphological characters share many formal characteristics with genes. Many are represented in the body as a single copy: only one brain, one nose, and one left arm. Each character can vary, but only one variant is found in the next generation (as discussed further below). It is my suspicion that the formal similarities between transmission genetics and phenotypic variation and inheritance reflect essential organizational features of organisms related to their evolution by natural selection (Wagner 1995). *Hence, homology of phenotypic characters and gene identity and transmission are parts of the same broader principle of life at different levels of biological organization.*

Duplicated Genes without Individuality

Before we return to patterns of morphological diversity, I want to briefly introduce another, more specialized genetic phenomenon that will be important later for drawing the analogy between gene evolution and morphological evolution. This is the case of gene families (i.e., groups of physically distinct genes that do NOT form individual lineages of descent). At face value this does not make sense in the context of the previous discussion because we argued that copies of a gene within the same genome can mutate independently and, thus, experience their own more or less independent evolutionary history.

Thus, it was a surprise when it was first found that there are clusters of genes in the genome that do not follow that pattern. For example, the genes for ribosomal RNAs occur in the genome in many copies, probably because of the need for large amunts of ribosomes for protein synthesis (Coen, Strachan et al. 1982). This gene family exists in most vertebrates and, thus, is likely an ancient feature of the vertebrate genome. One would expect then that the members of this gene family had diverged, as had other paralogous genes. Surprisingly, however, the different copies within an individual and species are much more similar than they are to rRNA genes from other species.

It is now clear that the members of this gene family, as well as those of some other gene families, evolve in concert; that is, they exchange information among one another so that the members of the family in the same species remain similar while diverging between species. The most common mechanisms to achieve this are frequent, unequal crossovers among members of the gene family, which homogenizes the sequence composition of the family, and gene conversion, during which one copy of a gene imprints its sequence on another member of the gene family. The result is that the members of this gene family, even though they are physically different segments of the DNA, evolve as a unit and, thus, form a single lineage of descent with modification.

An additional consequence of this example is that the question of whether a piece of DNA is a unit of evolutionary change or not is not determined alone by its physical organization but by the manner in which transmission and mutation processes work. Thus, we always have to consider the evolutionary dynamics of variation and selection in order to determine whether two physical pieces of the genome are biologically distinct entities or belong to the same lineage of descent. Similar reasoning will hold for some phenotypic characters that exist in multiple instances within the same body.

Character Identity and Character States

The discussion of gene identity in the previous sections showed that in genetics, we have to distinguish at least three different notions of sameness:

- Allelism: two genes in the same genomic context and having the same structure are the same allele. If two genes differ in their structures and are on the same locus, then they are different states (called alleles) of the same gene. Only one allele is transmitted from each parent to the offspring. Alleles compete for their presence in the next generation.
- Orthology: genes from different individuals or species in the same genomic context and most of the time having different structures. These genes form a lineage of descent by mutation and allele replacement, and they diverge only because of speciation events, which lead to different branches on the gene tree. Gene identity is based only on participation in the same lineage of descent, and not on similarity. Genes that form a lineage are the fundamental unit of evolution because their different states, alleles, compete for their representation in the next generation.
- Paralogy: genes that are related because they are derived from the same gene in an ancestor, but now occupy different genomic loci and form different lines of descent. Paralogous genes do not compete by natural selection; rather; they acquire unique functional responsibilities in the organism. They are no longer the same genes because they do not belong to the same unit of genetic evolution.

I will argue here that the same formal types of sameness apply to morphological characters. The failure to distinguish between these types of relationships among phenotypic characters is the cause for much confusion regarding homology. All of these relationships among genes represent broad-sense homology in the sense that all are based on genealogical continuity. They differ only in the kinds of events that connect them. Alleles are more or less direct copies of each other. Orthologous genes are related by direct descent and speciation, while paralogous genes are additionally related through at least one gene duplication event. A similar logic applies to entities at other levels of biological organization: autonomous organelles within cells, cell types, multicellular organs like limbs and brains, and perhaps even to functions and behaviors. Let us build the argument with the intuitively most accessible case: multicellular organs (i.e., morphological characters).

Typical morphological characters with clear homology among species usually come in a fixed copy number per individual. Each normal vertebrate has one brain, one head, two eyes, and, if any, never more than four limbs. Furthermore, these characters always come within the same organismal context; the head is at the anterior end of the body, the brain is within the head, and the forelimbs are attached to the shoulder girdle. These facts are uncontroversial because they are the basis for the classical phenomenological criteria of homology.

In addition, characters vary, as they come in different sizes, shapes, and colors (i.e., there are different *character states* for the same character).

However, in each organism, only one character state can be realized because each organism has only one copy of the respective character. Hence, character states replace each other between generations, just as alleles do. Over time, characters can form lineages of descent in which within different species different states of the same character are represented. For example, most tetrapods have two forelimbs; otherwise they have none. There is no wild-type tetrapod animal that has more than two forelimbs. Yet the forelimbs of different animals are modified (as described above) relative to the lifestyles of the respective animals. Hence, one can reconstruct a phylogeny of forelimbs that will follow the species' relationships, but describe the modifications they underwent due to selection or drift among alternative character states.

Of course, this analogy between genes and morphological characters is formal, and not material. The most important difference is that genes make direct copies of themselves. There is even material overlap, as each copy of a gene contains one DNA strand of the parental gene. In contrast, morphological characters are not direct copies of each other, but are built anew within each generation. The inheritance of characters and character states is more indirect and mediated through genes that orchestrate development. The relationships between the continuity of genes and the continuity of morphological characters will be discussed in the next chapter. Here I focus on the formal, phenomenological similarities, which guide my conceptual approach to homology.

Like genes, the identity of a morphological character is not tied to similarity; rather; it is tied to the historical continuity of descent. To illustrate this point, I want to introduce an example that will become important in the next chapter, in which we will discuss the genetic basis of character identity. Ancestrally (pterygote, or winged) insects have two pairs of proper wings associated with T2 and T3 (the second and third segments in the thorax, T) as is seen, as examples, in honey bees, grasshoppers, and most spectacularly in butterflies (figure 2.4A). In dipterans, there is only one pair of wings in the sense of a lift-producing appendage, which is localized on T2 (i.e., it is a forewing). There is a homolog of the hind wing on T3, but it is not shaped as a wing blade. Rather, it is a small, club-like appendage called the haltere (figure 2.4B). This appendage is a sensory organ, which helps maintain stability during flight, rather than being a lift-producing organ and has no shape similarity with typical wings.

Beetles also have only one pair of wing blades. However, this pair is associated with T3 and, thus, is a hind wing. The forewings of beetles have been transformed into a pair of highly sclerotized structures, called elytra, which function as protective covers for the abdomen (figure 2.4C). Again, elytra have very little similarity with typical wings, but are clearly homologous to forewings. Hence, butterflies, flies, and beetles all have two pairs of dorsal appendages that are homologous among species.

There is no doubt regarding the homology of elytra and halteres with the fore- and hind wings of four-winged insects, respectively, because beetles

FIGURE 2.4: Illustration of the concepts of characters and character states in insect wings. A) Butterfly with clearly recognizable forewing (FW) and hind wing (HW). Both are functionally "wings" (i.e., functionally adapted for flight, and signaling, in this case). B) In mosquitoes and flies, the forewing forms as a wing blade, but the hind wing is transformed into what is called a haltere, a sensory organ necessary for flight stabilization. C) In beetles, the flying wing is the hind wing, but the forewing is a protective cover, the elytra. Fore- and hind wings are character identities, and wing blade, halter, and elytra are character states of fore- and hind wings. (Source: Wagner, 2007, *Nature Reviews Genetics* 8:473–479.)

and flies are nested within a larger clade of winged insects, almost all of which have two pairs of wings, one pair on their T2 and one pair on their T3. Hence, homology (i.e., historical continuity) is the most parsimonious interpretation. There are the forewings that are shaped as flying organs in flies and butterflies, but are protective organs in beetles. And, there are the hind wings that are formed as functional wing blades in butterflies and beetles, but are sensory organs (halteres) in dipteran insects.

In morphology, we can distinguish between two kinds of entities. On the one hand there are *character identities*—for example, forewings and hind wings of insects. On the other hand there are various *character states* that insect wings can assume; the forewing can be a wing blade or an elytra or even a haltere (as in the enigmatic insect order Strepsiptera), and the hind wing can be a wing blade or a haltere. A wing blade can also have a different shape, structure, and color. The relationship between character identity and character states is the same as that for *gene identity* and *alleles* in genetics. There is even an analog of paralogous genes that, in morphology, is called serial homology or homonomy (i.e., morphological characters that exist in multiple copies in the same body like body segments, teeth of reptiles, or scales).

Characters and Character States: Who Is Who?

Teeth are the most durable remains of vertebrates and, thus, are the most abundant kinds of fossil in the fossil record. Much of paleonotology is based on comparisons of teeth and there are many theories that explain the evolution of teeth in great detail (Romer 1956; Fraser and Smith 2011). Mammalian teeth are among the most complex of teeth, particularly the molars

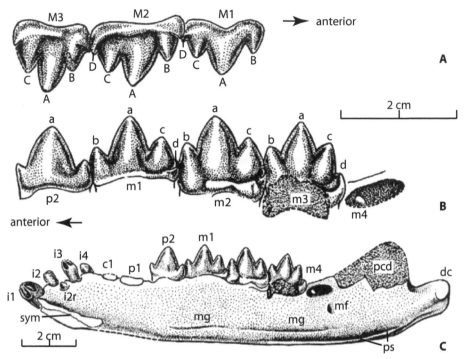

FIGURE 2.5: Teeth of the triconodont reptile *Jeholodens jenkinsi*. This is a stem mammalian with multi-cusped teeth. Note that each molar tooth has a prominent central cusp, which is tempting to compare to the single cusp of the typical reptilian tooth that consists of a simple cone. However, it is not clear whether the major and the minor cusps of these and other mammalian teeth are actually individualized and, thus, whether they are legitimate characters rather than features of the tooth, which is the character. Reprinted by permission from Macmillan Publishers Ltd: *Nature*, Ji Qiang, Luo Zhexi, and Ji Shu-an, 1999, *Nature* 398:326–330, copyright 1999.

and premolars, which consist of variable numbers of tooth cusps arranged in species-specific patterns. Most reptiles have only cone-shaped teeth that have only one cusp each.

The first stage during the elaboration of tooth shape in the stem lineage of mammals was the three-cusped tooth (a.k.a. the tritubercular molar) of animals aptly named *Triconodon*. In this tooth there is a main cusp in the middle and two minor anterior and posterior cusps (figure 2.5). It is tempting to identify the major cusp with the single cusp of the reptilian cone-shaped tooth. This was done by Cope in 1885 (Cope 1885) and named protocone by Osborn (Osborn 1888). From that starting point and onward, the challenge for paleontologists was to identify the homologies among all the cusps in mammalian teeth, which eventually resulted in a consensus view assembled in 1906 by Gidley called the "premolar analogy theory" (Gidley 1906).

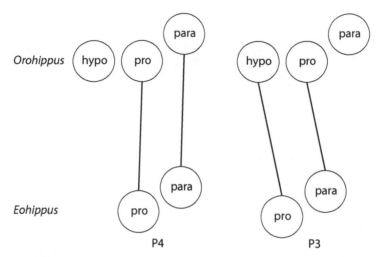

FIGURE 2.6: Example illustrating that tooth cusps are likely not individualized and do not form a unique line of descent. This example was analyzed by Van Valen in an influential paper in 1982. Van Valen compared the position of the lingual cones of premolars 3 and 4 (P3 and P4) in the basal horse species *Eohippus* and *Orohippus* (hypo = hypocone; pro = protocone; para = paracone). Note that in *Orohippus*, the configuration of cones is the same between P4 and P3, which led to giving the three cones the same names. In *Eohippus*, only two cones are present, called proto- and paracones. However, intermediate stages in evolution show that in P3, the hypo-and protocones of *Orohippus* correspond to the pro- and paracones of *Eohippus*, contrary to the situation in P4. Here, historical continuity and pattern similarity between P3 and P4 are in conflict. In particular, this comparison calls into question the serial homologies of the cusps between P3 and P4.

However, a seemingly minor detail during the evolution of horse teeth shows that the entire enterprise was conceptually flawed (VanValen 1982).

In horses, the premolar teeth had assumed the phenotype of molar teeth, a process called molarization. Initially, this process proceeded by adding cusps to the premolar teeth. In *Orohippus*, the number and the arrangement of cusps was very similar between molars and premolars. The problem, however, was that cusps that apparently corresponded among the molars and premolars in terms of position and shape in *Orohippus* arose in different locations (figure 2.6). Hence, cusps that correspond in location among teeth in the derived situation do not correspond to "the same" cusps during intermediate stages of evolution. Van Valen (1982) interpreted this observation as a breakdown of homology assessment among tooth cusps, which means that tooth cusps apparently lack the individuality to have historical continuity beyond that among very similar species.

What seems to have happened during horse evolution is that there was selection for increased grinding surfaces and each of the premolar teeth added cusps in different locations, depending on the details of its ancestral arrangement of cusps and its size. But beyond that, there is no special functional role that each cusp plays and they do not appear to be genetically individualized (see below). In our terminology, tooth cusps are not individualized characters, but teeth or tooth types are. Tooth cusps are the wrong level to attach statements of homology. The number and arrangement of tooth cusps is a character state of the homolog "premolar," but individual cusps are just parts of a unitary pattern.

Van Valen's interpretation was confirmed by recent work on tooth development (for a review of tooth development and its variational implications, see Jernvall and Jung 2000). Teeth develop from an epithelial mold formed by the ectoderm and filled with mesenchyme. The epithelial cells will produce enamel and the mesenchymal cells will produce dentin. The shape of the epithelial mold determines the shape of the final tooth.

Cusp development is initiated by a signaling center called the enamel knot, which secretes a set of signaling molecules: BMP, FGF, SHH, and WNT (Jernvall and Thesleff 2000). New enamel knots are initiated as the tooth germ grows, and at each new enamel knot, the same cocktail of genes is expressed. Hence, there is no molecular evidence that enamel knots and, thus, the cusps of the definitive tooth are developmentally or genetically individualized. Jervall and Jung (2000; abstract) write:

> It is unlikely that there is a simple 'gene to phenotype' map for dental characters. Rather, the whole cusp pattern is a product of a dynamic developmental program manifested in the activation of the developmental modules.

In contrast, different tooth types, incisors, molars, and so on, do express different transcription factor genes, a fact which is at least consistent with the notion that tooth types are developmentally and genetically individualized and, thus, the proper level of homology.[6]

The important lesson here is that not every morphological detail, even though it looks spatially well circumscribed, is necessarily an individualized character for which homology statements make biological sense. This is even true in those cases where similarity among species suggests correspondence among these elements, like between the single cusp of a reptilian cone tooth and the major cusp of the tritubercular molar. This important point was first brought to me (although I do not know whether he was in fact the first to make this point) by Brian Goodwin (Goodwin and Trainor 1983), even though the specific example Brian used to make his point turned out to

---

[6] This does not mean that tooth identity can't be lost, as for instance in the case of tooth whales. But it is not known what the expression and function is in whales of those genes that cause tooth identity in other mammals.

be flawed.[7] Sometimes, morphological "elements" are just parts of an overall pattern that represent a character state rather than being individualized characters themselves. Simply looking at a morphological pattern is not sufficient to assess character individuality.

Variational Modalities: More Than One Way of Being a Certain Character

In this section I want to consider examples of character state variation, which seems to suggest that we may need to introduce an additional concept to adequately describe how development shapes morphological evolution. Homologs are body parts that represent units of phenotypic evolutionary change that can persist for long periods of time. They are often found in large groups of organisms and sometimes can undergo quite radical transformations. For example, paired vertebrate appendages (fins/limbs) are characters that have existed longer than the most recent common ancestor of all gnathostomes (500 million years ago) and have undergone major transformations and re-organizations.

In our current scheme, however, limbs and fins are just two different character states (or actually different sets of character states; see below), because limbs are clearly derived from paired fins of fishes. Nevertheless, the transformation of fins to limbs was one of the most interesting of evolutionary transitions and has attracted much attention from paleontologists, as well as evolutionary and developmental biologists alike (Clack 2002). There is no question that this transition was a significant event with downstream consequences for the evolution of a group (tetrapods), as well as for the evolution of paired appendages themselves. Yet this transition has no conceptual standing in our current scheme of describing morphological evolution, because we do not have a way to distinguish between the fin-limb transition as a case of character transformation and the minor variations that distinguish a chimp hand from a human hand.

Of course, the most important question is whether such a distinction is biologically warranted. What is the difference between the fin-limb transition and other more common character state transformations, say size changes? I think that a distinction between these two cases of character transformations is warranted because the fin-limb transition involved developmental changes with major and, probably, irreversible variational consequences. A limb remains a limb, even though it is used for swimming, as in whales and dolphins,

---

[7] Brian Goodwin exemplified his idea with a model of pattern formation of limb skeletal elements, but it turns out that at least good portions of the limb elements are in fact developmentally and genetically individualized. Nelson, C. E., B. A. Morgan, et al. (1996). Analysis of Hox gene expression in the chick limb bud. *Development* 122: 1449–1466.

FIGURE 2.7: The mudskipper, *Periophthalmus barbarus*, is a teleost fish that lives out of water and uses its pectoral fins like limbs. Nevertheless, its fin anatomy is clearly that of a teleost fish with no resemblance to tetrapods. Photo by Bjørn Christian Tørrissen, bjornfree.com.

and a fin is a fin, even though it is used to "walk" on land, such as the pectoral fins of some gobies (e.g., mudskippers, *Periophthalmus*; figure 2.7).

However, what does it mean that a limb and a fin are different kinds of characters? They certainly are part of one and the same transformational series and, thus, are clearly homologs. Of course, one can make a list of dissimilarities between fins and limbs, but then any character change would be in the same category.

I think that there are four ways in which the fin-limb transition, as well as the origin of the teleost paired fins, is different from other, more minor character state transformations. Before we can argue this case, we need to provide a brief characterization of primitive fins, derived teleost fins, and tetrapod limbs, limiting the description to skeletal elements. Figure 2.8 gives an overview of the main gnathostome lineages. The elasmobranchs, with shark and chimeras and rays and the bony fish (osteichthyes; i.e., ray-finned fishes), and sarcopterygian fishes (lungfish and coelacanth), including the tetrapods, are the two sister clades of jawed vertebrates. Within the bony fish we have again two clades, the ray-finned fishes, Actinopterygii, and the fleshy-finned fishes, the Sarcopterygii, which include the tetrapods.

All bony fish "fins" have two kinds of skeletal elements: endoskeleton and dermal skeleton. The endoskeleton consists of cartilage or replacement bone

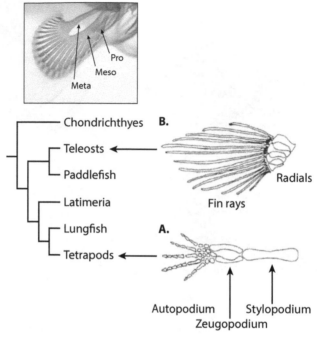

FIGURE 2.8: Teleost fins and tetrapod limbs are homologous but represent different varia-tional/character modalities. Note that the fin rays of the fins are not comparable to digits. Fin rays are dermal bones, whereas digits are endochondral long bones. Also, the proximal skel-eton in teleost fins is not comparable or even homologous to the proximal bones of the tetra-pod limb. The teleost pectoral fin has four radials that are all attached to the shoulder girdle, whereas the tetrapod limb has only one bone attached to the body and then a set of two distal bones, the radius and the ulna of the zeugopod. Basal fin structure is exemplified by the pecto-ral fin of a shark that has a "tri-basal" fin comprising a pro-, meso-, and metapterygium. The en-dochondral elements of the teleost fin are homologous to the propterygium, and the skeletal elements of the tetrapod limb are homologous to the metapterygium.

(i.e., bone that replaces embryonal cartilage). In contrast, the fin rays are dermal bones that develop at the interface between epidermis and dermal mesenchyme (Witten and Huysseune 2007). The fins of sharks and the basal lineages of ray-finned fishes have similar fin endoskeletons (i.e., a posterior branched element, the metapterygion, and anterior rod, shaped elements, called mesopterygion and protopterygion). This basic pattern is also found in basal ray-finned fishes, like the sturgeon and paddlefish (Metscher and Ahlberg 1999).

The transition to the two derived forms of paired appendages (i.e., teleost fins and tetrapod limbs) is characterized by complementary losses of certain endoskeletal elements. In its entirety, the teleost fin endoskeleton consists of four rods derived from the protopterygion. These bones are called radials in teleost fishes. In contrast, the entire endoskeleton of the tetrapod limb is derived from the metapterygion.[8] A consequence of this loss is that the limb, as well as other sarcopterygian fins, has only one element attached to the shoulder girdle, called the humerus or femur, respectively. In addition, limbs lack fin rays, but do acquire new elements, digits. A more detailed discussion of these events will be given in chapter 10.

The transition from fins to limbs seems to be rare. This is just a matter of historical record. As far as we know, only one lineage acquired limbs or paired appendages specialized for terrestrial locomotion. That might be a coincidence, although it probably is not. Furthermore, the transition from fin to limb is not explained alone by functional need. This follows from the fact that even the most terrestrial teleosts (*Periophthalmus*, mudskippers, some blennies, eels, and tropical catfish) do not acquire anything remotely similar to a tetrapod limb.

The fin-limb transition, as well as the origin of the teleost fin, is in three additional ways more fundamental than other character state transitions. First, it includes the irreversible loss of parts. Second, it includes the unique acquisition of novel parts (see previous paragraph and chapter 10). Finally, the development of endoskeletal elements proceeds in a fundamentally different way in teleost fins than in tetrapod limbs.

In limbs, the skeletal elements arise from mesenchymal condensations; that is, cells first aggregate and then differentiate into cartilage. Hence, during limb development, the individual elements form first, and then turn into cartilage. This mode of skeletogenesis is in contrast to the situation in teleost fins where the cartilage differentiates first and the skeletal elements individuate later (see below).

The evolution of the teleost fin proceeded via the loss of posterior skeletal elements (i.e., the loss of the metapterygion and the mesopterygion). The remaining endoskeleton elements were the radials. Early in larval life, the fin skeleton arises as a cartilage plate in which it is already functioning as a support structure. Later, this plate is subdivided by dedifferentiation of three stripes of cells to create four individualized cartilage rods, called radials. Hence, the teleost fin endoskeleton chondrifies first, and then individualizes the skeletal elements, whereas the tetrapod limb first forms individual elements by condensation, and then chondrifies. These two modes of development have consequences for the variational properties of limbs and teleost fins.

All teleost fins (~24,000 species, except the few that have lost paired fins) have four radials arranged in an anterior-posterior series. If present, all limbs have only one element proximally, up to two in the middle segment of the

---

[8] Distal of the radials are small nodular cartilages, called didal radials, which support the proximal end of the fin rays, but those are minor parts of the fin endoskeleton.

limb, and a variable number of digits and associated elements. These developmentally caused differences in variational tendencies ensure that a teleost
fin remains a fin, and a limb remains a limb, regardless of what its functional
roles are. Teleost fins never turn into anything resembling a tetrapod limb,
and a limb never turns into anything resembling a teleost fin, even when
the limb becomes functionally a fin, like the limbs of whales, plesiosaurs, or
mosasaurs (figure 2.9).

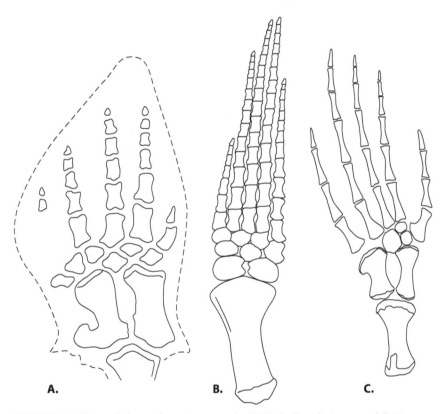

FIGURE 2.9: Flipper skeleton of aquatic tetrapods. A) Whale, B) a plesiosaur, and C) Moasaur.
These are functionally equivalent to pectoral "fins" in fishes. Even though these limbs are clearly
modified from their counterparts of terrestrial animals, they are easily recognizable as tetrapod
limbs. The conclusion is that the character modality "tetrapod limb" remains stable in spite of
reversals to ancestral functions that originally were performed by the "fish fin" modality. Hence,
radical transformations of characters result in the permanent occupancy of different sets of character states, despite a reversal of their selection pressure. [Redrawn from the following sources: A)
http://dita2indesign.sourceforge.net/dita_gutenberg_samples/dita_outline_of_science/html
/chapters/chapter_06.html B) http://frontiersofzoology.blogspot.com/2011/04/cfz-blog-on
-plesiosaurian-taniwhas.html C) http://www.geology.wisc.edu/~museum/hughes/images
/MosasaurFlipper.jpg.]

In a sense there is no question that teleost fins and tetrapod limbs are two different kinds of characters. Yet they are clearly homologs in that they represent a single tree of character diversification. But each has its own, non-overlapping sets of characters states (i.e., all the different shapes and functions limbs and fins can assume) and they never transition into each other. They are derived from a third set of character states represented in modern fauna by sharks, sturgeons, and other basal fishes. Otherwise, teleost fins and limbs are variationally isolated from each other (i.e., there do not seem to be mutations that turn a limb into a teleost fin and vice versa).

I propose calling these sets of character states two different "*Variational Modalities*" of the paired vertebrate appendage.

- Two or more sets of character states of the same character (homolog) are called *Variational Modalities* if they are non-overlapping and if transitions among character states within each mode are much more frequent than between modalities. It is expected that variational modalities differ in their developmental/functional organization and, thus, reflect mechanistically relevant differences among sets of character states.

By referring to relative magnitudes of transition frequencies, the vagueness of this definition is an easy target for criticism. Obviously, the term should be used only if it becomes a marker for a biologically interesting distinction, like that between fins and limbs, and where research into the mechanisms that cause these different modes of variability finds a mechanistic explanation, like those between fish fins and tetrapod limbs.

In order to show that the idea of variational modalities is not a special case for vertebrate appendages, I want to outline another example from flowering plants. Within the asterids, a clade of 65,000 species of flowering plants, the flowers have five fused petals, with a few exceptions. The typical (but not invariant; see below) arrangement of petals has two petals dorsally and three ventrally (figure 2.10). At the root of asterids, the flowers had radial symmetry (with five axes of symmetry), whereas bilateral flowers evolved at least eight times independently (Donoghue, Ree et al. 1998).

Bilaterally symmetric flowers can be classified according to the orientation of their petals. The most common pattern is two petals up and three down: a 2:3 pattern (Donoghue and Ree 2000) (figure 2.10). Other patterns also occur. Besides 2:3, the most common are 4:1 and 0:5. Transitions among these morphs are frequent. The 3:2, 1:4, and 5:0 patterns are also possible, but occur more rarely. It turns out that these patterns are conditional upon a radical change in flower development; namely, a 180 degree rotation of the petal (and other flower organs) anlagen such that there is one petal at the top, two on either side, and two below (figure 2.10). This character change is rare. Based

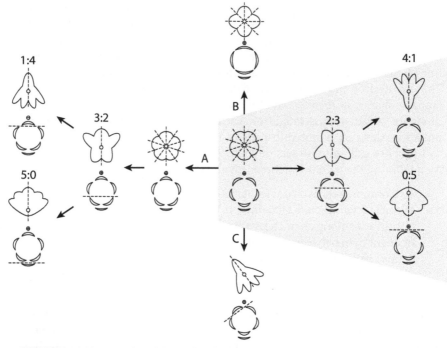

FIGURE 2.10: Variational modalities of pentameric flowers. The shaded area includes the most frequent petal configurations, 2:3, 4:1, and 0:5, where the first number is the number of dorsal petals and the second is the number of ventral petals. The other configurations, 3:2, 1:4, and 5:0 are accessible only after a rare transition (A), which inverts the dorso-ventral polarity of the flower. All flowers in this scheme are homologous, but represent different sets of character states that are mutually non-overlapping and accessible only through rare transitions. (B) and (C) indicate other rare transitions leading to different character states of flower symmetry (Donoghue and Ree, 2000, *Am. Zool.* 40:759–769).

on phylogenetic data, Donoghue suggested that it arose only twice, once in the ericad lineage (e.g., *Rhododendron*) and once in the lobelia lineage.

However, once the flower organs are rotated, the 3:2, 1:4, and 5:0 patterns are easily accessible character states (for more details, see Donoghue and Ree 2000). Hence, the asterid flowers exist in two variational modalities,[9] one in which the 2:3, 4:1, and 0:5 patterns are easily accessible and another in which the 3:2, 1:4, and 5:0 patterns are easily accessible, and transitions between these two sets of character states are rare. These two variational modalities

[9] There are some other sets of character states, like the loss of one petal leading to the four-fold symmetrical flowers of *Plantago* and an oblique flower in which one of the two dorsal petals becomes differentiated from the rest of the petals, as in Valerianacea (*Centranthus*) Endress, P. K. (1999). Symmetry in flowers: diversity and evolution. *Int. J. Plant Sci.* 160: S3–S23.

are characterized by a qualitative difference in flower development that leads to two non-overlapping sets of possible character states.

Character Identity and Repeated Body Parts: Serial Homology

The examples introduced above, insect wings and paired vertebrate appendages, also provide examples of serial homology. In most four-winged insects, the forewing and the hind wing are so similar and originate from different segments of the body that it seems likely that they are copies (i.c., similar to duplicated genes). The evolutionary history of insect wings, however, is not well understood and it is not clear if they originated from identically repeated dorsal appendages and then individualized later (which seems to be the most likely scenario) or if fore- and hind wings are similar due to convergence in structure because they serve the same function in most four-winged insects (i.e., flight).

The situation seems to be a little clearer in the case of paired fins in fishes. Most fishes have two pairs of fins, a pair of anterior pectoral fins and a pair of (primarily) posterior pelvic fins.[10] The fossil record indicates that these two pairs of fins did not originate simultaneously. The earlier vertebrates seem to have only anterior paired appendages, probably homologous to pectoral fins (see chapter 10). Only later in the fossil record are forms found that also have a posterior pair of appendages. The second pair of appendages probably evolved by re-deployment of the genetic information that evolved first in the anterior appendage. This is likely because of the similarity of gene expression patterns in the pectoral and pelvic fin buds. Hence, pelvic fins most likely originated from a duplication of pectoral fins by re-deployment of the genetic program for pectoral fin development.

Nevertheless, the anterior and posterior paired appendages have individualized, as can be seen in their divergent morphology and appendage-specific gene expression that determine fore- and hind fin/limb identity (e.g., *Tbx 4/5, HoxC6*, others). This scenario is formally equivalent to the origins of paralogous genes due to gene duplication and their subsequent individualization. To emphasize this analogy between serially homologous body parts and paralogous genes, one could call the former "paramorph" body parts.

But not all repeated physical structures in an organism are also individualized morphological characters like the forelimb and the hind limb (each of which is homologous to and derived from pectoral and pelvic fins, respectively). This form of serial homology is most prominent among cells

---

[10] In some highly derived teleost fish, the pelvic fins can be found in an anterior position, sometimes even more anterior than the pectoral fins. Nelson, J. S. (1994). *Fishes of the World*. New York, John Wiley & Sons, Inc. But this situation is a secondary modification, because all the basal ray-finned fishes have pectoral fins anterior and pelvic fins posterior.

and cell types. Most cell types in larger metazoan animals are represented by a large number of more or less identical copies. Red blood cells are perhaps the most notorious example of such a histological plurality. Only in small invertebrates will one find different cell types in one or a few copies per individual. Small invertebrates can show the phenomenon of species-specific cell number constancy, in which each individual of a species has the same number of cells.

For example, in small nematodes, the lineages of individual cell types are invariant among individuals of the same species and to some degree even among different species. But in vertebrates the typical situation is that of a plurality of occurrences of the same cell type in each individual with no individuality for each cell of a certain kind. Cells that belong to the same cell type evolve in concert (i.e., vary more between species than within species or individuals). Hence, an entire group of cells, which represent the same cell type, are a unit of evolutionary change (even though they are physically distinct entities) in the same way as a gene family that evolves in concert (i.e., ribosomal RNA genes comprise physically distinct genes, but are only one unit of molecular evolution).

This example shows that we have to distinguish between two forms of serial homology, one of which is that each copy of the same character has evolutionary individuality similar to paralogous genes. These characters can be called *paramorphs* (Minelli 2002; Minelli and Fusco 2005)[11] because of the similarity of this situation to genetic paralogs. On the other hand, there are serially homologous parts that lack individuality, like those genes that evolve in concert or cells of the same cell type. To my knowledge, the first to recognize this latter form of serial homology as distinct from paramorph characters was Rupert Riedl (Riedl 1978), who called this form of relationship "*homomorph*." Here, I will adopt the distinction between paramorph and homomorph body parts to indicate biologically distinct forms for serial homology.

The distinction between homomorph and paramorph body parts is not always straightforward and may require detailed developmental, genetic, and comparative data in order to decide. For example, many crustacean species have series of nearly identical limbs, all of which have the same qualitative parts but differ with respect to position and, often, size (figure 2.11). These limbs form a series of similar, but certainly not identical, appendages. Are these homomorph or paramorph body parts?

I believe that we do not know enough about these crustacean limbs to give an answer at this point. But there is new, surprising evidence regarding

[11] The notion of paramorph was introduced to explain the idea that body appendages in metazoans are duplications of the body axis. The utility of this term, however, goes beyond this specific hypothesized relationship between body axis and body appendages.

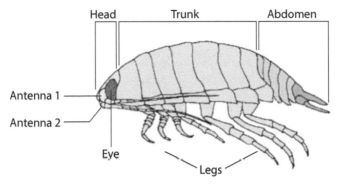

FIGURE 2.11: Body organization and appendages of an isopod crustacean are an example of serially homologous characters with unclear individualization. The limbs (pereopods) are qualitatively similar, but differ in size and shape. It is difficult to tell which ones are truly individualized and which are homomorphic. (Redrawn from http://www.quantum-immortal.net/other/crustacea.php.)

the individuality of digits in mammals that suggests that digits two to five in our hand are homomorphs, whereas digit one, the thumb, is individualized relative to the other digits (i.e., the thumb is a paramorph with respect to the other digits; see chapter 11).

At face value the digits of tetrapod limbs are typical serial homologs. They all comprise a series of small articulated long bones, the phalanges, and associated soft tissues. The phalanges form the movable digits that, in most cases, protrude from the palm. The most proximal phalanx articulates with another long bone that is longer than the typical phalanx, the metacarpal bone. The number of phalanges per digit is variable. In the ancestral five-digit hand of amniotes, the numbers of phalanges in digits one to five were 2, 3, 4, 5, 3 and the phalangeal formula for the foot was 2, 3, 4, 5, 4. In mammals, the consensus phalangeal formula is 2, 3, 3, 3, 3, which is also what humans have in their hands and feet. Hence, digit one always had two phalanges and was always different from the rest of the digits.

In addition, in terms of gene expression, the most distinct digit is digit one. For example, there are five different Hox genes that are expressed in the amniote hand: *HoxA13, HoxD10, HoxD11, HoxD12,* and *HoxD13.* Two of these are expressed in all the digits, *HoxA13* and *HoxD13.* In contrast, the remaining HoxD genes are expressed only in digits 2 to 5, namely, *HoxD10, HoxD11,* and *HoxD12.* Digit 1 is, thus, genetically a unique digit and, therefore, a good paramorph candidate.

This hypothesis was recently confirmed by an analysis of primate hand variation by showing that, in terms of its variational tendencies, digit 1 belongs to a different module than the remaining digits (Reno, McCollum et al. 2008). The remaining digits in mammals not only have the same morphology,

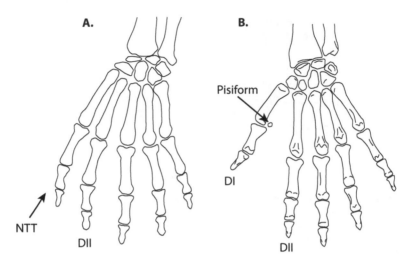

FIGURE 2.12: Homeotic transformation of the first digit into a digit II in the human hand. A) A non-opposable triphalangeal thumb (NTT) in which the first digit resembles the second. B) Normal human hand. Both are dorsal views with anterior to the left (left hand). (Source: Wagner, 2005, *ThiBS* 124:165–183.)

despite minor size differences, but are also a co-variational module. This data suggests that, in mammals, digit 1 and the collective of digits 2, 3, 4, and 5 are paramorphs (i.e., individualized serial homologs, while digits 2, 3, 4, and 5 by themselves may be a group of homomorphs that is, non-individualized serial homologs).

The hypothesis that digit I in mammals has a distinct identity was also confirmed by the existence of homeotic mutations that transform digit I into a digit II. In human genetics, this mutation is called "non-opposable tripha-langeal thumb (NTT)" (figure 2.12) (Joachimsthal 1900). This name indi-cates that in the mutant phenotype, the first digit (presumptive thumb) is no thumb at all, but has three phalanges like the rest of the digits and cannot be rotated toward the palm, as is typical for human thumbs.

What is remarkable about the NTT mutation, in contrast to other "tripha-langeal thumb" phenotypes, is that the similarity between the modified digit I and the remaining digits goes far beyond the number of phalanges. It also concerns details of the first metacarpal morphology that, in the NTT phe-notype, lacks the distal epiphysis (the point where the bone grows in length) and has instead a proximal epiphysis, as do digits two to five. In addition, the NTT "thumb" also lacks the proximal articulation surface that is typical for the thumb and which is necessary for the thumb's opposability. Also missing are the muscles and tendons that rotate the thumb in the wild type. If there is any convincing case of a total transformation of digit identity, then this is one. Clearly, the thumb represents a distinct morphological and developmental individuality that is controlled by a genetic switch.

It is highly unlikely that the same classification of homo- and paramorph digits as in humans and, perhaps, in most mammals, applies to all tetrapods. For example, the evolution of the posterior digits in theropod dinosaurs suggests that most, if not all, digits are paramorphs in this group, which includes the birds. In this group specifically, digits IV and V are lost and digits I, II and III are retained. These remaining digits also have quite specialized morphologies (Wagner and Gauthier 1999) and, in the chicken, the wing digits are also more distinct at the gene expression level than are the toes (Wang, Young et al. 2011). Hence, individuality of serial homologs is a characteristic that varies among clades.

Most likely, individualization is caused by or occurs in concert with functional specialization. For example, the serially homologous forelimbs and hind limbs are variationally more independent in mammals with distinct forelimb/hind limb modes of locomotion (e.g., flying or bipedal jumping), and are less so in quadrupedal mammals (Young and Hallgrímsson 2005; Young, Wagner et al. 2010).

Another character for which the question of individuality of serial homologs has been empirically addressed is the eyespots of nymphalid butterflies (Monteiro, Prijs et al. 2003). In many butterflies, a characteristic pattern element of wing pigmentation patterns is eye spots, which are characterized by concentric circles of pigmentation and a central "pupil." These pattern elements are often present in a series along the anterior-posterior wing margin and are clearly serially homologous (Nijhout 1991). They not only have the same color combination, but also express the same set of transcription factors and signals, like *distalless (dll)*, in the center and *engrailed* and *spalt* farther out from the center (Brakefield, Gates et al. 1996; Brunetti, Selegue et al. 2001).

However, they also show clear signs of individuality. For example, not all possible eyespots are found in all species and eyespots on the same wing can differ in size. These signs of individuality in interspecific variation are matched with evidence for individuality in the variation within a species. Monteiro and colleagues (Monteiro, Prijs et al. 2003) found X-ray–induced mutations in the butterfly *Bicyclus anynana* in which certain groups of eyespots are missing. For example, the wild-type hind wing has seven eyespots. Monteiro et al. found mutations with eyespots 3, 3+4, 1+2+3+4, or 1+2+3+4+7 missing. Clearly, individual eyespots and groups of eyespots can vary independently from others in terms of presence/absence.

Similarly, eyespots can vary independently in terms of size, as shown in inter-specific variation as well as by artificial selection experiments (Beldade, Koops et al. 2002). Interestingly, color composition does not seem to be independently variable among eyespots. An artificial-selection experiment on color composition did not yield differences among eye spots (Allen, Beldade et al. 2008), which suggests that individuality of serially homologous traits can be specific to certain aspects of the phenotype, like size and presence/absence, but not to color composition. Not surprisingly, individuality can be a matter of degree. In other words, serially homologous characters can be individualized with respect to one trait or property, like size, but not individualized with respect to another trait, say color composition.

Variable degrees of individuality among serial homologs point to the fact that individualized characters are spatiotemporally limited. This means that they have a more or less well-defined beginning in evolution (although sometimes hard to identify empirically) and, potentially, an end (i.e., characters can get lost). They also have a well-defined beginning during individual development of an organism, a fact that will be documented in the next chapter. Ancestrally, gnathostomes have two pairs of fins or limbs (limbs being a character state of paired fins), but there is nothing in a placozoan[12] animal that meaningfully could be called the homolog of a fin/limb.

Identifying the events in phylogeny when new characters arise and understanding the population genetic and developmental mechanisms of character origination is an important objective of evolutionary biology in general (see chapter 4). A consequence of the spatiotemporal limitation of characters is that they are individuals, and not sets or classes (Ghiselin 2005). These philosophical issues will be discussed in greater detail in chapter 7.

The idea that characters (i.e., individualized body parts) form lineages and, for that reason, can be analyzed like gene trees or species trees has been proposed by several authors. For example, R. Geeta[13] showed that monocot leaves form tree-like relationships based on similarity that follows the species trees. She took this result as evidence that leaves "are developmentally integrated, evolving entities" (Geeta 2003). In contrast, she also found that leaf primordia do not form a structure tree and concluded that there is no such entity as the "monocot leaf primordium."

Todd Oakley had argued that eyes and specialized parts of eyes act like replicators (Oakley 2003). Oakley and colleagues showed that a comparison of structure and species trees could also be used to identify structure origination events, and called their approach "hierarchical phylogenetics" (Serb and Oakley 2005). These examples are interesting because they suggests that rather conventional methods of comparative analysis can be used to test the idea of whether a particular anatomical part is, in fact, a homolog in the sense defended in this book (i.e., behaves like an entity that forms lineages of descent).

---

[12] Placozoans are a small group of marine invertebrate with very simple body organization that is considered representing the phenotype of a primitive metazoan animal.
[13] Professor Geeta publishes as R. Geeta; even the official University of Delhi website and her official faculty profile does not reveal what "R" stands for.

Character Swarms: Persistent Cases of Partial Individuality

When writing the chapter on skin appendages (chapter 9) and, in particular, the section on feathers, it became clear that there were cases that persistently lingered in an intermediate state between serially homologous, but individualized characters (paramorph), and homomorph characters. In the case of feathers, it is clear that groups of feathers have sufficient individuality to acquire distinct phenotypes, like the different types of feathers on the wing and the tail, as well as the feathers that cover the body. Nevertheless, variation and evolution show that even feathers of different types have a tendency to share derived character states, most likely because they share a large amount of genetic machinery. During evolution, they behave like a "swarm" of characters comprising individualized groups of characters, but varying to some degree in a concerted way, just like a swarm of birds comprises individuals although collectively they behave like a unit of motion.

This term was chosen for its analogy to the concept of "species swarms," which are collections of species that are hard to distinguish because they have not acquired full genetic individuality, but still share appreciable amounts of genetic variation and, thus, co-evolve in spite of the existence of a certain degree of differentiation. A more detailed discussion of character swarms is given in chapter 9.

Alternative Conceptualizations of Homology

Different conceptualizations of homology derive from different assumptions regarding the nature of organisms and from different research goals. The present account of homology assumes that organisms have structural organization, which includes parts that each constitute distinct developmental/evolutionary units (Brigandt 2007; Assis and Brigandt 2009), and that these units are typically those entities recognized in comparative anatomy (e.g., limbs and brains). Each of these units can come in different sizes and shapes, which are called character states. Yet these states are still variants of the same thing, a limb for example.

In contrast, alternative views do not make such assumptions and, consequently, deny that the distinction between characters and character states makes any sense (Bock 1977). These two views roughly correspond to the distinction between transformational and taxic notions of homology, respectively. An even more radical view, which in fact can even be traced back to Darwin according to some scholars, is that homology lacks any biological significance, a view that has been called the "residual conception" of homology (a term introduced by Amundson 2005 in reference to G. C. Williams; Williams 1992).

The residual view of homology is compatible with the taxic homology concept, but is not identical to it. The differences between the transformational and the residual views of homology are conceptually relevant and cannot coexist in the long run because they make different assumptions regarding the nature of organisms. These assumptions have to be confronted with empirical evidence and need to be resolved. Hence, ultimately the differences between the taxic and transformational views will be decided by empirical evidence.

The pragmatic aspects that shape different views of homology derive from different research programs. The present account developed within the joint research programs of evolutionary biology and developmental evolution, which seek to explain patterns of biological diversity through population genetic and developmental mechanisms. The taxic account of homology is motivated by the pragmatic needs of phylogenetic systematics. Distinctions motivated by pragmatic considerations are not necessarily in conflict and do not call for empirical resolution. The only criterion to distinguish between these conceptions is their usefulness. It might be that a concept arising in evolutionary biology may also be useful in phylogenetic systematics, or vice versa; yet, in principle, these alternatives can remain standing side by side as tools in different biological disciplines.

The Biologically Relevant Distinction: Historical Residues or Organizational Parts?

In order to highlight the difference between a purely historical explanation of homology and one that assumes that homology reflects the developmental organization of organisms, it is useful to compare the classical definition of homology by Owen with those used in evolutionary biology following Darwin. First, let us consider Owen's definition:

- Homologue . . . the same organ in different animals under every variety of form and function (Owen 1843, p. 379).

It is clear that Owen was thinking of body parts that could come in different forms and fulfill different functions. Thus, this definition implies that homology is not similarity; rather, it hypothesizes that organisms consist of parts that are *the same* in a deeper, but in Owen's account, some unspecified way that can come in different incarnations (i.e., "under every variety of form and function"). Sameness in Owen's conception is not tied to overall similarity, since similarity is declared irrelevant for sameness. This notion implies a distinction between characters, the things that are thought of as the same in different animals (and plants one may add), and character states (i.e., the different forms and functions of these parts).

Nikolaas Rupke (Rupke 1994) points out that for Owen, this view of homology was a solution to a practical problem. Owen was involved in a massive program of cataloging museum material and sought to unify the description of vertebrate skeletons. In Owen's time, these skeletal elements had different, disparate names in different groups of organisms. Furthermore, Amundson has argued that typology in nineteenth-century anatomy was mostly thought of as an empirical generalization that was pointing to a yet to be discovered causal principle to explain the patterns of disparity among organisms (Amundson 2005). Metaphysical ideas, linked to Platonic idealism and often associated with typology, were not central to this enterprise. Hence, Owen's concepts were part of an empirical, practical research program that, at its core, did not exclude evolutionary ideas (see Rupke 1994, p. 171).

The homology concepts expressed here and by others (e.g., Newman and Müller 2005; Assis and Brigandt 2009) have obvious affinities to Owen's definition, as seen in their insistence on the distinction between homologs (characters) and character states. The main difference between Owen and the more recent accounts is the idea that developmental biology will help clarify what Owen could only hint at—namely, the question in what way "organs" can be *the same* "regardless of form and function"? Hence, contemporary developmental accounts of homology imply and support a research program into questions about the developmental underpinnings of character individuality (see chapters 3 to 6) and the origin of characters (evolutionary novelties; see chapters 4, 5, and 6 as well as section II) (Amundson 2005).

Now let us turn to the homology concept derived from Darwin's views. Darwin himself apparently never provided a concise definition of homology (Mayr 1982). Thus, we use two definitions derived from the neo-Darwinian tradition (i.e., those of Simpson and Mayr).

- Homology is resemblance due to inheritance from a common ancestry (Simpson 1961, p. 78).
- Attributes of two organisms are homologous when they are derived from an equivalent characteristic of the common ancestor (Mayr 1982, p. 465).

The first notable difference between Owen's definition and those by Mayr and Simpson is that the latter do not talk of organs or any other biologically relevant units, but of "resemblance" and "attributes." In this context, the distinction between characters and character states becomes meaningless (Bock 1974; Stevens 2000). Also gone is the notion of sameness, since there is nothing that remains the same. Only residual similarities remain from the erosion of similarity with the common ancestor due to mutation, selection, and drift. The degree of similarity can be used as a guide to reconstruct the phylogenetic relationships among species, but beyond that it is of no theoretical relevance. Not surprisingly, then, is that in molecular biology, sequence similarity is sometimes expressed as "% homology," which implies that homology is a degree of resemblance due to derivation from a common ancestor.

This concept implies that there are no persistent causal factors that maintain homologous similarities; "they are merely the ancestral characters that happened by coincidence to survive (Williams 1992, p. 88)" (Amundson 2005, p. 240). For that reason, Amundson's characterization of this view as "*homology as residue*" hits the nail on the head.

Homology as residue has no theoretical significance other than as evidence for common ancestry. Homology is the "anticipated and expected consequence of common ancestry" (Wake 1999). "There is no reason to seek a naturalistic explanation for instances of homology, biologists should instead turn their attention to questions that can be resolved" (Wake 2003, p. 193). This view reduces the evolution of the phenotype to a conceptually irrelevant shadow of genetic change, and evolution becomes nothing more than changes in gene frequency.

The defenders of the residual view mostly point to two facts (Peter Gorfrey-Smith, personal communication). Their first argument is that morphological characters are not inherited, but are developmentally reconstructed in each generation (a fact I agree with and pointed out in 1989; Wagner 1989; although I certainly was not the first to notice). Only in the case of DNA and, hence, genes is it meaningful to speak of the same thing in different species because they are the product of a directly copying process. Conceptually, one may argue, the sameness of morphological or any other phenotypic character may be just an illusion. The defenders of a residual view can even point to the growing body of evidence from developmental evolution that shows that seemingly the same character can be realized by different genes in different species (Wray and Abouheif 1998).

These arguments are important and need to be taken seriously, although I do not want to address them fully here. Regarding the first argument, I want to point out only that, for natural selection, it does not matter whether an entity has been produced by copying or by a developmental process, as long as it is reliably present in each generation and contributes to the survival and the reproduction of that organism. The latter argument, the variability of developmental pathways, will be the center of attention in the next chapter (chapter 3). For now, I want to point out only that not all aspects of development are equally variable (see chapter 1) and that this pattern of differential conservation likely has a biological signal and we need to read this signal to resolve this conundrum.

The contrast between the developmental account (advocated here) and the view of homology as residue implies a very clear research objective—namely, to determine if the assumptions of the developmental account regarding the organization of development (see chapter 3 for details) are correct and whether they contribute to the observed patterns of morphological (phenotypic) disparity among species. The developmental account makes stronger assumptions about biological reality than the residual view and, thus, leads to testable predictions. These predictions and the manner in which they can be integrated into the research program of developmental evolution are the subject of the remaining chapters of this book.

Taxic Homology

The terms "taxic" and "transformational" were introduced by Niles Eldredge (Eldredge 1979) to refer to two kinds of evolutionary theories. Taxic theories are concerned with the origin of species richness (i.e., diversity), whereas transformational theories concern themselves with phenotypic and genetic change (also sometimes called anagenesis). In an influential 1982 paper, Colin Patterson adopted these terms to characterize different approaches to the homology concept (Patterson 1982). He defined homology as a shared derived character that characterizes a monophyletic group; that is, Patterson equated homology with the cladistic concept of synapomorphy.

- Homology = Synapomorphy (Patterson 1982, p. 29).

The idea of equating homology with the definition of monophyletic groups is widespread among cladistic systematists and is not due only to Patterson (Wiley 1975; Bock 1977; Platnick and Cameron 1977; Cracraft 1978; Nelson 1978). This notion of homology is clearly shaped to fit the needs of systematists—namely, to have a clear, workable definition of homology in the context of phylogeny reconstruction. As such, it is a legitimate clarification of the term that seems to be the best for this purpose, given that it is widely adopted in systematics. Any hypothesis that seeks to answer origin questions (i.e., questions regarding the origin of evolutionary novelties), however, is excluded from the purview of this concept.

For example, Patterson calls Gegenbaur's theory regarding the derivation of paired appendages from gill arches "empty" because within the extant diversity of vertebrates, gill arches and paired appendages characterize the same group—namely, gnathostomes. This might be so. Yet Gegenbaur's theory, however outlandish it may seem, is still a serious contender among the theories that seek to explain the origin of paired appendages. Recent developmental work has provided substantial support to Gegenbauer's theory because of the similarities in the developmental regulation of limb/fin development and the development of chondrychthian gill rays (Gillis, Dahn et al. 2009; see also chapter 10). Hence, Gegenbaur's theory may be empty within the cladistic framework, but is a rich source of inspiration for specific experimental work in developmental evolution.

Another consequence of equating homology with synapomorphy is that the distinction between characters and character states becomes unnecessary, and may even be a distraction. A monophyletic group can be characterized by a novelty (i.e., a new individualized body part like paired appendages or feathers); by a new character state, like "multilayered epidermis"; or by a certain amino acid in a protein. Both new characters and derived character states of a plesiomorphic character can be a synapomorphy and characterize a monophyletic group. Even the absence of a body part, like wings and limbs, can be a synapomorphy.

The refusal to accept the distinction between characters and character states is not without irony, because the standard form of presenting data for phylogenetic analysis forces the researcher to present a data matrix in which the rows represent species and the columns represent different characters, while the entries in the matrix are different character states (e.g., hind wings shaped as a wing blade or reduced to a haltere). There is a complex technical literature on whether this way of presenting data is necessary and whether it implies a distinction between characters and character states. But these questions do not need to concern us here. The point is that even the most technical approach to systematics in some form implies a distinction between different entities, as represented by the columns in the data matrix, and the different states, as represented by the entries in individual cells of the data matrix.

The conceptual irrelevance of the distinction between characters and character states is true for systematics because of its pragmatic needs, and not because body parts with different character states do not exist. Anatomical, variational, and developmental evidence shows that they do exist and that the distinction is necessary to understand the pattern of biological diversity and disparity at the phenotypic level. Hence, the transformational notion of homology advocated in this book is not in conflict with Patterson's taxic view because both are shaped to fit different research programs (Hennig 1966; Queiroz 1985; Donoghue 1992). Patterson even writes that, ". . . the transformational approach to homology may be more informative and more interesting than the taxic approach, a point [he] do[es] not deny" (Patterson 1982, p. 36).

The Organizational Homology Concept

A biological homology concept that can be perceived as an alternative to the view proposed in this book is the "Organizational Homology Concept" proposed by Gerd Müller, University of Vienna (introduced in Müller 2003; see also Love and Raff 2006). It also attempts to solve the problem of how characters can be "the same" in spite of differences in form, function, and even differences in development and genetic bases. It focuses on the role that individualized body parts play in the development and evolution of the organism rather than on notions of autonomy. There is much to be recommended in this view, particularly in those cases for which there is clear historical continuity of a body part and at the same time it is questionable whether there are dedicated gene regulatory networks underlying its development.

Müller lists his basic assumptions that guided his ideas about homology in seven premises:

1. Homologs are constant elements of organismal construction; they are independent of changes in form and function.
2. Homology signifies identity, not similarity.
3. Homologs are fixated by hierarchically interconnected interdependencies.
4. Homologs are developmentally individualized building units.
5. Homology denotes constancy of constructional organization despite changes in underlying generative (developmental/genetic) mechanisms.
6. Homologs act as organizers of the phenotype.
7. Homologs act as organizers of the evolving molecular and genetic circuitry.

Premises one and two are broadly accepted, and three is an explanation of why homologs can be conserved over long periods of evolutionary time. Premises four and five are concordant with my emphasis of developmental autonomy that was previously discussed in this chapter.

The main distinguishing feature of the organizational homology concept is encapsulated in premises six and seven. The idea seems to be that the very presence of a homolog acts like a boundary condition for the development of the remainder of the phenotype, as well as for the evolutionary opportunities of the underlying genetic/developmental machinery. Premises six and seven are, thus, related to premise three by explaining what kind of interdependency causes the "fixation" of a morphological character in a body plan.

These ideas are condensed in the following (preliminary) definition: *Homologs are autonomized elements of the morphological phenotype that are maintained in evolution due to their organizational roles in heritable, genetic, developmental and structural assemblies* (Müller 2003, p. 65).

This definition amounts to a "relational ontology" of body parts in which the existence and the identity of morphological characters are seen as being grounded in the (organizational) relationships that they have to other parts of the body. The emphasis on organizational/developmental relationships, rather than functional relationships, is dictated by the fact that identity of homologs is expected to be independent of function, as in the original definition by Richard Owen (see premise one).

As such, the organizational homology concept is not concerned with explaining how characters can be developmentally individualized and, thus, is complementary rather than competing with the ideas developed in this book. Although, it does address an aspect of the biology of morphological characters that is not discussed much—namely, the relationship between homologs and the body plan of which they are part. The organizational homology concept highlights the integration among homologs and any other aspects of the body plan.

A Case for Conceptual Liberalism

Concepts are mental tools, and tools are made to perform a given set of tasks. A hammer is made to exert strong forces on an object, either to drive a nail into wood or to smash a rock. A concept is also a tool. Its purpose is to bring order to a set of observations and to guide further study and practice (e.g., experimentation and technology). Even with the best tool, we should resist the temptation to overuse it (e.g, if we have a hammer, we should not consider all of reality as nails). This is particularly true for biological concepts, which always oversimplify reality. For example, I previously discussed the distinction between paramorph and homomorph characters. When making this distinction, I remained acutely aware that there are gradations of differences on which homomorph copies of characters blend into paramorphs. But the fact that sharp distinctions are impossible to make is not a problem, as long as a majority of cases clearly fall into one or the other category. It would also be foolish to abandon the species concept only because there are cases for which two populations are neither fully part of one species nor clearly separate species.

To the contrary, recognizing these situations in which the species concept does not fully apply leads us to a deeper understanding of the process of speciation. Studying populations that are in the process of speciation helps us understand how speciation works. Furthermore, these situations would not be recognized as worth studying if they did not stick out against the foil of the biological species concept. Hence, critical evidence regarding the process of speciation becomes recognizable only because it does not fit the species definition. Hence, in a paradoxical way, the situations that deviate from the species concept validate this concept because they help identify cases that deserve further study and from which we, in fact, learn a lot about the biology of populations.

Concepts, theories, and models are most valuable because reality does not always conform to them. This is a basic principle of scientific investigation. Scientific evidence for novel facts arises from discrepancies between well-supported theories and observation. Small deviations from the predictions of the Newtonian model of planetary motion provide evidence for the theory of special relativity. Without the Newtonian model, these observations would have no meaning and would not even be "deviations" from anything and could not lead us to deeper insights. Similarly, albeit less glamorous, is the use of the neutral theory of evolution that proposes that most molecular sequence change is due to drift to fixation of selectively neutral mutations.

This model makes strong predictions about the rate of nucleotide sub-stitutions over time (constancy of the rate regardless of population size; relationship between within-population variation and divergence between populations). Of course, everyone knows that real sequence evolution does not always conform to these predictions. But the reason we still teach this model is that it provides a yardstick to detect biologically significant facts. For example, the most widely used methods to detect natural selection at the molecular level work by identifying statistically significant deviations from the neutral model. In all the sciences, strong evidence drives discovery with models and discovers new things by finding discrepancies from these mod-els. This is not the same as falsification, in which one discards a model once deviations from the model are discovered. In contrast, models, concepts, and theories that reflect some aspects of reality, but fail in others, are essential tools for scientific discovery.

Viewed from this perspective, I do not expect the conceptual distinctions introduced and defended here to neatly and universally reflect reality. I do expect, however, that a significant part of biological diversity and disparity will be reflected in the distinctions between characters and character states, variational modalities, as well as between paramorph and homomorph char-acters. I also expect that there are cases that do not fit into these categories and I expect that these cases will be important for further study. These cases can teach us about the causes that lead to evolutionary novelties and novel body plans. If the concepts proposed here facilitate this process of discovery, then they will have fulfilled their purpose.

Sorting Patterns of Morphological Variation

In this chapter I have proposed a number of distinctions, many of which have been around by other names for more than a century, although they may not necessarily be familiar to all readers. For this reason, I summarize them here and provide a little "concept map" (figure 2.13) to offer an easier overview.

Let us begin with homology in the broad sense, which is just the idea that different body parts are, at some general level, "the same" without specify-ing whether these body parts are from the same individual or from different individuals or species. Then, we distinguish between corresponding body parts in different species or individuals, something that Owen called "special homology," and characters that are repetitions of the same part in the same individual, called "serial homology."

Continuing with special homology (i.e., correspondence of body parts be-tween species and individuals), we then have what one can call homology in the narrow sense, as exemplified by corresponding body parts of closely related species (i.e., the hands of a chimpanzee and a human). On the other

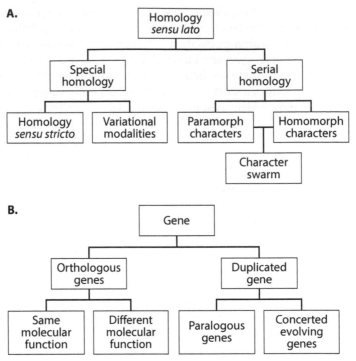

FIGURE 2.13: Concept map for the broader ideas on homology, both for morphological characters (A) and genes (B). See text for explanation.

hand, there are homologous parts that differ substantially in their variational tendencies and composition in terms of cell types and anatomical parts, which I call "variational modalities." The paradigm for variational modalities is paired teleost fins and tetrapod limbs that are homologous at the level of paired vertebrate appendages, but which have a very different developmental and evolutionary biology.

On the side of serial homologs, we need to recognize serially homologous characters that have acquired individuality and, thus, are two independent lineages of evolutionary modification. These are called "paramorph characters" in analogy to paralogous genes. On the other hand, there are body parts (e.g., cells of the same cell type) that lack any genetic individuality and are re-iterations of the same developmental program. Following Riedl (Riedl 1978), I like to call these "homomorph characters," meaning characters of the same shape; although, homomorphy is not a statement of similarity per se, but a statement of developmental genetic identity. In between fully individualized paramorph characters and completely identical homomorph parts are serially homologous characters that have a tendency to co-evolve by easily sharing character states. These are sufficiently individualized characters for which natural selection can specialize them for different functions, although derived character states are often shared. I call these "character swarms" for which the paradigm is feathers (see chapter 9).

To close this chapter, I want to emphasize that the purpose of this "taxonomy of terms" is neither phenomenological nor linguistic. The purpose is to clarify differences that, I think, are essential for a productive experimental and mechanistic research program aimed at understanding the evolution of complex multicellular organisms. The following chapters aim at moving us closer to this goal.

3

A Genetic Theory of Homology

"Zunächst müssen wir zugestehen, daß unstreitig dem alten
Homologiebegriff ein stark präformistisches
Moment innewohnt."[1]

(Bertalanffy 1936)

The concepts outlined in chapter 2 were intended to capture the major patterns of phenotypic variation: the existence of quasi-independent units of variation, of different character states, of different modes of variation, and of repeated body parts. These patterns have to be the product of both natural selection and the developmental mechanisms that engender the phenotype in each generation. Hence, the question arises, to what extent is there a mechanistic/developmental explanation for these patterns? Is the continuity of character identity subscribed by the continuity of genes? This is the point at which discussions regarding homology tended to become derailed in one way or another ever since the beginning of developmental biology with Hans Spemann in the early twentieth century (Spemann 1915). In the following sections I will review some of the relevant facts from developmental genetics and then propose a model that intends to explain these patterns and to serve as a guide for further research into the developmental evolution of morphological characters.

Why Continuity of Genetic Information Is Not Enough

In an influential paper published in 1982 entitled "Homology and Causes," Leigh Van Valen wrote: "All homologies involve a continuity of information. In fact homology can be defined, in a quite general way, as correspondence caused by a continuity of information. It now remains to see how this cause,

[1] First we need to acknowledge that the old homology concept has a strong preformistic tendency.

or kind of cause, produces the various relations we call homologies." (Van Valen 1982). This was the first clear description of the task that this chapter sets out to do. The first task, however, is to discuss whether or to what extent Van Valen's proposal to define homology, as correspondence caused by continuity of information, helps to identify experimentally testable explanations.

For a contemporary biologist, the first idea that the term continuity of information brings to mind is genetic information. Van Valen was quick to point out that he meant both more and less than DNA. On one hand, patterns of homology exist in cultures and in behavior that rely on non-genetic transmission of information. Homology and, thus, continuity of information is not limited to information transmission by DNA. Based on the limited data available at the time, he also intuited that different genotypes could produce the same phenotype and, thus, there could be homology without identity of genetic causes. The latter point is important for defining the agenda for this chapter, and I want to trace the history of this problem by focusing on some landmark publications (for a more extensive discussion, see Laubichler and Maienschein 2003).

The desire to explain the manifest continuity of characters by some underlying cause is as old as the homology concept itself. In fact, some historians argue that the concept of an archetype, proposed by pre-evolutionary biologists to explain patterns of morphological variation, has been a placeholder for some as yet to be determined causal principle (Rupke 1994; Amundson 2005). The first attempt to link homology to experimentally determined developmental mechanisms was that of Spemann, who is known for his discovery of embryonic induction.

Spemann was trained as a comparative vertebrate morphologist and, thus, was intimately familiar with the tradition of evolutionary morphology. When he and his collaborators discovered the inductive effect of the eye cup on the overlying ectoderm, which results in the formation of a lens, they tested this phenomenon in a variety of amphibian species. To their and everyone else's surprise, lens development could proceed in some species without an eye cup. Clearly, the lenses of different amphibian species are homologous, but their developmental causes (to the extent that one can assess them) are not.

As it turned out, lens-forming capacity is conferred upon the embryonic ectoderm in multiple steps, with the eye cup simply being the last tissue to provide an inductive signal (Henry and Grainger 1990; Grainger 1992; Lang 2004). Other sources of inductive signals include, for example, the heart-forming mesenchyme. The relative strength of these inductive signals varies between species so that, in some species, a lens can form only in the presence of an eye cup, while in some other species the other inductive signals are sufficient to initiate lens development. These and other experimental results led Spemann to examine the homology concept in a 1915 paper (Spemann 1915) in which he ultimately concluded that there was no way to explain homology with continuity of developmental mechanisms and, thus, gave up on seeking a connection between homology and experimentally verifiable causes.

Things did not improve with more knowledge regarding development. Another influential publication was De Beer's little book *Homology: An Unsolved Problem* (DeBeer 1971), in which he summarized all of the evidence since the time of Spemann that suggested a conceptual discontinuity between homology and mechanistic explanations of development. A particularly impressive example was lens regeneration in salamanders. It is relatively easy to remove the lens from an eye without doing too much damage to the rest of the eye. When one does this with salamanders, the lens can regenerate (figure 3.1). Of course, tissue formation and induction cannot be the same as during embryonic development, because the developmental conditions that existed during development cannot be reproduced. During development, the lens derives from the embryonic undifferentiated epidermis, which no longer exists in the adult. Rather, during regeneration the lens forms from the cells of the iris (figure 3.1), which corresponds to the rim of the eye cup and not to the embryonic epidermis. The regenerated lens is certainly "the same" organ as the one that developed during embryogenesis, but the developmental pathway seems quite different.

Lens regeneration fits into a larger pattern of variable developmental origins of cells for homologous organs. In general, homologous characters derive from homologous embryological origins and lead to the broad

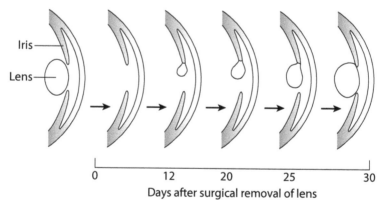

FIGURE 3.1: Lens regeneration in the newt. After the lens is removed from the eye of an adult newt, the lens forms from the iris. This mode of development is different from the developmental origin of the lens. During embryonic development, the lens derives from the ectoderm that covers the eye-cup. This was one of the most impressive early demonstrations that homologous body parts, in this case the lens, do not need to derive from the same developmental origin or have to follow the same developmental pathway. (Redrawn after: http://members .tripod.com/~Glove_r/Sheldrake.html.)

generalizations of comparative embryology, like the germ layer theory (Keimblattlehre) (e.g., Salvini-Plawen and Splechtna 1979; Hall 1998), according to which each organ and organ type derives from the same stereotypical embryonic cell layers (Keimblätter). These include the endoderm that leads to the gut and its accessory organs, such as (in vertebrates) the liver, the lung or swim bladder, and so on; the ectoderm that gives rise to the skin epidermis and the nervous system; and the mesoderm that gives rise to muscles and internal skeleton of vertebrates, kidneys, and so forth.

But this is not always the case and is most instructive for cases in which one adult structure is of hybrid origin in terms of its cell populations. Often, the contributions of different cell populations are variable. For example, the inner ear bone of chickens, the *columella*, consists of cells derived from the neural crest (and, thus, from the ectoderm) as well as from mesodermal cells. The relative contribution of either cell population can vary without consequence for adult morphology (Kontges and Lumsden 1996).

A pattern related to the variable origin of cells that contributes to a homologous character is positional variation. An example pointed out by Gavin de Beer (De Beer 1971) is the variable axial position of forelimbs. In newts, the forelimb derives from trunk segments 2, 3, 4, and 5, but in humans it derives from trunk segments 8 to 13. There is no shift of limb position during embryogenesis; rather, the differences are evolutionary. An even greater variation is obvious if we include birds and dinosaurs in our comparison.

Similarly, the regionalization of the body axis is highly variable. Can we speak of the thorax of the chick, derived from segments 15 to 21, as homologous to the thorax of a human derived from segments 8 to 19 as homologous? Historically they are, but embryologically they derive from non-corresponding parts of the body axis.

Even the mode of tissue formation can differ without affecting character identity. In vertebrates, bone can form in two very different ways. It is formed either as so-called replacement bone or as a membrane bone. Replacement bones develop from a cartilage precursor, which is later replaced by bone substance. Membrane bone develops from the ossification of connective tissue (membranes, tendons, etc.) without a cartilagenous intermediate (Hall and Miyake 1992; 1995). The mode of bone development is usually well conserved, so that most bones form as either one or the other in all species in which they are found (i.e., a bone usually "is" either a replacement bone or a membrane bone). In other words, the mode of bone formation usually is associated with bone identity.

But there are exceptions. For example, Bellairs and Gans (Bellairs and Gans 1983) pointed out that the orbitosphenoid, which is usually a replacement bone, is a membrane bone in amphisbaenian lizards. There is no doubt about the homology between the orbitosphenoid of amphisbaenians and other reptiles, yet their mode of ossification is fundamentally different.

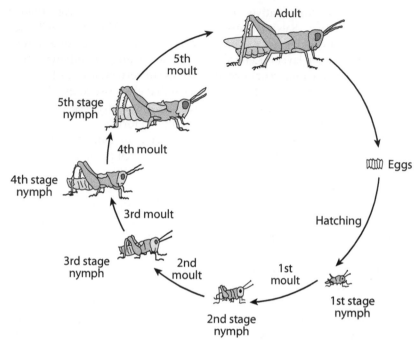

FIGURE 3.2: Life cycle stages of a hemimetabolic insect, the grasshopper. During grasshopper development, the different stages of post-hatching development are a step-wise transformation of the 1st stage nymph to the adult. For comparison, during the life cycle of a holometabolic insect, like the fruit fly (see figure 3.3), the larva differs radically from the adult (imago). The hemimetabolic life cycle likely was ancestral and the difference between the holometabolic and the hemimetabolic life cycle stages demonstrate that developmental can change radically but has little impact on the phenotype of the adult. (Redrawn from The Open Door Web Site: http://www.saburchill.com/.)

Another widely known fact pointing to developmental variation among clearly homologous characters is the variability of developmental modes in insects. Insects develop either directly through a number of larval and nymph stages that progressively become similar to the adult (imago). An example is the grasshopper (figure 3.2). In contrast, flies and butterflies and other more derived insects develop indirectly from a larval form, which is very dissimilar from the imago, through a radical metamorphosis (figure 3.3). Even early embryonic development is quite different between insect orders.

In flies, the segments are produced in a cascade of gene activity (see below), which leads to the simultaneous formation of all body segments. This mode of development is known as long germ development. In contrast, in primitive insects like thysanurans, the segments originate sequentially at the posterior end of the embryo in a budding zone (figure 3.4). This mode of development

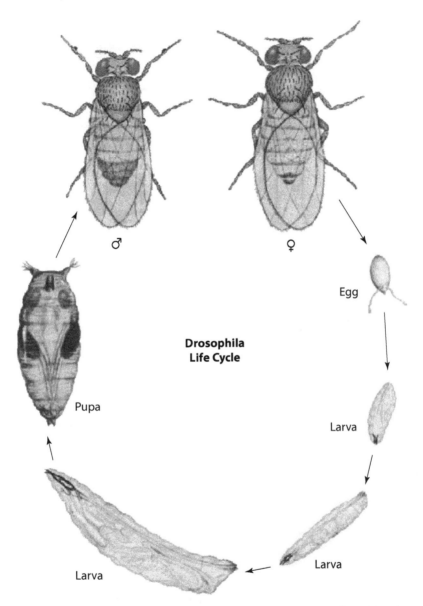

FIGURE 3.3: Life cycle of Drosophila, a holometabolic insect. Compare to figure 3.2 and note the remarkable differences between holometabolic and hemimetabolic life cycle stages. This is a reminder of how different the developmental stages among clearly related animals can be with little direct impact on the adult body plan. (Redrawn after Carolina Drosophila Manual.) Printed in U.S.A. © Carolina Biological Supply Company. Reproduction of all or any part of this material without written permission from the copyright holder is unlawful.

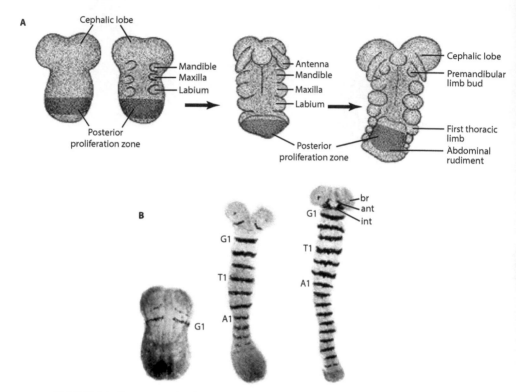

FIGURE 3.4: Short germ development. A) Embryonic development of *Petrobius*, a wingless insect (Thysanura). B) *engrailed* expression showing developing segment boundaries in beetle embryos (*Tribolium castaneum*). The embryo starts to form head segments, including the mouth parts of mandible, maxilla, and labium. Development proceeds by adding body segments at the posterior end of the embryo until all body segments are formed. Compare this mode of development with that of Drosophila in figure 3.5. (G1: first jaw segment, T1: first thorax segment, A1: first abdominal segment, br: brain, ant: antenna segment, int: intercalary segment). (From Scott Gilbert, 2010, *DevBio: A Companion to Developmental Biology*, 9th Edition, http://9e .devbio.com/article.php?ch=6&id=91.)

is called short germ development. Parasitic wasps have even more radically modified early development.

Clearly, all insects form a monophyletic group (i.e., a clade) and are derived from a common ancestor with the same basic body regions as modern insects, but their mode of development is quite different. The embryological evidence, however, does not tell us whether there is more continuity at the underlying genetic level than suggested by the dissimilarity of embryonic pathways.

The most definite evidence for discontinuity of genetic information for homologous characters has arisen since the 1990s from developmental genetics. In particular, the comparative work on insect segmentation, based on

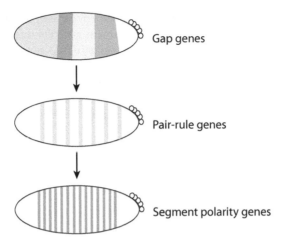

Gap genes

Pair-rule genes

Segment polarity genes

FIGURE 3.5: Main stages of Drosophila segment development. This is an example of so-called long germ development. The critical difference compared to short germ development illustrated in figure 3.4 is that all segments of the body form simultaneously in a sequence of three steps: gap genes specify broad regions of the body axis; pair-rule genes specify pairs of segments; and segment polarity genes finally specify the borders of the definite segments. This is a derived developmental pathway and illustrates that the insect body plan and major body parts have been maintained despite radical changes in early development.

the exquisitely detailed work on Drosophila segmentation, has been very influential (for an overview, see Davis and Patel 2002) (figure 3.5). Drosophila segmentation is, perhaps, the developmental process that has been analyzed in the greatest detail at the genetic and molecular levels.

In Drosophila, one can distinguish between three main phases of development based on the different genes that are active during these phases. First, there is a phase of gap-gene activity during which mutations cause the loss of broader regions of the anterior-posterior body axis. Examples of gap genes are *Knirps* (*kni*), *Krüppel* (*Kr*), and *Hunchback* (*hb*). Then there is a phase of pair-rule gene activity during which mutations lead to the loss of pairs of segments. For example, a loss of function mutation of the *ftz* gene results in the loss of parasegments (i.e., stripes of tissue that comprise the posterior half of each odd-numbered segment and the anterior half of each even-numbered segment). A mutation of the *eve* gene leads to a complementary defect. Subsequently, the segment polarity genes kick in to set up the limits of the definite segments. Mutations of these genes typically result in a loss of segment polarity (i.e., a loss of the differences between the anterior and posterior parts of segments).

It is important to note that each of the genes and gene classes mentioned above are absolutely necessary for Drosophila development. All loss of function mutations of these genes are lethal. Thus, it is reasonable to expect that a homologous set of genes also directs the segment development of other insects, despite the different modes of development noted above. This is, however, not the case. For example, the role of the pair-rule genes varies greatly between insect orders. In grasshoppers and other insect lineages, homologs of *eve* and *ftz* are known and are expressed during nervous system development, but do not participate in segment development (Patel, Ball et al. 1992; Grbic, Nagy et al. 1996). These genes have a conserved role in nervous system development, but not in segmentation.

The more comparative molecular work that is done on various species, the more examples of homologous organs with mechanistically different modes of development become known. Well-worked-out examples are emerging in nematode vulva development (Sommer and Sternberg 1994; Kiontke, Barriere et al. 2007) and differences between mouse and zebrafish somitogenesis (van Eeden, Holley et al. 1998). Ultimately, a comprehensive review of these differences may turn out to be synonymous with a review of all of zoology. *We can safely assume that some level of variation in the developmental mechanisms of homologous characters is the rule rather than the exception.*

Continuity of morphological characters is not subscribed by continuity of genetic information. This seems to be a pretty depressing situation if one thinks that homology should somehow be related to the evolution of development and motivates the notion that homology may be an illusion (i.e., rather than reflecting an important part of biological reality; see chapter 2 for discussion and references). However, on closer inspection, the pattern of developmental variation is not entirely random and might tell us something about how phenotype homology is reflected in genome evolution.

Lessons from the Variable Development of Homologs

In the previous section we mentioned a number of examples for which clearly homologous characters derived from different cell populations during embryogenesis or followed different developmental pathways to arrive at the same adult morphology. One possible reaction to this fact is to assert that homology is a meaningless concept (Wake 2003). That said, a more moderate reaction is possible—namely, to ask whether only certain aspects of the developmental process are relevant for homology while others are not. Here we will pursue the second option—namely, to assume, for example, that embryological origins are irrelevant for the developmental basis of homology and then ask what this implies for the search for the biological basis of homology.

If the embryological source of cells comprising a character is irrelevant for character identity, as suggested by a large body of comparative embryological evidence, then *the developmental processes leading to character identity have to commence in later developmental stages* than those that provide the cellular material. I think this suggestion is consistent with an elementary fact from developmental biology.

In many organisms, one can experimentally distinguish between the prospective fate of a cell population and cells that are determined to have a particular fate. The prospective fate of a cell population is what these cells will become during an undisturbed course of development. For example, two lateral patches of the embryonic "skin" of a newt (i.e., the ectoderm) have the prospective fate of becoming a lens, and the ectodermal cells at the future belly region become the outer layer of the skin, epidermis. This is what these cells become if the normal process of development proceeds undisturbed.

If development is disturbed, either by experimental manipulation (like removing the eye cup) or by mutations that prevent contact between the head ectoderm and the eye cup, then the prospective lens cells become epidermis (skin). If the eye cup is transplanted under the prospective belly epidermis, the prospective epidermis turns into a lens. Obviously, early in development, these cells are not committed to become either a lens or epidermis. Only later events seal the developmental fate of these cells.

These events are called embryonic induction and, in many cases, depend on chemical signals from neighboring cell populations. After induction, the cells become determined, which means that they pursue a developmental pathway largely independent of their environment, even when explanted into a tissue culture flask. Then the cells have "internalized" their fate and really are committed to become lens or epidermal cells, but not before.

I think that there is a simple lesson to be learned from these facts from developmental biology. Up to a certain experimentally detectable stage of development, characters do not exist. Otherwise, every adult character would have to be present in the zygote, which is obviously absurd. Thus, *developmental variation before the determination of the character is irrelevant for character identity*. Hence, the interspecific variation of those early stages of development should not worry us with regard to homology. This is not to say, however, that the embryological origin of a character's cells cannot be conserved to some degree and, thus, be used as a criterion to identify homologous characters. Of course, that is the case, as overall similarity is a heuristic criterion for finding homologous characters. However, for similarity we also know that it is conceptually not necessary for homology, as the homology of inner ear bones of mammals with the jaw joint of reptiles shows. Similarity and identity of embryological origin is, in fact, only a historical residue that points to homology, but is not necessarily part of character identity.

But is not the entirety of the developmental process eventually the cause of the adult morphology? How can we make a distinction between relevant and irrelevant parts of the developmental process? After all, if perturbed, any or most developmental events have effects on the adult morphology and, thus, clearly have causal importance! Shouldn't then all of development be considered part of the mechanistic basis of character identity (i.e., the cause of what a character is)? These are legitimate questions and it is necessary to dig deeper into the inner workings of an embryo to justify my suggestion: character identity arises at a certain stage of development and not earlier.

As mentioned above, in many (but not all; see below) cases cells assume their definite developmental fate through chemical signals from neighboring cells. Why then aren't the process of embryonic induction and the inductive signal part of the developmental causes of character identity, all the way back to the first inductive signal from the maternally provided factors in the egg cell? After all, they are all necessary during normal development to achieve the wild-type outcome of development. Again, the answer lies in a long known, but still surprising fact of developmental biology that had puzzled and stalled developmental biology for many decades during the mid-twentieth century.

Soon after the discovery of embryonic induction by Spemann and his collaborators, it was clear that the most likely material basis for an inductive signal was a molecule, rather than an electric or mechanical stimulus. After this was clear, the next goal of developmental biology was to identify these molecules or, at least, characterize their chemical nature. This step turned out to be extraordinarily difficult, but in an interesting way.

It was not that the biochemists of the mid-twentieth century were unable to isolate or at least characterize biological material. To the contrary, that was a time of extraordinary progress in biochemistry. The problem was biological. In order to know whether one has isolated, or at least enriched, a biologically relevant molecule, one needs a bioassay. This is a biological test that shows whether the molecule has the biological activity one wishes to explain. For example, it soon became possible to achieve neural differentiation in tissue culture, and it should have been straightforward, albeit difficult, to work toward isolating the neural inductor molecule. The real difficulty, though, turned out to be both conceptual and practical at the same time.

As soon as these *in vitro* differentiation assays were established, it became clear that "specific" developmental events, like neuronal differentiation, could be "caused" not only by specific biological agents, but also by artificial chemicals, and even simple pH changes. If this is the case, then we face the practical problem of proving that a certain fraction of biological material actually contains the natural inductive signal.[2] Alternatively, any material may simply induce differentiation as an artifactual effect. This technical difficulty had stalled progress in developmental biology for decades until the revolution of developmental genetics during the 1980s and 1990s.

---

[2] For example, when I was a student a famous case made the gossip circuit when an isolated inducing molecule turned out to be contamination from a resin used in columns for fractionating proteins.

The interesting conceptual lesson from the failure of biochemical attempts to isolate inductive signals is that, obviously, the information for the developmental fate of a cell is not contained in the inductive signal itself; rather, it is in the responding cells. It turns out that, during each stage of development, cells have a limited number of possible fates, and the inductive signals simply choose between them. If left unperturbed, most cells have a default developmental pathway that they will pursue, as for example ectodermal cells become skin if not told otherwise.

Inductive signals can be conceptualized as perturbations, which kick the cell into one or another intrinsically predetermined states, and this can also be done by non-specific perturbations, like increased Li^+ concentrations or non-physiological pH levels. This is the reason why it is so difficult to experimentally manipulate developing cells without causing massive artifacts. For example, in our lab we work with endometrial stromal cells. It is easy to cause the expression of "differentiation marker genes" like *prolactin* and *IGFBP* simply by treating these cells with chemicals that we use to deliver RNA or expression constructs into the cells. Even simply forcing the cells to take up some RNA or DNA can cause the cell to start differentiating, even though the appropriate physiological, hormonal signal is absent.

In the language of developmental biology, inductive signals are not instructive (i.e., do not contain information about the final product), but are only permissive (i.e., they trigger a process intrinsic to the cells themselves). The signal does not contain the blueprint of the character, as DNA does for proteins; rather, it is simply that: a limited, replaceable signal for a choice between a small number of alternatives. The alternatives, however, are intrinsic to the cells that receive the signal.

The information for character identity is within the cells that react to the signal and not in the inductive signal itself. Knowing that developmental signals do not contain essential information, but can be replaced by others quite easily in an experiment, makes it understandable why, during evolution, inductive signals could have changed quite readily, almost as readily as the embryonic source of cells that give rise to a body part. Hence, the biological basis of character identity can neither be tied to embryonic origin nor to developmental/inductive signals. This conclusion is consistent with the comparative developmental evidence that shows that both cellular origin and inductive signals are highly variable during evolution.

Homeotic Genes and Character Identity

Thus far I have argued that certain aspects of the developmental processes are not relevant for understanding character identity, but I have not yet made a positive statement regarding what *is* relevant for character identity. Here I will argue that the function of homeotic genes gives us a hint at what might be the developmental genetic basis of character identity. Essentially I will argue that the distinction between character identity and character states, as defended in chapter 2, is reflected in the genetic architecture of development in which character identity has a different genetic substrate than character states (Wagner, Chiu et al. 2000; Deutsch 2005; Wagner 2007).

The term homeotic was introduced by William Bateson (1861–1926) to identify variation in which "something has been changed into the likeness of something else" (Bateson 1894). Or, one may say, when one body part assumes the identity of another. The classical and best understood cases are the *Antennapedia* and the *Bithorax* mutations of Drosophila in which in the former the antenna is transformed into the likeness of a leg (figure 3.6A), and in the latter the haltere assumes the shape of a wing (figure 3.6B).

In many cases these transformations are partial, but clearly in the direction of the identity of another body part. In the previous chapter, the differences between forewing and hind wing and between wings and halteres were used to explain the distinction between characters and character states. It was this that Jean Deutsch (Deutsch 2005) used to point out that character identity has a genetic basis that is distinct from that of character states.

A bithorax mutation (i.e., a partial or complete transformation of the haltere into a wing) is caused by mutations in the *ultra bithorax* gene, *Ubx*. This gene encodes for a homeodomain containing transcription factor. There are two ways to interpret the role of the *Ubx* gene. One interpretation is that one could, quite legitimately, think of *Ubx* as a gene that determines the difference

FIGURE 3.6: Paradigmatic examples of homeotic mutations. A) *Antennapedia* mutation in which an antenna is replaced by a leg-like appendage. B) *Bithorax* mutation in which the haltere is replaced by a wing, resembling the forewing of a wild type fruit fly. (Source: http://forums.cannabis culture.com/forums/ubbthreads.php?ubb=showflat&Number=1176691;http://www.ucl.ac.uk /~ucbzwdr/teaching/b250-99/homeotic.htm.) Figure 3.6A reprinted by permission of F. Rudolf Turner.

between a wing, in the sense of a wing blade, and a haltere. After all, the loss of function mutation of *Ubx* results in a loss of haltere-like morphology and the recovery of a wing-like morphology. This mutation restores what looks like the four-winged phenotype of the ancestor of dipterans. It is possible to think of *Ubx* function as causing the difference between the four-winged ancestral morphology and the derived dipteran morphology in which the hind wing is shaped as a haltere. With this interpretation, *Ubx* is seen as a determinant of a derived character state (i.e., the haltere).

Another interpretation is that *Ubx* is a determinant of character identity (i.e., a factor necessary for the identity of hind wings), regardless of what the character state is (Wagner, Chiu et al. 2000). With this interpretation, the bithorax phenotype is a duplication of the forewing identity in the third thoracic segment (T3) of the fly. Either interpretation is consistent with the mutant phenotype in flies.

Is the distinction between these two interpretations "semantic" or biologically meaningful? In order to see that *Ubx*'s function is, in fact, to determine character identity and not character states, one has to reach for a broader comparison of *Ubx* function in insects.

The first evidence that *Ubx* may not be responsible for the character state of a haltere came from a study of *Ubx* expression in the butterfly *Junonia* by Sean Carroll's group (Warren, Nagy et al. 1994). Butterflies have four wings. If *Ubx* is, in fact, the determinant of the character state "haltere," then one would expect *Ubx* expression to be absent in the third thoracic segment. But this is not what Carroll and his collaborators found in *Junonia*. As in all other insects examined so far, the third thoracic segment expresses the same combination of genes as in Drosophila. Upon reflection, though, it is also clear that the hind wing of the butterfly is quite different from the forewing, and that the hind wing does have its own character identity; it requires some genetic factor to make it different from the forewing, even though it is a wing in the sense of a lift-producing organ.

The clearest evidence, however, for the fact that *Ubx* determines wing identity (i.e., forewing versus hind wing identity) rather than a haltere phenotype comes from a study of *Ubx* function in the beetle *Tribolium* (Tomoyasu, Wheeler et al. 2005). Remember that in flying beetles, the hind wing is the flying organ, unlike in flies in which it is the forewing. Hence, in beetles, the hind wing is a wing blade (at least in its fully expanded state of the wing, which is usually folded under the forewing). The forewing in beetles is a protective cover called the elytra. In their study the authors used RNA-interference (RNAi) to knock down the expression of *Ubx* and found that, in the absence of *Ubx* function, the normally blade-like hind wing was transformed into a second set of elytra (figure 3.7). This result can be interpreted only as the transformation of the beetle hind wing into the identity of the forewing.

 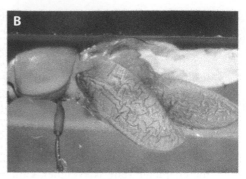

FIGURE 3.7: Comparison of the phenotype of a mutation in Drosophila (A) to a phenocopy of this mutation in the beetle *Tribolium*. The *Tribolium* phenotype was generated by knocking down *Ubx* gene mRNA. This is the same gene that is mutated in the Drosophila *Bithorax* mutation. What the *Tribolium* experiment clearly shows is that the loss of *Ubx* function results in a duplication of the anterior wing identity in the place of the posterior wing. This shows that *Ubx* is responsible for hind wing identity and not directly for the phenotype or character state of the "hind wing", for which "hind wing" means posterior dorsal appendage. The main conclusion drawn from this experiment is that genes that define character identity and character states can, at least in this case, be separated based on their mutational phenotype. (Wagner, 2007, *Nature Reviews Genetics* 8:473–479.)

Hind wings never had the shape of an elytra, which would be quite dysfunctional. *Ubx* determined hind wing identity, perhaps in conjunction with other genes, regardless of whether the hind wing had the shape of a wing blade or that of a haltere. Hence, *Ubx* function determines character identity regardless of character state (Wagner, Chiu et al. 2000; Deutsch 2005; Wagner 2007).

Loss of *Ubx* function results in the transfer of forewing identity upon the dorsal appendage of the third thoracic segment, regardless of what the character state is of the forewing. As a corollary, the genetic factors that make a hind wing a wing blade or a haltere have to be different from the character identity determining genes like *Ubx*. These character state determining genes act downstream of character identity genes.

A Model: Character Identity Networks

The realization that, in the case of insect wings, *Ubx* clearly confers character identity but not character state information per se can be condensed in a simple model of the genetic control of character development (figure 3.8). This model has three tiers of developmental roles for genes. The first tier is the determination of *positional information*.

This function is either fulfilled by the inductive signals among neighboring cell populations, by gene cascades that endow large areas of the embryo

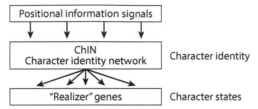

FIGURE 3.8: Cartoon model of the functional logic of metazoan development. It is proposed that there are three main classes of gene functions in character/body part development. First, at the highest level of this hierarchy are genes involved in determining the positional information of a group of cells that become the later character. This signal (or signals) is received and integrated by a gene regulatory network that relates the positional information signal to the effector genes that actually engender the phenotype. This relay network is called a Character Identity Network (ChIN) and its main function is to enable the activation of a position specific and organ specific developmental program.

with differential gene expression, or by creating repeated identical chunks of tissue, like the gap and pair-rule genes in Drosophila. All of the examples of interspecific differences in developmental mechanisms of homologous characters that were cited in the previous section were about genes involved in this developmental role. Variable inductive mechanisms that baffled Spemann and his colleagues are among these, as are the variable roles of pair-rule and gap genes in insect segmentation. Also, the well-documented differences in vulva development in nematode worms (Sommer and Sternberg 1994) have to do with cell-cell signaling leading to the determination of vulva-forming cells, as well as the mechanisms of somitogenesis that differ between mouse and zebrafish (van Eeden, Holley et al. 1998).

The next tier of gene function in this model comprises genes and gene regulatory networks that interpret the positional information signals and activate position-specific developmental programs. I call these gene networks *Character Identity Networks* (ChIN; Wagner 2007). In other fields these same gene regulatory networks are called "core regulatory networks" (CRN; e.g., Graf and Enver 2009). There is also a similar sounding concept, "kernels," which was proposed by Davidson and Erwin (Davidson and Erwin 2006).

However, the conceptual basis for ChIN or CRN is quite different from that of kernels *sensu* Davidson. Kernels are defined through their phylogenetic age. ChIN and CRN are defined by their biological function in determining character and cell identity (Wagner 2007; Graf and Enver 2009), regardless of their phylogenetic age. A ChIN is a ChIN, even if it might have originated yesterday. Hox genes, like *Ubx*, belong to a ChIN, but are not part of kernels as explicitly stated by Davidson and Erwin (Davidson and Erwin 2006).

ChINs form the interface between developmental signals and those genes that actually engender the morphological character during morphogenesis and differentiation. I argued above, following Jean Deutsch, that *Ubx* is an example of such a gene that confers hind wing identity. If there is a material locus of character identity, then it would have to be the ChINs. This implies that this layer of genetic control of development should be more conserved than positional information gene networks. This, in fact, seems to be the case, as I will explain below.

The next tier of this model of character development is the "realizer" genes that are controlled by the ChIN genes and are engaged in actually making the morphological product. These genes and their expression are necessarily variable, as they are responsible for realizing the specific character state; for example, a wing versus a haltere.

Hence, if there is any mechanistic basis for character identity and, thus, homology, it is likely to be within the continuity of Character Identity Network genes, which mediate between (phylogenetically variable) positional information signals and the realizer genes that determine the (variable) character state. ChIN gene function is also the point in the developmental program that translates the "abstract" positional information of early development into specific developmental individuality by controlling character-specific gene expression. They mediate the readout of the positional information and, thus, allow for the distinctness of one body part compared to others (i.e., they are the mechanistic basis of body part individuality both in individual development as well as in evolution).

During evolution, ChIN genes allowed for variational independence of body parts because they enabled differential gene expression and, thus, for the evolutionary optimization of gene expression and function in different body parts. In the next sections I will discuss several potential examples of ChINs to see whether they share common structural and functional characteristics.

Variation and Conservation of Segment Development

As already briefly mentioned above, in Drosophila the cascade of segment development comprises three stages: the stage of gap gene activity [e.g., genes *Knirps (kni)*, *Hunchback (hb)*, and others], which determines larger chunks of the body axis; the stage of pair-rule gene activity, such as *even skipped (eve)*, *paired (prd)*, *hairy (h)*, *runt (run)*, *fuzi tarazu (ftz)*, and others, which determines parasegmental units of the body axis (i.e., stripes that cover anterior and posterior halves of neighboring segments); and, lastly, the stage of segment polarity gene activity, which maintains the parasegmental borders and directs the execution of the developmental program to create the morphology of segments. Phylogenetic conservation varies between these different stages of segment development.

The activity of gap genes in Drosophila proceeds in the syncytial blastoderm of the early fly embryo. Pattern formation during this stage of development depends on the ability of transcription factors to diffuse without being impeded by cell membranes. Gap genes are expressed in distinct domains and their mutations result in the loss of larger contiguous groups of segments. However, the syncytial blastoderm is a derived feature of long germ development in insects. In this mode of development, all segments of the body axis are formed more or less simultaneously. Many insects and all other arthropods, however, follow a different mode of segmentation, so-called short germ development.

In short germ development, segments are formed sequentially at the posterior end of the embryo. Segment formation proceeds in a cellularized environment in which diffusion of transcription factor proteins is restricted. In short germ insects, gap genes are active during early development, but their expression pattern relative to the definite segments is different from that in Drosophila. Their expression domains are contiguous rather than restricted to distinct regions and, in most cases, are more anterior than their Drosophila counterparts (for an overview, see Damen 2007).

Knockdown (KD) experiments using RNAi reveal that these genes are involved in segmentation, but their expression domains do not correspond to the segments affected by KD. In *Tribolium*, for example, *Kr* is expressed in the domain of *eve* stripes 2 and 3, but with *eve* KD, stripes 4 and 5 are affected and the more posterior *eve* stripes do not form at all (Cerny, Bucher et al. 2005). Hence, gap genes seem to affect segmentation more indirectly than in Drosophila in which they directly regulate their pair-rule target genes.

In addition, in short germ insects, gap genes are engaged not only in segmentation, but also in segment identity determination (i.e., they have homeotic functions), a function that in Drosophila is largely restricted to Hox genes. Furthermore, a *Tribolium*-specific gap gene, *millepattes* (*mlpt*), has been identified (Savard, Marques-Souza et al. 2006), which does not seem to have a function in Drosophila segmentation.

In the non-insect arthropods (specifically the spider *Cupiennius* and the myriapods *Lithobius* and *Glomeris*), a gene regulatory network is involved in segmentation that is not involved at all in Drosophila segmentation—the Delta/Notch pathway (for a review, see Damen 2007). Delta/Notch has no role in segmentation either in flies or in *Tribolium*, but does have a role in the more basal lineage of cockroaches (Pueyo, Lanfear et al. 2008), and even annelids (Thamm and Seaver 2008). Delta/Notch is a signaling pathway that exists in all metazoans examined.

Notch is a transmembrane receptor of Delta, which is also a membrane integrated protein. Hence, Delta/Notch pathway signaling is limited to neighboring cells for which cell membranes that contain the ligand Delta abut cell membranes that contain the receptor Notch. It makes sense that this signaling pathway is used for segmentation in highly cellularized environments, such as those of spider and millipede growth zones (no data regarding Delta/

Notch activity in crustaceans appear to be available), but not in the syncytial environment of early fly development.

Clearly, the highest level in the segmentation cascade, gap genes' function, is quite variable among arthropods, even though the comparative data are still quite spotty. A scenario forwarded by Wim Damen (Damen 2007) explains the differences among arthropods. This suggests that Delta/Notch was the ancestral upstream activator of the segmentation cascade in arthropods. At the same time, gap genes functioned primarily to activate the Hox genes for determining segment identity. During the evolution of insects, the function of the Delta/Notch system was gradually replaced by the gap gene system, and Delta/Notch lost its function for activating Hox genes. This transition from Delta/Notch to gap gene control of segmentation was probably associated with the evolution of long germ development.

In Drosophila, the gap genes activate the pair-rule genes *eve, run,* and *h*; the so-called primary pair-rule genes. These are the genes that are expressed in segmentally repeated stripes. The primary pair-rule genes translate the nonperiodic pattern of gap gene expression into a periodic pattern and activate the secondary pair-rule genes *odd, opa, prd, ftz, ten-m,* and *slp.* Pair-rule gene activity seems to be involved in the segmentation of all arthropods, even though there are differences in their expression patterns and functional relationships.

For example, in the holometabolous insects, such as the cricket (*Gryllus bimaculatus*), grasshopper (*Schistocerca americana*), and the milkweed bug (*Oncopeltus fasciatus*), the expression and function of *eve* varies (Damen 2007). *Eve* is expressed in a segmental rather than a double segmental pattern in *Oncopeltus*, and has no segmentation function in *Schistocerca*. It is expressed in both segmental and double segmental patterns in *Gryllus*, and has paired function only in anterior segments, but not in posterior segments. Further, in *Gryllus* and *Oncopeltus, eve* also has gap function. In *Tribolium,* the set of primary pair-rule genes include *eve, run,* and *odd,* but in *Drosophila* they are *eve, run,* and *h*. Similar variability of expression and function can be found in other arthropod groups.

The layer of gene activity directly downstream of pair-rule genes in Drosophila segmentation are the so-called segment polarity genes, *Wingless, engrailed, Hedgehog, hairy,* and others (for a summary, see Wilkins 2002; Damen 2007). Their role is to establish the definite boundaries between neighboring segments and to control the execution of actual segment morphology development. The core of the segment polarity network involves reciprocal activation among cells that express the two signaling molecules *Wingless* and *Hedgehog. Wingless* signaling activates the expression of the transcription factor gene *engrailed (en)* that, in turn, activates *Hedgehog (hh)* gene expression, which signals back to *Wingless*-producing cells to activate *Wingless* expression (figure 3.9).

It turns out that this layer of the segmentation cascade is highly conserved in contrast to the variation at the higher levels of the segmentation hierarchy

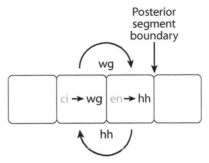

FIGURE 3.9: Example of a gene regulatory network that is rigidly associated with a character identity, the insect segment. This network is called the segment polarity network. In contrast to other processes underlying segment development in insects, the segment polarity network is conserved. Transcription factors are in gray, signaling molecules in black.

(e.g., in *Tribolium*; Farzana and Brown 2008). Expression of the Drosophila segment polarity gene homologs *engrailed* and *wingless* has been detected in segments from five or more insect orders and even in several crustaceans (Damen 2007). *Engrailed* expression has been found to be conserved in all four major extant arthropod clades, insects, crustaceans, myriapods, and chelicerates, and even in the "para-arthropods," like tardigrades (Gabriel and Goldstein 2007) and the velvet worms, Onychophora (Eriksson, Tait et al. 2009). These data suggest that this part of the segment polarity network is at least as old as arthropod segments (Damen 2007).

Repeated patterns of *engrailed* expression were even found in *Amphioxus* during somite formation and, thus, may have played a role in ancestral vertebrate somitogenesis (Holland, Kene et al. 1997). However, the relationship between arthropod and annelid segmentation and chordate somitogenesis remains unclear. Annelids belong to a major clade different from that of arthropods, the lophotrochozoans, whereas arthropods belong to the ecdysozoans. Comparative anatomy does not immediately lend support to homology between these two cases of segmentation.

Indeed, expression of "segmentation" genes *hh*, *Wingless*, and *en* in annelids is not consistent with a segment polarity network that is shared by arthropods and annelids (Seaver, Paulson et al. 2001), although the situation might be variable between species (Wedeen and Weisblat 1991; Prud'homme, de Rosa et al. 2003). Hence, it is unclear whether the common ancestor of all bilaterians had segments, and the similarities and differences between chordate, annelid, and arthropod segments for now do not result in a straightforward answer as to the homology of these cases of segmentation.

In any case, the essential elements of the segment polarity network seem to be conserved, at least within the arthropods (Damen 2007). It is clear that a segment polarity gene network is more conserved than is pair-rule gene function and the early pattern formation process prior to actual segmentation.

The main point here is that the gene regulatory network underlying segmentation is variable with respect to its phylogenetic conservation. Certain parts are highly variable, whereas others are more conserved. The segment polarity gene regulatory network seems to coincide in its phylogenetic distribution with that of arthropod segmentation. Hence, the segment polarity gene network is a good candidate for the arthropod gene regulatory network that determines the identity of arthropod segments as segments.

Eye Development and the *ey/so/eya/dac* (ESED) Networks

Drosophila eye development has been intensely studied for many years and provides an example of a gene regulatory network that could be paradigmatic for how character identity networks are organized. The eyes of Drosophila develop from two imaginal discs on each side of the head of the fly larva. Differentiation of the eye cells proceeds in a wave from posterior to anterior that is morphologically reflected in the so-called "morphogenetic furrow (MF)," which travels across the imaginal disc as the cells commit to their fates of making retinal cells and their support tissues. This means that the undetermined, proliferating cells are anterior to the MF and that determination and differentiation occurs posterior to the MF.

A critical step in eye development is activation of a gene regulatory network that we will call ESED; short for *eyless (ey), sine oculis (so), eyes absent (eya),* and *dachshund (dac).* These are four genes that constitute the core of the regulatory network of compound eye development in Drosophila. In addition, a paralog of *ey* has been found to be important, called *twin of eyeless* or *toy,* which is also essential for compound eye development. The names of four of these genes are different ways of saying "without eyes," based on the phenotypes of their mutants. This already indicates that these genes are jointly necessary for eye development.

Four of these, *ey, toy, eya,* and *dac,* are also sufficient to induce ectopic eyes if one of them undergoes targeted misexpression in the antennal imaginal disc. In the next paragraph I will summarize what is known about the function of these genes at the time when this chapter was written. There is a very extensive literature on these genes. However, I will not provide extensive references; rather, I direct the reader to two reviews from which I derived most of my information (Kardon, Heanue et al. 2004; Friedrich 2006).

Three of these genes, *ey, toy,* and *so,* encode for transcription factor proteins. The other two, *eya* and *dac,* encode for transcriptional co-activators (i.e., proteins that cannot directly bind to DNA but bind indirectly to DNA by binding to transcription factors[3]). In addition to being a transcriptional

---

[3] Although *dac* is suspected of harboring a DNA binding domain.

co-factor, the Eya protein is also a protein tyrosine phosphatase and, thus, part of signaling transduction machinery (Rebay, Silver et al. 2005).

During normal eye development, activation of these genes follows a simple linear hierarchy (dark broken lines with arrows in figure 3.10): *toy* is the first component of the network to be expressed; it activates *ey* that, in turn, is required for activating *so* and *eya*. *Ey* and *toy* are the most potent activators of ectopic eyes and are expressed in mutants that lack *eya* or *so*. *Eya* and *so* regulate each other; thus, in the absence of *eya*, no *so* is expressed, and in the absence of *so*, *eya* expression is reduced. The most downstream component of this network is *dac* that, during normal development, requires *ey* and *eya* to be expressed. But *ey*, *eya*, and *so* can be expressed in a *dac-/-* mutant background. In addition, a gene called *optix*, a paralog of *so*, is also essential in compound eye development and can induce ectopic eyes, but is expressed only during the proliferation of eye cells (i.e., anterior of the MF in the eye antenna disc; Seimiya and Gehring 2000).

Once they are expressed in the imaginal disc, three of these genes, *so*, *eya*, and *dac*, constitute a cross-regulatory network in which genes that are upstream in the developmental cascade are also regulated by "downstream" genes. Interestingly, *ey* is downregulated posterior to the MF and is, thus, not directly a part of the eye determination network, although it returns later in regulating opsin expression in receptor cells. That is to say that three genes in this network, *so*, *eya*, and *dac* are mutually interdependent in their role in eye development (figure 3.10).

For example, ectopic *eya* and *dac* can induce the expression of *ey*, and the ability of *eya*, *so*, and *dac* to induce ectopic eyes depends on the presence of an intact *ey* gene. In an *ey* mutant background, the ectopic expression of the

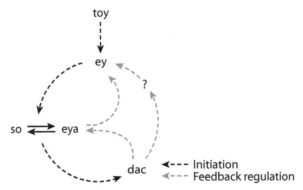

FIGURE 3.10: Character identity network of insect eyes. The pathway for developmental induction is shown in black and feedback loops are shown in gray. The feedback loops maintain the activity of this gene regulatory network. Like the segment polarity network, this core network is highly conserved among arthropod eyes. (Redrawn after Czerny et al., 1999, *Molecular Cell* 3:297–307.)

downstream genes is not sufficient to induce eye development. Hence, even though *ey* is the most upstream component in the developmental process, *eya*, *so*, and *dac* are equally critical for eye development and mutually maintain each other's expression. This dual action of eye determination genes is also reflected in the molecular mechanisms of their transcriptional regulation, as shown in the case of *dac*.

The transcriptional regulation of *dac* depends on two enhancers: one 5′ in intron 9 of the *dac* locus and one in the 3′ non-coding region (Pappu, Ostrin et al. 2005). The 5′ enhancer is activated by Ey and is likely the enhancer that activates *dac* transcription, while the 3′ enhancer is activated by the So:Eya protein complex and Smad, the transcription factor of the dpp signal that is responsible for maintaining *dac* transcription. There are different cis-regulatory elements for *dac* activation and for maintaining its expression.

In recent years, the molecular details of this network that was deduced from genetic data have been filled in. As expected, many regulatory interactions are due to classical transcription factor activities (i.e., through transcription factor proteins' binding to cis-regulatory elements). For example, the direct regulation of *so* by *ey* and *toy* was demonstrated by identifying a cis-regulatory element (so10) of the *so* locus. So10 binds Ey and Toy proteins at different binding sites for which Ey binding is important for compound eye development and the Toy binding sites are important for ocellus (i.e., accessory eye) development (Punzo, Seimiya et al. 2002).

In addition, however, it becomes clear that an essential layer of regulation is based on protein-protein interactions. For example, the Ey protein is not involved first in activating the other genes of the eye determination network, *so*, *eya*, and *dac*; rather, it forms a complex with two other transcription factors, homothorax (Hth) and tee shirt (Tsh). This Ey:Hth:Tsh complex inhibits the expression of the other eye determination genes and maintains the cells in a proliferative state. The switch from this function to eye determination and differentiation is mediated by a dpp signal, which represses Hth and liberates Ey to activate the downstream genes and results in the expression of So and Eya proteins. Eya is a transcriptional co-activator that has to bind to a transcription factor in order to be recruited to its target genes. Eya dimerizes either with itself or with So proteins. Only the Eya:So complex activates genes further downstream of this cascade.

When it is first expressed, however, So is dimerized with another cofactor, Groucho, and the Grou:So complex represses the expression of target genes. Eya disrupts the Grou:So complex and replaces Groucho to form the activating Eya:So complex. This complex then directly activates *dac*. As noted above, the *dac* gene has two enhancers, one that is activated by Eya:So and Ey and the other that is activated by Eya:So and Smad. Dac protein also forms a complex with Eya, although this is mediated by another transcriptional co-factor protein, CBP. These then lead to activating those genes that are directly involved in eye cell differentiation, like *atonal* and *gl*.

From the evidence summarized above, it is clear that the functional specificity of transcription factor genes depends not only on the interactions between transcription factor proteins and its binding sites on DNA, but importantly also depends on protein-protein interactions between transcription factors and transcriptional co-factors. This fact will be important when we consider the evolution of novel gene regulatory networks (chapter 6) and the involvement of similar sets of genes in different tissue types and organs (see below). First, however, we need to consider the evolutionary history and conservation of this network.

There are three levels of comparisons that are needed. First, we have to compare what we learned about Drosophila compound eye development with that of other Drosophila eyes (i.e., the development of paramorph eyes in Drosophila). Second, we need to compare the Drosophila compound eye with that of vertebrates, as the homology or non-homology of vertebrate and insect eyes is one of the most discussed issues in developmental evolution. Finally, we need to discuss the role of the ESED network in organs other than eyes.

Development of Paramorph Eyes in Drosophila

All of the work that was sketched in the previous paragraphs focused on the development of compound eyes (i.e., the eye of the Drosophila adult that is capable of image perception). In Drosophila and other insects, however, compound eyes are not the only light sensory organs. There are two other paramorph eyes. These are the ocelli of the adult and the larval eyes, called Bolwig's organ. All in all, the humble fruit fly has seven eyes, two compound eyes, three ocelli, and two larval eyes; the latter survive into adulthood as extra-retinal photoreceptors.

Morphologically, the simplest eye type is Bolwig's organ, which simply consists of a bundle of photoreceptors without supporting cells (figure 3.11). In the fly larva, the cephalopharyngeal skeleton acts as a makeshift light screen to provide some degree of directional light perception. In contrast, the compound eyes are the most complex; in Drosophila these include about 800 ommatidia containing a total of about 6,000 photoreceptors. Each ommatidium is a self-contained functional unit complete with receptor cells, pigment cells, and a lens. The ocelli are on the top of the head and contain up to 95 photoreceptors in a cup-shaped retina that is covered with a cornea and a lens (figure 3.12). All Drosophila eye paramorphs derive from the same embryonic anlage in the dorsal head neuroectoderm.

Phylogenetically, the oldest paramorphs are ocelli and lateral compound eyes. With the exception of myriapods, all arthropod clades have both median ocelli and lateral compound eyes and, thus, it is safe to assume that these two

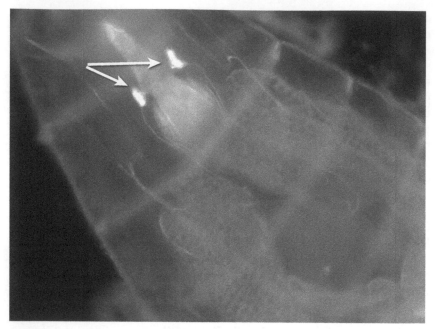

FIGURE 3.11: Bolwig's organ in a fly larva. This organ is a larval eye that persists into the adult as so-called HB-eyelets; see figure 3.12. (Image by permission from Elizabeth Kane, Harvard University. http://worms.physics.harvard.edu/images/hm_worm.gif.)

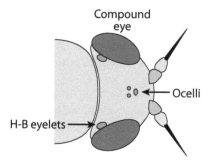

FIGURE 3.12: Three types of eyes on the adult fly head. Best known are the large compound eyes. Associated with the compound eyes are small eyelets, which are remnants of larval eyes (see figure 3.11). Finally, there are three ocelli on the top of the head.

paramorphs were present in the most recent common ancestor of arthropods. The differentiation of larval and adult compound eyes is a derived feature of holometabolous insects and is a consequence of the differentiation of early larval and adult morphologies. The simple structure of Bolwig's organ in Drosophila is, thus, likely the result of a secondary simplification (for more details see Paulus 1989; Oakley 2003; Friedrich 2006; Friedrich 2008). Hence, the serial homology of Bolwig's organ and adult compound eyes is quite well supported, and the serial homology of ocelli and compound eyes is likely because they share the same rhabdomeric photoreceptor cell type.

Now we want to ask what the similarities and differences are in the developmental control of paramorphic fly eyes. The most conserved aspect of eye development is the requirement for *eya* and *so* for eye determination and differentiation. Knockout genotypes of *eya* and *so* lack all three eye types, and these genes are essential for activating more downstream eye differentiation genes like *atonal*. In contrast, the requirement for the more upstream factors *ey* and *toy* varies among paramorphs.

Ey null mutants lack compound eyes but do have ocelli, whereas null mutants of *toy* have compound eyes but have distorted ocelli. As mentioned above, this difference in the requirement of *ey* and *toy* is reflected in the enhancer *so10* of the *so* locus (Punzo, Seimiya et al. 2002). Remarkably, Bolwig's organ development is unaffected even in *ey* and *toy* double deficient genotypes. This is surprising because Bolwig's organ and compound eyes are more closely related to each other than each is to ocelli (Friedrich 2006). A possible explanation for this is that *ey* and *toy* have functions both in eye cell proliferation and in activating the determination network.

Recall that Ey acts in a protein complex with Hth and Tsh to promote cell proliferation and is not expressed posterior of the MF. Only when *hth* is downregulated by the dpp signal is Ey liberated and activates *eya* and *so*. It seems that the role of *ey* is limited to those eyes that have a need for additional cell proliferation prior to differentiation (i.e., those eyes that require a large number of cells). In Drosophila, Bolwing's organ is highly reduced and neither *ey* nor *toy* is required for its development. Overall, these data show that there is no necessary link between eye development and *ey* activity. The only aspect of the eye determination network conserved in all three Drosophila eye types is *eya* and *so*.

The other member of the compound eye determination network, *dac*, is also not universally required for eye development. For example, a *dac* null mutant lacks compound eyes but has normal ocelli. A compound eye development requirement for *dac* does not even seem to be universal among insects. RNAi-mediated knockdown of *dac* in the milkweed bug *Oncopeltus fasciatus* does not result in a noticeable eye phenotype (Angelini and Kaufman 2004). Hence, it is possible that either *dac* was secondarily recruited into the eye determination network after the most recent common ancestor of the milkweed bug and flies or it was lost in the milkweed bug lineage.

The ESED Network in Vertebrate Eyes

Homologs of the ESED network genes exist in all animals, although they have different names. The orthologs of *ey* and *toy* in vertebrates are called *Pax6* and *Pax4*, the homologs of *so* and *optix* are called *Six1* to *Six6*, and *dac* is homologous to *dach*. Each of these gene families has different numbers of members in insects and vertebrates (table 3.1), and their gene family trees are complicated (Bebenek, Gates et al. 2004; Lynch and Wagner 2011). It was a great surprise, though, when it was discovered that the vertebrate ortholog of *ey, Pax6*, was also necessary for eye development in vertebrates, and that the Pax6 protein and the Ey protein were, to some degree, functionally equivalent in activating downstream eye determination network genes (Chow, Altmann et al. 1999). This discovery was used to support the notion that vertebrate eyes and insect eyes might, in fact, be homologous, despite their deep morphological dissimilarities. Thus, it is worthwhile to compare the similarities and differences between the vertebrate and insect eye determination networks and assess whether the ESED networks of Drosophila and vertebrates are, indeed, homologous.

Briefly, *Pax6, Six3, Six6, Eya1, Eya2,* and *Eya3, as well as Dach1* are all expressed in vertebrate eyes and are, to varying degrees, involved in eye development. In both mice and humans, homozygous mutants of *Pax6* display anophthalmia (i.e., lack of eyes), as do mutations of *Six6* in humans and knockdown of *Six3* in medaka fish. In contrast, *Six3* mutants result in microphthalmia, whereas *Eya1* mutant mice have no eye phenotype, although this mutation in humans results in cataracts. In *Xenopus, Eya3* knockdown results in a loss of anterior brain structures, including eyes.

Gain of function studies also support central roles for *Pax6, Six3,* and *Eya3* in vertebrate eye development. Misexpression in *Xenopus* of *Pax6* (Chow et al. 1999; Zuber et al. 2003; Chow, Altmann et al. 1999; Zuber, Gestri et al. 2003) and *Eya3* (Kriebel, Muller et al. 2007) and of *Six3* in medaka (Loosli, Winkler et al. 1999) results in ectopic retina and lens structures, similar to the ectopic eyes that are induced by misexpression of *ey* or *eya* in Drosophila.

Now let us examine whether the eye determination network in vertebrates is homologous to that in Drosophila. We will focus on the roles of *Pax6, Six,* and *Eya* genes. We look at *Pax6* because it was the reason for hypothesizing homology between vertebrate and insect eyes, even though the comparison of Drosophila paramorph eyes already shows that *ey/Pax6* is not essential for eye development—for example, not required for Drosophila larval eyes. In addition, we look at the role of *so/optix* related genes as well as *eya* homologs in vertebrates, because *so* and *eya* form the conserved core of the eye determination network in all three paramorph eyes in Drosophila.

As mentioned above, *Pax6* is essential for vertebrate eye development, as assessed by a mutant phenotype and the effects of knockdown experiments using medaka and *Xenopus. Pax6* misexpression is also sufficient to induce ectopic eye structures in *Xenopus* and induces other eye developmental genes, such as *Six3* (Zuber, Gestri et al. 2003). As with *ey, Pax6* has a role in cell proliferation in the eye field and anterior brain. Furthermore, *Pax6/ey* has a conserved function by regulating the expression of opsin genes, which probably was the ancestral function of *ey/Pax6* (Davidson 2006). Things become more problematic when we proceed to the homologs of the core eye determination genes in Drosophila (i.e., to the homologs of *so* and *eya)*.

There are three *so* related genes in Drosophila: *so, optix,* and *D-six4.* The latter is not expressed in eyes and, for that reason, has not been mentioned in this chapter. Both *so* and *optix* play essential roles in Drosophila eye development; *so* is induced by *ey* but *optix* expression is *ey* independent. In vertebrates, there are six *Six* genes (i.e., genes related to *so* and its paralogs). These are called *Six1* to *Six6*, with two *Six* semi-orthologs corresponding to each of the three Drosophila genes (table 3.1). The semi-orthologs of *so* are *Six1* and *Six2*. *Six1* is not expressed in the eye, and *Six2* is expressed only in the adult retina.

The main eye development genes from the *Six* family in vertebrates are the semi-orthologs of *optix: Six3* and *Six6* (a.k.a. *Optix2*). *Six3* is essential for eye development in mice and humans, *Six6* mutations can result in anophthalmia in humans and *Six6* (KO) in mice results in a loss of the optic nerve, but

Table 3.1. Vertebrate sine oculus homologs and eye development (summary after Donner and Maas 2004; Seimiya and Gehring 2000; Hu et al. 2008)

Drosophila Ortholog	Expression and Function in Fly	Vertebrate Paralog	Expression and Function in Vertebrates
so	Essential for eye development, expressed in optic lobe and differentiating eye; interacts with *eya*; expr. depends on *ey*	*Six1*	Not expressed during eye development
		Six2	Expressed in adult mouse retina; no eye induction effects; interacts with Eya protein
optix	Essential for eye development, but is not expressed in optic lobe and not in differentiating eye cells; does not interact with Eya protein; expr. independent of *ey*	*Six3*	Expressed during eye development; essential for eye development; relationship to *Pax6* variable; does not interact with Eya protein
		Six6	Expressed during eye development; essential for optic nerve in mouse; anophthalmia in humans; relationship to *Pax6* variable; does not interact with Eya protein
D-six4	Not expressed in eyes	*Six4*	Expressed in Xenopus lens; not expressed in eyes of mouse and zebrafish
		Six5	Weak expression in developing mouse lens and neural retina of the adult

not other eye structures. Even though *optix* is essential for Drosophila eye development, it is necessary for proliferation of the eye field rather than eye determination and differentiation, as it is not expressed in the eye-antenna disc posterior to the morphogenetic furrow (i.e., that part where determination and differentiation occur). However, the orthologs of *optix* in vertebrates play a role more similar to *so* (i.e., are involved in determination and differentiation), while the orthologs of *so* are not involved in eye development. Furthermore, *Six3*, an *optix* semi-ortholog, is regulated by *Pax6*, but *optix* itself is independent of *ey*.

It would be possible to reconcile these functional differences between the Drosophila and vertebrate eye networks if we assume that, ancestrally, the common ancestor of *so* and *optix* was already involved in eye development, and that after the duplication, the paralogs assumed different, complementary developmental roles in the Drosophila and vertebrate lineages. A gene lineage analysis, however, does not clearly answer this question, because *so* is found in basal metazoans, including sponges, but *optix* is not, and the gene tree reveals that the *so* clade could be paraphyletic (Bebenek, Gates et al. 2004). This means that what is called *so* in sponges could be the ancestral gene, which gave rise to the *optix* gene lineage at some time within the bilateria. Nevertheless, there is some moderate support for a *so* clade, which, if confirmed by further work, would imply that these paralogs already existed prior to the most recent common ancestor of all metazoans.

So plays its role in conjunction with *eya*, a function that is mediated by a direct protein-protein interaction between the So and Eya proteins. In contrast, the Optix protein does not interact with Eya, and neither do the vertebrate homologs of Optix, Six3, and Six6. Hence, it is most parsimonious to assume that Six-Eya interaction never played a role in vertebrate eye development. Rather, the eye determination networks were independently derived by building upon an ancestral role of *ey/Pax6* in opsin regulation. In the deuterostome (i.e., vertebrate) and in the protostome (i.e., Drosophila) lineages, different members of the *Six* gene family have been recruited into the gene regulatory network that controls the morphogenesis of eyes. Note that *optix* is not activated by *ey*, but *Six3* is activated by the *ey* ortholog *Pax6* (Zuber, Gestri et al. 2003). This also suggests independent recruitment of different *Six* gene family members into eye development. This scenario is also more consistent with the striking embryological and anatomical differences between insect and vertebrate eyes, which suggest independent evolution of eye morphogenesis.

The other element of the Drosophila eye determination network is *eya*, which is essential for eye development in the fly. In vertebrates, there are three paralogs, called *Eya1*, *Eya2*, and *Eya3*. Two of these paralogs, *Eya1* and *Eya2*, are not critically important for eye development. However, the third paralog, *Eya3*, is essential in *Xenopus* eye development, but its effects are independent of Six proteins because the Six3 and Six6 proteins do not interact with any Eya protein (Kriebel, Muller et al. 2007). Finally, the homolog of *dac* in vertebrates is not involved in eye development.

Overall, it is true that genes from some of the same gene families are involved in eye development in vertebrates as in insects, but the details of their molecular and functional interactions are so different that it is more likely that these gene regulatory networks were assembled independently during the evolution of eyes in the protostome and deuterostome lineages.[4] Most important, different gene family members of ancient pre-metazoan origin play similar roles in eye development in vertebrates and insects.

Non-homologous organs can use developmental genes from the same gene family, but their regulatory networks are likely to be non-homologous in the case of vertebrate and insect eyes. This is important if it is hypothesized that homologous morphological structures correspond to homologous gene regulatory networks (ChINs). A correlate of this assumption is that non-homologous organs should be regulated by non-homologous ChINs. This seems to be the case for vertebrate and arthropod eyes, as well as other organs (see next section).

ESED Networks in Organs Other than Eyes

In vertebrates, genes of the ESED network are active in many other organs, most notably in muscle development (Kardon, Heanue et al. 2004), kidney, and all organs derived from placodes (Schlosser 2005), such as the lens, the inner ear, lateral line organs, and others. Here, we briefly review the role and molecular function of a few of these organs to contextualize what has been discussed about eye development and evolution.

In vertebrates, skeletal muscle cells derive from the dorsal regions of somites. Members of the ESED gene families are expressed in these cells just prior to and while typical muscle differentiation markers are expressed (i.e., *MyoD* and *Myf5*). Specifically, *Pax3* and *Pax7*; *Six1* and *Six4*; *Eya1*, *Eya2*, and *Eya4*; and *Dach1* and *Dach2* are expressed in dorsal somitic cells. The functional relationships between *Pax3*, *Six1*, *Eya2*, and *Dach2* are known. *Pax3*, which is not an ortholog of *ey* but is related to the Drosophila gene *gooseberry*, positively regulates *Eya2* and *Dach2*, just like Drosophila *ey* regulates *eya* and *dac* in compound eye development. Furthermore, Eya2 synergistically functions with Six1 through a direct protein-protein interaction, just like the Drosophila Eya and So proteins.

Notably, *Six1* is a semi-ortholog of *so*, unlike the *optix* semi-orthologs *Six3* and *Six6* that are essential in vertebrate eye development. Dach2 also

---

[4] This scenario still does imply that photoreceptor cells are homologous between vertebrates and insects, or at least paramorph. This is likely given the wide taxonomic distribution of photoreceptor cells and the similar role of *Pax6/ey* in opsin expression. But see chapter 8 for more details on photoreceptor cell type evolution.

interacts with Eya2 in muscle development through a direct protein-protein interaction, just like their (semi/pro) orthologs Dac and Eya in Drosophila eye development. Recall that Dach does not play a role in vertebrate eye development. Hence, at the level of the ESED gene regulatory network, vertebrate muscle development is more similar to Drosophila compound eye development than Drosophila eye development is to vertebrate eye development. This is another reason to doubt the homology between insect and vertebrate eye development.

Placodes are regional thickenings of the embryonic epidermis that are precursors of many vertebrate-specific organs and characters.[5] The list of placode-derived structures is long and includes eyes, ears, scales, feathers, lateral line organs, and many more. Members of the ESED network are widely expressed in placodal cells. Specifically, *Six1/2* and *Six4/5* genes are expressed in all placodes. Members of the *Eya* family are widely expressed in cranial placodes, with the exception of *Eya3*, but some combination of *Eya* genes is expressed in all placodes, depending on the species. *Pax6* and *Dach* are also expressed in many, but not all, placodes.

Experimental work on *Six1* and *Eya1* suggests that in placodes, the ESED network regulates cell proliferation, prevents apoptosis, promotes cell shape changes, and induces cell differentiation cascades. In addition to expression, many of the molecular interactions among the ESED members are also relevant for placode development, as for example the protein-protein interactions among Eya and Six proteins and others. Members of the ESED network family of genes with more restricted expression among placodes are likely responsible for the differentiation of placodes for their specific roles.

Hence, it seems that these genes also play a similar abstract role by regulating the proliferation of organ precursor cells and the transition to cell fate determination and differentiation, as in eye development. It is also clear that the original members of the ESED network gene families were deployed in various combinations to regulate the development of non-homologous characters. Although more detailed study is required, it is plausible that different versions of the ESED network evolved for each unique character.

A Note on Similarities Among Gene Regulatory Networks

From the comparative biology of the ESED network it is clear that certain genes and proteins tend to be involved together in quite different contexts, like eye development, and muscle and ear development. Why is it that during evolution, time and again the same sets of genes work together, rather than some more random collection of genes? This is surprising when the organs

[5] This section on placodes is based on the review by Gerhard Schlosser: Schlosser, G. (2005). Evolutionary origins of vertebrate placods: insights from developmental studies and from comparisons with other deuterostomes. *J. Exp. Zool. Part B (Mol. Dev. Evol.)* 304B: 347–399.

and tissues are unrelated, like eye and muscle. It is also clear that the functional details differ between different tissue types. Nevertheless, the same suspects are often expressed and function together.

Should we take the co-expression and functional interactions of genes in different tissues as evidence of homology of the respective tissues? For example, are we forced to think about a serial homology between muscles and eyes? Certainly there is a temptation to do so, and in some instances the homology of tissues and cells does correspond to shared gene regulatory networks. The degree to which similarity of gene regulatory networks is indicative of homology depends on how unlikely it is to find two genes that interact in two unrelated tissues. Hence, we have to ask what determines the likelihood of two genes functioning together in a particular context?

The biological role of a gene is, to a large degree, determined by its cis-regulatory elements and the interactions of its gene products. All of these regulatory elements, cis-regulatory as well as protein-protein interaction motifs, are likely to be shared among duplicated genes. This is particularly true if paralogs are generated by a whole-genome duplication. Both paralogs are initially identical and have the same cis-regulatory and protein-protein interaction domains. Consequently, it is also likely that, at least initially, the duplicated genes are regulated by the same upstream genes. If two genes are ancestrally regulated together, then it is likely that the paralogs of both genes are also expressed in the same cells.

Furthermore, regardless of the mode of duplication, if two proteins functionally interact through a protein-protein interaction, so will all their paralogs, at least initially. Protein-protein interactions do change over time, as for example the interaction between Eya and So. The So protein in Drosophila binds to Eya, while its paralog Optix does not (Kenyon, Li et al. 2005; Kenyon, Yang-Zhou et al. 2005). If duplication of the genes does not occur through a whole genome duplication, but through a local tandem duplication or retrotransposition, then the protein-protein interactions are still inherited by the paralogs, even though the cis-regulatory elements might not.

There are also instances where members of the ESED network are expressed and probably function totally independently of other members of the network. Examples are *dac* and *Dach1* in the Drosophila wing and the vertebrate limb, respectively. *dac* and *Dach1* are expressed in these organs, but no other member of the ESED network is. Similarly, *Six1* is expressed in skeletal precursor cells from the sclerotome, but no other ESED member is.

From these elementary considerations it is clear that once a set of transcription factors interact, it is likely that their paralogs also interact, even in novel functional contexts. This is the case because paralogs often inherit the molecular elements that mediate the functional interactions. Hence, similarity of a gene regulatory network of some tissues, even one in which multiple genes are involved per se, is not strong evidence for homology. Genes that ancestrally interacted are still likely to do so even in novel developmental roles, because these functional interactions do not need to be evolved from scratch once they exist in an ancestral protein pair.

The Role of Protein-Protein Interactions

While much emphasis is placed on the role of cis-regulatory elements in regulating gene activities, the examples discussed above also show that gene regulation critically depends on protein-protein interactions. The role of protein-protein interactions is not limited to the generic fact that transcription factors have to interact with an initiation complex to either inhibit or enhance transcription, but is critical for the functional specificity of the transcription factors and their co-factors. Recall that the So protein inhibits the expression of eye regulatory genes if it is associated with Groucho, whereas it enhances their expression if it is associated with Eya. This is a theme that is somewhat neglected in evolutionary biology, given the emphasis on cis-regulatory element evolution (see chapter 6 for a more thorough discussion of this issue).

Here, I want to point out only that protein-protein interactions among transcription factors and their evolution are likely key elements in the determination and evolution of character identity. To support this point, I will discuss two examples in some detail. One involves the identity of a mammalian cell type, the endometrial stromal cell, and the other is flower organ identity.

Gene Regulation in Endometrial Stromal Cells

In placental mammals, the development of the fetus depends on communication between the mother and the fetus through the placenta. The placenta is a composite organ comprising a fetal part and its maternal counterpart. The maternal part is formed by the endometrium, the inner lining of the uterus. The endometrium includes a superficial epithelium and an underlying stroma (in addition to uterine glands and blood vessels). In most placental mammals, and also ancestrally, the fetus breaches the endometrial epithelium and invades the stroma. In other words, the fetus acts like a parasite, and the natural reaction of the maternal body would be to attack the invader with fatal consequences for pregnancy.

Hence, a new function had to evolve, which is primarily performed by the endometrial stromal cells—namely, suppressing an immune reaction and inflammation, as well as limiting the invasion of the fetal placenta into the wall of the uterus. In humans and many other primates, in order to perform this function, the endometrial stromal cells undergo differentiation in preparation for the arrival of a fetus; this differentiaiton is called decidualization (Emera, Romero et al. 2011). It is called decidualization because, eventually, if no conceptus arrives, these cells will undergo cell death and will be cast off during menstruation. Because of its importance for human fertility, gene regulation in endometrial stromal cells has been intensely studied and has yielded important insights into the mechanisms of gene regulation and its evolution (e.g., Gellersen and Brosens 2003).

A paradigm for gene regulation in endometrial stromal cells is the expression of prolactin (PRL). PRL is mostly known as a hormone produced by the pituitary in the brain of most, if not all, vertebrates. In placental mammals, PRL is also expressed in some white blood cells, the ovary, and decidualized (differentiated) endometrial stromal cells (ESC). In primates, gene transcription in the ESC is controlled by an alternative promoter about 6 kb upstream of the translational start site. Transcription from this alternative promoter is controlled by two kinds of signals.

One is progesterone, a steroid hormone that reaches the uterus via the bloodstream. Progesterone acts through a nuclear receptor, PR-A, which binds progesterone and then enters the nucleus as a transcription factor. However, activated PR-A is not sufficient to cause PRL expression by ESCs. In addition, a number of transcription factors are needed, which are mostly activated through another signaling system, PKA/c-AMP. In cells, c-AMP levels increase in response to activation of a G-protein coupled receptor activated by other hormones and paracrine factors, such as luteotropic hormone, relaxin, prostaglandin E2, and others. These signals lead to the upregulation, activation, and nuclear translocation of transcription factors like C/EBP-/, Stat5, and FoxO1a. All of these transcription factors also bind to the decidual PRL enhancer. Together with PR-A, these proteins form a complex that is necessary for activating PRL transcription.

The idea is that the formation of this complex of transcription factors activated by different signaling pathways is a molecular mechanism for integrating multiple signals into one coherent transcriptional response (Gellersen and Brosens 2003). In other words, different signals contribute to activating different transcription factors that, together, form a complex or cooperate in other ways to jointly regulate the target genes. Hence, transcription factor protein-protein interaction is a mechanism for integrating signals so as to elicit a coherent gene regulatory response. Thus, it is likely that the evolution of transcription factor protein functions played a critical role in the evolution of gene regulation, at least in higher eukaryotes.

Floral Organ Identity and Obligate Hetero-dimerization

The genetics and development of floral organ development is, perhaps, among the best understood cases of organ identity determination. The development and evolution of flowers will be discussed in greater detail in chapter 12. Here it will serve as a paradigm for the role of protein-protein interactions in the development of switch-like, qualitative decisions during development.

Briefly, the four different flower organs of typical eu-dicot flowers—sepals, petals, stamens, and carpels—are determined by a combinatorial code of transcription factor expression, which consists of three classes of genes called A-, B-, and C-class genes (Causier, Schwarz-Sommer et al. 2010), as follows: A = sepal identity; A+B = petal; B+C = stamen; and C = carpel identity. Here we focus on the role of B-class genes, which are represented by two genes called *pistillata (PI)* and *APETALA3 (AP3)* in *Arabidopsis*, and *DEFICIENS (DEF)* and *GLOBOSA (GLO)* in the snapdragon (*Antirrhinum*). I will follow the description of the molecular biology of these genes in *Antirrhinum* as summarized in Lenser et al. (2009).

The expression of both B-class genes is jointly necessary for petal and stamen identity. Their expression is initially caused by external signals, but then is maintained by a positive feedback loop that eventually restricts the expression of these genes to the second and third whorl of the flower anlage. The auto-regulatory function, as well as the regulation of downstream target genes depends on the obligatory formation of a heterodimer complex: DEF:GLO. This complex then becomes the active transcription factor for maintaining the expression of both genes. The effect of this dependency on the formation of a complex is that it is mutationally difficult to dissociate the expression domain of these genes and their target genes. Computational studies have shown that this is particularly robust to both mutational perturbation and stochastic noise; thus, it is a classical mechanism for the canalization of organ identity (Lenser, Theissen et al. 2009).

The Transcriptional Logic of Transcription Factor Complexes

If the expression of target genes depends on a transcription factor complex consisting of transcription factors from different signaling pathways, then the formation of the complex acts like an "AND" gate, which requires the activation of two or more inputs to elicit a certain output—for example, PRL expression in endometrial stromal cells. In other words, the genes that translate multiple extracellular signals into a coherent qualitative response do so by forming a transcription factor complex, which contains the transcription factors that respond to these different external signals.

The alternative to the cooperative activation of target genes through the formation of an obligatory transcription factor complex is that different transcription factors bind independently to a promoter/enhancer and then act additively. Independent contributions of different transcription factors would result in a graded response of cells in terms of their gene expression to various combinations of external factors. A graded response, however, is different from the all-or-nothing response that is typical of cell differentiation and organ identity determination. Usually, during normal development, a cell is either a neuron or a glial cell, but not something in between.

The qualitative nature of the differentiation response of cells, as well as that of tissues and organs, is what leads to the concept of character and cell identity (i.e., the subject of this book)—that is, the biological nature of character identity across species and its evolution. If in fact, as suggested by the examples discussed above, cell and character identity depend on the integration of multiple signals and transcription factor activities through the obligatory formation of transcription factor complexes, then protein-protein interactions and their evolution are essential for the molecular mechanisms that underlie both the phenomenon of character identity (homology) as well as the origin of novel phenotypic characters (i.e., novelties).

Characteristics of Character Identity Networks

In this chapter we discussed in some detail gene regulatory networks for arthropod segmentation and eye development. In both cases, we identified a core of regulatory relationships that are much more highly conserved than other aspects of development. In both cases these genes play a role in translating positional information signals into specific morphogenetic activities of cells and, thus, are the interface between "abstract" positional information and the specific activities of cells that lead to the final product of development. There are also some dynamical properties that distinguish these networks from other gene regulatory relationships and signal transduction cascades.

While the genes in ChINs represent a hierarchy of causal interactions during development (i.e., there is an orderly sequence of gene activation events), *eventually the network becomes self-regulatory wherein members of the network mutually sustain each other's expression and act synergistically on downstream genes.* Members of these networks are, thus, not only a link in a linear chain of causation, but eventually form a functional unit in which developmental causality is realized at the level of the network rather than at the level of the single gene. This is reflected in the empirical fact that members of a ChIN are jointly necessary for organ development, and some of its members individually are sufficient to initiate morphogenesis and differentiation. The latter property depends on the details of the molecular mechanisms and is not universally realized.

For example, *so* is certainly part of the core of the Drosophila eye determination network, but individually it is not sufficient to induce ectopic eye structures, as are other members of the network like *eya*. The reason probably is that the So protein is already present in prospective eye tissue, but in a complex with Groucho, which makes it a transcriptional repressor. Only if So is recruited into a protein complex with Eya will it become part of an activator complex. Hence, it is easy to understand why *eya* is sufficient to induce eye development but *so* is not, even though both are part of the core eye identity network. Given this background, I want to propose a preliminary characterization of a *Character Identity Network* (ChIN).

A ChIN is a set of genes with the following characteristics:

- The genes form a cross-regulatory network of gene regulation in which the members of the network sustain each other's expression.
- The gene network is historically associated with the existence of a morphological character of a certain identity. That is, the homology of the morphological character and that of the network are co-extensive.
- The members of the network are jointly necessary for the development of the morphological character, and some of the network members are also individually sufficient to trigger the morphogenesis and differentiation of the character.[6]
- The members of the network also jointly repress the development of an alternative character or cell identities by repressing or disrupting the activities of alternative ChINs.
- The members of the network synergistically control the activity of downstream genes that underlie the developmental activities of those cells that make up the character. This synergism often requires the obligatory formation of transcription factor complexes through protein-protein interactions. One can call the protein complex of the core regulatory genes the Core Regulatory Complex (CRC), and the functional cooperation mediated by the CRC is, perhaps, the cause of the evolutionary cohesion of the ChIN.

This is a slightly cartoonish view of how real gene regulatory networks actually work. For example, some gene products have roles prior to the activation of the core determination network by associating with other proteins, and actually suppress the expression of other members of the network, as is the case with So in Drosophila eye development. Also, the degree of mutual cross-regulation might be variable. However, this characterization implies that genes in a ChIN are likely to be co-adapted and causally cohesive entities, which may also explain their evolutionary conservation.

With all its manifest limitations, the ChIN model is not meant to be an exact picture of reality. Thus, it is not a definition; rather, it is a characterization that is meant to function as an idealized image. This is an idealization that might be useful for understanding the evolution of morphological characters, to explain their historical continuity, and the origin of morphological characters (i.e., evolutionary novelties). I will pursue the latter question in the next chapters.

---

[6] This ability to induce ectopic cells, tissues, and structures is, of course, conditional upon the competence of the cells to react to the gene product and the availability of the tissue topology necessary to realize the morphogenetic process.

4

Evolutionary Novelties: The Origin
of Homologs

In chapter 2 the homology concept was introduced as reflecting a pattern of phenotypic evolution, one for which species that descend from a common ancestor tend to retain corresponding parts, like brains and limbs. The idea is that parts of organisms retain historical continuity and that they undergo (quasi-) independent modifications. It was also noted that historical continuity of parts has its limits in that tracing the history of a body part into deep evolutionary time always leads to a point where the traces of the body part disappear. It was concluded that homologs (i.e., body parts with historical continuity) have definite beginnings in the history of life.

There are no teeth or limbs in *Trichoplax*[1] and there were none in the most recent common ancestor of *Trichoplax* and vertebrates. But teeth did originate at some time in the lineage from ancestral chordates to gnathostome vertebrates. During these periods in evolutionary history, new homologs originated. At those times, these body parts were novelties (i.e., they were units of evolutionary transformation that were different from anything that had come before them). The evolutionary processes that led to the origin of body parts will be considered in this chapter. But before we seek to explain character origination, we have to make sure that it makes good scientific sense to introduce a concept such as "novelties." In fact, there are good reasons to avoid such a concept (e.g., cut down on semantic noise) if the conceptual distinction between novelties and other evolutionary events has no consequences that matter.

---

[1] *Trichoplax adherens* is the anatomically most primitive free-living animal. It essentially consists of a bag made of a simple epithelium and a few mesenchymal cells in its interior. There is no definite body axis or any inner organs.

Modes of Evolution

Life in all its forms is complex and fantastically multifaceted. There is an incredible diversity of species out there, with 24,000 species of bony fish alone, about 350,000 species of described beetle species, and a diversity of microbes that science has not even been able to approximately estimate, let alone actually count. There is also a staggering complexity of molecular and functional mechanisms that sustain the lives of these many species. Adaptation is the term used to describe the exquisite fit between the physiology and structure of organisms and their environments. There is also an amazing disparity among the major forms of life; even within the limited realm of animals, some life forms do not seem to have anything in common with other life forms. It takes a trained biologist to list the commonalities between a human and a cnidarian or a sponge. It is safe to assume that, to date, no single theory has adequately explained all these aspects of biological diversity. To claim otherwise is akin to the fable of the blind men and the elephant.

The story of the blind men and the elephant probably originated in India and can be traced back at least to Jainism, one of the oldest known religions. There are various incarnations of this story, but all share the idea that humans are like blind men who are asked to discover the true nature of an elephant. Because they are limited to what they can feel with the palms of their hands, each can only report on a small part of the elephant. After feeling one of the legs, one says the elephant is like a pillar, another calls it a tree branch after examining the trunk, and one calls it a rope because he got hold of the tail. Various forms of this story differ in what follows after the different discoveries by these blind men, but often these differences of opinion lead to violent fights among them that may or may not be resolved by a seer who tells them about the relativity of their views. Unfortunately for nature, which is the elephant that science examines, there is no seer who can tell us what the elephant really is like. The best we can do is to garner reports regarding as many aspects of the elephant as possible and continue listening to each other.

Darwin's original explanatory model was entirely focused on selection, which led to many problems in explaining the origin of species (Mayr 1942; Mayr 1987). The point was that the origin of species turned out to be a problem of a different kind from character transformation and adaptation. Character transformation can be explained by natural selection, at least to some extent. However, among sexually reproducing species, at least, speciation is an entirely different problem; namely, the question is not how the species characters became different, but how a population can divide into two, each of which then becomes an independent line of descent. The answer is that any process that leads to reproductive isolation, regardless of how, can lead to a new species.

There are many mechanistically different ways for how this can happen, from local adaptation, to hybridization and genome duplication, and non-adaptive divergence in geographic isolation. Hence, the problem of speciation is conceptually a totally different kind of problem than is adaptation, and one that needs a different research program to address it than that used for adaptation. This was the main intellectual achievement of the New Synthesis in evolutionary biology. Now every student of evolution knows that adaptation and speciation are different kinds of processes and that neither implies the other. Here, I propose a similar move—namely, that it is conceptually and in terms of research programs necessary to distinguish between the evolution of adaptations and the origin of novelties (Müller and Wagner 1991; Newman and Müller 2001; Wagner and Lynch 2005).

What Is the Difference between Adaptations and Innovations?

On the most general level, one can say that an *innovation is the origin of a novel body part that may serve a novel function or specialize in a function that was already performed in the ancestral lineage*, but without a dedicated organ. Clearly, these novel characters contribute to the survival of the species that possess them and, thus, have adaptive value. Why then could this process not be subsumed under the adaptationist research program? To examine whether this is a viable approach, let us first reflect on what is required to explain an adaptation.[2]

It is widely accepted that adaptations are traits or features of an organism that owe their existence to the action of natural selection (Futuyma 1998). The agenda of the adaptationist program is to explain the evolution of features that enhance the survival and reproductive success of individuals (Mayr 1983; Stearns 1986). It is well established that natural selection is the cause and, thus, the explanation of adaptive traits. Putting aside all the complications that result from historical contingencies and the many problems in actually demonstrating the role of natural selection, there is a common denominator to all attempts to demonstrate adaptations. This is the assumption that natural selection is causally sufficient to explain adaptations.

If natural selection is a sufficient cause (given certain boundary conditions) of adaptive traits, then the outcome of adaptive evolution has to be reproducible. Consequently, repeated evolution of the same or similar traits under similar ecological or functional conditions is seen as evidence in favor of adaptation.[3] Similarly, the ability to evolve a certain trait in the laboratory,

---

[2] The argument developed in this section was first published in Wagner, G. P. and V. J. Lynch (2005). Molecular evolution of evolutionary novelties: the vagina and uterus of therian mammals. *J. Exp. Zool. (Mol. Dev. Evol.)* 304B: 580–592, and the text following is a modification of that text.

[3] The fact that some apparent morphological novelties can originate multiple times in a clade has been used to challenge the very notion of evolutionary novelty (Eberhard 2001), but the evidence in this case is far from conclusive (Wagner and Muller 2002).

either by experimental evolution or artificial selection, is a good argument that natural selection is, in fact, the cause for this trait.

Finally, the ability to predict the outcome of evolution from functional optimality considerations is also a strong argument in favor of adaptation (Orzack and Sober 2001). The goal of the adaptationist research program is to identify the functional and ecological factors that caused these fitness differences that led to the selection of this trait (Mayr 1983; Orzack and Sober 2001).

Natural selection requires the availability of heritable variation, and demonstrating the heritability of the trait under consideration is a standard part of the adaptationist research program (Stearns 1992). But the dependency of natural selection on heritable variation adds nothing to the explanatory force of adaptationist research programs, because heritable variation does not determine the outcome of the evolutionary process (Amundson 1989). The crucial causal factor remains natural selection.

The availability of heritable variation can be thought of as a boundary condition rather than the cause of adaptation (Sterelny 2000). But there are certain traits for which the presence of heritable variation is not guaranteed, even in large outbred populations. For example, there is no heritable variation in *Drosophila* for directional asymmetry in the location of ocelli (Maynard-Smith and Sondhi 1960), and it is unlikely that one can demonstrate the presence of selectable variation for flight feathers in an alligator population (the closest living relative of birds). Nevertheless, flight feathers evolved from archosaur epidermal scales (Prum and Brush 2002, and see chapter 9). This leads to the question of how feathers arose, with their radically different morphology compared to a scale, and whether natural selection is a sufficient explanation of their origin. This is the problem of evolutionary novelty.

The agenda for the study of novelties is to explain the origin of characters. New characters open up new functional and morphological possibilities to the lineage that possesses this character. There is no question that such traits exist. Feathers have utility for various new functions that were not served by the ancestral epidermal scales, like heat insulation, flight, and communication (see chapter 9 for more details). Feathers also have an increased range of possible morphologies compared to epidermal scales. For example, selection for the shape of crocodile epidermal scales is not likely to result in branched pennate appendages. The reason is that the range of morphologies characteristic of feathers requires a quite specific set of developmental features, which feathers have but that scales do not (Prum 1999; Harris, Fallon et al. 2002; Yu, Wu et al. 2002; and see chapter 9 for more detail).

Most likely, feathers evolved first for heat regulation and assumed more sophisticated functions later (Prum and Brush 2002). Hair in mammals is another epidermal appendage that also plausibly could have originally evolved for heat insulation, but never gained the ability to support flight, as do feathers. Mammals, of course, did evolve flying species. In fact, the second largest group of mammals is bats with 925 described species (Koopman

1993). But bat flight is based on expanded skin folds between the fingers, and not skin appendages.

The reasons for the distinct morphological and functional versatility of hairs and feathers are developmental. Feathers and hairs differ in their modes of development (see chapter 9 for details) with vast implications for their variational properties (Prum 1999). The derived developmental pathways create different opportunities for further adaptive change in one character (e.g., feathers, but not in others, like scales or hair).

From the sketchy example above, as well as those examined in chapter 1, it is clear that developmental pathways can promote or limit the ability to realize certain morphologies. The question then is whether these novelties are just some special adaptations or whether their study requires a different research program. Certainly novelties can be adaptive because they most likely increase the fitness of those organisms that have them (although, this is not necessarily the case; see Gould and Lewontin 1978). But is that all? Does natural selection explain the most important consequence—namely, the new variational properties that come with the novelty?

There are a number of features characteristic of novelties that make it unlikely that the adaptationist program will give us satisfactory answers. For one, novelties are rare. One can assume only that the necessary genetic variation does not arise at a sufficiently high frequency in natural populations, or it requires a specific genomic predisposition in order to occur (Wagner 2001). In contrast, genetic variation for other traits arises at a high level of regularity; for example, genetic variation in abdominal bristle number in *Drosophila* (Falconer and Mackay 1996), as well as other traits (Roff 1997).

This phenomenological fact is supported by the discovery that novelties are often realized by "genetic rewiring" of developmental genes rather than by tweaking of existing regulatory relations (Davidson 2001; Wagner and Lynch 2010). It seems that new regulatory interactions arise at a much lower rate than quantitative variation of existing regulatory interactions; but this remains an open empirical question. Furthermore, the specific new potential of a novelty can hardly be "seen" by natural selection that originally selected the new trait. Natural selection is shortsighted by taking hold only of immediate benefits. Thus, it is unlikely that natural selection can provide a satisfactory account of the fact that feathers turned out to be able to support flight, whereas hair did not.

Revisiting the Conceptual Roadmap: Which Way to Novelty?

It is relatively easy to place an intuitive handle on the notion of novelty by pointing to examples of what we mean by novelties: feathers in the evolution of birds (theropod dinosaurs, to be exact; see chapter 9); the origin of flowers (see chapter 12); and insect wings. This is the way language works. Intuitive generalization and repeated use by the members of a language community can lead to a stable system of meaning and word use. However, this mode of fixing the meaning of a term is not sufficient for science.

Science is not only a language game among individuals, but also includes Nature, and our conversation with Nature requires us to know what questions to ask her so that she will answer us. In a less poetic vein, we need to decide what kind of information we need and what kind of experiments to do to learn about the processes that produce novelties in evolution. Similarly, the intuitive notion of a species is perhaps as old as any human language, since distinguishing between different kinds of plants and animals was, and is, essential for human survival.

But a proper understanding of speciation required sophisticated knowledge of population genetics in order to arrive at a research program that penetrated the inner workings of how species originated. In this section I want to propose a perspective of what novelties are in order to determine what kind of research questions we need to ask to, eventually, gain a deeper understanding of the processes that generate novelties during evolution and ultimately may lead to an understanding of how and why complex organisms arose.

The first thing to note is that each of the paradigmatic novelties—feathers, flowers and insect wings—were novelties at the time they arose and are now perfect exemplars of homologs (i.e., they represent clades). Most likely, they originated only once (with qualifications) and are now transmitted along lines of descent. Consequently, the notion of homology and novelty are two sides of the same coin (Müller and Wagner 1991). What we think about the nature of homologs will inform us about what kind of processes are able to create new homologs (i.e., novelties). Therefore, I will start the discussion of the biology of novelties with the definition:

> *Morphological novelty is a structure that is neither homologous to any structure in the ancestral species nor serially homologous to any part of the same organism.* (Müller and Wagner 1991)

I will discuss alternative concepts later.

Based on what I wrote above, this definition is conceptually expedient, but also comes with a cost. It depends on how well we understand and agree upon the nature of homology, which is an assumption not worth betting good money on. For better or worse, however, the usefulness of any homology concept also depends on its ability to spawn productive research programs and, thus, the notions of novelty and homology are tied to each other in terms of their scientific utility.

The notion of homology proposed in chapter 2 fundamentally depends on the distinction between character identity and character states. Character identity refers to that part of the organism that forms a lineage of descent with modifications (i.e., the thing that corresponds among different species like the forewing of a bee and that of a dragonfly). Character states are the different shapes and forms that a character can assume. I argued that homology should be associated with character identities rather than character states.

It follows, then, that *the origin of an evolutionary novelty is the evolutionary process through which a novel character identity arises. In other words, an evolutionary novelty originates when a part of the body acquires individuality and quasi-independence.* With the acquisition of these attributes, the body part starts to be a unit of phenotypic evolutionary change and potentially a homolog, if the descendent species proliferate and form a clade.

If we consider the origin of novelties as the process by which a body part becomes individualized, then it is also clear that the study of adaptation and that of novelties have to be different research programs. The study of adaptation tries to understand how fitness-enhancing features evolved, given the environmental boundary conditions inherited by the species from its ancestors. The study of novelties tries to understand how the organism differentiates into quasi-independent parts.

The explanation for adaptation is natural selection. We are not yet sure what the explanation for novelties is. Almost certainly, natural selection will be part of the answer. Yet the goal of novelty research is still different from the study of adaptation. Similarly, the answer to the question of how species arise involves natural selection, in part, but we know now that natural selection is not sufficient to explain speciation. In order to explain speciation, we have to invoke additional mechanisms, like geographic isolation, hybridization, and others. I suspect that the origin of novelties also requires natural selection as well as additional mechanisms, but what they are will have to be determined by more empirical research.

In addition to the conceptual reasons for distinguishing between adaptations and novelties, there is increasing evidence that the molecular genetic mechanisms involved in adaptations and innovations also tend to be different (Wagner and Lynch 2010). Novelties likely require large-scale reorganizations of the gene regulatory network. Gene regulatory network reorganization involves, for example, the creation of novel cis-regulatory elements. In contrast, adaptive modifications often involve only the modification of existing cis-regulatory elements. For that reason, transposable elements play a greater role in major evolutionary transitions and innovations than in small-scale modifications (Feschotte 2008; Oliver and Greene 2009).

Another Layer of Complexity: The Origin of Variational Modalities

The above definition for evolutionary novelty is pleasing, at least if one values logical purity. It clarifies the notion of novelty by linking it to the notion of character identity. However, in spite of its logical virtues, it has a serious defect; namely, it ignores the complexity of biological evolution. By narrowing the notion of evolutionary novelty to the origination of character identity, it excludes the very paradigm I used to introduce the distinction between novelties and adaptations: the origin of feathers. The problem is that feathers are derived from scales (i.e., from pre-existing structures: epidermal scales).

Based on that fact, the origin of feathers can be seen as a character state transformation rather than the origin of a novel character identity. Also, one of the intuitive paradigms of evolutionary novelties, the origin of limbs in tetrapods from paired fins, falls into the same category. Here also the novel trait, tetrapod limb, is derived by a transformation from pre-existing structures: fins. This too is a character state transformation, but still it appears significant and the derived state is novel and noteworthy.

We encountered this problem already in chapter 2 when we reflected on the nature of character identity. There we recognized that the intuitive notion of character identity combines two distinct biological situations: different individualized characters (i.e., different character identities *sensu stricto*) and different variational modalities. Differences in character identity reflect the individuality of different body parts, both developmentally as well as evolutionarily—for example, the difference between the brain and the liver. Variational modalities reflect "deep" differences between instances of the same character.

The paradigm for the latter is the difference between limbs and teleost fins, as discussed in chapter 2. Variational modalities differ in the anatomical composition of the character (e.g., there is no part in a teleost fin that corresponds to digits), and they also differ in development with consequences for the variational opportunities and constraints of the character. Hence, variational modalities are non-overlapping sets of character states with rare or difficult transitions between them. The latter provision gives a variational modality its notion of "evolutionary individuality," as characters tend to remain within one or another variational modality (set of character states) over longer periods of evolutionary time. Because of that, variational modalities represent different "kinds" or "personalities" of the same character identity.

Clearly, character identity and variational modalities are different biological phenomena and need to be distinguished. Consequently, the origin of a novel character identity (the origin of paired appendages in early vertebrates) and the origin of a novel variational modality (limbs from fins) also represent different kinds of evolutionary events. Thus, it is necessary to distinguish between two forms of novelties. Conveniently, one may call them Type I and Type II novelties.

> *Type I novelty*: the origin of a novel character identity. The paradigms are the origin of the vertebrate head and the origin of the insect wing.
> *Type II novelty*: the origin of a novel variational modality. The paradigms are the origin of the tetrapod limb from paired fins and the origin of feathers from epidermal scales.

These two forms of novelties imply different research questions. To explain Type I novelties, one has to address the origin of gene regulatory networks that underlie character identity (see chapter 3 for the notion of character identity networks, or ChINs). Explaining Type II novelties requires explaining the different developmental modalities that lead to different and persistent variational tendencies. The paradigm for the latter kind of explanation is Richard Prum's theory of feather origins based on the distinct mode of feather development as compared to other skin appendages (Prum 1999; Prum and Brush 2002).

It must be mentioned that it was previously recognized that the term "evolutionary novelties" covered a biologically heterogeneous territory and led to a similar "typology" of evolutionary novelties (Müller 2010). Gerd Müller introduced a distinction between three types of novelties, but the conceptual origins were different from those introduced here. For example, the notion of character identity, as introduced in chapter 2, does not play a prominent role in Müller's scheme. For Müller, the difference between the origin of a physical part (e.g., a new cartilage) and the differentiation/individuation of pre-existing parts was conceptually more important than the notion of character individuality. Thus, differentiation of existing parts and the origin of novel parts were different types of novelties for Müller, while for me they are just two different ways to get a new character identity.

Thus, the typologies introduced by Müller and those introduced here reflect different emphases in terms of the empirical questions. Müller was interested in the origin of novel physical parts as a consequence of thresholds during developmental pattern formation (Müller 1990), like the origin of novel skeletal elements. In contrast, for me the emphasis is on understanding the genetic underpinnings of character identity and variational modalities. And, this is how it should be: different concepts reflect different priorities in research programs.

Phenomenological Modes for the Origin of Type I Novelties

To get an overview of what kind of evolutionary events a theory of novelty origination has to explain, we can turn to a list of apomorphic morphological characters that are characteristic of some higher clades. These include for example the major clades of mammals (i.e., mammals themselves), the monotremes (platypus and echnida), and the therians, which include the marsupials (opossum and kangaroos) and the placental mammals (table 4.1). From a list of apomorphic characters it is clear that some of these apomorphies are not novelties, even in the broadest sense of the word, as for example the loss of characters (the loss of nuclei in erythrocytes, the loss of teeth in monotremes, etc.), or simple shape changes (e.g., bent cochlea).

Table 4.1. Apomorphic characters of the major mammalian taxa classified according to their evolutionary transformation underlying their origin (after Müller and Wagner 1991, Table 1, pp. 238–239; data after Ax 1984).

Type of Change	Apomorphic Character	Plesiomorphic Character	Taxon
Loss of elements	Erythrocytes without nuclei	Erythrocytes with nuclei	Mammalia
	Teeth absent	Teeth present	Monotremata
	Coracoid absent	Coracoid present	Theria
	No oil droplets in retinal cones	Oil droplets in retinal cones	Placentalia
	Marsupial bones absent	Marsupial bones present	Placentalia
Shape change	Lateral temporal skull opening	No lateral temporal skull opening	Mammalia
	Cochlea bent	Cochlea straight	Mammalia
	Cochlea coiled	Cochlea not coiled	Theria
	Penis simple	Penis forked	Placentalia
Differentiation of repeated elements	Teeth heterodont	Teeth homodont	Mammalia
	Qualitative differentiation of cervical and thoracic vertebrae	Gradual differences between cervical and thoracic vertebrae	Mammalia
	Some hairs specialized as whiskers	No whiskers	Theria
New elements	Marsupial bones	—	Mammalia
	Hair		Mammalia
	Muscular diaphragm		Mammalia
	Lips and facial musculature		Mammalia
	Glans penis		Mammalia
	Thigh glands		Monotremata
	Pseudo-vagina		Marsupialia
	Corpus callosum		Placentalia
	Vagina		Placentalia
Change of context	Centrum of first vertebra fused to second		Mammalia
	Secondary jaw joint	Primary jaw joint	Mammalia
	Angular (tympanic) fused to temporale	Angular part of lower jaw	Mammalia
	Yolk sac attached to uterus	Yolk sac not attached to uterus	Theria
	Separate opening of gut and urogenital sinus	Common opening of gut and urogenital sinus (cloaca)	Theria
	Scapular origin of supracoracoid muscle	Coracoidal origin of supracoracoid muscle	Placentalia
Fusion/integration of plesiomorphic characters	Secondary palate	Separate maxillary processes and palatines	Mammalia
	Nipples	Dispersed external orifices of milk glands	Theria

Others are potential examples of how novel body parts originate: a) differentiation of repeated elements (e.g., distinct tooth types from homodont reptilian teeth); b) de novo origination of elements that have no trace in the ancestral species (hair in mammalia or the corpus callosum in placental mammals); and c) new elements arising from the fusion of plesiomorphic elements (centrum of first vertebra fused to second vertebra, which is derived in mammals, or the origin of nipples from the clustering of the openings of milk glands in Theria, whereas the milk glands are diffuse in monotremes, which presumably is ancestral). Another group of characters is relational, as for example the attachment of the supracoracoid muscle to the scapula rather than to the coracoid.

This list of modes of origination of novelties is both more inclusive and more restrictive than those of other authors. For example, Newman and Müller (Newman and Müller 2001) would not include the differentiation of repeated elements in their notion of novelty, mostly because the repeated elements from which a novel character differentiates predates the "novel" element; hence, the novel element seems to be not "truly" novel. Although this argument makes some intuitive sense, I will discuss below the reason for including this mode of origination in my list.

This list can also be considered as too narrow. For example, Brian Charlesworth (Charlesworth 1990) argued that essentially every mutation with a phenotypic effect is a novelty because it is "something new." This latter point of view, of course, rejects the notion of novelty as a distinct concept and subsumes all evolutionary change driven by natural selection as adaptation. Hence, the limits of what to include in the list of novelties is informed by the theoretical perspective of the particular author. The goal of this section, then, is to justify my choice of examples of novelties. Furthermore, the distinction among the different modes of origination may point to different causal scenarios for the origination of novel characters and, thus, may be useful.

Differentiation of Repeated Elements

The differentiation of repeated elements is, perhaps, one of the most important modes of evolution that can lead to new elements (Remane 1952; Riedl 1978; Weiss 1990). The list includes the differentiation of teeth into different tooth types in mammals, the differentiation of arthropod body segments into the body regions of insects or crustaceans, the differentiation of hair into whiskers and body hair in therians, any case of a novel cell type, and many more. In chapter 2 we discussed the distinction between simple repeated elements, like two red blood cells in a given individual (homomorphic characters), and characters that arose as individualized instances of repeated elements. The latter have been called "paramorphs" to distinguish them from the repetition of identical elements (see chapter 2, as well as Minelli 2002; Minelli and Fusco 2005).

The notion of paramorph characters acknowledges the fact that differentiation of repeated elements can lead to body parts that follow their own history of descent with modification, even though ancestrally they arose from identically repeated (homomorphic) elements. Paramorph characters, thus, are truly novel in the sense that they represent novel quasi-independent body parts and are, therefore, different homologs. For example, with a few exceptions (e.g., toothed whales), mammals have differentiated teeth. Two of these tooth types are incisors (figure 4.1) and molar teeth (figure 4.2).

Incisors are usually blade-shaped frontal teeth, whereas molars are usually broad grinding teeth with multiple cusps. In some lineages, the incisors assume some extreme modifications, as in elephants, where they become protruding tusks (figure 4.1C) and are no longer any good for biting. Similarly, in male narwhales, one incisor becomes a long straight tusk (figure 4.1B),

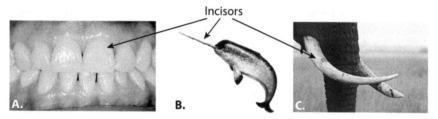

FIGURE 4.1: Shape variation of mammalian incisors demonstrating the evolutionary individuality of incisors relative to other teeth. A) Human tooth row, B) narwhale "horn," C) elephant tusks.

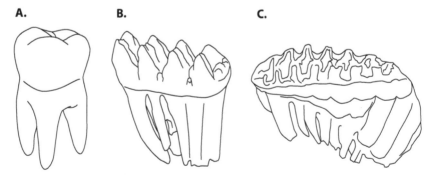

FIGURE 4.2: Mammalian molar tooth variation. A) Human third molar, B) mastodon molar tooth, C) African elephant molar tooth. Compare to figure 4.1 to appreciate the distinct evolutionary individualities of incisors and molars.

the function of which is uncertain. At the same time, molar teeth were being modified for their own end; for example, in elephants they became high crowned (so-called hypsodont) teeth that are specialized for grinding abrasive plant material (figure 4.2B and C). Clearly, incisors and molars had their own largely independent history of evolutionary modification.

For that reason, paramorph characters are included in my list of novelties. They are novelties, regardless of the fact that the "material" from which they differentiated, the repeated elements like segments and homodont teeth, predate the origin of the character identity itself. The reason is that the criterion for a body part to be a named entity (i.e., a homolog) is evolutionary individuality and not where the cells for the character come from. Also, if the definition of a Type I novelty is the origin of a new homolog, then the origin of individuality has to be the decisive criterion for what counts as a novelty. This argument has consequences for what we need to explain mechanistically in order to understand novelties.

Paramorph characters can also lose their individuality if the functional reason for their existence ceases. For the tooth types in mammals, this is what happened in the toothed whales (figure 4.3). In killer whales and dolphins, for example, the tooth row consists of a variable number of homodont single cusp teeth, which are superficially similar to the simple teeth of most reptiles. Hence, homologs have a limited lifetime and it does not make sense to call the first two or three teeth of a dolphin "incisors" and some more proximal teeth "molars." There is no basis for this. Hence, the demise of a paramorph tooth type is either the total loss of teeth, as in monotremes, or the loss of individuality.

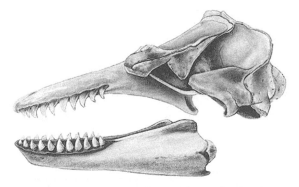

FIGURE 4.3: Skull of a killer whale. Note homodont tooth row of uniformly cone-shaped teeth. This condition is secondary, as whales are placental mammals and, thus, derived from an ancestor with distinct tooth types. Illustration by Katherine Zecca, Alaska Fisheries Science Center, NMFS, NOAA.

De Novo Origination

Completely novel physical elements are those for which there is no trace in the ancestral phenotype. For mammals, examples include hair, marsupial bone, corpus callosum, and many more (see table 4.1). Among these, the corpus callosum (CC) seems to be the least problematic case (figure 4.4). The corpus callosum is a massive tract of nerve fibers that connects the two telencephalic brain hemispheres. It exists only in placental mammals, but is not the only such fiber tract. In placentals, there is also a smaller fiber tract, called the *commissura anterior* (CA). It travels through the *lamina terminalis*, which embryologically corresponds to the anterior tip of the neural tube. This CA is also present in marsupials and is somewhat larger than in placentals.

The main difference between the CC and the CA, and which makes the CC a novelty, is that, embryologically, the CC fibers bridge the gap between the two forebrain hemispheres rather than traveling through the *lamina terminalis or any other pre-existing embryological structure*. Thus, the development of the CC requires breaching the integrity of the embryonic brain surface so that nerve fibers can travel from one forebrain hemisphere to the other. Lack of the CC is a malformation in humans that does not seem to have serious consequences. In fact, in these individuals, more fibers are found in the CA than in normal individuals, as is the case in marsupials. Hence, the CC is primarily a new passageway for nerve fibers, most of which would otherwise find an alternative route to cross to the other brain hemisphere through the CA.

Nevertheless, morphologically and developmentally the CC has no counterpart in the outgroups of placental mammals. It is an invariant species

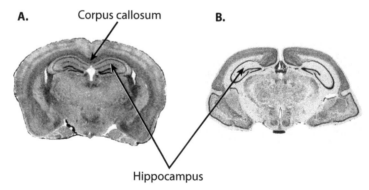

FIGURE 4.4: The corpus callosum is a massive tract of nerve fibers that connects the cells of the two cerebral hemispheres. The corpus callosum is an innovation of placental mammals. A) Coronal section through the mouse brain showing the corpus callosum. B) Coronal section through a marsupial brain without a corpus callosum (*Antechinus flavipes*). [A: http://www.mbl.org/images/coronal.jpg; B: http://brainmuseum.org/specimens/dasyuromorphia/marsupmouse/sections/238marsupmouselg.jpg.]

characteristic among placental mammals. As human pathology shows, it can be lost quite easily, but lack of the CC never becomes a species character in placentals and, hence, is likely to be maintained by natural selection.

Another group of de novo characters are novel bones or cartilages. An example in mammals is the marsupial bone that attaches to the pubis and functions in supporting the brood pouch. Much developmental work has been done on a novel bone in birds because of its accessibility to experimentation in chickens. This is the cartilage/bone that connects the fibula with the tibia in bird legs, the so-called *syndesmosis tibiofibularis* (STF) (Müller 1989; Müller and Streicher 1989; for details see chapter 5). The STF is a consequence of the capability of amniote connective tissue to react to mechanical stimuli with the production of cartilage and, ultimately, ossification of this cartilage. This fact led Gerd Müller to propose his side-effect hypothesis for the evolution of novel characters (Müller 1990).

The idea is that novel structures arise from cellular reactions to changing epigenetic environments, as for example the mechanical forces that act on connective tissues in fetal life, which change for some other possibly adaptive reasons. For example, a reduction in the size of the fibula changes the force distribution on the developing fibula and, thus, introduces a novel cartilage. Hence, the novel structure, here the STF, is not directly the focus of natural selection; rather, it is a developmental side effect of some other change, namely the reduction of the distal fibula. I will discuss this theory further in chapter 5.

Another often cited example of a novel structure is mammalian hair. In fact, hair is found only in mammals. But one can make a plausible argument that hair is more a paramorph of other skin appendages than a complete de novo formation. Hair is one of several skin derivatives that share developmental mechanisms during their early differentiation. Hair, epidermal scales, the filiform papillae of the tongue (Dhouailly and Sun 1989), feathers, and even teeth share the same mode of early development (Chuong, Widelitz et al. 1996). They all derive from an interaction between the ectoderm and the underlying mesenchyme, which form a close association and exchange molecular signals that are similar or identical between these different forms of skin derivatives, such as Wnt 10, BMP-2 and -4, TGFb, Shh, and others (Widelitz, Jiang et al. 2000; Botchkarev and Paus 2003; Chuong and Homberger 2003; Wu, Hou et al. 2004; Mikkola 2007).

Hence, within an individual mammal, hair, mammary glands, scales (if present), lingual papillae, and even teeth could be paramorph structures (i.e., all derived from an ancestral skin derivative). What the closest paramorph of hair is exactly is not clear, but an argument can be made for a derivation from keratinous scales through the intermediate stage of mechanosensory bristles, as are found, for instance, in agamid lizards (Pinkus 1922). Alternative scenarios will be discussed in chapter 9.

It requires some deep knowledge regarding development and evolution of a structure to decide whether it is a de novo formation or a paramorph. Nevertheless, I suspect that de novo origination, like that of bones and cartilages, requires somewhat different developmental mechanisms than the origin of paramorph characters through differentiation. Thus, it might be useful to maintain the distinction between differentiation of repeated elements and de novo origination as suggested by the classification of novelties by Gerd Müller (2010).

Novelties from the Fusion of Ancestral Characters

Fusion is the inverse of the first mode of origination that we have discussed here: differentiation. Whereas in differentiation a character originates from gaining individuality among repeated elements, fusion is the process during which individualized characters lose their individuality and form a novel, more inclusive unit that is then the new character, or during which repeated elements are united into a larger unit. For example, the second vertebra in mammals differs from those of other vertebrates, as it includes a part of the first vertebra. Specifically, the body of the first vertebra is fused to that of the second and, thus, forms a new element, and the first vertebra is now a bony ring without a body (figure 4.5).

Another example is the origin of nipples in mammals. Mammary glands are modified hair-associated glands that, in monotremes, are diffusely distributed over the chest of the female (see chapter 9 for more details). The milk that is secreted is collected by the breast hair and taken up by the young by licking at the breast hair of the mother. In higher mammals, the openings

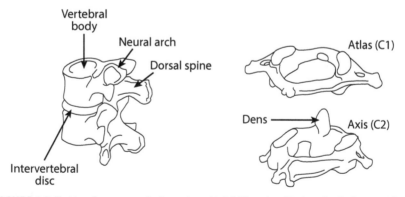

FIGURE 4.5: First two human cervical vertebrae (right). Note that C1, the atlas, lacks a vertebral body, as is seen on the right side of the other vertebrae. The body has been fused to the second cervical vertebra (C2) and is visible there as the "dens" of C2. This configuration is a derived feature of mammals.

of the milk glands are united into a unitary structure, the nipple or a teat, and milk can be obtained by the young without the aid of hair. Hence, the evolution of the mammalian breast includes the origin of a paramorph hair-associated gland and the subsequent fusion of these elements into a unitary structure.

Another non-zoological example of innovation through character fusion is the origin of the angiosperm flower. In gymnosperms, the male and female organs are separate, even when they occur on the same plant. The first critical step during the evolution of the angiosperm flower was the unification of the reproductive organs by expressing both male and female organs on the same shoot axis and then integrating subdenting leaves as the perianth into the flower (see chapter 12 for more details). Hence, the flower was the product of character integration that occurred in at least two steps: the integration of male and female organs and the integration of sterile leaves to become the perianth.

Compared to the other two modes, character fusion is less often investigated as a mode of innovation. It is not yet clear how prevalent this mode of novelty origination is and what kinds of processes are involved, although in the case of the flower, some deep insights into the relationship between the molecular biology of flower development and character integration have been garnered (see chapter 12).

From Phenomenology to Explanation

I argued above that Type I novelties could arise in three different phenomenological modes: through differentiation of repeated elements; de novo origination; and the integration or fusion of ancestral elements. In any case, the end product has to be an individualized body part that was not previously there. By implication, any evolutionary process that produces a novelty has to create the developmental genetic machinery that makes an individualized body part during ontogeny. Hence, explaining the evolutionary origin of novel body parts means to explain the origin of the genetic regulatory network that directs the development of this novel body part.

Explaining the evolution of the genetic regulatory network requires looking at three different issues. For one, we need to consider what evolutionary forces favor the origin of novel body parts and, thus, the evolution of the underlying gene regulatory network. This issue will be discussed in the remainder of this chapter. Then we need to look at the natural history of development and ask, at what level of the developmental hierarchy do modifications occur that could lead to evolutionary novelties? This will be the subject of chapter 5. Finally, we need to ask, what kind of mutations contribute to the evolution of gene regulation and what is their relationship to the origin of novelties? This question will be discussed in chapter 6.

The first question, what are the evolutionary forces and scenarios that can lead to the evolution of novel characters, includes factors like the role of environmentally induced phenotypes versus the role of mutations in creating novelties (discussed in chapter 5). What kinds of selective forces play roles in the evolution of character individuation? Is the evolution of canalization a likely factor in the origin of novel characters?

How Natural Selection Creates Character Individuality

The various modes of origination of Type I novelties, whether from individualizing repeated elements or from the creation of a novel skeletal element or a pigment spot, all ultimately require an evolutionary process that endows them with variational, genetic, and developmental individuality. Only with this individuality can the body part be a quasi-independent unit of phenotypic evolution. To evolve this individuality requires the developmental and genetic mechanisms that are discussed in subsequent chapters. In addition, we need to understand the population genetic mechanisms that favor character individuality. However, understanding these population genetic mechanisms is surprisingly difficult.

Population genetic theory is best when it explains either the selection of a mutation, given or assuming that it has an additive genetic effect on fitness, or when it explains the response to directional selection in terms of the mean phenotype of a quantitative character, as exemplified by the "breeder's equation": $R = h^2 * S$.[4] Population genetics is at its most sophisticated when population effects and life history characters directly determine the fitness of mutations. Examples are the theory of sex-ratio evolution (Fisher 1930) and the dynamics of genomic conflict, as for example between the paternal and the maternal genomes in the cases of live-bearing animals and flowering plants (Haig 1993).

Unfortunately, population genetics is least satisfying when it comes to explaining variational properties, like the evolution of mutation rates, canalization and robustness, and modularity. This is not to say that there is not extensive theoretical and empirical work on these subjects; rather, the explanatory value of these approaches is still quite unclear. Here I want to summarize what is known in each of these fields and explore its potential role in explaining the origin of novelties.

---

[4] R is the so-called selection response (i.e., the difference between the mean value of a quantitative trait between two consecutive generations). The symbol h^2 is the heritability (i.e., the slope of the regression line between the average parental phenotype and the offspring phenotype). S is the selection differential, which is the difference in mean value within the parental generation before and after selection (see Falconer and MacKay 1996: *Introduction to Quantitative Genetics*. Essex, England, Longman).

Explaining Modularity: Selection for Variational Individuality of Body Parts

Modularity is a concept that is motivated by the goal of understanding the evolutionary origin and significance of phenotypic organization (a.k.a. body plans). Rudy Raff introduced this in his influential book *The Shape of Life* (Raff 1996) to conceptualize the fact that the development of multicellular organisms is partitioned into quasi-independent processes and parts, as with the development of the limb bud and of the brain. Raff's criterion for modularity is context independence during development and, thus, his concept should be properly called developmental modularity. For example, under the right experimental circumstances, a frog embryo can be induced to form a limb on odd parts of its body. But even in these unusual locations, the limb bud differentiates into an anatomically well-organized limb (Hinchliffe and Johnson 1980; Müller, Streicher et al. 1996).

We previoulsy discussed another case of developmental modularity; namely, the fact that ectopic expression of the gene *ey* can induce the development of well-formed and even light-sensitive eyes on the legs and wings of fruit flies (Halder, Callaerts et al. 1995). I think that it is not coincidental that the units of development that Rudy Raff had in mind when he introduced his modularity concept correspond closely to the morphological units I am concerned with here (i.e., morphological characters or homologs).

In the same year, and independently, Wagner and Altenberg (Wagner and Altenberg 1996) introduced a corresponding concept: variational modularity. This idea is that clusters of phenotypic traits co-vary due to the pleiotropic effects of genes, but are relatively independent of other such clusters. That is to say, traits that belong to the same variational modules tend to be affected by mutations of the same genes, whereas traits from different variational modules tend to be affected by mutations at different genetic loci.

Finally, there are various modularity ideas that originated in molecular biology, which have to do with the structures of molecular networks (Hartwell, Hopfield et al. 1999). For example, gene regulatory networks have been found to have some modular organization in the sense that there are groups of genes that are co-expressed and are, most likely, regulated by the same set of upstream regulatory genes (Li, Sun et al. 2009). There are groups of proteins that form clusters in a protein-protein interaction network and may have common functions and evolutionary history (Fraser 2006). There are also signal transduction networks that seem to function in a similar way in various contexts and, thus, may be self-contained functional units (i.e., modules). An example is the hedgehog (Hh) signaling pathway, which includes a stereotypical set of components, like the receptor patched, the membrane-bound protein smoothened, and the transcription factor Gli/ci. This pathway plays a role in very different biological contexts from brain and limb

development to feather development in vertebrates and to segment formation and wing patterning in insects.

The relationships among these various notions of modularity are somewhat intuitive, although they are not entirely clear (see discussion in Wagner,
Pavliček et al. 2007). For example, developmental modularity is certainly related to differential gene expression and, thus, it is plausible that developmental modularity is related to modular gene regulatory networks. But it
is also likely that there are multiple modular gene regulatory networks that
are active in each developmental module—for example, at least one for each
cell type in a developmental module and different co-regulatory modules for
different cellular functions, like the ribosomal gene network and the network
underlying different metabolic pathways. Hence, my guess is that a developmental module contains multiple gene regulatory modules and multiple
modular signal transduction networks.

The relationship between variational and developmental modularity
is also not one-to-one. Two developmental modules can express the same
genes and, thus, may be part of the same variational module. For example, a
limb bud is paradigmatic of a developmental module as used by Rudy Raff to
introduce this concept. But the left and the right limb bud express the same
genes and, thus, are also highly correlated in their variation. The left and the
right forelimb belong to the same variational module, but each is an independent developmental module.

It seems plausible that forms of modularity have the following relationship to each other: a variational module is also a developmental module that,
in turn, contains multiple molecular network modules. For a part of the body
to be a homolog in the sense developed here, variational modularity is necessary. Hence, we will focus on the likely evolutionary causes of variational
modularity.

Evolution of Variational Modularity

Variational modularity is an idea that dates back to Olson and Miller (Olson
and Miller 1958), who pointed out that the correlations among quantitative
phenotypic characters are highly structured. These correlation patterns seem
to reflect a mixture of developmental and functional relationships. That is
to say that functionally related traits tend to have higher correlations than
do functionally unrelated traits. In addition, serially homologous traits, like
upper arm and upper leg lengths, tend to have higher correlations than do
unrelated traits (Van Valen 1965).

Traits from the same developmental gene expression territory are also
more strongly correlated than are traits that derive from embryonic territories
with different gene expression patterns (Nemeschkal 1999; Reno, McCollum
et al. 2008). For example, the digits of the hand differ in the set of Hox genes
that are expressed during their development. Specifically, digit 1, the thumb,

expresses only two Hox genes, *HoxA13* and *HoxD13*, whereas digits 2 to 5 additionally express *HoxD12* to *HoxD9*. Reno and colleagues showed that the morphological correlations among digits 2 to 5 were much higher than the correlations between digit 1 and all other digits both in terms of population variation and evolutionary differences. In primates, the thumb shows more independent evolutionary change relative to the other four digits than each of digits 2 to 5 relative to each other. Hence, Reno et al. identified two variational modules in the primate hand: digit 1 and one module comprising digits 2 to 5.

The pattern of phenotypic correlations corresponds to the patterns of pleiotropic effects of genes. For example, Kenny-Hunt and colleagues (Kenney-Hunt, Wang et al. 2008) mapped the genes that influenced 70 skeletal traits and identified 105 loci that affected these traits in a cross between two mouse inbred strains. They found that traits that showed strong phenotypic correlations also tended to be affected by the same genes. Hence, variational modules are likely to be sets of traits affected by the same genes and, probably, also are clusters of traits that could co-evolve because they have a very similar genetic basis.

What then determines the pattern of pleiotropic effects and, thus, the pattern of co-variation among traits? A hint of how correlation patterns and patterns of pleiotropic effects might evolve comes from a study of limb variation by Young and Hallgrímsson (Young and Hallgrímsson 2005).

Young and Hallgrímsson studied the variation of limb bones in six mammalian species. In most of these species they found the familiar pattern with correlations between serially homologous characters. The femur (upper leg) and humerus (upper arm) were correlated, but the humerus was less correlated with the tibia (lower leg), the pattern first described by Van Valen (Van Valen 1965). In addition, elements in the same limb were also more correlated, probably due to functional constraints.

What is interesting, though, is that the correlation between serially homologous traits breaks down in species with specialized forelimbs and hind limbs. In the bat, the serially homologous elements of the autopod (hand/foot) and the zeugopod (lower leg/lower arm) are uncorrelated, whereas the corresponding elements in the gibbon and macaque are quite strongly correlated. It seems that correlations between serially homologous traits are ancestral, but that functional specialization can lead to a loss of correlation (i.e., an increase in variational and, probably, also genetic individuality; figure 4.6).

The same pattern has been found in primate evolution in the transition from quadrupedal primates to bipedal humans (Young, Wagner et al. 2010). This is interesting because it might be a hint of how natural selection can lead to the individuality of body parts. As discussed above, individualization of serially homologous body parts is a major mode for the evolution of novel characters and, thus, the selective forces that can produce this individuality are important for explaining novelties.

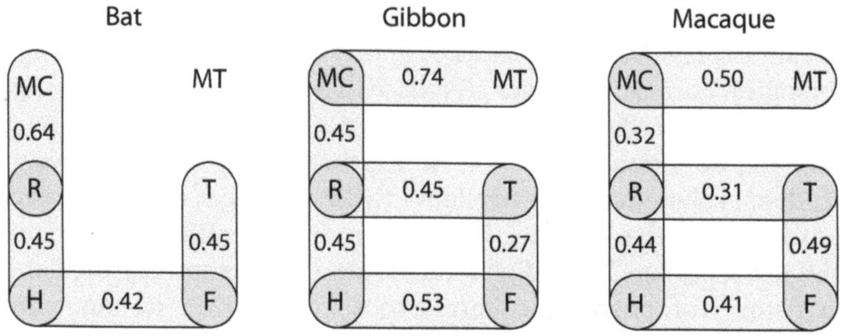

FIGURE 4.6: Forelimbs and hind limbs become less integrated (less correlated) if they specialize for different functions. These diagrams summarize the correlation patterns between limb long bones of a bat and those of two primates, the gibbon and the macaque. Note that the correlations between the radius and tibia as well as metacarpals and metatarsals present in gibbon and macaque are missing for the bat. H = humerus, R = radius, MC = metacarpal, F = femur, T = tibia, MT = metatarsal. (Redrawn after Young and Hallgrímsson, 2005, *Evolution* 59:2691-2704.)

Functional specialization of characters implies that the selective forces that act upon them will be different. Not only will there be different optimal phenotypes selected for different tasks, but selection to change one character will not usually coincide with selection to change the other character. For example, for bats, the environmental conditions that require a modification in wing shape or size, say due to altitude or higher temperature and, thus, lower air density, will not necessarily coincide with environmental changes that require modification of the feet, say for clinging to different substrates. Functionally differentiated characters will rarely experience directional selection at the same time and of the same kind. And, if one character needs to change, but the other needs to stay the same, this should benefit from low genetic correlation among these characters (Wagner and Altenberg 1996).

This is the case because selection on one of two genetically correlated characters will lead to a change in the unselected character, a phenomenon called "correlated selection response." This means that selection on one character may lead to a loss of adaptation at a genetically correlated character. If these two characters often experience directional selection independently of each other, then a decrease in correlation will be beneficial. This seems to be a reasonably intuitive idea, although it turned out to be surprisingly difficult to model this process (for a review, see Wagner, Mezey et al. 2005). One of the first successful attempts to simulate the evolution of variational modularity was the study by Kashtan and Alon (2005) in which they used logical circuits as a model of the genotype.

A logical circuit consists of elements that take two or more inputs and transform them into one output according to some rule. The inputs and outputs are binary, either 0 or 1 as in a digital computer, and the rule can be a logical (Boolean) function. A genome then consists of a number of these logic elements and the connections among them. Mutations change the connections among the elements and selection among mutant genotypes proceeds according to a given goal. The goal for the network is to produce a certain output for each possible input configuration.

For example, their circuit had four inputs: x, y, z, and w. The network was selected to calculate the following logical function: G1=$((x \text{ XOR } y) \text{ AND } (z \text{ XOR } w))$ (see figure 4.7). When the authors selected for this goal, the network evolved many different possible solutions (i.e., networks that could calculate the function G1). In this experiment, the evolved networks were almost always non-modular.

In another experiment, the authors periodically changed the goal function from G1 to G2 = $((x \text{ XOR } y) \text{ OR } (z \text{ XOR } w))$. In this case, the networks always evolved modularity, in the sense that there were sub-circuits dedicated to calculating the functions shared between G1 and G2, $(x \text{ XOR } y)$ and $(z \text{ XOR } w)$, and another part that represented the variable part of the function: either the AND or the OR function connecting $(x \text{ XOR } y)$ and $(z \text{ XOR } w)$. Hence, if the fitness function was modular, that is, if there were aspects that remained the same and others that changed, then the system evolved different parts that represented the constant and the variable parts of the environment (figure 4.7).

This example was intriguing because it overcame some of the difficulties of earlier attempts to simulate the evolution of variational modularity (Wagner, Mezey et al. 2005), although it did use a fairly non-standard model of a genotype-phenotype map: logical circuits. In a second example, Kashtan and Alon (2005) used a neural network model with similar results. Hence, the questions arise, how generic are these results? and can one expect that similar processes occur in real life?

Modeling the evolution of modularity became significantly easier after a kind of genetic variation was discovered by quantitative trait locus (QTL) mapping in the lab of James Cheverud at Washington University called "relationship QTL," or r-QTL for short (Cheverud 2001; Cheverud, Ehrich et al. 2004; Pavlicev, Kenney-Hunt et al. 2008). An r-QTL is a genetic locus that affects the correlations between two quantitative traits (i.e., their variational relationship and, therefore, "relationship" loci). Surprisingly, a large fraction of these so-mapped loci are also neutral with respect to the character mean. This means one can select on these "neutral" r-QTLs without simultaneously changing the character mean in a certain way.

It was easy to show that differential directional selection on a character could easily lead to a decrease in genetic correlation between characters (Pavlicev, Cheverud et al. 2011). Of course, it is not guaranteed that each and every population has the right kind of r-QTL polymorphisms, nor is it yet clear what kind of genetic architecture allows for the existence of an r-QTL.

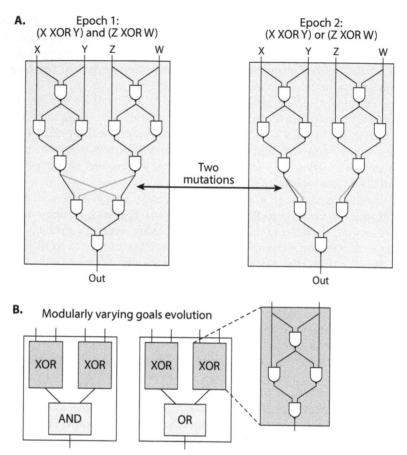

FIGURE 4.7: Evolution of logical networks during computer simulations of network evolution. A) Two networks that evolved to adapt to two different goals. Both networks are modular and have the ability to quickly switch between performing different tasks with only two mutations. B) Goal functions varied between epochs used to evolve networks in A). Each symbol in the wiring diagrams represents a not-AND function. (Redrawn from Kashtan and Alon 2005, *PNAS* 102:13773-13778.)

Nevertheless, these findings make it plausible that differential directional selection can enhance the genetic/variational individuality of traits and, thus, may play a role in the origin of evolutionary novelties by selecting for variational individuality.

It must be added, though, that there has been relatively little research in this area and that we will need to see more to determine whether we understand what is going on here, if anything. In particular, one difficulty is the mathematical modeling of gene interaction (epistasis), because the details of an epistasis model determine the outcome of the evolution by natural selection (Hansen 2006). One result shows that natural selection increases or decreases mutational variance, depending on whether the average epistatic effects are positive or negative (Hansen, Alvarez-Castro et al. 2006). This means that the genetic architecture is more determined by the genetic architecture that we start with than by the nature of the selection forces that act upon it. In other words, the evolution of a genetic architecture could be arbitrary with respect to selection.

Variational Modularity and Evolvability

The kind of selection discussed in the last section points to a connection between the evolution of modularity and that of evolvability. Evolvability is the capability of populations to respond to directional selection, and evolution of evolvability suggests that genotypes differ in this capability. Recent reviews of this idea are provided by Pigliucci (2008) and Wagner and Draghi (2010). The idea that evolvability itself can be subject to natural selection is controversial (Lynch 2007), although there is increasing evidence from simulation studies that evolvability is an evolvable trait under quite generic conditions (Masel and Bergman 2003; Kashtan and Alon 2005; Salazar-Ciudad 2007; Wagner 2007; Crombach and Hogeweg 2008; Draghi and Wagner 2008; McBride, Ogbunugafor et al. 2008; Draghi and Wagner 2009; Wagner 2010; Pavlicev, Cheverud et al. 2011). Modularity is one way how the evolvability of a genotype could increase and, thus, selection for modularity under directional selection could be seen as a special case for the evolution of evolvability.

The idea was that if different variational modules could be dedicated to performing different adaptive functions, then these functions would be evolvable without interference from other functional constraints. Thus, variational modularity is thought to contribute to evolvability, which is largely the consensus view. In fact, Kashtan and Alon (2005) pointed out in their study discussed above that the evolution of modularity was associated with a major increase in evolvability and, thus, selection of evolvability was likely the driving force for the evolution of variational modularity.

Interestingly, though, Thomas Hansen (Hansen 2003) found that evolvability was maximized not when modules were totally uncorrelated, but if about 16% of the genes were pleiotropic and the remainder had modular effects. This is the case if the number of genes is constant and the average effect of a mutation on a trait is also constant. Under these assumptions, if all the genes that affect only one trait are reduced, then the total amount of genetic variation available is less and, thus, evolvability will be less as well. The reduction in evolvability is greater than the gain of evolvability by reducing pleiotropy; thus, there is an intermediate optimum. Hence, even by assuming that modularity evolves to maximize evolvability, one would not predict that the correlations between functionally independent traits would be zero. Rather, the prediction would be that there are pleiotropic genes, although they make up only 16% of all genes that affect the trait. The exact details of what degree of pleiotropy is optimal depends on factors that determine the amount and kind of genetic variation and, thus, it is difficult to make predictions for each particular case (Pavličev, Cheverud et al. 2010).

An intuitive model may make the connection between functional specialization, modularity, and differentiation more clear. Consider the origin of the endometrial stromal cell (ESC) in placental mammals as an example. This cell is part of the uterus lining and negotiates with the invasive fetus to establish the placental interface between maternal and fetal tissues. It is not yet clear how this cell type arose. Evidence suggests that it either derived from an immune support cell (the follicular dendritic cell, FDC) or a myofibroblast, both of which are derived from the bone marrow (Oliver, Montes et al. 1999; Garcia-Pacheco, Oliver et al. 2001; Muñoz-Fernández, Blanco et al. 2006; Aghajanova, Horcajadas et al. 2010).

In either case, the ESC would derive from a cell type that is associated with and underlies all the epithelia of mucosal surfaces of the body, such as the mouth, the gut, and the genital tract. Both cell types are involved in wound healing, which makes sense since the fetus also invades maternal tissues and, thus, creates a wound in the female genital tract. The difference, however, between a reaction to an injury in the gut, say, and the invasion of a fetus into the placenta is that the injury caused by a fetus is tolerated and managed rather than healed as long as pregnancy lasts.

Unlike the FDC of the gut, the ESC in the uterus expresses genes that result in suppressing inflammation, and maintaining and managing the invasion of the placenta. Hence, the evolution of placentation implies the acquisition of target genes by the FDC in the uterus that allow pregnancy to be maintained. If they were expressed in the rest of the body, however, the same genes would be deleterious because they would suppress the inflammation that is necessary to combat pathogens. Recruitment of these new genes will nevertheless occur as long as the advantage created by their expression in the uterus is greater than the damage they can do to other parts of the body.

However, the very fact that there are deleterious pleiotropic effects of these target genes' acquisition by the FDC induces a selection pressure to express these genes specifically in the uterus. Hence, the negative pleiotropic effects of adaptations to invasive placenta immediately lead to a direct selection pressure for differential gene expression. Gene expression limited to the uterus will increase fitness by the same amount as their pleiotropic effects have decreased it. Direct selection for modularity and, thus, character identity is possible if the function that is selected for is antagonistic to other functions of the body.

Other Models for the Evolution of Variational Modularity

There are other classes of models that can explain the evolution of variational modularity (for a review, see Wagner, Pavlicev et al. 2007). Two of these are instructive because they show how selection on properties other than evolvability can also produce variational modularity. One class comprises scenarios in which modularity directly affects fitness. In another class, modularity is a side effect of selection for robustness.

Modularity is an abstract property and, in order for it to affect fitness, it needs to interact with other organismal traits. One proposal for which modularity could be considered a more or less direct target of selection refers to constraints to adaptation (Leroi 2000). In this proposal, modularity is selected if it breaks a developmental constraint and, thereby, makes adaptive phenotypes accessible that would otherwise be genetically unattainable. For example, one scenario in which modularity would directly improve fitness is when ontogenetic development of the optimal phenotype itself is aided by a modular organization of development. This has been shown in two artificial-life studies; one used network learning as a model of ontogeny (DiFerdinando, Calabretta et al. 2001) and the other simulated artificial ontogenies of robots under the control of gene regulatory networks (Bongard 2002). However, one can also view these models as an extreme case of selection for evolvability, as breaking the constraint increases evolvability and is selected because it allows or facilitates evolution of an adaptive trait.

It has long been known that artificial neural networks can learn certain tasks by implementing simple rules for modifying the strength of neural connections. This ability to learn, however, is impaired if the same neural network is asked to learn two different tasks: a phenomenon called neural interference or cross talk. A paradigm for this problem is learning to recognize an object on an array of receptors (e.g., a "retina") and to determine where the object is located on the retina: the so-called "what" and "where" problem. Neural interference can, of course, be eliminated if the two tasks are learned by two non-overlapping sub-networks.

DiFerdinando and colleagues (2001) showed that a genetic algorithm would spontaneously evolve a modular network dedicated to the two functions. In this simulation, the genetic algorithm was limited to evolving the general architecture of the network (i.e., which neuron was connected to which other neuron). A learning algorithm was then used to determine the strength of the connections (i.e., the actual function was learned/developed during the ontogeny of the individual, while the architecture evolved by mutation and selection). Hence, in this case, modularity evolved because it aided learning the task during the lifetime of the individual and, thus, directly improved the fitness of the phenotype.

Another artificial-life model that pointed in the same conceptual direction was the study by Josh Bongard who evolved robots with an artificial ontogeny (Bongard 2002). He also showed that modularity of the developmental regulatory network evolved in those lineages that most successfully evolved functioning robots. Hence, it is possible that genetic, developmental, and variational modules can evolve because they aid the function of the phenotype. They can do that by regulating physiological and developmental processes necessary to maximize fitness in a given environment.

Another way that selection can drive the evolution of a trait is by correlated selection responses. In this case, selection acts on a character that affects fitness. But any character that is genetically correlated with this adaptive character will also change, even though the trait may not affect fitness. For example, selection on body size can lead to changes in skull shape if there is a genetic correlation between body size and skull shape. Hence, skull shape would evolve under the influence of natural selection, even if skull shape may not directly affect fitness and, thus, is not the direct target of natural selection.

Similarly, it is now thought that genetic robustness might evolve as a correlated effect of selection for environmental robustness (Wagner, Booth et al. 1997; Meiklejohn and Hartl 2002). There are some theoretical results that suggest that something similar could be the case with modularity. Specifically, it has been proposed that there might be a correlation between modularity and robustness and, thus, modularity could arise as a correlated selection effect of the evolution of robustness (on robustness, see Andreas Wagner 2005). One argument for this mode of evolution derives from a computational study of RNA secondary structure evolution (Ancel and Fontana 2000). This kind of model has been called "differential erosion of interactions" under stabilizing selection.

In a seminal paper, Lauren Ancel and Walter Fontana (2000) investigated the evolution of RNA secondary structure using a computational model of RNA folding. In their model they allowed for environmental variation in the form of alternative, sub-optimally folded secondary structures, and showed that three variational properties increased during evolution: environmental and genetic robustness as well as modularity of secondary structure elements. In short, it was shown that the directly selected property was environmental robustness, which in that model was synonymous with thermodynamic stability. Further, it was shown that genetic robustness and modularity were correlated with environmental robustness and evolved as correlated selection responses. In this case, modularity was assessed by the independence of the melting profile of different secondary structure elements and the independence of a secondary structure element from its sequence context.

It makes sense that modularity and robustness are correlated in RNA secondary structure. Modularity limits the structural consequences of genetic and environmental perturbations on parts of the secondary structure. The presence of a secondary structure element does not depend only on the presence of complementary base pairs, but is also influenced by nucleotides outside of the sequence that folds into the stem-loop region. This is the case because nucleotides outside of a specific structure element determine whether there are alternative secondary structures that compete with the formation of that element.

Ancel and Fontana showed that this dependency on the sequence context decreased with increasing robustness of the overall secondary structure and, thus, reduced the overall impact of mutations on the secondary structure element. This result says that evolution of robustness led to fewer and weaker pleiotropic effects. Because these results were based on a biophysical model of RNA folding, it is likely that this property is at least generic for RNA secondary structure. Similar correlations between genetic robustness (designability), thermodynamic stability, and the existence of protein domains have been suggested by Li and collaborators (Li, Helling et al. 1996) and have been supported to some extent by comparative computational studies of protein evolution and structure (Rorick and Wagner 2011).

The evolution of modularity as a consequence of selection for robustness, as exemplified by the RNA simulation studies, can be understood as a case of differential elimination of pleiotropic effects (Wagner, Mezey et al. 2005). The effect of a mutation increases with the number of characters that it affects (Wagner, Kenney-Hunt et al. 2008). Hence, robustness can increase by reducing the number of affected characters per mutation, for example, by limiting mutational effects to a phenotype module. One can consider these mutation effects as "soft" because they can be eliminated by changing the genetic background. On the other hand, the effects of mutations at nucleotides that are part of the structure do not depend on the genetic background. One can call these effects "hard" because they cannot be eliminated. It is plausible that hard effects are those that occur within the structure elements and soft effects are indirect effects that arise from nucleotide positions outside of the element.

This consideration can be generalized by a model called the "differential erosion model." The erosion of rocks can lead to highly structured landscapes. Soft rocks are washed away more quickly than are hard rocks, as was the case for Devil's Tower in Wyoming that was generated from the erosion of soft rock surrounding the solid lava core of a volcano. Similarly, if the indirect effects of mutations are more readily suppressed by selection for robustness than are the effects on the core process in which the gene product is directly involved, then modularity can evolve as a consequence of the evolution of robustness by differential erosion of indirect (soft) effects.

This would imply that genes that are involved in a functional module should also have more limited effects on fitness than those genes that are hierarchically above the modules and regulate and control modules. Furthermore, one would predict that functional interactions within a module (i.e., those mechanisms underlying "hard" genetic effects) are phylogenetically more stable than are those effects that go beyond individual modules.

Explaining Functional Specialization

Thus far we have dealt with variational properties, modularity, correlations, and robustness, but have ignored a fundamental question. As a rule of thumb, and I do not know whether there is really any exception from this rule, different characters usually are dedicated to different functions. However, it is not immediately clear why this should be so. After all, why should functional specialization of organs even have to evolve, given that single-celled organisms still exist and, in fact, make up by far the greatest fraction of biodiversity, as one and only one cell fulfills all the functions necessary for a productive life (reproduction, ingestion, metabolism, excretion).

There are still more forms of microbial life than there are multicellular forms, so it is hard to argue that being multicellular and functionally differentiated is intrinsically better. Of course, there has to be something cool about functional specialization, otherwise we humans would not partake in it. But that statement is not much help, since it is no more informative than saying that we live in the best of possible worlds. Certainly there have to be some special circumstances that have to be met to trigger the evolution of functionally specialized organs, even though less specialized organisms are otherwise evolutionarily stable. A general theory to answer this question was proposed by Claus Rüffler and collaborators (Rueffler, Hermisson et al. 2012).

The one special example of functional specialization that received serious attention from theorists was the differentiation into soma and germline. The soma-germline differentiation caught the eye of biologists because it implied that some cells gave up their individual fitness "for the greater good." In organisms with germline segregation, only some cells, the germline cells, can contribute their genes to the next generation, whereas the majority of cells are genetic dead ends. Hence, from a population genetic point of view, the existence of terminally differentiated somatic cells is surprising because it is a type of cellular "altruism."

As it turns out, the specialization of cell types into those with somatic and reproductive functions is likely to be the result of a trade-off between these two functions. In protozoans, such as the green alga *Volvox*, there is a trade-off between locomotion and reproduction. For small planktonic algae, survival depends on locomotion, the ability to stay in the water column rather than sinking to the ground or floating to the surface. These organisms' locomotion depends on the actions of their flagella, which just happen to use a part of the cellular machinery that is necessary for cell division: the centriole. As a consequence, these cells have to temporarily give up locomotion when they reproduce (i.e., if the cells divide).

Time spent reproducing is time that is not spent to ensure survival. The solution of this conflict is that some cells give up their personal reproduction and dedicate their lives to locomotion, or what is more generally called somatic functions, whereas another set of cells focus on reproduction while relying on the somatic cells to carry them around. These latter cells are generally called germline cells. Hence, the question is, under what circumstances is such a division of labor advantageous (i.e., when is the fitness of a colony with germ-soma differentiation greater than the fitness of an undifferentiated colony in which all cells invest in somatic as well as reproductive function)?

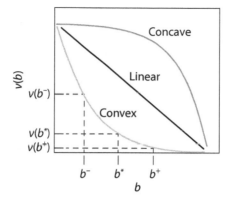

FIGURE 4.8: Trade-off relationships between viability, v, and fecundity, b, in colonial algae. For a strictly convex function, $v(b)$, if we take a particular point, say b^*, and two points equidistant below and above b^*, say b^- and b^+, respectively, then $v(b^-) + v(b^+) > 2\,v(b^*)$. This means that a loss of a contribution to fecundity is more than compensated for by the increase in viability. This implies that there is an advantage for specializing one part of the colony for fecundity and the other for viability functions. In contrast, a concave performance function implies that there are diminishing returns on investment in either component and specialization of cells for the two fundamental functions of fecundity and viability is not produced by natural selection. (Redrawn after Michod, 2006, *PNAS* 103:9113–9117.)

A fairly general theory of this was proposed by Richard Michod (Michod 1999; Michod 2006), who showed that colony fitness benefited from differentiation if the trade-off curve between reproductive and viability functions was convex (figure 4.8). A convex trade-off means that a decrease in, say, viability is compensated for by more than a proportional increase in fecundity, and vice versa.

This style of modeling was extended to the general problem of body part differentiation by Claus Rüffler and collaborators (2012). Within a fairly general mathematical framework it was found that there were generally three factors that favored functional differentiation by natural selection: position effects; accelerating performance functions; and synergistic effects among characters (figure 4.9). Position effects result from a situation in which one body part is inherently better positioned to perform a certain function. For example, the appendages of an arthropod that are close to its mouth are more likely to contribute to feeding than are more posterior legs.

The second condition, accelerating performance functions, corresponds to the factor identified by Michod (2006) for the case of soma-germline differentiation discussed above. Finally, synergistic interactions among characters describe a situation in which the total performance of two characters is higher than the sum of the contributions of each character by itself. This fairly general framework is likely to facilitate our understanding of how and under what circumstances functional differentiation can and will evolve.

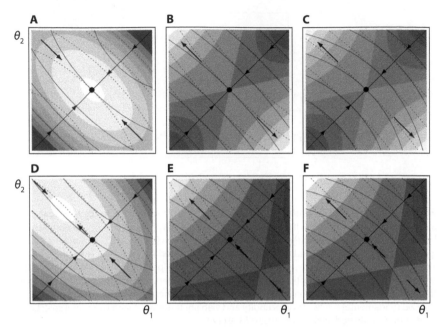

FIGURE 4.9: A model for the evolution of functional specialization according to Rueffler and colleagues (2012). This model considers two phenotype modules contributing to two functions that contribute to fitness. Two situations are distinguished: two equivalent modules (A–C) and two non-equivalent modules (D–F). Contours of the fitness landscape are indicated by shading, with lighter shades indicating higher fitness. Solid circles indicate the locations of fitness maxima in the constrained trait (i.e., the situation when the two modules cannot diverge). The expected direction of the evolutionary dynamics is indicated by arrows. The solid and dashed curves depict iso-performance curves of the underlying performance functions F1 and F2, respectively. A) Iso-performance curves are convex for F1 and concave for F2, indicating that functional differentiation decreases performance in both tasks. Thus, this point is a fitness maximum. B) Iso-performance curves are concave for F1 and convex for F2, indicating that functional differentiation increases performance in both tasks. Thus, this point is a saddle point of the fitness landscape. C) Iso-performance curves are concave for both F1 and F2, indicating that functional differentiation increases performance for task 1 and decreases performance for task 2. In this example, the increase in performance in task 1 is sufficiently large to outweigh the decrease in performance in task 2 such that the point is still a saddle point of the fitness landscape. For plots D, E, and F it was assumed that module 1 had an intrinsic advantage in contributing to task 2, whereas module 2 had an intrinsic advantage in contributing to task 1. Each plot in the lower row represents a perturbation of the corresponding plot in the upper row. In D, nonequivalence of modules moves the fitness maximum above the diagonal, whereas in E and F, nonequivalence moves the saddle point below the diagonal. In both cases, selection favors specialization of module 1 for task 1 and of module 2 for task 2. Note that extrema of the fitness landscape correspond to points where iso-performance curves for F1 and F2 are tangent to each other. (Source: Rueffler et al., 2012, *PNAS* 109:E326–E335.)

Explaining Robustness and Canalization

Canalization is the idea that natural selection could lead to stabilizing the optimal wild-type phenotype against the effects of mutations and environmental perturbations (Waddington 1957; Schmalhausen 1986). The rationale for this idea is that stabilizing natural selection acts against deviations from the optimal phenotype and, thus, should also select those genotypes that are less prone to express deviant phenotypes in response to genetic or environmental variation. This idea was the subject of intense research during the 1950s and up to the early 1970s, but was largely abandoned until the mid-1990s when new molecular and modeling techniques led to a renaissance in this field.

Before we go into the technical details, it is worth noting that Waddington's inspiration to conceive his theory of canalizing selection was very closely related to the subject area of this chapter: the origin of morphological novelties. Waddington did not express it in those words, but he was impressed by the qualitatively distinct nature of cell types and thought that this fact required an explanation (Waddington 1957). Because variation is the basic principle of Darwinian evolutionary theory, it seemed odd to him that cell types should present such stable, and qualitatively different, phenotypes. One could imagine a situation for which cell types do not exist. Rather, there would be only a continuum of cell phenotypes that blend into each other through quantitative variations in their expressions of various proteins.

In fact, intermediate cell types that are not found in say, mammals, are known to science. For example, cnidarians, and even echinoderms have cells that are truly intermediate between epithelial cells, which usually specialize in forming barriers,[5] and muscle cells, or so-called epithelio-muscle cells (figure 4.10). However, even such apparent oddities usually are well-defined cell types within their respective species. Waddington, thus, reasoned that natural selection should be able to enhance the distinctness of different cell types and, at the same time, endow them with stability against mutations and environmental effects. This is what he originally called canalization.

As mentioned above, the basic idea is simple: stabilizing selection acts against deviations from the optimal phenotype and, thus, should favor genotypes that are less sensitive to mutations and environmental variation. But as it is often the case in science, when researchers take a closer look, reality turns out to be more complex than the initial idea.

For one, there are technical difficulties, because canalization is a variational property that is hard to measure with standard transmission genetic

---

[5] For example, barriers between the internal and the external space of an organism, like the skin or the gut epithelium, or the barrier between the inner space of blood vessels and the body cavity, so-called endothelial cells.

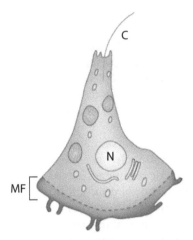

FIGURE 4.10: Epithelial muscle cell from a cnidarian larva. This cell has the functional characteristics of both an epithelial cell and a muscle cell and suggests that specialized cell types are derived from cell types that performed multiple functions. The sister cell type model suggests that new cell types arise from sub-functionalization of multifunctional cells like the epithelial muscle cell. MF = muscle fibers, C = cilium, N = nucleus. Reprinted by permission from Macmillan Publishers Ltd: *Nature Reviews Genetics,* Arendt, 2008, *Nature Reviews Genetics* 9:868–882, copyright 2008.

techniques (see, for example, Rendel 1967 for early attempts to get a handle on this issue). Even modeling the process of canalizing selection was done surprisingly late, because it involved gene interactions (epistasis) and genotype by environment interactions that were challenging to model and required new modeling techniques. The first computer simulation study that demonstrated the possibility of canalization evolving from natural selection was published in 1996 by Andreas Wagner (Wagner 1996).

In that paper, Andreas Wagner introduced a modeling framework for simple models of gene regulation that subsequently became the standard model for studying the evolution of gene regulation by computer simulation. The first analytical study of natural selection that used population genetic theory was published one year after Andreas's study (Wagner, Booth et al. 1997) and also required developing novel modeling approaches. The main result from these early studies was that environmental canalization should be easy to evolve, but genetic canalization was more difficult.

The reason is that selection for genetic canalization depends on the presence of genetic variation because, otherwise, selection can't "see" the effect of genetic robustness. Yet stabilizing selection not only selects for canalization, it also eliminates genetic variation. The trouble is that stabilizing selection is more efficient for eliminating genetic variation than is selecting for canalization. The end result is that, paradoxically, strong stabilizing selection is predicted to be less effective in selecting for canalization than is moderately strong selection (Wagner, Booth et al. 1997). Of course, this scenario depended on the assumption that the amount of genetic variation was maintained by mutation-stabilizing selection balance. The situation might be less problematic if one considers situations in which the amount of genetic variation in a population is maintained at a higher level than by mutation, as for example by hybridization or high mutation rates, a fact that was already recognized by Ernst Mayr in an essay on evolutionary innovations (Mayr 1960).

One way out of this conundrum is to assume that the evolution of genetic robustness is the indirect consequence of selection for environmental robustness. As mentioned above, the rationale is that selection for environmental robustness is predicted to be easier and if there is a positive genetic correlation between environmental and genetic robustness (i.e., if genotypes that are environmentally robust are, on average, also genetically robust) then selection for environmental robustness will also produce genetic robustness as a side effect. This idea received a boost from a study described above by Ancel and Fontana (2000). Ancel and Fontana showed that genetic robustness evolved, but as a statistical consequence of selection for environmental stability. In the case of RNA secondary structure, a positive correlation between environmental and genetic robustness was due to the fact that both were related to the thermodynamic stability of the molecular structure. Whether such a correlation is also generally true for organismal phenotypes other than RNA and protein structure is not clear. Both the empirical and the modeling evidence are ambiguous.

Another modeling approach was brought to bear by Erik van Nimwegen and colleagues in a paper published in 1999 in which they used genotype space models to prove the existence of selection for mutational robustness (Nimwegen, Crutchfield et al. 1999). Ofria, Adami, and Lenski used the power of computational "artificial life" models to show the same thing (i.e., selection can lead to the selection of a less fit but more robust genotype if the mutation rate is high;[6] Wilke, Wang et al. 2001).

Another difficulty with the population genetic theory of canalization is that the population genetic consequences of gene interaction are not well understood, as they depend on the model of gene interaction that is used. For example, it has been shown that with certain models that are more general than those used in Wagner et al. (1997), stabilizing selection can also lead to de-canalization; that is, stabilizing selection can make the genotype more sensitive to mutations (Hermisson et al., 2003). Similarly, the same kind of model predicts that under directional selection, for which one would

---

[6] This fact was already published by Schuster and Swetina in 1988 using analytical mathematical methods. This paper was about ten years ahead of its time and, as a consequence, was largely ignored. In fact, I knew of this result at the time I was in Vienna and I heard Schuster talk about this result, but I also did not understand its significance.

intuitively predict de-canalization, canalization can evolve (Hansen 2006; Hansen, Alvarez-Castro et al. 2006; Hansen 2013). Another unexpected result was that the gene regulatory network model introduced by Andreas Wagner (Wagner 1996) could evolve canalization even without any selection on the phenotype itself (Siegal and Bergman 2003).

It turned out that, in these models, there was indirect selection against gene regulatory networks that displayed stable oscillations of gene expression rather than a stable fixed point of gene expression. This indirect selection was sufficient to produce canalization for any phenotype (Siegal and Bergman 2003). All of this means that there is no simple relationship between the mode of selection and the consequences for the variational properties of the genotype, as assumed by Waddington.

On the empirical side, the difficulties are also considerable. The main objective in demonstrating genetic canalization is to demonstrate that genotypes differ in their sensitivity to mutations. The difficulty with standard transmission genetic approaches is that they are intrinsically incapable of distinguishing between differences in the amount of genetic variation present in a population and differences in the sensitivity of the phenotype to the effects of mutations. At the time of Waddington, the standard empirical approach to demonstrate differences in canalization was to introduce a major mutation into a wild-type genotype. A paradigm of this approach was the effect of the *Tabby* mutation on the number of whiskers in mice.

Mice usually have a fairly constant number of whiskers on their snout and with very little variation between individuals. The *Tabby* mutation decreases the average number of whiskers and also increases the variation of whisker number. Artificial selection showed that this variation was heritable and, thus, presumably genetic. The interpretation of this result was that the wild type was canalized and that the mutation had resulted in a new phenotype outside the "range of canalization"; thus, formerly "hidden" genetic variation became visible.

This was the standard approach for studying genetic canalization until molecular techniques became available. The problem, however, is that this approach is based on a flawed assumption. Hermisson and Wagner (2004) showed that any character, whether canalized or not, will show this result if it is under stabilizing selection and if the genes are epistatic. Alternative empirical approaches were not available until the mid-1990s.

The first paper that used an alternative approach was by Stearns and Kawecki (Stearns and Kawecki 1994), who measured the effects of P-element insertions on various life history traits. They argued that traits more closely related to fitness, and presumably under stronger selection, were more robust than other traits. This was the first attempt to get a better handle on measuring genetic canalization using modern genetic techniques. The problem was that different traits could have different genetic target sizes (i.e., depend on different numbers of genes), so that this comparison may be confounded (Houle, Morikawa et al. 1996).

In 1996, Greg Gibson showed that Drosophila genotypes that were more sensitive to ether for producing the bithorax phenotype had an allele at the *Ubx* locus that resulted in lower Ubx protein concentrations in the haltere imaginal disc (Gibson and Hogness 1996). This was the first insight into the mechanisms for a classical case of environmental canalization. The first evidence that phylogenetically more conserved characters are also more canalized came from a computational analysis of the secondary structure of RNA virus genomes by Andreas Wagner and Peter Stadler (Wagner and Stadler 1999). They used computational methods to estimate the mutational robustness of virus genomes and found that phylogenetically conserved secondary structure elements were more robust to nucleotide changes than those that were phylognetically variable. Note that this is not logically necessary, because a character can be phylogenetically conserved due to stabilizing selection, even though it is as mutable as a less conserved trait.

Finally, in 2005, Paul Turner and his colleagues showed that viruses evolved lower mutational robustness if they were placed under a regime in which natural selection was weaker (Montville, Froissart et al. 2005). This little survey showed that molecular genetics, computational chemistry, and experimental evolution techniques led to a number of original new approaches for detecting and measuring genetic and environmental robustness. Overall, their results suggest that genomes do in fact acquire some degree of robustness when they are under sustained stabilizing selection and may lose it when selection is relaxed.

This is remarkable given that the situation is not that clear if we look at the problem based on the first principles of population genetics (see above). This means that either there is a strong, pervasive correlation between environmental and genetic robustness so that genetic robustness evolves easily as a consequence of selection for environmental robustness (i.e., as a correlated selection response), or the genetic architecture of real organisms is more favorable for the evolution of genetic robustness by direct selection than what is assumed in some of the theoretical models.

When we return from this excursion into a rather arcane branch of population genetics to the topic of this chapter, morphological novelties, and the original inspiration for Waddington (the evolution of distinct cell types), it is clear that the work on genetic canalization is quite divorced from both of these concerns. In the hands of population geneticists (including my own work on this subject), the idea of canalization became either a problem of the variability of quantitative traits, say bristle number, or of variation in fitness itself. However, there is an emerging consensus that organisms are canalized at the systems level (see the book by Andreas Wagner; Wagner 2005). Even though much detail still needs to be worked out, work since the mid-1990s suggests that natural selection can lead to both phenotype canalization and genetic modularity. Both may turn out to be key processes for the origin of novel body parts or individualized characters.

Natural Selection and the Origin of Novelties: A Roundup

While there is no serious question regarding whether natural selection plays a role in the origin of novelties, at least not in my mind,[7] specific scenarios and explanations are difficult to come by. The problem is that if we frame the objective of a population genetic theory of novelties as explaining the origin of quasi-independent parts of the phenotype (often dedicated to a specialized function), then we run into a number of difficult and still unsolved problems in evolutionary genetics: robustness, modularity, specialization, and evolvability. I described these areas of research, possibly somewhat disproportionately to the taste of some, because in recent years, novelties have become predominantly seen as a problem of developmental genetics. But the emphasis on gene regulatory networks, molecular evolution, and developmental mechanisms obscures the fact that we also face difficult problems at the population genetics level to understand the evolutionary forces that shape individualized body parts.

To illustrate how easy it is to underestimate the difficulty of this problem, I want to close this chapter with an anecdote. When I started the real work of what would become the Wagner et al. paper in *Evolution* with the title "A population genetic theory of canalization" (Wagner, Booth et al. 1997), my thought was that I would more or less directly be able to address the origin of modularity and the origin of individualized body parts. To get going, however, I first looked at what mathematical population geneticists would do—namely, consider the case of stabilizing selection on a single character, and asked whether robustness/canalization can evolve. As mentioned above, this idea had been around since the 1940s and 1950s, although at the time no serious population genetic model of genetic robustness had been published.

I thought that this part of the project would become a footnote in the paper that I really wanted to write, namely, a paper on modularity. As I worked on it, it soon became clear that a footnote would never provide sufficient space to deal with this problem. Ultimately, this "footnote" grew into

---

[7] Alternative positions are discussed, some more credible than others. The one most strongly based on population genetics is that of Michael Lynch, who prefers a model of genome evolution in which the driving forces are drift and mutational biases and variations in stabilizing selection due to variations in population size. Others assume that developmental mechanisms and plasticity are sufficient to explain the origin of novelties, as for example the models by Newman and Müller. Newman, S. A. and G. B. Müller (2001). "Epigenetic mechanisms of character origination," in *The Character Concept in Evolutionary Biology*, ed. G. P. Wagner. San Diego, Academic Press: pp. 559–580.

a 50+ page manuscript that the journal *Evolution* was generously willing to consider despite its enormous size. After that paper, I thought that the population genetic theory of canalization was done; after all, I had started with the first principles of population genetics and showed what could happen.

As the reader now knows from the sections above, things quickly became more complicated with the discovery of indirect selection for robustness by Siegal and Bergman, and the paradoxical selection for canalization and de-canalization discovered by Hermisson and Hansen. Now my conclusion is that we are far from understanding the issue for which I planned only a foot-note in the mid-1990s. Major progress became possible with computational models (van Nimwegen, Andreas Wagner, Hansen, Ofria, Adami, Draghi, Ancel and Fontana, Hogeweg, and others). These models are at least more complex, if not more realistic, than the analytical population genetic models I started out with.

We may now be in a situation in which computational modeling is ahead of our ability to really understand what is going on inside these models, and how to frame a theory around these results. Or, it may be that we need an-other step up in the complexity of computational models of evolving gene regulation and development to capture what is going on when novel body parts arise. That is to say, in politics and in science, even if tempting, it is unwise to declare "mission accomplished" too quickly, and it is intrinsically hard to say when a mission is, in fact, accomplished. Only because this re-search is summarized in a book does not mean that it is completed.

Finally, I want to point out that major progress in molecular and developmental evolutionary biology is now making it possible to consider a comprehensive theory of evolutionary innovation. The boldest attempt at this goal has been proposed by Andreas Wagner from the University of Zürich, Switzerland, in his recent book *The Origin of Evolutionary Innovations* (Wagner 2011). I recommend this book to the reader who is interested in a broad perspective on the mechanisms of evolutionary innovations.

5

Developmental Mechanisms for
Evolutionary Novelties

When analyzing the development of any body part, one is confronted by a canonical set of themes: many, if not all characters develop from an interaction with the environment; organ rudiments start to form in response to certain inductive signals that tell the embryo where and when to form a certain body part; once organ-specific development is initiated, the field of cells becomes spatially organized, often through the action of discrete signaling centers; and gene regulation is organized in a way that ensures character-specific execution of the developmental program. Given their role in the development of virtually all body parts, each of these developmental mechanisms has the potential to play a role in the evolutionary origin of novel body parts. Here I want to discuss their role in evolutionary novelties from an organ-level, developmental point of view. In the next chapter, a genetic, DNA-centered point of view will be assumed by focusing on the molecular genetic side of these processes.

The Environment's Role in Evolutionary Innovations

When various lines of evidence in the nineteenth century clearly pointed to the fact that species are mutable, based on evidence from comparative anatomy, geographic distribution and variation, paleontology, and other fields, phenotypic plasticity was the only known phenomenon by which the phenotypes of organisms could be modified. Thus, it was tempting to assume that these modifications were incipient evolutionary modifications that became heritable and resulted in species differences. Even Charles Darwin was the proponent of such a theory (Geison 1969). Later, some of the bloodiest battles in the history of biology were fought over this idea (Koestler

1971)—literally bloody because these included suicide (e.g., Paul Kammerer) and political persecution (Lysenko).

It appears that a similar fight is about to begin, as can be seen by the publication of a number of books and the discovery of environmentally induced gene imprinting (Schlichting and Pigliucci 1998; Hall, Pearson et al. 2003; West-Eberhard 2003; Gilbert and Epel 2009). Of course, the standard position in evolutionary biology is that permanent evolutionary modifications arise from spontaneous genetic changes (mutations) without the helping hand of the environment, and that the environment is limited to a role as the selective agent that drives natural selection. Randomness of mutations means that the frequencies and kinds of genetic changes that occur are independent of what is useful for an organism living within a given environment. There is ample empirical evidence that mutations are, in fact, random in that sense.

On the other hand, in mainstream evolutionary biology, phenotypic plasticity is seen as an adaptation to unpredictable environments (Stearns 1989; Scheiner 1993), but not as a driver of evolutionary change, as are mutations and selection. In this view, the reaction norm (i.e., given a certain genotype, the relationship between an environmental factor and the resulting phenotype), is, itself, a character that undergoes evolutionary modification by mutation and selection (Stearns 1989). But counterarguments have also been mounting, particularly during recent years.

There are three types of arguments in favor of an active role of phenotypic plasticity. Some authors point to the fact that environmental modifications can be transmitted between generations, even without a DNA sequence change. This became most abundantly clear with recent research on the inheritance of environmentally induced DNA modifications, like DNA methylation (Bird 1993), but is not limited to that mechanism.

Other forms of direct phenotype heritability, also called epigenetic inheritance, are known and include language and other maternal or paternal effect mechanisms. In the case of language it can be shown that the evolutionary dynamics are similar to those of genetic inheritance in that they can be understood with the same types of differential equations as those used in population genetics (Nowak 2006).

Another set of facts are the phenomena of genetic assimilation and genetic accommodation. One speaks of genetic assimilation when selection on an environmentally induced phenotype leads to the genetic heritability of this phenotype. The most likely explanation of this phenomenon is the selection of alleles that phenocopy the induced phenotype. These alleles can become selected if the frequency of the environmentally induced phenotype is not 100%, but varies among genotypes. Selection then favors more sensitive genotypes and, eventually, those genotypes that do not need the environmental trigger to produce the phenotype.

There is an extensive literature devoted to the importance and interpretation of these phenomena, which do not need to concern us here in great detail. What we need to specifically discuss is what role phenotypic plasticity may play in the origin of novel body parts. Proponents of an active role for plasticity in evolutionary change summarize their importance by four points (see the summary by Gilbert and Epel 2009):

- *The derived phenotype is not random*: usually, organisms react with one or a small set of phenotypic states to changes in the environment. The result of the environmental change is often integrated and detailed rather than chaotic or random.
- *Plastic phenotypes can become genetically fixed*: this has been demonstrated repeatedly in the lab, but is also likely to have occurred in nature. The latter is likely if the environmentally induced phenotype of one species or population is similar to the genetically fixed phenotype of a closely related species or population. For example, the heat shock–induced wing pattern of the central European subspecies of the butterfly *Aglais urticae* resembles that of the Sardinian subspecies, while the cold-shocked wing pattern resembles that of the Scandinavian subspecies (Goldschmidt 1938).
- *If the phenotype is induced first by the environment, then the number of individuals with the new trait can be greater than one*: because the environment can affect all individuals in a local population, the environmentally induced phenotype will be present in more than one (or a few[1]) individuals, as would be expected if the novel trait arose through mutation. Hence, the chance of phenotype fixation is much greater than when the novel phenotype first occurs as a result of a single allele.
- *Integration of the new trait with the remainder of the organism does not necessarily require additional mutations*: as an organism reacts with plasticity to changes in the extra-organismal environment, so do traits to their intra-organismal environment (i.e., to other traits in the same organism). Often, this plasticity is ancestral and tested by natural selection to enhance the chances of an adaptive outcome. This dramatically reduces the number of independent mutations that are necessary for the origination of a complex adaptive trait. For example, the size of the liver in mammals is directly regulated by the amount of skeletal muscle in the body. A mutation that

---

[1] Even in the case of a random mutation, the number of individuals affected by it does not need to be equal to one. A larger number of individuals can be affected if the mutation occurred in a germline cell a few generations prior to meiosis. If this is the case there will be more than one gamete carrying the new mutation and more than one offspring potentially can have the mutant allele (Woodruff and Thompson 1992, *Journal of Evolutionary Biology* 5: 457–464).

increases the muscle mass does not require another mutation to adapt liver size to meet the increased metabolic demands of more muscle cells.

These points affect any new trait, whether it is a modification of an existing character or whether it leads to a novelty in the sense used here. Nevertheless, these points address a number of problems when one attempts to explain novel characters.

One problem often cited is the question of whether an organism with a novel phenotype will find a partner for sexual procreation. A novel phenotype might be incompatible sexually with the ancestral type. For example, mating among gastropods is facilitated if the two partners have the same shell coiling. A single individual with alternative shell coiling will have a hard time producing offspring. But if the environment causes a higher number of alternative-coiling morphs, the chances of producing a new generation of differently coiled snails are much greater and so is the chance of eventually fixing this phenotype within the population. In general, because environmentally induced novel phenotypes will arise in more than one individual, any degree of assortative mating will enhance the chance for the eventual fixation of this character, regardless of whether or not the new phenotype is compatible with the ancestral wild type.

Naturally, examples of environmentally induced traits that qualify as modifications of existing characters are more numerous than those that can potentially become a novelty (i.e., a new individualized body part or leading to a new variational modality). One example is the induction and genetic assimilation of skin callosities. Local keratinization of the amniote skin can arise from localized mechanical stimulation of the skin. But it can also be congenital, as is the case for the ostrich that hatches with calluses at those sites on the body where the animal will touch the ground when resting (e.g., Waddington 1957).

This is certainly a striking potential example of a genetically assimilated trait and one that could be a novel individualized part of the body. However, because there is only one species of ostrich left, it is difficult to say when this genetically assimilated character arose and whether this character behaves like a homolog (i.e., was inherited among species and then evolved its own variations), even though this is likely.

The question we need to ask here is whether an environmentally induced and genetically assimilated callosity would be expected to have the developmental and genetic individuality necessary for it to be an individualized body part. The answer is that we do not know because, to my knowledge, these types of questions have not been investigated experimentally. But we can ask what the currently accepted model for genetic assimilation implies with respect to this question.

Current thinking, which dates back to Waddington (Waddington 1942), is that cells that start to spontaneously keratinize more strongly than in other parts of the skin evolved (through natural selection) a sensitivity to the local conditions in the parts of the body where they develop. For example, it could be that the skin over the sternum and the pelvis (i.e., where the callosities of the ostrich develop) is already stretched and bent more than that of other parts of the body in the embryo, and that this becomes a sufficient signal for these cells to keratinize. It could also be that there are other factors that induce local keratinization and that they may be related to regional differences in vascularization or some unequal distribution of signal molecules produced by the dermis. Given our current understanding of skin development, some scenario of this kind is plausible.

Development of a callus does not require that the cells that form the callosity are different in their developmental potential than any other part of the skin, and they do not need to execute a developmental program that is different from those of any other epidermal cells. Nevertheless, because there is almost certainly some genetic variation in the amount and the spatial distribution of those factors to which skin cells will react, the size and shape of the callosity will display some amount of additive genetic variance. If true, one can view this as evidence of incipient individuality, and that natural selection is able to modify these callosities.

Variational individuality of the kind necessary for adaptive modification in response to natural selection does not require a full-blown character identity network, as discussed in chapter 3. Variational individuality is probably a precursor of developmental individuality for which a group of cells has its own distinct developmental regulatory network that executes a unique developmental program that, as soon as it is triggered, is largely context insensitive.

Release of Cryptic Genetic Variation

The relationship between genetic and environmental changes is also complicated by the fact that environmental changes often result in the release of cryptic genetic variation. That is, genetic variation that was neutral with respect to the phenotype wthin the old environment becomes expressed as phenotypic variation within the new environment. The release of genetic variation due to changes in the environment or due to a major mutation has traditionally been seen as a sign of genetic canalization of the wild type in the old environment (Waddington 1957).

As discussed in the previous chapter, however, the release of cryptic genetic variation is a generic property of genotypes that display non-linear relationships between genetic and/or environmental factors and the phenotypes (Hermisson and Wagner 2004). When a character is under stabilizing selection for some extended period of time, some genetic variation will accumulate that is neutral within this environment. If, then, the environment changes and at least some of the formerly neutral genetic variation is expressed within the new environment, the genetic variance of this character will be greater after the environmental change than it was before. It has been shown that the genetic variance always increases, even if the genetic canalization[2] remains the same. A metaphor may explain this fact.

One can think of new mutations as dust that falls at a constant rate onto the floor. Stabilizing natural selection can then be viewed as a vacuum cleaner that removes this dust from the floor. Genetic robustness is then like a rug through which the dust can settle, but which prevents the vacuum from removing it. This means that the "dust" is not seen by natural selection and is, thus, "neutral." An environmental change can then be viewed as a change in either the size or the location of the rug.

If, however, the rug is shifted, the degree of genetic canalization remains the same (i.e., it still covers the same floor area in terms of square feet), but because its position has changed, it reveals some of the dust that had accumulated under the rug. This means the genetic variation that is now visible to selection has increased, even though the area of the floor that is covered by the rug remains the same. Hence, a change in the environment will always lead to a temporary increase in the genetic variance of the phenotype, regardless of whether or not genetic canalization remains the same.

The best understood mechanism for cryptic genetic variation that can be released in a new environment is the heat shock protein systems involving Hsp90 (Nathan, Vos et al. 1997; Rutherford and Lindquist 1998; Queitsch, Sangster et al. 2002). Heat shock proteins are those that associate with other proteins and aid in stabilizing their structures. Generic heat shock proteins are specifically expressed at higher temperatures in order to protect the organism from this environmental insult. However, there are also Hsp's that are expressed under normal temperatures, like Hsp90.

The function of Hsp90 is to stabilize proteins that generically have to be structurally unstable, like those proteins that change their structures in response to small molecules (e.g., in signaling pathways). If the efficiency of the Hsp90 protein is decreased, either by introducing a mutation in the Hsp90 gene or by a pharmacological agent (e.g., geldanamycin), one observes a release of cryptic genetic variation. This can also occur in response to any kind of environmentally induced stress, such as when the amount of Hsp90 becomes saturated with destabilized proteins.

Released cryptic variation more or less affects all parts of the phenotype (Nathan, Vos et al. 1997; Rutherford and Lindquist 1998; Queitsch, Sangster

---

[2] One can measure the genetic canalization by the mutational variance, V_m (i.e., the amount of genetic variance that enters a population by mutation per generation, assuming a constant mutation rate). The larger this value is, the lower will be the genetic robustness or canalization of the character.

et al. 2002). There are complex dynamics between those environmental and genetic factors that affect the phenotype. This blurs the directionality of causation between genetic and phenotypic change. The phenotype is changed due both to the presence of an environmental stimulus (e.g., heat shock) and genetic variation.

The release of cryptic genetic variation is often viewed as a mechanism that increases evolvability in the face of environmental change (Rutherford 2000; Rutherford 2003). This is plausible because, as Rutherford has shown, the released amount of genetic variation is not burdened with strong, unconditionally deleterious pleiotropic effects and, hence, at least some of this genetic variation is potentially adaptive (Milton, Huynh et al. 2003; Milton, Ulane et al. 2006).

The question that concerns us here, however, is whether the release of cryptic genetic variation enables the evolution of novelties. No empirically based answer appears to be available at this point because this question has, to my knowledge, not been addressed with data. However, one way to address this issue is to ask whether environmentally induced stress can reveal or enable the kind of genetic variation that is required to restructure a gene regulatory network.

The question of what kinds of mutational mechanisms are responsible for rewiring gene regulatory networks will be addressed in chapter 6. There it will be argued that one of the major sources of genetic variation in the structure of gene regulatory networks is the activity of transposable elements. Transposable elements generically carry a number of transcription factor binding sites that, when inserted into a genomic region, can have an effect on the expression of nearby genes. In fact, it has been found that the rate of transposition of some transposable elements is responsive to stress (Strand and McDonald 1985).

For example, the Drosophila transposable element Copia can be induced by heat stress because it has a heat shock–like promoter in its long terminal repeats. Whether this mechanism and the phenomenon of stress-induced transposition activity is a general fact is not clear. But at least this type of mutation that is critical for the evolution of the structure of gene regulatory networks is potentially stress inducible and, thus, environmentally regulated.

Where Does the Positional Information for Novel Characters Come From?

During development, cells need some signal that "tells them" when and where in the embryo they should organize to form a certain part of the body. This signal can be intracellular, such as inherited cytoplasmatic factors that were asymmetrically distributed in the egg cell and were then passed on differentially to the daughter cells. Often the signal is emitted from a neighboring cell population, a phenomenon called embryonic induction. Or, positional information can be derived from differential gene expression arising from pattern formation in the precursor field from which the body part develops. The latter occurs during Drosophila body axis development when differential gene expression in the syncytial blastoderm endows the cells with positional information along the anterior-posterior axis of the embryo.

Positional information signals are absolutely critical for the development of an organ or body part. Any knockout of an inductive signal results in the loss of the body part in affected individuals. For example, the loss of gap gene function in Drosophila results in the loss of a group of segments. Thus, it is justified to consider the inductive or the positional information signals as *the* or at least *a* cause for the development of the body part. After all, such a signal is, in a specific organismal context,[3] necessary for character development and, in a certain operational sense, also sufficient (i.e., in certain situations, applying the signal is sufficient to trigger the development of the character).

Thus, an inductive signal and organ development can be seen as having a cause-effect relationship with each other. Therefore, it is tempting to speculate that the evolution of positional information signals may also have been the evolutionary cause for the origin of the body part. After all, I just agreed that they can be viewed as "*a*" cause for the development of the body part. Why then should the developmental cause not also be the evolutionary cause for the origin of the character? With a few exceptions, to be discussed in the next section, however, this line of reasoning seems to be flawed.

The main reasons why this idea is flawed were discussed in chapter 3, in which the irrelevance of positional information signals for the homology of characters was explained. Briefly, the outcome of a developmental process is determined by the developmental predisposition of the cells that react to an inductive signal, and not by the inductive signal itself. For this reason, inductive signals and other positional information mechanisms are replaceable[4] and, therefore, are not closely linked to character identity. More important, there are directly relevant empirical facts that contradict the idea that the evolution of positional information is a factor in the origin of novel body parts.

An example that illustrates that positional information signals can be much older than the characters that depend upon them is the relationship between the zone of polarizing activity (ZPA) in the paired appendages of vertebrates and the development and evolution of digits. Limb development

---

[3] This qualification is necessary because inductive signals can be mimicked by artificial perturbations and are thus not strictly necessary in a broader sense.

[4] The replacability of inductive signals extends to both experimental intervention as well as evolutionary changes.

and the ZPA are a paradigm for both developmental biology and evolutionary biology and will be discussed with respect to various aspects in this and other chapters.

The ZPA is a signaling center that was first discovered because of the effect of transplanting cells from the posterior edge of the chicken wing bud to an anterior location of another wing bud.[5] These transplants often result in a mirror image duplication of the digits in the hand. It was later discovered that the ZPA is a group of cells that produce a molecular signal, a protein called Sonic HedgeHog (Shh), which plays a critical role in determining the number and the identity of digits (Riddle, Johnson et al. 1993). A complete knockout of the *Shh* gene (*Shh-/-*) results in the complete loss of digits in the hand and to only one digit in the foot of the mouse (Chiang, Litingtung et al. 2001). It may even be inaccurate to speak of the loss of digits in the hand and the foot because, in fact, all autopod structures seem to be absent in *Shh-/-* mice.

What is interesting, though, is that digits are the actual innovation of the fin-limb transition (see chapter 10). It is interesting to ask: what role, if any, did the evolution of the ZPA have for the origin of digits? It turns out that all vertebrates that have paired appendages (i.e., bony fish, with tetrapods being technically part of bony fish clade, as well as the sharks) have a ZPA (i.e., a population of cells at the posterior margin of the fin bud that expresses *Shh*). Based on this taxonomic distribution of the Shh-expressing cells, it is clear that at least the most recent common ancestor of all jawed vertebrates had a ZPA.

Even though *Shh* is essential for the development of a derived character (the digits of tetrapods), the ZPA is much older than the tetrapod clade. The ZPA cannot be a factor for the origin of digits. Digits arise based on an existing signaling system, including the ZPA and other signaling centers. Digit development, thus, depends on *ancestral* signaling systems rather than originating from a novel signaling center. These signaling centers had a passive role during the origin of digits. The functional locus of the genetic changes that resulted in the origin of digits had to be in the gene regulatory network that reacts to these ancestral signals—Shh for example.

A classic example for the principle explained in the previous paragraph is the induction of skin structures on the face of an amphibian larva. This example dates back to the early years of experimental embryology during the 1920s and 1930s. I cite this example here because I want to show that this principle was already apparent in the classical experimental embryological literature.

In 1932, Hans Spemann and Oscar E. Schotté published a paper in the journal *Die Naturwissenschaften* (Spemann and Schotté 1932) on their

---

[5] This experiment works only with wings, but not with the foot. The likely reason is that the wing has a reduced number of digits and, thus, an amount of uncommitted mesenchymal cells that are competent to form digits.

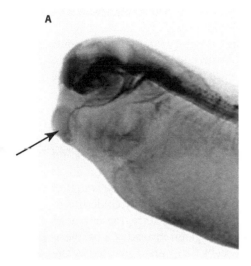

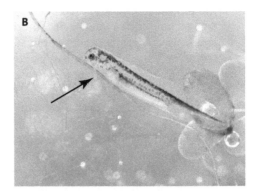

FIGURE 5.1: Larval adhesion organs of amphibians. A) Anuran larva with cement gland (arrow). B) Salamander with "balancers." These two structures are not homologous, but they are still induced in salamanders and frogs by the same mesenchymal signals. This example shows that the same positional information signal can be recruited into the development of unrelated characters. Figure 5.1B is reproduced by permission of Mark Aartse-Tuyn.

transplantation experiments in which they had transferred epidermis between frog and newt larvae. They showed that the inducing signal was the same between frogs and newts, but the morphological results were determined by those cells that received the signal.[6] Anuran and urodele larvae differ in the kinds of "ancillary organs" that young larvae may develop (figure 5.1). I call ancillary organs those specialized structures that help the hatchling to survive until their limbs or body axis matures to a degree so that the larva can control its body position. The anuran tadpoles produce adhesive discs, also called "suckers" or cement glands, while the larvae of some urodeles produce elongated "balancers." Balancers are transient structures that may have multiple functions, including keeping the larva upright for feeding

---

[6] A series of experiments regarding the development of balancers had already been published in 1925 by Ross G. Harrison at Yale University in the *Journal of Experimental Zoology*, in which Harrison, among others, described an experiment during which *Ambystoma punctatum* epidermis was transplanted onto *Rana sylvatica* embryos. In these chimeric larvae, imperfect but clearly recognizable balancers developed. Hence, one set of experiments that made the Spemann/Schotté paper so famous had already been published seven years prior to when the 1932 Spemann/Schotté paper appeared. The 1925 Harrison paper, however, was not cited in the Speman/Schotté paper in 1932. This is a curious fact since Schotté must have been in contact with Harrison when his paper was published, because he started working in the Harrison lab in 1932 and stayed until 1934 as a Sterling Research Fellow (Lilversage 1978).

until the tail musculature has matured. They also may play a role as adhesive organs and sensory organs (Crawford and Wake 1998).

Spemann and Schotté performed a number of transplantation experiments with toad and newt embryos. In these experiments, they first transplanted the prospective belly ectoderm ("skin") of each species into the prospective face region of the same species and showed that, during this stage, prospective belly epidermis could participate in the formation of facial structures, including balancers or cement glands and teeth, depending on the species. This result suggested that other tissues of the developing oral region induced the naïve belly epidermis to form balancers or cement glands in urodeles and anuran larvae, respectively.

What was really surprising, though, was that xenoplastic transplantation experiments also worked. For example, if the prospective belly epidermis of a frog embryo was transplanted into the prospective facial region of a newt embryo, the newt larva would develop a cement gland instead of a balancer. This meant that the reactive cells (i.e., those cells that were instructed by the inductive signal) determined the outcome—in this case, the transplanted frog cells on the newt face. It also followed that the inductive signal in the newt could induce the frog cells to form the anuran cement gland. Reciprocal experiments also worked; that is, transplantation of a prospective newt belly epidermis onto a frog's prospective facial region resulted in a frog larva with a balancer. These were not isolated observations, as they created a cottage industry of work with a large amount of replication and confirmation (Holtfreter 1935; Rotmann 1935; Holtfreter 1936; Schotté and Edds 1940).

The molecular factors that are responsible for inducing these facial structures are best understood for the cement glands of *Xenopus*. From work with *Xenopus*, it is clear that the same inductive signals that cause the formation of the stomodeum (mouth) are also involved in inducing the cement gland (see, for example, Wardle and Sive 2003).

To my knowledge less, if anything, is known about the molecular factors that induce balancer development in the newt, although a few speculations based on the available data are plausible. First, both newts and frogs develop mouth openings, as do all other deuterostome animals. If the inductive signals in *Xenopus* for stomodeum and cement glands overlap, it is plausible that the signals for stomodeum development in the newt are also used to induce balancers. This assumption would explain the Spemann/Schotté/Harrison experiment; newt epidermal cells could form balancers on a frog face.

From what is known about the frog, the molecules implicated in these signals are pretty generic, including BMPs, Wnt, FGF, and various hedgehog signals (Wardle and Sive 2003). What determines the position of cement gland development is the combination of inductive and inhibitory signals that overlap in a specific region of the face (Bradley, Wainstock et al. 1996; Sive and Bradley 1996). The overlap of these signals with those that induce the mouth opening makes it plausible that a similar or even the same configuration of signals is present around the prospective mouth opening of a newt larva and that the balancer-forming cells react to the same molecular signals as the cement gland–producing cells in the frog (although in a different combination, as the facial positions of these structures are not the same).

To put these results into an evolutionary context, it seems plausible that both anurans and urodeles have a similar set of signals that cause their (homologous) mouth openings and that in the frog and newt lineages, these signals were independently co-opted into inducing derived (non-homologous) facial structures: the cement gland and the balancers. Of course, this scenario assumes that the common ancestor of frogs and newts did not have a facial structure that was homologous to both balancers and cement glands.

There is agreement in the literature that balancers and cement glands are not homologous (Duellman and Trueb 1986; Crawford and Wake 1998). This conclusion is based on embryological and phylogenetic evidence. For example, Lieberkind (1937) showed that cement glands and balancers were derivatives of different branchial arches. Balancers developed on the mandibular arch and cement glands on the hyoid arch.

The comparative evidence arguing against the homology of cement glands and balancers has multiple layers. First, there are no urodele larva with cement glands and no anuran larva with balancers (Schotté and Edds 1940), which is consistent with independent origination of these structures. Within urodeles, balancers are known from two clades that are not closely related. Balancers are found in Hynobiidae and in the clade formed by the Salamandridae and Ambystomidae. The most parsimonious scenario, using the phylogenetic hypothesis of Frost et al. (Frost, Grant et al. 2006, p. 115, Figure 53), suggests two independent origins within the urodeles (figure 5.2), which implies that the most recent common ancestor of urodeles did not have balancers or similar organs.[7] In addition, no balancers are known for Australian lungfish larvae (e.g., Kemp 1986), and the adhesion gland in the African lungfish is found more caudally than those of frogs. Finally, there are no comparable structures in the egg-laying caecilians, which do not have aquatic larvae (e.g., Dunker, Wake et al. 2000). Hence, the available comparative and embryological evidence is consistent with the assumption that

[7] The possibility favored by Crawford and Wake (1998) that the balancers in Hynobiidae and Salamandridae are homologous is also not excluded based on this evidence. Assuming a single origin and multiple losses has only two more steps (one gain and three losses) than the most parsimonious scenario. Whether the assumption of a single origin (Dollo) implies balancers in the most recent common ancestor of Urodela, however, critically depends on the phylogenetic hypothesis. The phylogenetic hypotheses of Frost et al. (2006) and Wiens et al. (2005) imply balancers in the root node of Urodela, while the phylogenetic hypotheses of Larsen and Dimmick (1993) and Weisrock et al. (2005) suggest an origin of balancers within the urodele clade. In the latter case, the balancer could not be homologous to cement glands, as there would be no historical continuity between balancers and anything ancestral to the most recent common ancestor of urodeles.

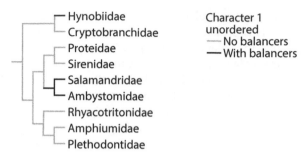

FIGURE 5.2: Phylogenetic distribution of balancers in urodeles suggests two independent origins of balancers, which makes it unlikely that cement glands in anurans and balancers in urodeles are homologous. (Tree after Frost et al. 2006).

urodele balancers and anuran cement glands are independent innovations in urodeles and anurans, respectively.

The model that results from these examples (e.g., fingers and the ZPA, as well as balancers and cement glands), suggests that novel structures seize upon existing signals in a part of the embryo or larvae to direct the development of a novel structure to its appropriate position. The role of positional information can best be described as passive or enabling, but not as active or specific. Novel structures can evolve without significant changes in the signals that, in the derived state, are their inducing agents. There is, however, an important exception: a possible role of mechanical stimuli for inducing the differentiation of skeletal tissues. This is the topic of the next section.

Derived Mechanical Stimuli and the Origin of Novelties in the Avian Hind Limb Skeleton

In this section, I want to describe the empirical basis for a model of the origin of novelties that emphasizes novel inductive stimuli as the mechanistic basis for the origin of novel structures: the side-effect hypothesis (Müller 1990; Newman and Müller 2001). This idea was based on a number of papers that explained the development and evolutionary origin of a skeletal structure in the bird and theropod lower leg: the *syndesmosis tibiofibularis*. This is a cartilaginous or bony crest of the tibia that connects the fibula of bird legs to the tibia (figure 5.3).

A bird's leg is characterized by an elongated tibia (shin bone) that has elements of the tarsus incorporated at its distal end and, for this reason, is called *tibiotarsus*. Compared to the *tibiotarsus*, the *fibula* is reduced, in particular at its distal end where it has completely lost its distal epiphysis and does not participate in the tarsal joint. In 1959, a French embryologist, Armand Hampé in

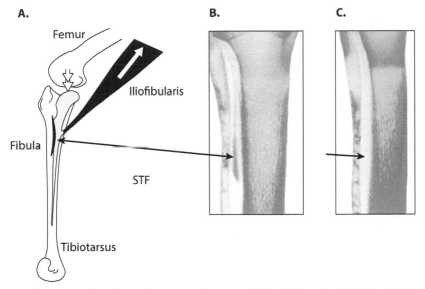

FIGURE 5.3: Example of a novel anatomical structure induced by mechanical stimuli during development. The acronym STF indicates the *syndesmosis tibiofibularis*, a connective tissue bridge between the tibia and the fibula in the bird leg. A) The distal end of the bird fibula is reduced, but the proximal end remains as the insertion point for the strong iliofibularis muscle. To compensate for the force derived from the iliofibularis muscle, the fibula is attached to the tibia through the STF. B) The cartilage anlage of the STF in normal development. C) In animals paralyzed in ovo, the STF does not develop, which shows that mechanical stimulation by the iliofibularis muscle is necessary for STF development. [A) After Müller and Streicher 1989, *Anatomy and Embryology* 179:327–339. B & C from Müller 2003, *Evolution and Development* 5:56–60, reprinted by permission of John Wiley and Sons.]

the Laboratory for Experimental Embryology of the CRNS in Paris, published a series of experiments in which he inserted a mica barrier between the presumptive tibia and fibula cell populations and observed a phenotype in which the tibia and the fibula had equal lengths (figure 5.4). This was a startling result because it looked like experimental re-creation of a "reptilian"-like configuration (i.e., an experimentally induced atavism; Alberch and Alberch 1981; Hall 1984). This result was enthusiastically embraced by the nascent community of developmental evolution researchers during the 1980s for whom developmental mechanisms that could create or re-create phylogenetic transformations were the focus of much attention. Hampé's experiments were frequently cited in reviews, but were never followed up.

A young anatomist at the University of Vienna Medical School, Gerd Müller, finally repeated Hampé's experiments and published a detailed account of these effects in the *Journal of Evolutionary Biology* in 1989 (Müller 1989). What Müller found was that the "elongation" of the fibula in this experiment was, in fact, only relative because the experiment had caused a length

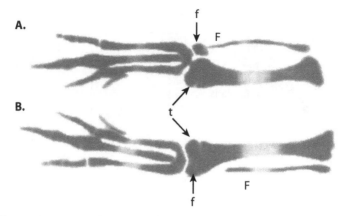

FIGURE 5.4: A) Experimentally altered chicken leg. Note that the fibula (F) is longer than in the control (B), and the fibulare (f) remains separate from the tibia. B) Normal chicken leg. Redrawn by permission after Müller, 1989, *J. Evol. Biol.* 2:31–47. Copyright © 2002, John Wiley and Sons.

reduction of the tibia rather than an elongation of the fibula. A careful description of the effects of this experiment revealed a number of traits that, in fact, were atavistic. Among these was a loss of the *syndesmosis tibiofibularis* (STF), which is an ancestral character for birds.

Following up on this result, Müller developed his "side-effect hypothesis" for the origin of evolutionary novelties (Müller 1990). For this reason, it is worth considering the merits of this idea in some detail, not only because of its relevance to the issue of novelties, but also because the "epigenetic" view of Müller's account differs in outlook from the more genetic view proposed in the remainder of this volume.

What Müller realized was that, while the distal part of the bird fibula was greatly reduced, the proximal part was never lost. A possible functional reason for this is that the proximal fibula in extant birds is still the insertion point of an important muscle, the *iliofibularis* muscle (MIF). The idea is that as the fibula loses its distal anchorage at the tarsal joint, the force from the pull of the iliofibularis muscle has to be compensated by some other structure. This structure is the STF in modern birds and theropod dinosaurs that connects the remaining proximal portion of the fibula to the massive tibiotarsus (Müller and Streicher 1989; Streicher and Müller 1992). If this scenario is correct, then the most parsimonious mechanism that produces the STF would be the mechanical stress from the MIF that directly causes the development of the STF.

In support of this scenario, Müller and Streicher found that the ontogeny of the STF followed the typical sequence of a pressure-induced ossification in a tendon. First, there was a fibrous connection between the tibia and the fibula in which cartilage formed later. In turn, the cartilage was replaced by an osseous outgrowth from the tibia. Consistent with a model that this sequence of events depended on the spatial arrangement of the bone and, thus, the mechanical distribution of stress, the formation of the structure stopped at early stages when the distance between the tibia and fibula was experimentally increased (Müller and Streicher 1989; Streicher and Müller 1992). Chick embryo motility did start at the time of STF formation (Wu, Streicher et al. 2001) and paralysis experiments resulted in the loss of the STF (Muller 2003) (figure 5.3B and C).

From this evidence it is clear that the STF belongs to a class of skeletal characters whose development depends on a mechanical stimulus. For example, the human kneecap develops in response to pressure in the tendon that runs over the knee (Muller 2003). The formation of a cartilage in response to pressure is an ancestral reaction norm of amniote connective tissue and is not the cause for the origin of novelties. In these cases, the evolutionary event that caused the origin of a novel structure was changes in the distribution of mechanical stress during the development of the skeleton. In turn, these mechanical changes are due to evolutionary changes in the sizes and shapes of bones.

This well-supported scenario is the inverse of that explained in the previous section for which the inductive stimulus was ancestral and the derived characters were due to a derived reaction to the stimulus. In the case of the STF and similar structures, the reaction norm is ancestral and the derived appearance of the mechanical stimulus (pressure) in a new location is the actual causal agent for the origin of the novel structure.

From the perspective developed in this book, however, the Newman/Müller (Newman and Müller 2005) scenario for the origin of novelties is incomplete, as it does not explain the eventual developmental/genetic individualization of the novel structure. It is possible that these types of structures individuate by two different modes. Perhaps there is adaptive pressure to make their development more reliable and, thus, independent of the mechanical stimulus. This would be a case of selection for developmental robustness (see chapter 4 for a discussion of this idea). Another possibility is that the structure is selected to assume a specific shape, which would increase its functional performance in one way or another. This scenario could also lead to a character-specific developmental program that would lead to the developmental/genetic individuality of the structure. This would make the epigenetically induced structure into an individualized character rather than simply the outcome of an ancestral reaction norm.

The Origin of Character Identity Networks

In chapter 3, the ChIN concept was introduced as a conserved gene regulatory network that translates the positional information signals into the expression of genes and results in the character's phenotype. ChINs enable the differential gene expression that is necessary for developing individualized body parts. Consequently, the evolutionary origin of ChINs might be best understood as an adaptation that prevents the expression of genes that are adaptive in one organismal context, but that are harmful in another context. In chapter 4 it was argued that the immunosuppressive function of endometrial decidual cells during pregnancy in placental mammals would likely be deleterious if expressed in other parts of the body, say in the gut or the nasal cavities, where infection is a constant threat. Immune suppression, however, was essential as soon as the placenta became invasive in the stem lineage of placental mammals.

Because ChINs are the mechanistic interface between positional information signals and the development of the phenotype, an explanation of their origin must include three components:

- How does the appropriate set of target genes come under the control of the gene regulatory network of the cell or the cell population?
- How is the expression of target genes linked to those signals that ensure differential expression within the appropriate spatial and temporal context?
- How does the gene regulatory network acquire the quasi-autonomy that is characteristic of individualized body parts and cell types?

Although I was unable to come up with a scenario that could explain all these aspects, because there seemed to be too many possibilities, this set of questions will frame my discussion on the examples in the second part of this book in which I will discuss a number of examples of the evolution of character identity. Here, I will limit myself to a few qualifying remarks.

The acquisition of a unique set of target genes is certainly central to the evolution of some novel cell types; however, this might not be universal. For example, the evolutionary origin of a novel skeletal element of a vertebrate, as described in the previous section, probably used the set of cell types that were already used in cartilage and bone development. Thus, the acquisition of target genes might not be an important aspect of skeletal element identity as it is in other cases. Also, the phrase "acquisition of target genes" might suggest an active recruitment of genes under the control of a set of signals. Yet this also does not need to be the case, even in novel cell types. For example, it seems plausible that the evolution of a specialized cell type, a muscle cell say, from a more generalist cell, say an epithelio-muscle cell, might result from the differential suppression of gene expression rather than from the acquisition of target genes (see chapter 8 for more details).

The relationship between differential gene expression and signals is essential for understanding character identity. Signals can directly modulate or initiate gene expression in order to match an organism's physiological status to its environmental conditions. The most famous example is the induction

of metabolic enzymes by the presence of nutrients, as for example the induction of lactose-processing enzymes due to the presence of lactose. Other
examples are the induction of heat shock proteins by elevated temperatures
and many more.

In these cases of physiological modulation, the signal directly controls
gene expression, and the response, in most cases, depends on the sustained
presence of the signal. Alternatively, a signal can act as a trigger whereby the
signal sets in motion a process that, ultimately, no longer requires the signal
to continue. This type of signal–target gene relationship seems to be more
frequent or, at least, more typical of developmental gene regulation.

Developmental signals initiate a developmental cascade of events that unfolds largely independently of the signal. Trigger signals are often difficult to
identify experimentally because the effect of triggering signals can easily be
mimicked by non-physiological perturbations (see chapter 3 for a discussion
of this fact). This type of gene regulation is also the basis for the surprising
effects of "master regulatory genes" in which the incorrect expression of a
single gene can cause a complex developmental program to be deployed (e.g.,
Pax6 and Drosophila eye development).

In any case, in contrast to the modulating signals involved in physiological
adaptation, these stimuli cause the expression of genes that can sustain differential gene expression in the absence of the signal itself. Hence, the evolution
of a ChIN involved establishing regulatory relationships that could maintain
differential gene expression, even in the absence of the initiating signal. One
example was extensively discussed in chapter 3: the cross-regulatory relationships between the regulatory genes underlying Drosophila compound
eye development. In this case, the core mechanism is a small network of
cross-regulatory genes that mutually sustain each other's expression and
which jointly regulate downstream target genes.

In recent years, a model system for the origin of a ChIN has emerged
from the lab of Artyom Kopp at the University of California, Davis (Kopp
2011). This model is the origin of sex combs, a character on the first leg pair
of Drosophila males that is restricted to a subset of Drosophila species. These
results will be discussed in chapter 6.

The Evolution of Novel Signaling Centers

The ChIN concept was introduced in order to contrast the developmental
mechanisms of differentiation to those of positional information. Because
many positional information signals are chemical in nature and often produced by localized cell populations (i.e., signaling centers), it may appear
that ChINs cannot include cell-to-cell signaling or the activity of signaling centers. But this cannot be true because ChINs direct the execution of

organ-specific developmental programs. These programs, however, have to include pattern formation (i.e., the spatial regulation of gene expression). But spatial gene regulation is dependent on cell-cell signaling, as shown for example in the putative arthropod body segment ChIN discussed in chapter 3, the so-called segment polarity network. The segment polarity network includes at least two signaling pathways, the hh (hedgehog) and the wg (wingless) pathways, to set up segment boundaries. Clearly, ChINs have to include signaling pathways and the relationship between positional information and the role of signaling in ChINs needs to be further clarified.

I propose distinguishing between positional information and pattern formation in the following way. A signal is a positional information signal if its role is to trigger the activity of a ChIN or, more generally, to trigger a self-organizing developmental cascade. A chemical signal will induce pattern formation if it is part of the local self-organization network of a group of cells that develop into an organ rudiment. The distinction between positional information (PI) and pattern formation (PF) is, of course, not genetic because the same signaling pathways can be used during either activity. Furthermore, the distinction is relative and hierarchical.

For example, the role of the ZPA and Shh in determining digit identity is that of a PI signal for digit development, as argued above in the section on the role of PI in the origin of novel characters. On the other hand, in chapter 10 it will be argued that the ZPA is probably part of the ChIN of the vertebrate paired appendages (fins and limbs; recall that limbs and fins are character states of the same homolog, the paired appendage in vertebrates). Hence, the same signaling center can provide PI to one character (digits) and be part of the ChIN of a higher level entity, the limb or fin, which may or may not include digits.

In other words, if a character contains individualized characters as parts, like the limb does with digits, then parts of the ChIN of that character can be a PI signal for its component parts (digits). With this in mind, we now consider what is known about two signaling centers that arose in the context of two novel morphological characters: the butterfly eyespot and the turtle carapace.

The Butterfly Eyespot Organizer

Butterfly wing eyespots develop around a signaling center in the wing imaginal disc, which expresses the transcription factors Dll, spalt, engrailed, cubitus interruptus (and thus has signs of hh signaling) and Smad, which is part of the TGF-B signaling pathway (Brunetti, Selegue et al. 2001) (figure 5.5). This center seems to emit two diffusible signals, Wnt and dpp, which belong to the TGF- family of signals. These signals activate a cross-regulatory network of genes that is responsible for the placement of the different pigments on the wing surface.

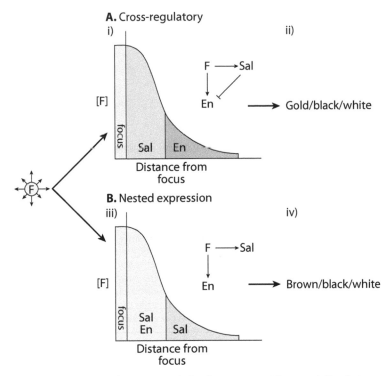

FIGURE 5.5: Two gene regulatory networks leading to concentric transcription factor expression patterns that cause the expression of different pigments around the signaling center, the eyespot organizer. (Redrawn after Brunetti et al., 2001, *Current Biology* 11:1578–1585).

Specifically, around this focus one finds *Dll* and *sal* expression, which induces the expression of black pigment, and around it one finds a ring of *en* expression, which induces the deposition of a gold-colored pigment. Outside the gold ring there is another ring of dark pigment, although the transcription factor that induces this pigment deposition is not known. This scenario explains the situation in the butterfly *Bicyclus anynana*. There is experimental evidence that the exclusive expression of dark- and gold-colored pigment is ensured by a mutual repression between *sal* and *en*. This interaction seems to be lacking in *Precis coenia*, in which the gene expression domains and the pigment rings overlap.

There are currently two theories for the evolutionary origin of the eyespot gene regulatory network. Both trace their origins back to the components of the ancestral gene regulatory network of the insect wing. One that was proposed by Keys and colleagues (Keys, Lewis et al. 1999) made the argument that the eyespot signaling center was derived from the signaling center at the anterior-posterior compartment boundary of the wing (figure 5.6). In the

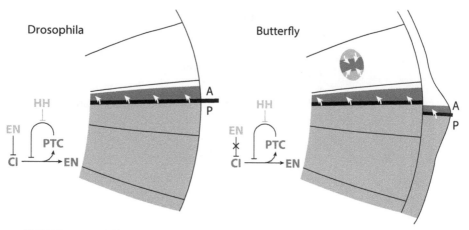

FIGURE 5.6: Model for the origin of the eyespot signaling center in butterflies after Keys and collaborators (1999). Left: signaling network responsible for the anterior-posterior compartment boundary in the wing. Right: nearly the same network is active in the eyespot organizer, except for an inhibitory link between *engrailed* (*en*) and *cubitus interruptus* (*ci*). Hence, this model suggests that the eyespot organizer gene regulatory network evolved by co-option and modification of an ancestral gene regulatory network, which in this case was the anterior-posterior compartment boundary network. (Redrawn after Keys et al., 1999, *Science* 283:532–534.)

insect wing, the posterior compartment is characterized by the expression of the transcription factor en, which induces the expression of the signaling molecule hh. The effect of the hh signal in the posterior compartment is suppressed because en inhibits the expression of the proximal effector of hh signaling, *cubitus interruptus* (*ci*).

However, at the border of the posterior compartment, the hh signal diffuses into the anterior compartment in which hh signaling is executed because there is no en to suppress *ci*. The action of *ci* then leads to the expression of the dpp signal, which creates an elongated signaling center just anterior of the *en* expression domain (i.e., anterior of the posterior wing compartment). The dpp signal, in turn activates (among other genes) the transcription factor gene *spalt* (*sal*).

Keys et al. reported the expression of many of the components of the a-p signaling center in the developing eyespot and concluded that the evolution of the eyespot resulted from the recruitment of the anterior-posterior signaling network by removing the inhibition on hh signaling in the presence of en. This was a necessary assumption because eyespots also develop in the posterior wing compartment, where *en* is expressed. This model explains many features of the eyespot gene regulatory network, but it fails to account for the fact that the eyespot signaling center also includes parts of the proximo-distal signal with the expression of wingless (*wg*) and *Dll*.

In a later paper, Reed and Serfas (Reed and Serfas 2004) linked the evolution of the eyespot gene regulatory network to Notch/Delta signaling involved in the development of wing veins. Notch is a cell surface receptor for its ligand Delta (also cell surface bound). Ancestrally, expression of Notch was linked to the determination of inter-vein tissue. In addition, during eyespot development, Notch is expressed in a stripe that proximally ends in a node of Notch expression is the first molecular sign of eyespot development. Comparative data further show that eyespot Notch expression is derived from a stage during which Notch expression is confined to a stripe midway between neighboring veins. These data are consistent with the morphological observations that mid-vein patterns of pigmentation are phylogenetically older than eyespots and may be the morphological precursors of eyespots.

A comparison of the models proposed by Keys and collaborators (Keys, Lewis et al. 1999) and that of Reed and Serfas (Reed and Serfas 2004) shows that both provide evidence that a novel character recruits signaling systems and gene regulatory networks that are ancestrally deployed in the wing. It is too early to say whether the Notch/Delta and the en/hh systems play different roles in eyespot development. Intuitively it seems plausible that the en/hh including Dll and wg systems are part of the ChIN, which causes both pattern formation and deployment of the pigment pathways, while the Notch/Delta plays the role of a PI signal because it does not seem to be engaged in the later deployment of the actual pigmentation pattern. If the latter is true, then the evolution of the eyespot included the recruitment and re-patterning of an ancestral PI signal.

The Turtle Carapace

The turtle body plan is one of the great intellectual challenges that comparative anatomy has provided to evolutionary biology. Turtles have such a weird anatomical organization that its origin seems to defy Darwinian logic (Rieppel 2001) (figure 5.7)—that is, the idea that the transition between body forms is gradual and adaptive during each step. The reason that this idea is a problem with turtles is that not only is their body encased in armor, a feature they share with other vertebrates (like the crocodiles or the armadillo), but the way in which the shell is anatomically organized is unique and perplexing.

The dorsal armor, the carapace, consists of a number of different bony plates of diverse embryological origins. The midline is formed by a row of so-called neural plates, which are attached to and derived from the dorsal spines of the vertebrae. The main body of the carapace, however, is formed by the so-called costal plates, which are fused with and derived from the ribs. The real surprise is the relationship between the ribs and the shoulder girdle (e.g., the scapula, or shoulder blade). In all other tetrapods, the scapula

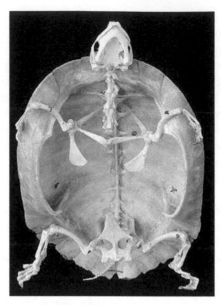

FIGURE 5.7: A major body plan innovation that has puzzled biologists for centuries. In the turtle, the shoulder girdle is located inside the carapace, which is derived from the rib cage. This arrangement is the inverse of what is seen in all other amniotes, in which the shoulder blade is external to the rib cage. It is difficult to conceive of a gradual transformation scheme to derive the turtle body plan from the ancestral and usual amniote body plan. Reproduced by permission from Kuratani et al., 2011, *Evolution and Development* 13:1–14. © 2011 Wiley Periodicals, Inc.

lies external to the rib cage and, hence, the scapula lies dorsal to the ribs. In turtles, however, the scapula lies ventral to the ribs and, of course, inside of the carapace, which is formed by the ribs.

Here is the problem. There are two known configurations: one has the scapula outside and dorsal to the ribs, like in all tetrapods except for turtles, and a second has the scapula ventral to the ribs. This is the problem for the Darwinian way of thinking: what evolutionary sequence can explain the origin of the turtle body plan in a series of small steps when the scapula can be only either outside or inside of the rib cage? What would the intermediate steps be? Matters are not helped by the fact that the fossil record does not reveal any intermediate morphologies. Turtles seem to be the ultimate hopeful monsters—a sudden, radical deviation from the ancestral body plan without any plausible or documented intermediate forms. What further confounds the situation is that even their phylogenetic affiliation with fossil forms is difficult to assign.

A number of scenarios have been advanced to explain the origin of the turtle carapace, although we cannot do justice here to this body of literature. However, what is understood to some degree is the developmental basis of

Chicken　　　　　　　　Turtle

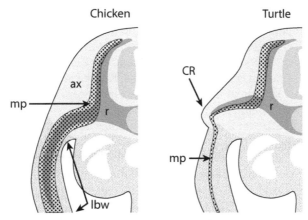

FIGURE 5.8: In the chicken, as in all other amniotes except for turtles, the ribs (r) grow ventrally from the axial compartment (ax) to the body wall. In the turtle, rib growth is arrested at the axial level in association with a signaling center, called the carapacial ridge (CR). Mp = muscle plate, lbw = lateral body wall.

the carapace. This research has shown that carapace development is guided by a unique signaling center: the carapacial ridge (CR). This is the reason why carapace development is discussed here.

Unlike the so-called ribs of teleost fish, the ribs of amniotes arise from the somites and then grow ventrally to encase the inner organs of the thorax. In contrast, in turtles, the ribs never turn ventrally; rather, they grow sideways and penetrate the dermal layer of the prospective skin. In this way the ribs, in a manner of speaking, grow "over" the developing shoulder girdle and create the odd topological relationship between the scapula and ribs. The derived growth trajectory of the ribs in turtles seems to be caused by an attractive signal from a derived signaling center: the previously mentioned carapacial ridge (figure 5.8).

This ridge forms the periphery of the future carapace and resembles, in some respects, the distal part of the limb bud (Burke 1989). The distal extremity of the limb bud is also formed by a signaling center: the so-called apical ectodermal ridge (AER). The AER is a ridge of elongated, pseudostratified epithelium that produces mostly fibroblast growth factor FGF8 signaling molecules. The AER forms at the boundary of the dorsal and the ventral compartments of the vertebrate embryo in response to a mesodermal FGF10 signal. This is essential for the distal outgrowth of the limb bud.

The carapacial ridge (CR) also forms at the boundary of the dorsal and the ventral compartments and is characterized by an epithelial ridge like the AER. Underneath the CR, the mesenchyme produces FGF10 (Loredo, Brukman et al. 2001) and displays early expression of Msx (Vincent, Bontoux et al. 2003).

There is also evidence for Wnt signaling (Kuraku, Usuda et al. 2005). But the molecular similarities between a limb bud and the CR remain limited. For example, there is no evidence of FGF8 expression in the epithelial CR cells and the obligatory interaction between the AER and the ZPA has no analog in the CR (Loredo, Brukman et al. 2001; Nagashima, Kuraku et al. 2007).

In a manner similar to the case of the eyespot organizer, the CR signaling center is a mix of ancestral and unique features. As with many signaling centers, the CR arises at a compartment boundary and recruits existing signaling pathways (FGF10 and Wnt). However, the collection of signals that is expressed and their anatomical organization are different from any other signaling center yet described (e.g., a lack of FGF8, Shh, and BMP4 expression; Kuraku, Usuda et al. 2005). Hence, we must conclude that novel signaling centers are put together from elements of ancestral signaling centers, but are uniquely configured into a novel gene regulatory network.

Experimental work suggests a two-step model for the development of the costal plates of the carapace. First, the FGF10 signal from the CR mesenchyme results in a dorsolateral rather than a ventral growth direction of the developing ribs (Burke 1989; Cebra-Thomas, Tan et al. 2005). Second, in concert with the perichondrial ossification of the ribs, a powerful BMP signal induces the ossification of the dermal tissue surrounding the ribs (Cebra-Thomas, Tan et al. 2005). Hence, the costal plates are extensions of rib perichondrial ossification, and are not dermal bones (Kaelin 1945; Rieppel and Reisz 1999).

Ablating the CR results in a ventral growth of the turtle ribs, at least according to some authors (Burke 1989; Cebra-Thomas, Tan et al. 2005). That is, the ancestral situation and insertion of an FGF10-loaded bead in a chicken embryo results in a turtle-like growth pattern of the chicken rib (Cebra-Thomas, Tan et al. 2005). Hence, it seems that the origin of an FGF10-emitting signaling center is sufficient to explain the derived growth pattern of turtle ribs. This is a very satisfying scenario, as it seems that the origin of the CR would be sufficient to explain the salient features of the turtle body plan. However, this emerging consensus has been challenged by a series of more recent discoveries.

Two recent discoveries support an alternative evolutionary scenario for the origin of the turtle carapace. One was the discovery of a fossil turtle that lacked a carapace, and the other regarded the developmental function of the CR. As mentioned above, in the past, one obstacle to a deeper understanding of turtle evolution was that there were no fossils that would help us understand the transition from the canonical amniote *Bauplan* to the turtle configuration. Until recently, the oldest known turtle fossil, *Proganochelys*, already had a fully formed carapace and, thus, did not help us to understand its origin. In 2008, the description of a new fossil, *Odontochelys,* was published (Li, Wu et al. 2008), which had a fully formed plastron but no carapace. Interestingly, its ribs were already dorsally arrested (i.e., they did not bend ventrally to contact the sternum.) Oddly, its ribs converged toward the

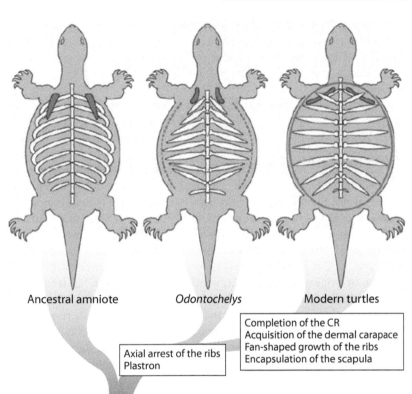

Ancestral amniote   *Odontochelys*   Modern turtles

Axial arrest of the ribs
Plastron

Completion of the CR
Acquisition of the dermal carapace
Fan-shaped growth of the ribs
Encapsulation of the scapula

FIGURE 5.9: Model of carapace evolution according to Nagashima and Kuratani (2009, 2011). The first step was the dorsal arrest of rib growth, as represented by *Odontochelys* and illustrated in figure 5.8. In *Odontochelys*, the ribs point toward the mid-trunk region, an arrangement reminiscent of the effect of experimental cauterization of the carapacial ridge (CR). Note that the shoulder blade was anterior of the ribs at this stage of evolution. CR expansion then led to a re-orientation of the ribs to more fan-shaped pattern that, at the anterior end, resulted in situation such that the shoulder girdle was covered by the ribs. Redrawn with permission from Nagashima et al. 2009, *Science* 325:193–196. Reprinted with permission from AAAS.

mid-trunk region rather than fanning out like the ribs of modern turtles do (figure 5.9). Because of this arrangement, the shoulder blade was neither outside nor inside the rib cage; rather, it was anterior, like in early stages of turtle development.

The curious arrangement of the ribs in *Odontochelys* also provides hints for the evolutionary function of the CR. Using cauterization experiments, Nagashima and colleagues from Shigeru Kuratani's lab at the RIKEN Institute for Developmental Biology, Kobe, Japan, showed that removing the CR resulted in a convergent growth of the ribs (Nagashima, Kuraku et al. 2007), similar to the situation in *Odontochelys*. In addition, removing the CR did not affect the dorsal arrest of the ribs (contra Burke 1989 and Cebra-Thomas et al., 2007). Hence, regardless of what the early function of the CR may have been, at least a late function of the CR is to direct rib growth in a divergent fan-shaped pattern.

Putting this observation together with the fossil morphology of *Odontochelys*, as well as embryological evidence regarding muscle development, Hiroshi Nagashima, Shigeru Kuratani, and colleagues (Nagashima, Sugahara et al. 2009) came up with a new scenario for the origin of the turtle carapace, called the folding theory (Kuratani, Kuraku et al. 2011).

According to the folding theory, the first step was the dorsal arrest of rib growth, but not the formation of the carapace (figure 5.9). The developmental cause of dorsal arrest is unknown at this point. This stage of evolution is thought to be represented by *Odontochelys*. At this stage, there might have been an incomplete CR involved in dorsal arrest, although evidence regarding the role of CR in dorsal arrest is contradictory. The next step was the expansion or origination of the CR, which resulted in (as in modern turtles) the fan-shaped expansion of the dorsal ribs. With the anterior move of the anterior ribs, the scapula became covered by the ribs, which resulted in a folding of the body wall and an internalization of the shoulder girdle.

This folding can still be seen in the arrangement of the deep muscles attached to the shoulder blade, which are the early-developing muscles (Nagashima, Sugahara et al. 2009). Late-developing limb muscles acquired novel attachment sites and, thus, are hard to compare between turtles and other amniotes. Hence, the CR results in the anterior growth of the anterior ribs and to the internalization of the shoulder blade. The origin of the bony elements of the carapace was likely a later step after the internalization of the shoulder girdle.

Overall, the carapace of the turtle is a complex novelty that involves the origin of novel skeletal elements through the interaction of periosteal rib ossification and dermal ossification, the dorsal arrest of rib growth, and the origin of a novel signaling center, the CR. This results in the fan-shaped arrangement of ribs and the infolding of the body wall together with the shoulder girdle. Hence, even though the carapace incorporates ancestral elements, like ribs, it is a novel developmental and an evolutionary unit that is integrated through novel regulatory interactions, like those between the CR and the embryonic ribs, the ribs and the dermis, and others.

The Developmental Biology of Novelties: Reflections

The development of a multicellular organism results from complex interactions between the environmental and intraorganismal signals and the cell and tissue autonomous reaction norms of those cells that form a character or body part. Not surprisingly, the evolutionary changes in the developmental mechanisms that contribute to the origin of novel body parts can also be complex and multifaceted.

Only a few tentative patterns can be gleaned from the fragmentary evidence we have regarding the developmental biology of evolutionary novelties. The most important seems to be that the majority of those evolutionary changes that resulted in novelties are those that affect the cell and tissue autonomous reaction norms rather than the inductive signals that induce these reactions. This conclusion arises less from positive evidence regarding the evolution of gene regulatory networks that underlie organ-specific autonomous developmental pathways; rather, this suggestion derives mostly from the well-established variability of positional information signals and the fact that PI signals tend to be older than the induced structures (e.g., ZPA and digits, balancers and cement glands, even butterfly eyespots). One exception to this rule is mechanical stimuli that can induce the differentiation of skeletal elements in amniotes. In this case, the driving force is changes in the distribution of mechanical stress during fetal development rather than novel cell types or novel tissue–specific reaction norms.

In any case, the greatest challenge for future research will be to understand the origin of gene regulatory networks that endow groups of cells with the ability to maintain differential gene expression and enable the quasi-autonomous execution of an organ-specific development "program," even when the initial inductive signals are absent. In other words, the main problem is the origin of the character identity network (ChIN).

6

The Genetics of Evolutionary Novelties

"Auf diese Weise könnten wir also dahin definieren, daß "homolog" Organe sind, die [. . .] unter entsprechenden organisieren-
den Beziehungen entstanden sind."[1]

(Bertalanffy 1936)

In chapter 3 the ChIN concept was introduced: a conserved gene regulatory network that translates positional information signals into the expression of those genes that result in a character's phenotype. ChINs enable the
differential gene expression that is necessary to develop individualized body
parts. ChINs are the most conserved part of the gene regulatory network that
underlies character development and, thus, are most consistently associated
with manifest character identity. Here we want to discuss which molecular
mechanisms might be involved in the origination of novel gene regulatory
networks (and, thus, ChINs) and what these mechanisms imply for the origin of novel characters.

Gene regulation is a complex process that includes all the steps involved
in making the information stored in DNA manifest in cellular activity. This
includes all of those processes that influence the ability of DNA to be regulated
by certain factors, such as chromatin modification and DNA methylation; to
the actual transcription of DNA; processing of an RNA transcript into mature
mRNA; those processes that affect the translation of mRNA into a protein; and,
finally, protein modifications by enzymes that add additional residues to the
protein, such as the addition of carbohydrates (i.e., glycosylation) and the addition of phosphate groups (i.e., phosphorylation). Cellular activity can be influenced by factors that act on each one of these processes and, thus, the evolution
of gene regulation has to be considered at all these levels of gene regulation.

Nevertheless, most work on the evolution of gene regulatory networks has
focused on the evolution of transcriptional regulation (Davidson 2006) and,

[1] In this way we could define as homologous those organs that arise from corresponding organiz-
ing relations [translated by Günter P. Wagner].

more recently, the role of non-coding RNA for the translation and stability of mRNA. Given the enormous complexity of these processes, this emphasis seems to be misplaced, although there is a good reason for this focus. The expression of all protein-coding genes has to involve transcription and translation, while the other processes that influence gene function are not always necessary. There is no gene expression without transcription. However, protein modification is not necessary, even though it occurs in many instances and is important when it does occur. Keeping in mind that our understanding of the evolution of gene regulatory networks is incomplete and imperfect, we will focus on those aspects of gene regulation about which we know most at this time.

The transcription of a protein-coding gene requires the recruitment of a complex of proteins that contain RNA-polymerase II, the enzyme that can create an RNA molecule from a DNA template. This complex is called the initiation complex and that part of DNA to which it binds is called the promoter. The likelihood of initiating transcription is further influenced by the presence or absence of transcription factor proteins that bind to other parts of DNA. These other DNA parts are often within 200 kb of the promoter and, thus, can be quite far removed from the actual gene.

Collectively, all DNA parts that affect transcription of a protein-coding gene on the same DNA strand are called cis-regulatory elements (CREs). Transcription factors that bind to these elements can either enhance or reduce transcription. CREs other than the proximal promoters have been called either enhancers or repressors, depending on whether their presence increases or decreases the rate of transcription. But this distinction is not that useful, as the effect of such a CRE on transcription often is dependent on the set of proteins that form the complex on DNA. Some of the proteins that determine transcriptional activity do not even bind to DNA—so-called transcriptional co-factors. Thus, *the functional effect of a CRE is not intrinsic to the DNA sequence of the CRE* and is not even determined by the nucleotide sequence per se. For this reason, we will call all those CREs that are *not* proximal promoters "enhancers," regardless of whether the proteins that bind to the CRE at any given time enhance or repress transcription.

To parse the complex problem of the evolution of transcriptional regulation into manageable pieces, I will focus on two complementary topics: the evolution of CREs (i.e., those modifications of DNA sequences that create opportunities for transcription factor proteins to bind to DNA) and the evolution of the transcription factor proteins themselves.

Evolution of cis-Regulatory Elements

There have been book-length treatments of this topic (Carroll, Grenier et al. 2001; Davidson 2001; Wilkins 2002; Davidson 2006; Stern 2011). I recommend that the interested reader consult these excellent summaries of the vast literature of gene regulation in development and evolution. The purpose of the brief review provided here is to explore which processes we need to consider in order to investigate the evolutionary origin of ChINs. To this end I will first summarize studies that have been done on two paradigmatic cases of CRE evolution. One of these cases was studied by David Stern's laboratory (when at Princeton University) regarding the evolution of the bristle pattern on the first instar larvae in Drosophila. The second case concerns the evolution of wing pigmentation, also in Drosophila, that originated from Sean Carroll's laboratory at the University of Wisconsin. Finally, I will discuss the role of transposable elements in re-wiring gene regulatory networks.

Shaven Baby and the Evolution of Bristles in Drosophila Larvae

Drosophila larvae are wormlike, headless, and legless organisms that display a rudimentary external segmentation that is visible in the arrangements of bristles of various sizes (figure 6.1). In the first instar larva of *Drosophila melanogaster,* the dorsal and lateral surfaces of the larvae have fields of small denticles, which are called trichomes. The function of trichomes is not known, but they are probably maintained by natural selection because they are found in many, but not, all Drosophila species. For example, *Drosophila sechellia* lacks these trichomes, as do *Drosophila ezoana, -borealis, -lacicola,* and *–montana.*

David Stern's laboratory set out to investigate the genetic basis of these morphological differences among Drosophila species (Sucena and Stern 2000; Sucena, Delon et al. 2003; McGregor, Orgogozo et al. 2007). Their starting point was the discovery of a gene, called *shaven baby* (*svb*), which was necessary and, to some extent, sufficient for trichome development. In fact, this gene has two names, *svb* and *ovo*, because of its two principal functions, one in trichome development and the other in ovary development. The name *svb* derives from its role in trichome development and *ovo* from its role in ovary development.

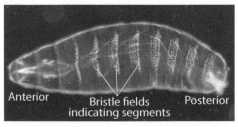

FIGURE 6.1: Drosophila larva in which bristle rows indicate body segments. Bristle patterns are an important model system for the study of developmental evolution. (Source: http://www.mutagenes.net/gene-expression-2/l-m.html.)

Here I will continue to call this gene *svb*. This gene encodes a zinc-finger transcription factor that regulates the morphogenesis of the cytoskeleton in epidermal cells so that they develop protrusions on their apical surface. These eventually are covered by cuticle and make up the bristles of a trichome. It has been shown that *svb* is always expressed in cells that form a trichome. If *svb* expression is suppressed, a trichome will not form.

The variation between species in terms of the presence or absence of trichomes on particular parts of the body surface is correlated with the presence or absence of *svb* expression in these areas. Genetic data showed that these phenotypic differences were due to cis-regulatory differences at the *svb* locus rather than changes in the transcription factor amino acid sequence of *svb* itself. The *svb* locus has multiple CREs: three alternative promoters, two of which are used in the ovary to produce different isoforms of OVOA and OVOB proteins, and one upstream promoter to produce the *svb* transcript (Andrews, Garcia-Estefania et al. 2000). Furthermore, a detailed dissection of the *D. melanogaster svb* locus has shown that *svb* expression in the dorsal and lateral epidermis of the first instar larva is controlled by three enhancers, each of which directs *svb* expression to a subregion of the wild-type *svb* expression domain (McGregor, Orgogozo et al. 2007; figure 6.2).

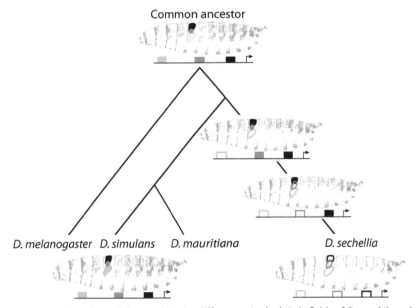

FIGURE 6.2: Genetic model to explain the differences in the bristle fields of *Drosophila sechellia* and *D. simulans*. Expression of the gene *svb* is controlled by three enhancers. Each enhancer drives expression in different spatial domains of bristle development. Stepwise loss of enhancer activity likely caused the phenotype of *D. sechellia*. Reprinted by permission from Macmillan Publishers Ltd: *Nature*, McGregor et al. 2007, *Nature* 448:587–590, copyright 2007.

Detailed studies of the orthologous regions of the *Drosophila sechellia* genome showed that all three of these CREs have changed to produce the *sechellia* phenotype. Hence, even though a single "allele" explains the difference between the *D. melanogaster* phenotype and that of *D. sechellia*, the evolutionary changes involved mutations at three different portions of the *svb* locus, which affected three functionally modular CREs. Thus, an individual mutation most likely had smaller effects than the substitution of a *D. melanogaster* allele with that of *D. sechellia*. Most likely, similar changes must underlie the parallel changes in the other species that have lost both, trichomes and *svb* expression (McGregor, Orgogozo et al. 2007).

It is clear that changes in the CREs of the *svb* locus played a pivotal role in the evolution of trichome patterns, rather than mutations in the many other genes that are also necessary for trichome development, like the genes in the Wnt and epidermal growth factor signaling pathways, which regulate *svb* expression. David Stern suggested that *svb* derived its privileged role from its gatekeeper function, which integrates the positional information from the Wnt and EGF signals, and directed the remodeling of the cytoskeleton, which ultimately creates the phenotypic trait: the bristles of the trichome (Stern and Orgogozo 2008). Thus, *svb* has a function similar to that suggested for ChINs.

Except, in this case, there is no direct evidence that *svb* is part of a mutually cross-regulatory network, like a ChIN, as is the case in the eye determination network (see chapter 3). Thus, it seems to be a character identity gene, rather than a network (if indeed *svb* acts alone, which would be unusual). What is odd about this idea, though, is that the same transcription factor also plays a role in ovary development and, thus, there have to be some coregulators or cofactors that differ between epidermal cells and ovary cells to explain the different roles *svb* plays in each cell type. Of course, it is possible that the different target genes *svb* activates in these cells are determined by differences in these cells' chromatin status.

One thing worth noting is that the evolutionary transitions in the Drosophila group are all from the presence of a trichome to its absence; hence, they are all cases of character loss. Because *svb* is a key control element for the development of a denticle in an epidermal cell, mutating this locus is the most efficient means to eliminate trichomes without any obvious consequences for other aspects of the phenotype. This example clearly shows that morphological differences can be caused by changes in the CREs, which are facilitated by their modularity.

What we do not learn from the published literature that I am aware of is how *svb* came under the control of its upstream regulatory factors and how it acquired its role in directing epidermal cell morphology. We can say only that the evolution of this role included the origin of an alternative promoter, different from the two used in the ovary, and that the *svb* promoter is further influenced by three enhancers that regulate expression in different regions of the body. This was a large amount of sequence change in order to make an epidermal denticle.

Wing Spots

Another paradigm of cis-regulatory evolution is the evolution of the anterior-distal spot of black pigment (melanin) in the wings of some Drosophila species—for example, *D. biarmipes* (figure 6.3). This trait is absent in *D. melanogaster, –pseudoobscura, -gunungcola,* and *-guanche,* but is present in *D. biarmipes,* and *–tristis* and many other species. In a taxon sample of 29 Drosophila species, Prud'homme and colleagues (Prud'homme, Gompel et al. 2006) found that the phylogeny suggested that this spot had evolved two or three times and was lost at least five times (figure 6.4). In two papers, Carroll's lab investigated the genetic basis for the presence or absence of this wing spot in four different species pairs (Gompel, Prud'homme et al. 2005; Prud'homme, Gompel et al. 2006).

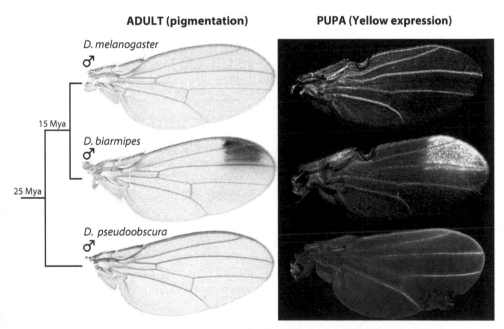

ADULT (pigmentation)   **PUPA (Yellow expression)**

D. melanogaster ♂

15 Mya

D. biarmipes ♂

25 Mya

D. pseudoobscura ♂

FIGURE 6.3: Development of a simple novelty, the anterior distal wing spot of *D. biarmipes*. This novelty arose in the *melanogaster* group. The location of the adult wing spot is preceded by a co-extensive domain of yellow expression. This shows that the regulation of yellow is critical for understanding the evolution of this pigmentation spot. Reprinted by permission from Macmillan Publishers Ltd: *Nature*, Gompel et al., 2005, *Nature* 433:481–487, copyright 2005.

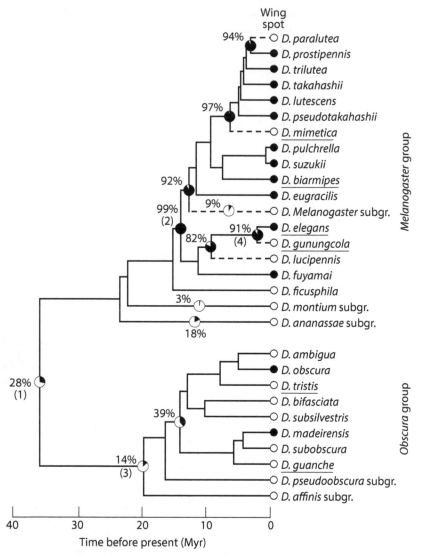

FIGURE 6.4: Evolutionary history of wing spots in the *Drosophila melanogaster/obscura* groups. Wing spots originated at least twice in this group and were lost multiple times. Reprinted by permission from Macmillan Publishers Ltd: *Nature,* Prud'homme et al., 2006, *Nature* 440:1050–1053, copyright 2006.

The black pigment in Drosophila is melanin that is produced from 3,4 dihydroxyphenylalanine (L-DOPA) by the gene product of the *yellow* (*y*) locus, while DOPA is obtained from tyrosine by hydroxylation that produces dihydroxyphenylalanine. The biochemical function of the Y protein is not known, but genetic evidence shows that in Drosophila Y is essential for the production of DOPA-melanin (Wittkopp, True et al. 2002; Wittkopp, Vaccaro et al. 2002). *Y* is expressed at low levels all over the wing of *D. melanogaster* and other "spotless" Drosophila species, which gives the wing a light pigmentation shade. In *D. biarmipes,* as well as in all "spotted" Drosophila species examined, *y* is strongly expressed in the anterior distal area of the wing where the pigment spot forms. Carroll and colleagues compared two cases of spot loss within the *D. melanogaster* group.

In this group of species, the wing spot was homologous, based on the phylogenetic reconstruction of Prud'homme and collaborators (Prud'homme, Gompel et al. 2006). In one study, promoter comparisons showed that the existence of the *D. biarmipes* spot was due to changes in the CRE that, in *D. melanogaster,* drove low-level *y* expression all over the wing (Gompel, Prud'homme et al. 2005). The orthologous genomic region was responsible for the loss of the wing spot in *D. gunungcola* when it was compared to the closely related spotted species *D. elegans.* This result would be expected, given that the wing spots of *D. elegans* and *–biarmipes* are homologous.

Because *D. elegans* and *–gunungcola* are closely related, it was possible to identify a small set of nucleotide substitutions that included the actual nucleotide substitutions that led to the loss of the anterior distal wing spot. This showed that between two to seven nucleotide substitutions were sufficient for *y* expression loss in the anterior distal domain of the wing (Prud'homme, Gompel et al. 2006). Presumably these nucleotide substitutions abrogated the binding sites for one or more activating transcription factors; however, the identity of these transcription factors is not known as I write these lines.

Regulation of *y* was also investigated in the spotted *D. tristis* species, which has a spot that is independently derived from that found in the *D. melanogaster* group. The regulation of *y* was also compared to ancestrally spotless *D. guanche.* It turned out that the enhancer that drove *y* expression in the wing spot of *D. biarmipes* and *–gunungcola* did not contribute to spot melanization in *D. tristis.* Rather, the spot enhancer of *D. tristis* was in the intron of the *y* gene. The same enhancer in spotless *D. guanche* drove *y* expression and, thus, pigmentation along the wing veins.

Hence, the evolution of novel spot enhancers was due to a modification of ancestral enhancers, rather than the evolution of a new enhancer from scratch. Furthermore, for non-homologous wing spots, the enhancer derived from modifications of different ancestral enhancers, but which drove *y* expression in some other part(s) of the wing.

In both cases, the wing spots in *D. tristis* and those found in the *D. melanogaster* group (but not in *–melanogaster* itself), the novel domain occupied more or less the same location on the wing. This convergence was probably due to the recruitment of the same set of transcription factors into driving higher *y* expression in parts of the wing. For *D. biarmipes* it was shown that spot expression was due to a combination of both activating and repressing transcription factors. One repressing transcription factor was identified: *engrailed (en)*.

en is expressed in the posterior compartment of the insect wing and had acquired a function in the spotted members of the *Drosophila melanogaster* group to suppress *y* expression in the posterior wing compartment. But the activator that drove higher expression in the distal part of the wing was not identified. Thus, the authors proposed that insect wing pigmentation patterns evolved from connecting *y* expression or the expression of other genes in the pigment biosynthetic pathway to an ancestral and largely conserved "trans-regulatory landscape"—that is, a pattern of partially overlapping expression domains of a set of transcription factors.

According to this theory, the combinatorial possibilities of choosing different sets of transcription factors more or less determines the range of possible pigmentation patterns in fly wings. This model has been called the "Christmas tree model" (so named in Wagner and Lynch 2008). Pigmentation patterns are like ornaments on a Christmas tree, which are hooked up to a pre-existing set of twigs on the tree. The ancestral expression domains of transcription factors correspond to the twigs of the Christmas tree, and the patterns of pigmentation to the Christmas tree ornaments.

The Christmas tree model is appealing in its simplicity, but may not reflect the essential structural complexity of wing spot evolution. Recent work in the Nicolas Gompel and Benjamin Prud'homme labs in Marseille, France, shows that the origination of the wing spot phenotype was likely a two-step process (Arnoult, Su et al. 2013) (figure 6.5). Arnoult and collaborators found that in *D. biarmipes,* the expression of *yellow* as well as other enzymes necessary for pigment production was under the control of *Dll*. In spotless species, like *D. melanogaster* or *D. annanasse,* however, *Dll* could not induce the expression of pigments. Furthermore, late *Dll* expression varies between species with wing spots and this variation corresponds to differences in the pigment patterns.

According to Arnoult and collaborators, the first step in wing spot evolution likely was the stepwise evolution of a regulatory module in which the expression of the genes required for pigment production came under the control of *Dll* in its original location. The second step was the diversification of the wing spots once the co-regulatory module was under the control of *Dll* and for which changes in the expression pattern of *Dll* were sufficient to change the shape of the wing spot.

If this scenario is correct, as the authors of this study point out, then the origination of the wing spot as a character is a different kind of process than the subsequent diversification of the wing spot. This model clearly maps onto the conceptual distinction between the origin of a character (novelty = origin of a *Dll*-controlled co-regulatory module) and the evolutionary modification of the character (different character states—e.g., different spot shapes).

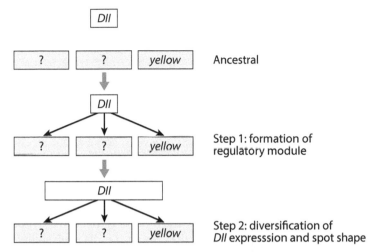

FIGURE 6.5: Two-step model of wing spot evolution according to Gompel and Prud'homme (Arnoult, Su et al., 2013). Ancestrally, there were no regulatory links between *Dll* and the pigmentation genes, including *yellow*. The first step was the formation of a co-regulatory module under the control of *Dll*. In the second step, once *Dll* expression was sufficient to deploy a pigmentation module, variation in the spatial expression pattern of *Dll* became the mechanism for wing spot shape evolution.

Are Novel Pigment Spots Novelties, and Why Does It Matter?

In the context of the origination of these Drosophila wing spots, a legitimate question can be raised as to whether these wing spots are Type I novelties in the sense developed in chapter 3 (i.e., novel elements of the phenotype). Specifically, it is interesting to reflect on the status of the anterior distal wing spot relative to our distinction between characters and character states. Is it more useful to consider the wing spots as characters or should we consider them as traits of a character state of the wing, and does it matter?

Certainly these wing spots have historical continuity, at least within the *D. melanogaster* group of species, and it is plausible that they can have different character states of their own. For example, if we consider them as characters, then the amount of pigment produced would represent a character state. The latter might be regulated by the number of activating transcription factor binding sites in the *y* spot enhancer. From a formal point of view, these spots can be treated as characters, like eyespots on butterfly wings. As it turns out, the answer depends on what model of wing spot development and evolution is correct, the classical Christmas tree model or the new Gompel/Prud'homme model described above.

Under the Christmas tree model (i.e., that pigmentation is linked to a pre-existing trans-regulatory landscape of transcription factor expression), the developmental pathway consists of only the polymerization of DOPA to produce black melanin. The identity of the spot and its location are determined by an ancestral set of transcription factors that are expressed in this part of the body (i.e., the trans-regulatory landscape, TRL, of the wing, as it is called by Carroll and colleagues[2]).

According to this model, the wing spot is a trait that is tightly integrated with other aspects of the wing, and its variational autonomy is limited to the amount of pigment that is deposited. If the theory of the TRL is correct, then any shape changes in the wing spot would have to come about by changes in the expression patterns of the transcription factor in the TRL itself. Thus, it is likely to affect other aspects of wing development and has quite limited autonomy/modularity. Under this model, it seems more appropriate to consider the wing spot as a trait of a character state of the wing, and the wing as the actual character. The reason why this matters is that it is a hypothesis regarding the evolutionary potential of the eyespot (i.e., its evolvability as an individualized part of the phenotype).

In contrast, under the Gompel/Prud'homme model, the wing spot has gained developmental and variational autonomy through the evolution of a regulatory module controlled by late *Dll* expression. The results presented by Gompel and Prud'homme and their collaborators suggest that late *Dll* expression gained independence from the ancestral TRL of the wing and, thus, emancipated the morphology of the wing spot, to some degree, from the developmental architecture of the wing and acquired quasi-independence, which makes it useful to recognize it as a distinct character.

In this context, it is informative to compare models of pigment spot development on the Drosophila wings with those of butterfly eyespots discussed in chapter 5. Butterfly eyespots are also pigment deposits on an insect wing, but are considerably more complex. They deposit three types of pigments in a highly spatially regulated manner that is typically, but not always, in concentric rings. The shapes of eyespots are not limited to the simple round phenotype; rather, they can assume complicated variations such that one can have almost an alphabet using only naturally occurring eyespot shapes (Nijhout 1991).

Eyespots develop from a signaling center, which expresses the transcription factors Dll, spalt, engrailed, cubitus interruptus (and, thus, has signs of hh signaling), and Smad. This center seems to emit two diffusible signals, Wnt and dpp. These signals activate a cross-regulatory network of genes that are responsible for placement of the different pigments on the wing surface (see chapter 5 for more details).

---

[2] If we consider the spot a character, then the area that ancestrally expresses the transcription factors also has to be a character, although a cryptic one, because there is already a distinct identity to this part of the wing, which is only made visible through pigmentation. But if we do that, then each wing domain comprising some possible combination of transcription factors expressed in the wing and which potentially can define a unique expression domain of some downstream target gene like *yellow*, also has to be considered a distinct character.

It is important to observe that regulation of pigment production in the butterfly eyespot is removed from the direct influence of the ancestral wing TRL by at least two layers of gene regulation. One is the establishment of the complex cross-regulatory network of the eyespot signaling center (for details regarding this network, see Evans and Marcus 2006). The two signal molecules transiently expressed by the focal signaling center result in the expression of transcription factors, which then establish the pattern of pigment deposition. Variation in this network is responsible for some of the variations in eyespot size, shape, and position (Brunetti, Selegue et al. 2001; Monteiro, Glaser et al. 2006).

Different serially homologous eyespots can also acquire developmental and variational individuality to acquire their own phenotypes and, thus, evolutionary fates (Monteiro 2008). Hence, the high evolutionary potential of eyespots and their ability to become paramorph (individualized) pattern elements depends on the existence of a gene regulatory network that is inserted between the conserved ancestral wing TRL and the factors controlling pigment formation. To some degree, this is also implied in the Gompel/Prud'homme model of Drosophila wing spots.

Hence, the existence of a ChIN emancipates the pigment spots from the tyranny of the ancestral TRL and makes the eyespot a (quasi-) independent character. In contrast, according to the Christmas tree model of wing spot development, the distal pigment spots of Drosophila would truly be a trait of the wing, rather than a character in their own right. Note that the distinction made here between characters and pattern elements of a wing character state is relevant for explaining the different evolvabilities of the distal wing spots of the fly and butterfly eyespots.

The idea that butterfly eyespots are individualized parts of the wing pattern is consistent with a recent finding from Monteiro's lab (Oliver, Beaulieu et al. 2013). This was that eyespots likely originated from a group of pigment spots on the ventral posterior wing and then were transferred to other parts of the wing as packages. The ability to translocate to other parts of the body requires that the developmental pathway have a high degree of modularity and autonomy, as would be required for it to be considered a character. For these reasons, the distinction is not merely semantic, but is relevant to the explanation of phenotype diversity.

What this detailed discussion of wing spot models and eyespots also reveals is that traits and characters are connected by a continuum (figure 6.6), as is everything else in biology. Wing spots assuming the Christmas tree model would have limited autonomy, probably mostly limited to the amount of pigment produced. If the model of wing spot development proposed by Sean Carroll is correct, then the shape and the location of the wing spot is dictated by the pre-existing trans-regulatory landscape.

In contrast, butterfly eyespots have an "internal regulatory anatomy" to themselves through the arrangement of different pigments relative to each other (dark and golden rings, for example) and they can develop in different places and combinations on the wing. In fact, the emancipation of hh signaling from repression by engrailed is a key insight of the model for the origin of eyespots proposed by Keys and collaborators (Keys, Lewis et al. 1999). Eyespots are still part of the wing surface, but are not physically separated parts of the body, as are many typical organs like the limbs, the brain, or the liver. At the other extreme are simple descriptive features, like the spatial relationships among scales or dermal bones in the skull, which do not have a biological existence of their own.

A natural reaction to the existence of a continuum between "traithood" and "characterhood" (figure 6.6) would be to question the value of this distinction. But this attitude would also imply having to give up the species concept, or the gene concept, or any other biological term. The reason why these continua do not render as useless the concepts that they connect is that these continua are closed intervals, which means that they have definite and clearly different endpoints. Two populations of the same species when compared to two populations from clearly separated species represent different biological

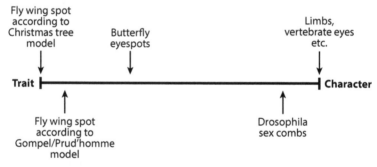

FIGURE 6.6: The continuum between "trait" and "character." At the extreme end of the spectrum, a trait is, for instance, a pigment's spot on an insect wing that is entirely defined by the ancestral trans-regulatory landscape (TRL) of the wing. Such wing spots lack developmental independence from their anatomical location. According to the model of Gompel and Prud'homme (Arnoult, Su et al., 2013), these wing spots can be independent if they are a regulatory module with late *Dll* expression at its top. If the wing spot in fact is due to a pigmentation module controlled by *Dll*, it would be independent of the remainder of the TRL. Butterfly eyespots have even more individuality by inserting more layers of autonomous local regulation between the positional information signals and the expression of the phenotype. Sex combs have a simple core regulatory network (ChIN) of two genes, *Scr* and *dsx*, which makes them quite independent of their location and allows them to change location (i.e., be expressed on different parts of the leg; Kopp, 2011). The reason why I did not place it all the way to a fully individualized character on this scale is that there is (to date) no evidence for co-adaptation between the Scr and Dsx transcription factor proteins. Co-adaptation among the members of a core gene regulatory network greatly enhances the historic stability of a ChIN. At the extremes on this scale are the classical anatomically, developmentally, genetically, and variationally individualized body parts of comparative anatomy.

realities, as does the difference between a trait and a fully individualized organ. Continua show that these things, species and characters, can evolve in a stepwise fashion, but do not deny the fact that the products of evolution (i.e., the existence of species and organs) are important facts of biology.

Sex Combs: The Origin of a ChIN

There is one example of an evolutionary novelty that occupies a sweet spot in terms of research accessibility. This is the so-called sex comb: specialized bristles on the forelegs of *Drosophila melanogaster* males. Sex combs occupy a sweet spot because they are phylogenetically relatively young and, thus, there are many extant species that represent the ancestral character state; and they occur in a genetic model organism (Drosophila) and, thus, are well understood in their genetic and developmental underpinnings and are accessible to experimental manipulation like few other examples discussed here. Artyom Kopp's lab at the University of California at Davis has made this system its focus to understand the genetics of this evolutionary novelty. The summary provided below is based on a review by Kopp (2011), including the references therein, as well as a key research paper by Tanaka and collaborators (Tanaka, Barmina et al. 2011).

In *Drosophila melanogaster,* sex combs are specialized bristles on the first tarsus segment of the forelimbs of males. These structures are found in the clade comprising the *melanogaster* and *obscura* groups in the genus *Drosophila* (figure 6.7). They are used during courtship and copulation, although their use varies between species. Sex combs probably arose independently in the relatively distantly related *Lordiphosa* flies.

Sex combs (SCs) are serial homologs to the transverse bristle rows (TBR) and are derived from the most distal TBR on the first tarsal segment of the foreleg. The modifications involve thickening of the bristles and rotation of the TBR into a longitudinal position through a morphogenetic movement of the initially transverse bristle row. SCs exhibit a wider variety of species-specific modifications in terms of bristle orientation and the number and size of teeth, and can even extend to the second tarsal segment, as for example in the case of *D. ficusphila* and others (figure 6.7).

SCs have all the hallmarks of a newly derived and clearly individualized structure. They vary independently from the other TBR, the serial homologs from which they are derived, and fulfill a new function in mating (TBRs are used to clean the eyes). Hence, it is interesting to ask, what are the genetic underpinnings of SCs and how do they relate to the developmental genetics of species without SCs?

In *D. melanogaster,* SCs develop on the anterior ventral distal tarsus, and near the distal joint on the first leg of males. Hence, the sites of SC development have a segmental, leg, and leg axis and within-tarsal spatial identity.

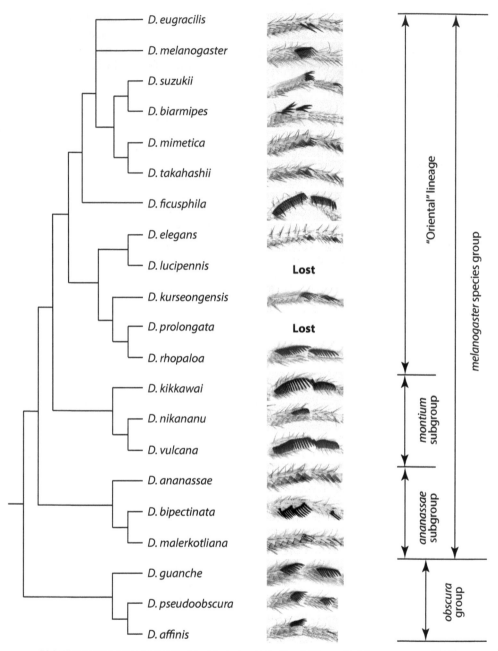

FIGURE 6.7: Taxonomic distribution of sex combs in the *melanogaster/obscura* group of fruit flies. Sex combs are a novelty on the forelegs of males in the *melanogaster* group of fruit flies. They evolved from a transverse bristle row that ancestrally existed on the legs of both males and females. Redrawn with permission from Kopp, 2011, *Evolution and Development* 13:504–522, figure 2. © 2011 Wiley Periodicals, Inc.

Consequently, SC development is influenced by and depends on all the signals that define this spatial identity. The Hox gene *sex combs reduced (Scr)* with its evocative name defines the segmental identity of the SC location, *wingless* defines the ventral location, and the absence of *engrailed* defines the anterior position. The identity of the most distal tarsal segment is defined by co-expression of *Dll, rotund (rn)*, and *dac*, as well as low *bric a brac (bab)* expression. The location close to the tarsal joint is probably regulated by Notch signaling (for references, see Kopp 2011). The sex-specific development of SCs depends on the activity of *double sex (dsx)*, the sex-determination locus in Drosophila.

Dsx is expressed in both males and females, but differs in their splice isoforms, with *dsxM* expressed in male tissue and *dsxF* in female tissue. Sexual phenotypes in Drosophila are cell autonomous, which means that in experimental and natural variants, cells with male and female phenotypes can coexist in the same animal. What is interesting, though, is that in the wild type, the sex-specific isoforms of *dsx* are expressed only in those cells that show sex-specific characteristics. Parts of the body that are not sexually dimorphic do not express *dsx*. Consequently, cells that contribute to SCs express *dsxM*, but the other parts with TBRs that are sexually monomorphic do not.

Kopp and collaborators showed that the development of SCs depended on and was co-extensive with the co-expression of *dsx* and *Scr*. During development, *Scr* is active first as the gene that defines segmental identity. Then, *Scr* in conjunction with location identity conferring signals activates *dsx*, which in turn feeds back to *Scr* expression (figure 6.8). *dsx* and *Scr* form a positive feedback loop, sustaining each other's expression, and become independent of the positional identity signals. In corresponding locations on the legs of species without SCs, no *dsx* is expressed. Furthermore, size and location variations of SCs among species are tightly correlated with the extent of *dsx* and *Scr* co-expression (Tanaka, Barmina et al. 2011). From this and more data, Kopp and collaborators concluded that the key event in the evolution of SCs was the origin of positive feedback between *dsx* and *Scr*.

Based on published data, the *dsx-Scr* network qualifies as a ChIN. Both genes are jointly necessary for SC development and they form a positive feedback loop to maintain each other's expression. To my knowledge, this is the first example for which the derivation of ChIN in a defined functional and phylogenetic context can be inferred. Outgroup species lack this positive feedback loop, while the derived character identity is rigidly associated with the co-expression of the *dsx-Scr* network. Because of the wealth of knowledge on Drosophila development, it is possible to create a speculative scenario for the origin of SCs. Needless to say, there is no direct evidence for this scenario. Nevertheless, it illustrates how, at least in principle, a novel ChIN may arise. Thus, these speculations may be useful for designing research on the origin of Type I novelties. They also illustrate how the transition from local

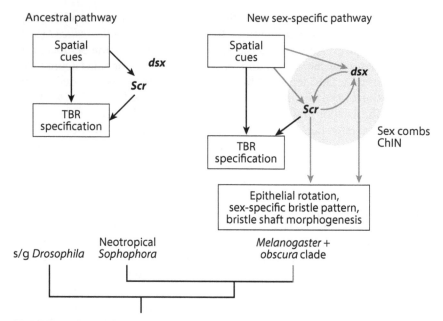

FIGURE 6.8: Origin of the sex combs character identity network from *dsx* recruitment and the formation of a positive feedback loop between *Scr* and *dsx* (modified after Figure 3 from Kopp, 2011, *Evolution and Development* 13:504–522).

modification to the origin of novel character identity may have occurred. For these reasons, these speculations are summarized below.

A Speculative Scenario on the Origin of a Simple ChIN

The speculation provided here is based on the following considerations. First, it is clear that the ancestral gene expression pattern for the origin of SCs was based on what is known from outgroup species (see Kopp 2011 and Tanaka, Barmina et al. 2011). Second, we want to propose steps in evolution that plausibly had a phenotypic effect so that natural selection could latch on to them. Third, this intuition is also informed by research on genes that affect the variational relationships among characters (Pavličev, Norgard et al. 2011).

Both ontogenetically and phylogenetically, the expression of *Scr* is primary. *Scr* is involved in labial and pro-thoracic segment identities that clearly preceded the origin of SCs. The signals that define the unique positional information for future sites of SC development are also ancestral and developmentally fundamental. Hence, the unique positional information, from which SCs evolved, precedes the origin of SCs consistent with arguments in chapter 5 about the relationship between positional information and the evolution of novelties.

The first event in the evolution of SCs could have been selection to modify the TBR most distal in the first tarsal segment, perhaps due to sexual selection. The positional identity of this TBR was already defined by ancestral positional information (PI) signals, and it is likely that the first heritable variation of the pre-SC TBR was through the variable modification of *Scr* target gene expression by these PI signals. It would have made sense that these PI signals were sex-specific because of the selection pressure for TBR modification; thus, the recruitment of *dsx* expression as a modifier is plausible. This suggestion, however, requires an additional assumption: the recruitment of *dsx* expression had some selectable phenotype.

For example, if *dsx* ancestrally had target genes in the bristle development pathway, then the recruitment of *dsx* could have had a phenotypic effect that fed the selection pressure to modify the distal TBR. Hence, the initial recruitment of *dsx* as a new target gene for *Scr* could have been just another way to affect the phenotype of the TBR. In addition, this pathway made these modifications sex-specific.

Because *dsx* is a transcription factor, it could evolve to affect the expression of *Scr*. Even this step likely had a quantitative effect on the phenotype of the distal TBR, because positive feedback probably increased the expression levels of both *Scr* and *dsx*. In *D. melanogaster*, quantitative variation in *Scr* and *dsx* activity is correlated with size variation in the SCs (see references in Kopp 2011). However, once a positive feedback loop was established, the activities of *dsx* and *Scr* became independent from the inductive signals that initiated this process.

This is the point where positional identity, which is dependent on the identity of the positional information signals, *transitions to character identity*, which is maintained by the positive feedback loop between the two organ identity genes. What this scenario implies is that character identity could originate from a seed for positional uniqueness derived from a modification of PI signals through the evolution of a network among transcription factors, which initially served only the local modifications of a serially homologous set of characters. Selection for the reinforcement of local modification can lead to character identity through the coincidental evolution of regulatory links among transcription factors.

Initially, this regulatory link had quantitative, modifying effects. But if these transcription factors link up in a positive feedback loop, then we have the origin of character identity. The latter step can also be favored by selection for variational independence of the derived modified character.

This scenario, though completely speculative, shows how the origin of a qualitative character identity can be linked to and originate from selection on quantitative variation. Selection for more and more quantitative differences can, under the right circumstances, lead to the eventual individualization of two serially homologous characters and, thus, to a Type I novelty. Whether it works that way needs to be tested with appropriate model systems.

Origin of Novel cis-Regulatory Elements: Transposable Elements

Logically, there are two ways by which new cis-regulatory activities can arise: through modification of an existing CRE and by the origin of a novel CRE. The examples on the evolution of pigment patterns discussed above showed that novel cis-regulatory activities, like pigment deposition in a part of the wing, often can be due to modification of a CRE which ancestrally had a function in the same general body region (e.g., pigmentation along the wing veins). However, there are also examples of cis-regulatory elements that have essential functions in the derived species, but have no counterparts in outgroup species (i.e., most likely truly novel CREs).

For example, the gene for prolactin (PRL) is expressed in the pituitary in all vertebrates. PRL is a polypetide and is a secreted signaling molecule that acts as a hormone with a wide variety of functions for regulating behavior and other physiological functions. In the brain, PRL is transcribed from a proximal promoter, which is regulated by the transcription factor Pitx1. In humans, apes, and monkeys, however, PRL is also expressed in the uterus, specifically, by endometrial stromal cells (Emera, Casola et al. 2011). In the uterus it plays a key role in maintaining pregnancy by inhibiting the expression of enzymes that break down progesterone (a steroid hormone essential for maintaining pregnancy) and by suppressing pro-inflammatory signals (e.g., IL-6).

PRL is not expressed in the uterus of a pregnant opossum, nor is it expressed in the homologous parts of non-mammalian female reproductive tracts in chicken, salamander, or fish (Lynch, Tanzer et al. 2008). Hence, uterine PRL expression is a derived feature among the eutherian mammals and independently evolved at least three times: once in primates, once in rodents, and once in afrotherians (Emera, Casola et al. 2011). In each case, uterine PRL expression is controlled by a promoter other than that used in the brain.

In the stromal cells of humans and other primates, PRL is transcribed from a promoter that is about 6 kb upstream of the transcriptional start site used in the brain (Gellersen, Kempf et al. 1994). No trace of the alternative promoter or of the additional enhancer is present in the genome of any non-eutherian mammal or any other more distantly related animal (Lynch, Tanzer et al. 2008). Hence, PRL expression in the uterus is controlled by a truly novel CRE, rather than by modification of an ancestral CRE.

Most interesting, though, these two novel CREs are both derived from two independent transposable element (TE) insertions upstream of the PRL locus. The uterine enhancer is a MER20, which has been inserted at the PRL locus in the stem lineage of eutherian mammals (i.e., after the most recent common ancestor of marsupials and eutherians and before the most recent common ancestor of eutherians). The alternative promoter is a MER38, which is a class 1 long terminal repeat transposable element and is found in primates and rodents (Emera, Casola et al. 2011). Interestingly, in the elephant, the uterine PRL promoter is produced by an alternative promoter that is about 10 kb upstream of the coding sequence and is derived from another transposable element (LA-L2), which exists only in afrotherians.

Furthermore, mouse transcripts of the PRL ortholog have a promoter even more upstream (70 kb) from a CRE derived from a transposable element called MER77 (Emera, Casola et al. 2011). This and much more evidence (briefly summarized below) supports an old theory that was originally proposed by Britten and Davidson (Britten and Davidson 1971) and in some form goes back to Barbara MacClintock. This theory proposed that transposable elements may play a major role in the evolution of gene regulatory networks.

The idea of transposable elements as a major class of agents for gene regulatory evolution is based on the fact that all transposable elements harbor transcription factor binding sites and sequences that are very close to perfect transcription factor binding sites (Rebollo, Romanish et al. 2012). These binding sites most likely evolved due to a function in the life cycle of a transposable element itself, rather than as adaptations to aid functionality in the host genome. If a transposable element with a certain transcription factor binding site inserts close to a protein-coding or even a non-coding RNA gene, then it provides a ready-made transcriptional regulatory element for the gene.

In addition, the upstream region of a gene is a preferred insertion site for transposable elements because this part of DNA tends to be more accessible than other parts of the genome. If such a transposable element inserts into a number of genes, then all these genes may immediately fall under the control of a transcription factor for which the transposable element provides a binding site or the right combination of binding sites. In this way it is conceivable that a large number of genes can come under the control of a novel upstream regulator in a short amount of time.

Supporting evidence for this model is mounting, in part from the comparative analysis of genome sequences (Feschotte 2008; Oliver and Greene 2009; Emera and Wagner 2012; Rebollo, Romanish et al. 2012). The first genomic evidence was the demonstration that many ancient transposable elements in the human genome are under stabilizing selection, as inferred by the rate and pattern of nucleotide substitutions (Silva, Shabalina et al. 2003;

see also, Lowe, Bejerano et al. 2007). Hence, it is likely that these elements fulfill some function that natural selection maintains. Because the majority of these transposable elements no longer transpose (i.e., are dead as transposable elements), this stabilizing selection most likely relates to the fitness of the host (i.e., they fulfill a useful function in the human genome).

Evidence for this is rapidly proliferating. For example, Jordan and colleagues found that 25% of experimentally characterized human promoters contained transposable elements (Marino-Ramirez, Lewis et al. 2005). This is almost certainly an underestimate, because old transposable elements are likely to have lost all the sequence signatures of their former existence as transposable elements due to mutation and drift. One-fourth of the DNAse I hypersensitive sites in human CD4+ T cells, which is characteristic of a CRE, overlap with annotated transposable elements due to their accessibility to transcription factors, and most of these sites are primate specific.

Bourque and colleagues (Bourque, Leong et al. 2008) have shown that binding sites for specific transcription factors identified by chromatin immunoprecipitation are highly over-represented in certain clade-specific transposable elements. Furthermore, transposable elements in the neighborhood of regulated genes have been shown to have transcriptional activity in reporter assays. This shows that transposable elements may contribute to clade-specific features of a gene regulatory network. There are many more reports with similar evidence (for reviews, see Feschotte 2008; Oliver and Greene 2009; and Rebollo, Romanish et al. 2012).

Particularly interesting and more specific evidence has been provided by David Haussler's lab (Wang, Zeng et al. 2007). Wang and colleagues looked at experimentally verified binding sites of the transcription factor p53 in the human genome. They found that these are highly over-represented in certain classes of endogenous retroviruses (ERV), in particular, transposable elements in the LTR10 and MER61 families. These families are primate specific and were transpositionally active at the time when New World monkeys split from the primate lineage. More than one-third of the experimentally verified p53 binding sites were in ERVs, and a subset has been shown to be transcriptionally active in reporter experiments. In the primate lineage, it seems that the evolution of the p53 transcriptional gene regulatory network has been greatly modified by retroviral element invasion in a clade-specific manner.

Comparative sequence analysis of mammalian and Drosophila genomes has also revealed that transposable elements not only play a role in the origin of novel CREs, but can assume a variety of functions, from modifying splicing patterns of mRNA to create alterative protein products, to being the origin of regulatory non-coding RNA, and even the origin of new classes of transcription factor proteins (for a review, see Feschotte 2008). For example, the paired domain, which characterizes our good friend *Pax6* (a.k.a. *eyeless*), is a protein derived from a TE-encoded protein, a transposase (Breitling and Gerber 2000).

The End of Uniformitarianism in Evolutionary Genetics: The Conceptual Impact of the Role of Transposable Elements in Gene Regulatory Network Evolution

Although the idea that transposable elements may play a major role in gene regulatory network evolution is old (Britten and Davidson 1971), and evidence in favor of this model is expanding, it seems to be time to reflect on how these facts should change our view of evolution. I want to devote a few lines to reflect upon the implications of transposable element–induced mutations. I think that the evidence that transposable elements play a major role in the evolution of gene regulatory networks affects various uniformitarian ideas that are broadly accepted in evolutionary biology.

Uniformitarianism is a set of ideas that claims that past events were caused by the same types of processes and factors as those we can observe today. This idea played a major role in the history of geology and also greatly influenced modern evolutionary thought through the influence of Lyell on the young Charles Darwin (see Mayr 1982). Uniformitarianist thinking also expresses itself in the assumption that all of evolution is based on the same kind of genetic variation that we find in any extant population—basically, nucleotide substitutions and minor insertions and deletions. Transposable element–mediated genetic variation, however, does not fit the bill.

Transposable elements are usually lineage or clade specific, and their effects depend on the sequence that a transposable element brings to the genome. Some transposable elements, like class 1 retrotransposons, are most likely derived from RNA viruses and, thus, can be viewed as accidental infection of a genome. Also, the virus that infects one lineage differs from the virus that infects another lineage. It is not clear from whence other transposable elements derive, like the DNA-based transposable elements, but these are also lineage specific.

We saw above that the transcriptional effects of a transposable element insertion depend on the transcription factor binding site that a transposable element brings to the genome. Thus, the genetic variation that arises in one lineage will be quite different from that arising in another lineage that is infected by a transposable element harboring another kind of binding site. It is not far-fetched then to think that the evolutionary fate of a lineage is strongly influenced and different from that of other lineages, in part, because of the nature of genomic parasites that infect its genome at any point in time.

This means that the response to natural selection is influenced not only by the actual selection pressures and the adaptive past of the lineage, but also by the kind of genetic variation arising in its genome, which is in part dictated by the accidental acquisition of different retroviruses and other genomic parasites. Mutational processes are not uniform across lineages, and they are also not uniform over time. Lineage invasion by a transposable element is an episodic process. There is a well-defined point at which a certain transposable element enters a lineage and is transpositionally active for only a limited amount of time until the genome figures out defenses against this parasite. There is a limited time window during the evolution of a lineage in which a particular transposable element contributes to the genetic variation of the species. Because transposable elements can cause major gene regulatory rewiring, the period of genetic innovation is also limited and, thus, one must assume that phases of enhanced rates of innovation are highly episodic.

The disruptive negative effects of transposable elements can be felt in any cell type and with respect to any physiological function. However, their potentially beneficial effects of providing novel CREs are not uniform. As Britten and Davidson have argued, the transposable element–derived CREs are predisposed to affect gene expression in cells that express transcription factors for which binding sites are ancestrally present in the transposable element. Hence, the potential to rewire a gene regulatory network is highly biased toward regulation in cells that already express some of the transcription factors which the transposable elements can bind.

It might not have been an accident that advanced placentas evolved from that mammalian lineage that was infected by a MER20, rather than by any other therian lineage (Lynch, Leclerc et al. 2011). MER20s brought with them binding sites for transcription factors that were probably already expressed in the cell types from which endometrial stromal cell derived, which include binding sites for FOXO1 and C/EBP-b and pre-binding sites for AbdB-related Hox proteins.

The potential for innovation in gene regulatory networks is likely to be non-uniform across lineages, different in epochs during phylogeny, and different among cell types. This implies that studying the current genetic variation in an extant population might not give us an adequate picture of the kind of genetic variation responsible for past evolutionary events—in particular, those events that involve major gene regulatory network reorganization. This makes assessing evolvability with population genetic means highly questionable, at least with respect to the likelihood for major transitions.[3] Only a historical comparative approach using inferences from comparative genomic data can provide the information necessary to understand past evolutionary events—in particular, genetic events related to innovations. This

[3] Population genetic methods to estimate evolvability are appropriate for assessing the ability of a population to respond to environmental changes and its ability to adapt within the limits of its given gene regulatory network. But extrapolation to macro-evolutionary evolvability, or potential to innovation is not possible. Hence, there is a limit to the extent to which a micro-evolutionary mechanism can be extrapolated to the macro-evolutionary scale.

also means that innovation is a different kind of process than is adaptation, which is usually studied within populations at the micro-evolutionary level.

The Role of Gene Duplications

Almost as soon as genes and chromosomes were established as the building blocks of the hereditary material, it became known that genome and genome duplications may have played an important role in evolution (Kuwada 1911; Tischler 1915; Bridges 1935; review and references in Taylor and Raes 2004). It was also recognized that gene duplication and the functional redundancy that it entailed may have provided mutational opportunities that did not exist prior to the duplication event.[4]

In other words, it was hypothesized that gene and genome duplications could increase, even though temporarily, the evolvability of a lineage and, thus, offered opportunities for change—in particular, innovations that may have been inaccessible to the ancestral lineages. In the second half of the twentieth century, this idea was most often identified with Susumu Ohno, who gave this view a convincing voice in his 1971 book, *Evolution by Gene Duplication*. This book emphasized the potential role of gene duplication in major evolutionary transitions, specifically in the origin of vertebrates (Ohno 1970).

At the descriptive level (i.e., the assertion that gene and genome duplications are important features of evolution), Ohno and his predecessors have been fully vindicated. In some lineages, gene duplications occur at a rate that is comparable to that of nucleotide substitutions (Lynch and Conery 2000). For example, Lynch and Conery calculated that there are hundreds of gene duplications in each human gamete. The fact that genome duplications played a major role in structuring the vertebrate genome, as hypothesized by Ohno, is also broadly accepted (Vanderpoele, Vos et al. 2004).

In contrast, the role that gene duplication plays in phenotype evolution, specifically with respect to the origin of novelties, remains controversial. The reason for this probably is that gene duplication can have many different consequences. Also, examples for which gene and genome duplications may be related to innovations are rare, and instances for which a case can be made regarding a causal connection between gene duplication and innovations are even rarer. More support exists for the idea that gene and genome duplications can play a role in speciation (i.e., the origin of genetic incompatibilities among sexually reproducing populations; Lynch and Force 2000).

---

[4] My thinking about the role of gene duplications has, naturally, been strongly influenced by the work on this topic done in my lab and that of collaborators. For this reason, the review of evidence regarding the role of gene duplication will be heavily biased toward work from my lab. This is not meant to be a comprehensive or even fair summary of the vast literature on gene duplication and I apologize for ignoring many important contributions.

There is also broad support for the idea that duplicated genes can be maintained by sub-functionalization (i.e., by a complementary loss of functions among paralogs; Force, Lynch et al. 1999). At the same time, there is vociferous criticism of the view that gene duplication is a factor in the evolution of novelties (Carroll 1995; Lynch 2007). The reason why the idea that gene duplication plays a role in the origin of novelties is not going away, in spite of this criticism, is a persistent empirical pattern that seems to link gene and genome duplications to major transitions in evolution.

More often than not, when an ancient gene duplication is traced to its phylogenetic origin, it is found to be associated with either a major change in body organization or an adaptive radiation of a major clade. Examples include the origin of vertebrates (Vanderpoele, Vos et al. 2004), the origin of crown group teleosts (Taylor, Braasch et al. 2003; Crow, Stadler et al. 2006), and the evolution of flowering plants (Irish and Litt 2005; Melzer, Wang et al. 2010). These examples and many more are suggestive of a causal role for gene and genome duplications during the evolution of novelties, like flowers or the vertebrate body plan. I will try to briefly review the evidence for this association in the next section.

First, I want to acknowledge that there is also strong evidence against such a connection. There are many documented genome duplications without any apparent innovations or radiations in their wake, in particular in amphibians (e.g., the genus *Xenopus*) and in fish (e.g., sturgeons, salmonids). There are also many examples of major transitions that are not associated with a genome duplication: the origin of amniotes, the mammals and birds as examples. Is the association between gene duplications and innovations and adaptive radiations just a perceptional artifact based on too limited data or is the connection there, but the causality is different from what has been assumed?

Here we will argue that this association is real, even though it is difficult to establish an appropriate null model to test this assertion statistically. What I will argue, however, is that the causality is the other way around. If there was a period of extensive adaptive radiation that coincided with gene duplications or a genome duplication, then the probability of maintaining paralogs was greater than if the duplication occurred during an adaptively less active period.

In this section we have summarized a few suggestive cases for an association between gene duplication and the evolution of novelties and adaptive radiations. However, the case made is not based on a fully factorial data set to prove association in the statistical sense. Such a study has yet to be done. What makes these few examples so evocative, though, is that gene duplications often affect exactly those genes that are known to be involved in body plan development, organ identity, and novel characters.

Vertebrate Hox Gene Clusters

Although evidence had been mounting since the 1960s and 1970s that genome duplications had occurred during the evolution of vertebrates, the possibility that these events could be related to the origin of the vertebrate body plan got a major boost with the discovery of Hox genes in vertebrates (Schughart, Kappen et al. 1987). Hox genes encode for homeodomain-containing transcription factor proteins, which were originally discovered because of their role in determining segment identity in Drosophila. In Drosophila, there are eight Hox genes with homeotic function in two clusters on the same chromosome, and all with a highly conserved sequence motif: the homeobox.[5]

Stunningly, and highly suggestive, it was found in the mouse that there are 38 Hox genes in four clusters on four different chromosomes (figure 6.9). These clusters are called Hox-A, Hox-B, Hox-C, and Hox-D that are on mouse chromosomes 6, 11, 15, and 2, respectively. Genes in corresponding positions in each of these clusters are more similar to each other than to other genes within the same cluster. This is most parsimoniously explained by the assumption that an ancestor of vertebrates probably had only one cluster with up to 14 genes and that this cluster was subsequently duplicated to yield the four canonical clusters found in most gnathostomes, including the shark, and the basal bony fishes *Latimeria* and the bichir *Polypterus* (Amemiya, Powers et al. 2010; Raincrow, Dewar et al. 2011; figure 6.10).

This scenario is also supported by the fact that a close relative of vertebrates, Branchiostoma (Cephalochordata), has only one Hox cluster with 14 Hox genes (Ferrier, Minguillón et al. 2000) and that some vertebrates have paralog group 14 Hox genes (Powers and Amemiya 2004).

The currently accepted scenario envisions a nice progression in the Hox gene inventory in parallel with the evolution of body plan complexity (figures 6.10 and 6.11). This scenario is that basal chordates have one cluster with up to 14 Hox genes (Ferrier, Minguillón et al. 2000), almost twice as many as Drosophila. The four canonical Hox clusters likely arose early on in vertebrate history and coincided with two genome duplications (Ruddle, Bartels et al. 1994). Even without a specific idea in mind, the question remains, what was the role of these gene duplications during the coincidental evolution of the vertebrate body plan?

Another suggestive event of Hox cluster duplication is linked with the origin of the largest group of extant vertebrates: the teleosts (about 24,000 species). In 1998, Amores and colleagues presented convincing evidence that zebrafish had seven instead of the four canonical Hox clusters (Amores,

---

[5] The *homeobox* is the 180 nucleotide long DNA sequence that encodes for the 60 amino acid long *homeodomain* of the Hox protein.

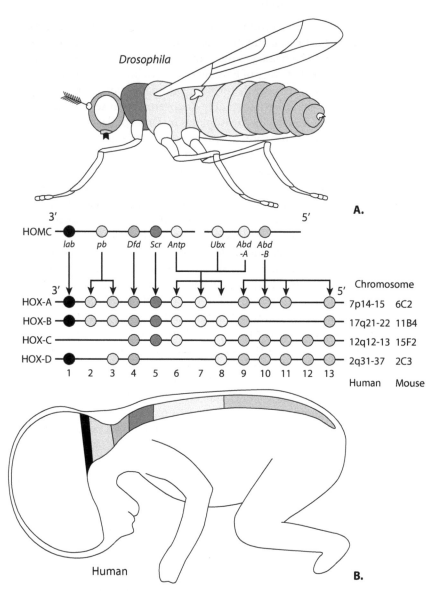

FIGURE 6.9: Hox gene clusters in Drosophila and mammal. A) In Drosophila there is one cluster of seven homeotic genes that determine segmental identities along the body axis. B) Mammals and many other vertebrates have four clusters with 38 Hox genes or more. The expression pattern along the body axis is similar to that in the fly.

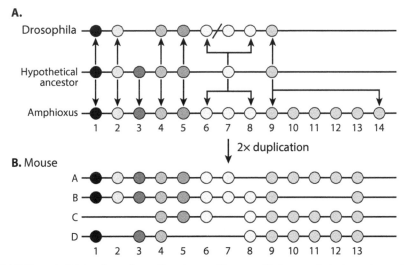

FIGURE 6.10: Origin of vertebrate Hox clusters. The most likely scenario is that prior to the origin of vertebrates, Hox genes multiplied by local tandem duplication and gave rise to the Hox gene cluster found in virtually all bilaterians. The four-cluster situation of most vertebrates was then achieved by two rounds of whole-genome duplication. This scenario explains why genes with corresponding positions in the four Hox clusters are more similar to each other than to other genes in the same cluster. Reprinted by permission from Macmillan Publishers Ltd: *Nature Reviews Genetics*, Prince and Picket 2002, *Nature Reviews Genetics* 3:827–837, copyright 2002.

Force et al. 1998). This event virtually coincided with the origin of teleosts, rather than earlier in ray finned fish phylogeny (Crow, Stadler et al. 2006).

Even though these associations between Hox cluster duplications and the origin of major vertebrate groups is suggestive, it has to be pointed out that this association is not very strong (Donoghue and Purnell 2005). For example, the majority of teleost species are in clades that arose long after the most recent common ancestor of teleosts, the Cypriniformes and the Perciformes, not to mention that the timing of the two vertebrate-specific duplications is quite loose and the relationship between them and any event in body plan evolution is also rather loose. A somewhat stronger correlation between retained paralogs and adaptive radiations has been shown in plant evolution for which major extinction events and the origins of retained paralogs are correlated (Van de Peer, Fawcett et al. 2009).

Evolution of Transcription Factor Proteins

In chapter 3 the importance of protein-protein interactions among transcription factors was discussed as a mechanism to confer regulatory specificity.

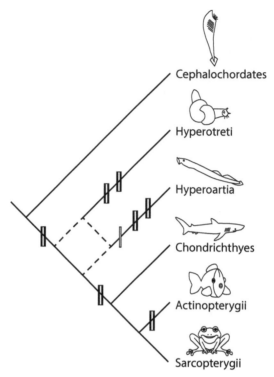

Cephalochordates

Hyperotreti

Hyperoartia

Chondrichthyes

Actinopterygii

Sarcopterygii

FIGURE 6.11: History of Hox cluster duplications in vertebrates. The most likely scenario is that one Hox cluster duplication occurred before the most recent common ancestor of vertebrates and a second after the divergence of extant cyclostomes. There was a third Hox cluster duplication coincident with the origin of teleost fishes, the largest group of extant vertebrates. In addition, the two extant cyclostome lineages underwent independent duplications [after Figure 9 in Prohaska et al., 2006, Chapter 10 in Papageorgiou (ed.), *HOX Gene Expression*, Landes Bioscience].

This is also the manner in which the input from different signaling pathways is integrated into a unitary response by a target gene and, thus, is the molecular basis for the difference between modification of the same and a qualitatively different entity, like a cell type. If this view of transcription factor function is correct, then the origin of Type I novelties has to include changes in the transcription factor proteins in order to evolve novel functional specificities and to integrate the output of signaling pathways into qualitatively new characters.

However, there has been substantial resistance within the developmental evolution community to the idea that transcription factors evolved in a functionally relevant way (Davidson and Erwin 2006; Prud'homme, Gompel et al. 2007; Carroll 2008). There are good reasons for why transcription factor evolution has been seen as unlikely. Therefore, I want to begin my discussion of transcription factor evolution with some of these reasons.

The most powerful evidence for a conserved biochemical role for transcription factors and other developmental genes derives from experiments in which the transcription factor from one species, say mouse, can replace the function of the homologous transcription factor in another, distantly related, species, like the fruit fly. To my knowledge, the first experiment of this kind was done in the laboratory of Bill McGinnis, then at Yale University. McGinnis compared the fly Hox gene *deformed* (*Dfd*) with its human semi-ortholog[6] *HOXD4* (McGinnis, Kuziora et al. 1990).

Dfd has a positive auto-regulatory function, which means that the Dfd protein enhances the transcription of its own mRNA. McGinnis showed that the expression of *Dfd* mRNA could be caused by the HoxD4 protein from mouse. This means that this specific biochemical function of Dfd had been conserved since the time of the most recent common ancestor of mammals and flies. This was a very important discovery, which has been backed up by many subsequent experiments with other transcription factor comparisons, mostly between Drosophila and mouse for technical reasons.

However, the picture quickly became more complicated because, in experiments that followed McGinnis's discovery, it was found that mouse genes often could replace their fly homologs only in a subset of their natural functions. To my knowledge, the first example was a comparison of the Drosophila *tinman* gene with its human homolog, *Nkx2.5* (Park, Lewis et al. 1998; Ranganayakulu, Elliott et al. 1998). Both genes are involved in heart development and some of the *tinman* target genes could also be activated with the Nkx-2.5 protein from mouse, but not all. A target gene that Nkx-2.5 could regulate was *FascIII*, but it could not regulate *eve*, *zfh-1*, or *D-MEF2*. Similar examples have accumulated since then, most notably evidence for the functional non-equivalence of the Hox protein UBX from Drosophila, brine shrimp, and velvet worm (Grenier and Carroll 2000; Galant and Carroll 2002; Ronshaugen, McGinnis et al. 2002).

Another reason for the idea that transcription factor proteins might not have evolved was certainly also that at the same time that evidence for functional conservation was accumulating, there was also success with identifying many cis-regulatory changes that were responsible for morphological differences between species. In each case, the responsible transcription factor was found not to have been changed. Examples include differences in the bristle pattern in the femur of the third leg of Drosophila, which could be traced back to changes in the CRE of Ubx, but with no changes in the Ubx amino acid sequence (Stern 1998); the loss of pelvic spines in freshwater stickleback populations, which was traced back to the CRE of the gene *Pitx1*

---

[6] *HoxD4* is a homolog of *Dfd*, but not an ortholog, because there are paralogs in the mammalian genome that also derive from *Dfd*, namely, *HoxA4*, *HoxB4*, and *HoxC4*.

(Shapiro, Marks et al. 2004; Chan, Marks et al. 2010); and the examples of cis-regulatory evolution discussed earlier in this chapter.

The experimental and comparative evidence for the conservation of transcription factor function was finally integrated into a theory that transcription factor proteins are not sufficiently modular to allow for adaptive modification, whereas CREs are often highly modular and, thus, do allow for genetic variation of high specificity and low pleiotropy. Pleiotropic effects reduce the chances that a mutation will have a net-positive effect on fitness. Thus, genetic elements that are highly pleiotropic are less likely to contribute to adaptive change than are genetic elements that are less pleiotropic. However, a literature survey of protein evolution suggests that protein modularity has been underestimated (Lynch and Wagner 2008; Wagner and Lynch 2008). Here, I will discuss only two classes of protein variation that escape the trap of pleiotropic effects and, consequently, also evolve rapidly and readily in transcription factor proteins.

Transcription factor proteins function through protein-DNA and protein-protein interactions. They form protein-DNA complexes that interact with the transcriptional machinery. A protein-DNA interaction is usually mediated through highly conserved DNA binding domains and these evolve slowly, if at all. Initially it was found that there are sequence motifs that mediate protein-protein interactions and that are highly conserved, as for example the so-called pentapeptide YPWM of some Hox proteins.

Recently, however, evidence has accumulated that transcription factor protein-protein interactions are often mediated through short linear motifs (SLiMs), usually only 3 to 10 amino acids long with only 2 or 3 amino acids that are essential for the protein-protein interaction (Neduva and Russell 2005). Most important these SLiMs are often found in natively unstructured regions of the protein, which liberates these motifs from structural constraints (Fuxreiter, Tompa et al. 2007). All these factors—small size, lax sequence specificity, and a lack of structural constraints—contribute to their evolvability and, thus, to the evolvability of transcription factor protein-protein interactions. In contrast to structured domains, it was found that SLiMs were poorly conserved and easily evolved independently in different proteins.

A paradigmatic example of the evolvability of SLiMs involves the variations of the transcription factor protein fushi tarazu (Ftz) (Löhr, Yussa et al. 2001; Heffer, Shultz et al. 2010). The *Ftz* gene was originally discovered in Drosophila due to its role in segmentation (fushi tarazu is Japanese and means "insufficient segments"). This gene is part of a homeotic gene complex, the Hox gene cluster, although in Drosophila it has no homeotic function (i.e., has no function in determining segment identity). The role of Ftz in segmentation requires the physical interaction between Ftz and a nuclear receptor (called Ftz-F1). The interaction between Ftz and Ftz-F1 is, on the side of Ftz, mediated by the SLiM LXXLL.

A comparison of Ftz function across insect orders revealed that its role in segmentation was likely to have been derived. There was no evidence for its segmentation function in the grasshopper *Schistocerca*. In contrast, in the grasshopper, the Ftz ortholog has homeotic function, as expected from its location in the Hox gene cluster. It also lacks the nuclear receptor interaction SLiM, LXXLL. Instead, it has an interaction motif, YPWM, which mediates its interaction with Extra denticle (Exd), as do many other Hox proteins (figure 6.12). The homeotic function of some Hox proteins, including grasshopper Ftz, requires physical association with Exd. Interestingly, the Ftz protein of the flour beetle, *Tribolium castaneum*, has both SLiMs and also has both homeotic and segmentation functions.

Putting the functional and structural information into the phylogenetic context of these three species (figure 6.12) suggests that, ancestrally, Ftz was

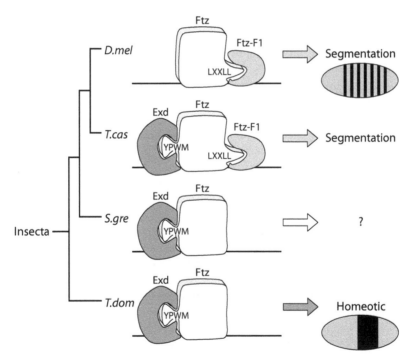

FIGURE 6.12: Evolution of protein-protein interactions with the Hox gene *Ftz* in insects according to Löhr and Pick (2005). Ancestrally, *Ftz* had homeotic function like most other Hox genes, depending on its interaction with Exd. During the evolution of insects, *Ftz* acquired a nuclear receptor interaction motif, LXXLL, and an *Ftz*-F1 interaction necessary for its derived function in segmentation. In Drosophila, the Exd interaction was lost and *Ftz* became limited to a segmentation function. *D.mel, Drosophila melanogaster; T.cas, Tribolium castaneum; S.gre, Schistocerca gregaria; T.dom, Thermobia domestica.* Redrawn by permission from Merabet and Hudry 2011, *BioEssays* 33:499–507. Copyright © 2011 WILEY Periodicals, Inc.

a homeotic Hox protein, homologous to the Hox3 paralog group in other animals. Later on in insect phylogeny, Hox3 acquired the LXXLL SLiM and a role in segmentation. Finally, in the dipteran lineage, Hox3/Ftz lost its YPWM motif along with its homeotic function and became a gene specialized for segmentation. This is a beautiful example of the evolvability of transcription factor proteins.

Just like linear motifs, simple sequence repeats (SSRs) are evolutionary labile and often variable in length between species. Therefore, they have been called "evolutionary knobs" that fine-tune transcription factor function. SSRs are microsatellite-like amino acid repeats that are particularly abundant in those proteins that regulate gene expression and evolve rapidly (Karlin and Burge 1996; Mar Albà, Santibáñez-Koref et al. 1999; Young, Sloan et al. 2000; Alba and Guigo 2004). For example, the rate of amino acid insertion and deletion in the glutamine-proline (QP)-repeat of the *Alx4* gene (also discussed below) is 2.7 indels per 100 million years, nearly 3 times the rate of duplicate gene fixation, 5 times the average nucleotide substitution rate, and 10 times the rate of novel exon formation.

SSRs have also recently been implicated in generating morphological divergence. In a study of SSR variation in 17 developmental genes among 92 dog breeds, Fondon and Garner found extraordinarily high levels of variation in the length of SSRs between different breeds (Fondon and Garner 2004). Furthermore, they found evidence that repeats were fixed in breeds more rapidly than expected under drift, and that these repeats were more diverse than would be expected under a neutral model. While most of these variations were minor changes in repeat length (usually two or three amino acids) five genes (*Six3, HoxA7, Runx2, HoxD8,* and *Alx4*) were found to have large expansions or contractions in SSRs in coding regions (Fondon and Garner 2004).

Although the function of most of these repeats is unknown, previous developmental and biomedical studies with mice and humans suggested that mutations in *Alx-4* could result in phenotypic effects. The Great Pyrenees dog breed, for example, is usually homozygous for the *Alx-4* Δ17aa mutation. An official characteristic of this breed is an extra toe on both hind feet (bilateral rear, first-digit polydactyly). All four Great Pyrenees dogs with bilateral polydactyly examined by Fondon and Garner (Fondon and Garner 2004) were homozygous for the *Alx-4* Δ17aa mutation, while a single individual that lacked these extra dewclaws and the other 88 breeds had neither polydactyly nor the 17aa deletion in *Alx-4*. This form of polydactyly is similar to that observed in *Alx-4* knockout mice.

Amazingly, deleting the QP-repeat, which is reduced in size in the Great Pyrenees, in mice specifically abolished ALX-4 binding to the cofactor LEF-1, which drives target gene expression in the limb bud (Boras and Hamel 2002). These data indicate that repeat length variations can have functional effects and suggest that changes in repeat length can alter gene expression by modulating protein-protein interactions between transcription factors.

Will a change in the length of simple amino acid repeats always lead to negative pleiotropic effects? Biomedical studies of repeat expansion diseases, a class of genetic diseases caused by expansion and contraction of SSRs, suggest that SSRs might have extremely few pleiotropic effects. For example, expansion of a polyalanine repeat in *HoxD13* by 7–14 residues causes synpolydactyly, a dominant developmental limb deformity characterized by duplication of fingers, and webbing between fingers (Goodman, Mundlos et al. 1997; Kjaer, Hedeboe et al. 2002; Anan, Yoshida et al. 2007; Zhao, Sun et al. 2007). No other organs or tissues are affected. Combined with the data that SSRs can mediate protein functions, these results suggest that changes in the length of repeats can have specific functional consequences without globally affecting protein function. Thus, one of the primary postulates of the cis-regulatory paradigm, namely that transcription factor proteins are strongly constrained in their potential to contribute to morphological change because of negative pleiotropy, is not supported by data from the function and evolution of SLiMs and SSRs.

As described above in the section on the role of gene duplication, we found that one teleost paralog of the gene *HoxA13*, namely *HoxA13a*, has acquired a new function in the zebrafish lineage. Intriguingly, the two paralogs *HoxA13a* and *HoxA13b* differ not only in their rates of nucleotide substitutions but also in their indels. *HoxA13b* underwent only deletions as compared to the unduplicated ancestral sequence, whereas *HoxA13a* underwent only insertions. Furthermore, a Morpholino®-mediated knockdown of *HoxA13a*, but not of *HoxA13b*, caused loss of the yolk sac extension (Crow, Amemiya et al. 2009). The yolk sac extension is a larval feature of zebrafish and its relatives and is an evolutionary novelty in the cyprinid clade (Virta and Cooper 2009; Virta and Cooper 2011). Together, the statistical and experimental evidence suggests that *HoxA13a* acquired a new developmental function and that indels might have played a role in the origin of a developmental novelty.

Finally, I want to briefly summarize a study that provided the first and most unambiguous link between an evolutionary novelty and a changed transcription factor protein function. This concerns the origin of the gene regulatory network of endometrial decidual cells: those cells that negotiate the interface between the fetal and maternal tissues at the site of implantation in placental mammals (Lynch, Tanzer et al. 2008). A small handful of transcription factor genes has been shown to be essential for decidual function in humans. Among these are two Hox genes, *HoxA11* and *HoxA10*. For *HoxA11* it was found that, in the lineage to mammals, the amino acid sequence underwent an increased rate of evolution (Chiu, Nonaka et al. 2000). Subsequently, with more data, it was found that, indeed, most of these changes had occurred coincidentally with the evolution of placentation (Lynch, Roth et al. 2004).

One function of *HoxA11* is to regulate the expression of the hormone prolactin in decidual cells. When *HoxA11* proteins were tested for their capability to up-regulate a reporter gene from the enhancer used by decidual cells, the so-called decidual enhancer, it was found that only proteins from placental mammals could up-regulate reporter gene expression. Neither the protein from opossum, the platypus, and the chicken, nor the reconstructed protein of the most recent common ancestor of opossum and humans could up-regulate the reporter gene from the decidual enhancer. However, the reconstructed protein from the most recent common ancestor of placental mammals had the same regulatory capability as the mouse and human proteins. Hence, it is clear that, coincident with the origin of the decidual gene regulatory network, the biochemical function of the HoxA11 protein changed to allow for PRL up-regulation in decidual cells during pregnancy.

Similarly, it was found that another key decidual transcription factor, CEBP-beta, which stands for CCAAT/enhancer binding protein-beta, had also changed its function coincident with the evolution of invasive placentation (Lynch, May et al. 2011). In this case, the biochemical mechanism is known. Human CEBP is the target of GSK3-beta kinase, which results in the phosphorylation of a serine. This results in a conformational change of the protein, which makes its DNA binding domain accessible for interactions.

This critical serine residue is absent in ancestral CEBP proteins. Instead, there are two different phosphorylation sites on a few N-terminal residues. These two serines have been replaced with alanines and, thus, are not available to be phosphorylated. Interestingly, this replacement led to a change in the functional response to phosphorylation. In the phosphorylated state, the ancestral protein is inactive, whereas the derived protein is activated. There are a number of available molecular pathways for transcription factors to be able to change their activities through evolution of their amino acid sequences: new interaction motifs, as in ftz; the expansion of amino acid repeats (Alx); or changes in phosphorylation sites, as in CEBP.

The Complexity of Transcription Factor Function: Molecules or Agents?

Research into the molecular functions and evolution of transcription factor proteins revealed that the classical model of transcription factors as conserved tools in the developmental toolbox underestimated their functional complexity. I will explain this point by contrasting two types of natural entities: molecules and agents. The point here is not that transcription factors are not molecules or that agents have powers that somehow go beyond what molecules can do. Rather, the point is that each category—molecule and agent—implies a different conceptual scheme. Thus, I will argue that transcription factor proteins are more complex than the classical view of molecules and, for that reason, are more similar to agents. Nothing of what follows is really new to molecular biologists, but has not yet fully informed the thinking in evolutionary genetics. For that reason, it might be useful to point out in what way transcription factors are more complex than simple molecules.

If one thinks about molecules based on a background in classical chemistry, like what I learned in chemical engineering school, one thinks of a configuration of atoms. This configuration has a core (e.g., a chain of carbon atoms) and some reactive residues hanging off of it, like carboxyl or hydroxyl groups. The kinds, the numbers, and the arrangements of these reactive groups define the chemical properties of a molecule. The "chemistry" of a substance, in particular that for small organic molecules, is defined by its reactive groups and the reactivities of the carbon "body" (whether it is aliphatic and saturated or aromatic, i.e., cyclic and unsaturated).

This model can, to some extent, also be applied to proteins and transcription factors. They have been found to have specific "active domains" that mediate certain biochemical functions, like DNA binding, nuclear localization, ligand binding, and enzymatic activity. One can think of these as biochemically active elements, just like the reactive groups in a small organic molecule.

Things got more complicated with the discovery of so-called allosteric effects, in which a ligand changes the conformation of a protein and determines whether a certain biochemical activity occurs or not—for example, enzymatic activity. In transcription factors and other regulatory molecules, allostery is usually called *internal regulation*. This is a significant step because it means that the biochemical activity of a protein is not defined by the presence or absence of certain domains; rather, there is context-dependent regulation of what kind of biochemical activity is available. The point here is that accumulating evidence suggests that evolutionary changes in transcription factor function can be achieved by evolutionary changes in the internal regulation of transcription factor activity rather than by changing the basic biochemical capacities of the protein (DNA binding, enzyme catalysis, etc).

If, as all evidence shows, internal regulation is a major part of transcription factor function and evolution, then the formal structure of transcription factor function is very similar to the formal model of animal behavior. The basic idea of how animal behavior or any other biological entity functions is tripartite. There are sensory functions that gather "information" from the environment. This sensory information is then fed to a module that is dedicated to information processing, and the output of this module is signals that activate or inhibit certain "motor" functions, whether they are locomotive or biochemical (e.g., secretory and other functions; figure 6.13). The same basic functions are performed by most, if not all, transcription factor proteins.

They sense the environment by ligand binding or by being subjected to modifications by kinases or other protein-modifying enzymes. Internal regulation of protein activity then "decides" which of the protein's biochemical

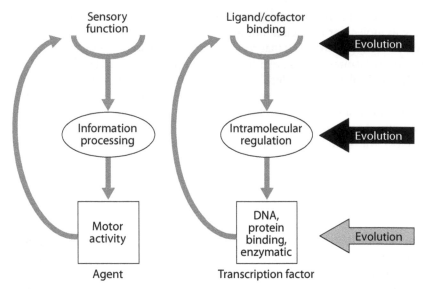

FIGURE 6.13: Transcription factor proteins are, at an abstract formal level, comparable to agents. An agent, like an animal, can be conceptualized as performing three distinct functions: sensory perception, information processing, and motor activity. All models of animal behavior can be accommodated in this basic model. Recent research into the function and evolution of transcription factor proteins suggests that transcription factors also perform these three basic functions. Sensory function is performed through ligand binding, cofactor binding, and protein modification. Information processing occurs through intra-molecular regulation, as for example by auto-inhibitory peptides that hide or expose other parts of the protein. Finally, the "motor" activities are performed by the domains of the transcription factor dedicated to the basic biochemical functions of a protein, like DNA binding, enzymatic activity, and so on. It turns out that the sensory and information processing parts of transcription factors are preferred targets of evolutionary change. The arguments that originally suggested that transcription factors could not evolve are valid with respect to motor functions, but not with respect to the sensory and information processing functions of transcription factors because the latter are more context sensitive. Changes in these functions would be expected to have limited pleiotropic effects and, thus, were more likely to have been advantageous than changes in the "motor" functions.

capacities are deployed in a certain cell or cell state. For example, many transcription factors can both increase or decrease the transcription rate of target genes. An example from my lab is the "house transcription factor" HoxA11 that has two repressor domains, one in the homeodomain and one in the Ala-repeat region, but also has an N-terminal activation domain. Whether HoxA11 will promote activation or repression depends on its own regulatory state and interaction partners. The point here is not that this is the case, which is widely known in molecular biology, but that these internal regulatory mechanisms are a preferred target of evolutionary change.

The first example from our lab that forced us to rethink the way that transcription factors evolved was a study by Kathryn Brayer, who studied the evolution of the protein-protein interaction between HoxA11 and Foxo1a (Brayer, Lynch et al. 2011). Briefly, she found that this protein-protein interaction evolved in the stem lineage of mammals, that the interaction occurred with the homeodomain of HoxA11, and that amino acid substitutions in HoxA11 were responsible for the evolution of the derived protein-protein interaction.

What was unexpected, however, was that there was not a single amino acid substitution in the homeodomain of HoxA11 at that time in phylogeny when the protein-protein interaction evolved! The model we came up with to explain these results was that the ability of the homeodomain to interact with Foxo1a was ancestral and may even extend to other Fox proteins (Foucher, Montesinos et al. 2003), but that other parts of the HoxA11 protein, outside the homeodomain, regulate whether this interaction takes place (figure 6.14). Hence, it seems that the protein-protein interaction between HoxA11 and Foxo1a does not arise from a new protein-protein interaction domain; rather, it occurs through a change in the internal regulation of protein-protein interactions. CEBP evolution that was discussed above (Lynch, May et al. 2011) is also a case of evolutionary changes to internal regulation for which phosphorylation decides whether or not and to what degree the DNA binding domain of CEBP is available.

The importance of the idea that internal regulation is a target of natural selection is that, unlike the biochemical activity domains, changes to intramolecular regulation are likely to be more context-specific and, thus, less

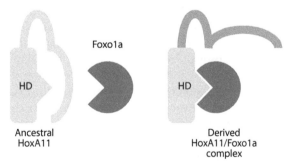

Foxo1a

| HD | | HD |

Ancestral
HoxA11

Derived
HoxA11/Foxo1a
complex

FIGURE 6.14: Evolution of a protein-protein interaction between HoxA11 and Foxo1a through intra-molecular regulation. Foxo1a binds to HoxA11 at its homeodomain. Ancestrally, HoxA11 and Foxo1a did not interact, even though the homeodomain was identical to that in derived HoxA11 proteins that can interact with Foxo1a. The evolutionary changes that allowed the evolution of this protein-protein interaction occurred N-terminal of the homeodomain and most likely affected the availability of the interaction surface of the homeodomain to Foxo1a (from Brayer et al., 2011, *PNAS* 108:E414–E420).

likely to have negative pleiotropic effects. It is true that mutations in the DNA binding domain are unlikely to be adaptive, even though the DNA binding preference of transcription factor does evolve (e.g., in fungi). But in complex organisms the potential for their evolution is low. However, the functional specificity of transcription factors and other regulatory molecules is not defined solely by their sets of biochemical activity domains; rather, specificity is also defined by context-specific regulation of these biochemical activities. Evolution of the "sensory" and "information processing" functions of transcription factors is the answer to the question of how transcription factors with pleiotropic biological roles can evolve (figure 6.13).

The Evolution of miRNAs

For a long time it was thought that the main players in molecular biology were DNA and proteins, with mRNA, rRNA, and t-RNA as their handmaidens to translate DNA into proteins. In recent years, however, it became clear that this picture was but a tiny segment of biological reality. With the advance of genomic technology and the discovery of genes without protein products, a vast world of novel RNA-based biological agents was and is being discovered. It turns out that of all the DNA in the human genome, only 2%–3% of this DNA codes for proteins, while more than 50% is transcribed into some sort of RNA (Kapranov, Willingham et al. 2007). Among these RNAs are a large variety of regulatory RNAs with an equally wide array of regulatory functions. Among them, the best understood non-coding RNAs are the micro-RNAs (miRNAs for short).

miRNAs are about 22 nucleotides long and are produced from longer RNA transcripts, which form a hairpin structure and are then processed to become a mature miRNA. Their main mode of action is to bind to the 3'-UTR (untranslated region) of an mRNA and prevent its translation into protein. miRNAs are expressed mostly during the later stages of development (Wienholds, Kloosterman et al. 2005) and seem to have an effect on stabilizing cell differentiation, as well as in buffering the phenotype against random variations arising from stochastic gene expression (Peterson, Dietrich et al. 2009).

miRNAs can have quite dramatic effects on the canalization of a morphological phenotype. For example, Li and colleagues (Li, Wang et al. 2006) showed that variations in the numbers of sensory bristles in Drosophila were strongly controlled by miRNA9a. The numbers of sensory bristles have been studied for decades using quantitative genetic methods—in particular, the number of sensory bristles on the scutellum[7] of the fly's back. The number

---

[7] The scutellum is a small triangular shield on the back of the fly corresponding to the third segment of the fly thorax.

of scutellar bristles is usually highly stereotypical in a species and serves as a paradigm for a canalized trait. This trait was studied by James M. Rendl, who published a little book about this research and became the main reference on canalization in the pre-molecular era of genetics (Rendel 1967).

The formation of a sensory bristle depends on the expression of a transcription factor called *senseless* that determines the ability of an epidermal cell to commit to sensory organ development. miRNA9a targets the 3'-UTR of *senseless* mRNA and controls its abundance in the cell. Li and colleagues showed that a loss of miRNA9a function resulted in higher variability in the number of sense organs that were formed. In contrast, miRNA9a overexpression resulted in a severe loss of sensory bristles. Hence, miRNA9a controls both the average number and the variability of sense organ number.

miRNAs also have an intriguing pattern of evolution. Most miRNAs, once originated and integrated into a gene regulatory network, no longer change to any extent. Thus, it is possible to reliably identify miRNAs by sequence comparisons, even though their sequences are short. In addition, they seem to rarely get lost once they are established, which makes them ideal phylogenetic markers.[8] In 2006, two groups independently showed a striking pattern of miRNA innovation in metazoan phylogeny: Peter Stadler's group at the University of Leipzig in Germany (Hertel, Lendemeyer et al. 2006) and Kevin Peterson's group at Dartmouth College in New Hampshire in the United States (Sempere, Cole et al. 2006).

These groups showed that the pattern of miRNA acquisition is highly non-uniform and seemingly focused on periods of major body plan innovations (figure 6.15). For example, in vertebrate phylogeny there were two major bursts of miRNA acquisition, one at the origin of vertebrates, when 41 new miRNA families were generated, and one at the origin of mammals, when 63 new miRNAs were acquired (Heimberg, Sempere et al., 2008). In fact, Sempere and colleagues pointed out that, to date, the number of miRNAs is the closest genomic correlate of phenotype complexity, as measured by the number of cell types (figure 6.16). This means that an increase in the number of cell types is associated with an increase in the number of miRNA species in the genome. This correlation makes sense, given the role of miRNAs in stabilizing cell types in terms of their gene expression patterns.

Most likely, as in the case with gene duplications, the association between phenotype innovations and miRNA numbers is caused by selective retention.

---

[8] Indeed it seems that deeper phylogenetic relationships that are notoriously difficult to reconstruct with conventional sequence comparison methods are being resolved with miRNAs. The reason is that normal gene sequences continue to evolve after a lineage split and, thus, the phylogentic signal can erode by later evolution. In contrast, miRNAs stay put and, thus, are like molecular fossils identifying related lineages. The only drawback is that miRNA inventories are expensive to determine and some of the data is based on the lack of certain miRNAs in certain species, which can always be a detection artifact.

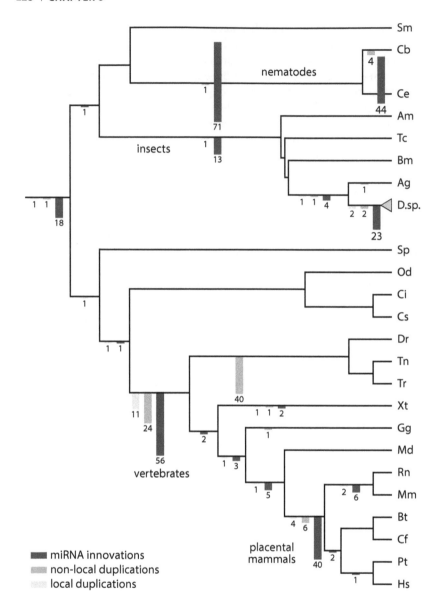

FIGURE 6.15: Phylogenetic reconstruction of miRNA innovation events during animal evolution. miRNA innovations are associated with major evolutionary transitions, such as the origin of vertebrates and placental mammals. Hs: *Homo sapiens*; Pt: *Pan troglodytes*; Cf: *Canis familiaris*; Bt: *Bos taurus*; Mm: *Mus musculus*; Rn: *Rattus norvegicus*; Md: *Monodelphis domestica*; Gg: *Gallus gallus*; Xt: *Xenopus tropicalis*; Tr: *Takifugu rubripes*; Tn: *Tetraodon nigroviridis*; Dr: *Danio rerio*; Sp: *Strongylocentrotus purpuratus*; Ci: *Ciona intestinalis*; Cs: *Ciona savignyii*; Od: *Oikopleura dioica*; Dsp: *Drosophila sp.*, Ag: *Anopheles gambiae*, Tc: *Tribolium castaneum*, Am: *Apis mellifera*, Bm: *Bombyx mori*, Ce: *Caenorhabditis elegans*, Cb: *Caenorhabditis briggsiae*, Sm: *Schistosoma mansoni* (after Hertel et al., 2006, *BMC Genomics* 7:25).

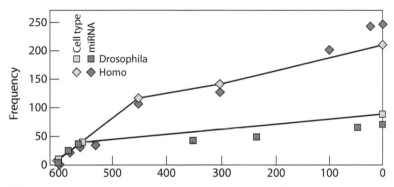

FIGURE 6.16: Correlations between the numbers of cell types of a species and the numbers of miRNA families in its genome. The increase in miRNA families closely tracks the increase in body plan complexity as measured by the number of cell types (after Sempere et al., 2006, *JEZ-B* 306B:575–588). Note: horizontal axis = millions of years before present.

Given the number of novel miRNAs found in closely related Drosophila species (Hertel, Lendemeyer et al. 2006; Sempere, Cole et al. 2006), it seems that miRNAs are continually created and preferentially retained when novel cell types or organs evolve. miRNAs probably arise from the prodigious amounts of non-coding RNA produced in many, if not all, animal genomes.

Much of this RNA does not need to have specific functions, but is either a side product of large polycistronic transcripts or the result of spurious transcription. These transcripts can quite readily evolve imperfect hairpin structures by coincidence. Once RNA exhibits a hairpin of the appropriate size, it then becomes subject to the enzymatic machinery that makes a potential miRNA from it. Hence, large amounts of spurious or junk non-coding RNA transcription provide a supply of potential novel miRNAs that can be utilized if the biological need arises.

A Material Difference between Innovation and Adaptation?

The genetic mechanisms that were described here are a collection of exotic mutations: new cis-regulatory elements from transposable elements; novel transcription factor functions; and new miRNAs. It seems that rewiring a gene regulatory network, as required for the evolution of a morphological novelty, uses a quite different set of mechanisms than usually associated with adaptive changes that is, changes in enzyme activity and gene expression due to small changes in cis-regulatory elements. This distinction hints at the possibility that the difference between adaptation and innovation is not only conceptual, but that the conceptual difference might be mirrored by a difference in the molecular mechanisms. It is far from clear whether this distinction will hold up, because there are still only a limited number of cases of innovations that are understood at the molecular level. However, one should at least not prematurely dismiss this possibility.

The possibility of a mechanistic difference between adaptation and innovation is also interesting because the characteristics of the genetic mechanisms may explain the phenomenology of innovations; innovations tend to be rare and episodic and result in a phenotype that tends to be canalized in its major features. As discussed above, one of the main characteristics of mutations by transposable elements is that they are episodic and specific to certain lineages. Mutations caused by transposable elements are most prevalent after the infection of a genome by a new retrovirus or any other new transposable element.

Similarly, genome duplications also temporarily open a window of evolvability by releasing constraints on gene evolution, and the maintenance of duplicated genes is often associated with body plan innovations. There is also a tendency for maintaining novel genetic elements with the origin of morphological novelties: new genes, new cis-regulatory elements, new miRNAs, and probably many others. Transcription factor protein evolution is likely necessary for the evolution of novel functional specificities, and miRNAs are involved in canalizing phenotypes once they have arisen.

Hence, the conceptual uniqueness of innovations (i.e., the origination of a novel cell type or of a quasi-independent body part) as compared to adaptations (i.e., the modification of existing body parts and physiological processes) may require a set of mutational mechanisms that can radically rewire gene regulatory networks and stabilize/canalize the phenotypic product of these changes. If further research supports this idea, then the conceptual distinction between adaptation and innovation will be linked to and grounded in the distinctness of the underlying molecular mechanisms.

7

The Long Shadow of Metaphysics on
Research Programs

It should thus be clear that metaphysical preconceptions
profoundly influence the course of scientific investigation.

—Michael Ghiselin, 1969, p. 53

A chapter on "metaphysics" would seem to be out of place in a book on a scientific subject. The vast majority of these books can do without this, thank you very much, although the shadow of metaphysics hangs over all of them, even if often unrecognized. This is because science and philosophy have a two-way relationship with respect to their influence. On one side, scientific discoveries affect how we see the world and ourselves and, thus, also affect how our worldview is cast into concepts and categories, like matter and mind, causality and chance, or factual and theoretical ideas. On the other hand, the prevailing philosophical viewpoints in a community influence what is considered a proper way of pursuing science.

For example, reflections about the mind of God have no place in modern scientific discourse, while Newton and to some degree even Einstein felt that they had to invoke ideas about what goes on in the mind of God. Is chance really a part of the world ("God does not play dice"; Einstein), or is it just an admission of our ignorance regarding highly complex forms of causation? For biologists, philosophical discussions about the "nature of species" and the "nature of higher taxonomic categories (genera, phyla, etc.)," are closer to home, as these ask, in what conceptual mold shall we cast our empirical knowledge about biological diversity and variation? Do higher taxonomic categories mean anything other than a way to economically summarize descriptive data about species?

Such a position is called nominalism by asserting that taxonomic categories are human-made distinctions for our own convenience, but are otherwise meaningless. Or do they reflect an objective reality, as for example, phylogenetic relatedness? Similar issues arise in discussions about homology.

Thus, a book on homology should provide answers, albeit preliminary, to questions like what is homology? and what are homologs? These questions raise the following ones: what kind of conceptual category shall we use to summarize our factual and theoretical insights into the structure, variation, development, and evolution of body parts? Are homologs nominal kinds as simple arbitrary summaries of phenotypic structure and variation—a position I clearly reject; or are they natural kinds[1]—a position that is highly controversial?

If they are natural kinds, then are they of the same sort as chemical elements, like gold or sodium? Shall we venture to give a formal definition of these concepts, or shall we choose a more flexible way of talking about these issues? Before we dive into these largely unsettled questions, let us reflect a little more about the relationship between philosophy and science.

Metaphysics as the Sister of Science

Metaphysics essentially means "behind physics," originally simply meaning the book written by Aristotle that was placed behind his book about physics, which happened to be his book on what we would now call philosophy of science or philosophy of nature. Thus, metaphysics is the natural extension of the theoretical reflection that starts as soon as humans venture beyond the most rudimentary descriptive and practical knowledge about nature. This is so because any significant scientific insight has implications for broader philosophical issues.

What we know about the structure of the brain and how its components work has implications for our notions of the self, the mind, or spirit, or whatever words we use to refer to our mental existence. The fact that we currently think of the brain as an elaborate computing device greatly affects what meaning we attach to these words. Similarly, what we learn from chemistry and atomic physics about the structure of matter has implications for how we think about what keeps the world together at its core (forces between electrons and protons, rather than a world spirit), and for what we are willing to contemplate as a causal explanation.

One of these views, which was strongly shaped by the physics and chemistry of the nineteenth century, is reductionism (Brigandt and Love 2012). This is the view that the laws that govern the atomic and sub-atomic particles are the only source of causal explanations, meaning that anything in the world is nothing more than the macroscopic manifestation of the interactions among these particles. Everything, in this context "everything" really means everything, from the storm front that moves over our town as I write these lines to the love I feel toward my children.

---

[1] The notion of natural kinds refers to the naturalness of groupings of objects, and not whether the objects themselves are natural.

On the other hand, philosophical generalizations, once they are established and widely accepted, also stabilize certain scientific theories and practices, sometimes against the pressure of empirical evidence. For example, the reductionist bias in much of the sciences encourages a focus on microscopic parts and molecular mechanisms, even though emergent phenomena are equally evident and in need of study on their own. Pattern formation and many other forms of non-linearity can lead to cooperative phenomena that can make the reductionist promise seem vacuous.

The reductionist promise, as I see it, is the idea that once we understand the laws that govern the components of a system, we can deduce or at least understand the behavior of any aggregate of these elements. This promise cannot be delivered, however, if the behavior of the system strongly depends on the historically contingent structure of the system, as is the case with organisms. I think there is no way around this , since any reductionist explanation will have to assume certain facts about how cells and organisms are set up, which are not themselves part of a reductionist explanation; they are historically contingent and, thus, not derivable from the laws of atomic physics. Of course, within the confines of historically contingent facts, like the structure of the eukaryotic cell and the anatomy of the human brain, reductionist explanations work brilliantly (Craver 2005; Bechtel 2010; Brigandt and Love 2012).

The same two-way street exists between evolutionary biology and the philosophy of science. As briefly sketched out in chapter 1, twentieth-century evolutionary biology had a distinct ontology that focused on genetic variation and population processes. This theory-specific ontology forced broader metaphysical revisions, which will be summarized in the next section, leading to philosophical categories like that of "individual" in contrast to classes and sets. These philosophical ideas, in turn, stabilized the paradigm of how evolutionary biology is done.

These metaphysical commitments work regardless of whether these ideas explicitly enter the day-to-day decisions of the practicing scientist. For these ideas to be effective, and in the way that they are, it is sufficient that the opinion leaders in the field are influenced by them. At any one time in history, at least some of the opinion leaders were deeply interested in these philosophical issues, as obviously was the case for Ernst Mayr (Mayr 1982) who dominated evolutionary thinking for the better part of the twentieth century.

The reason why philosophical issues have to be part of this book is that many of the biological facts, problems, and theories regarding homology are inconsistent with the philosophical positions that grew from twentieth-century evolutionary biology. I want to make these positions explicit here so that I can address them for what they are—namely, philosophical generalizations from a style of scientific research, rather than self-evident truths or logical necessities. They are neither self-evident truths nor logical necessities. If not addressed explicitly, they remain in the subtext of discussions of biological substance and make scientific communication even more difficult than is necessary.

An example of the power of philosophical ideas in scientific discourse is the denunciation of certain scientific positions as "typological" and, thus, incompatible with the "true" understanding of evolution (Amundson 2005). Denunciation as "typological" excluded large parts of paleontology, comparative anatomy, embryology, and developmental biology and even phylogenetics and taxonomy[2] from the evolutionary synthesis as the mainstream paradigm of evolutionary biology. It is probably not a coincidence that developmental evolution first took hold in developmental biology rather than in evolutionary biology.

The strictly anti-typological character of evolutionary biology of the mid-twentieth century has never had much influence on the rest of biology. In particular, molecular biology of the last century was deeply committed to an ahistorical and typological thinking. Jacques Monod's dictum "what is true of *E. coli* is also true for the elephant" is perhaps the quickest way to make this point. Of course, we know that there are many things that are true of *E. coli* that are not true of elephants and humans. For example, the *E. coli* chromosome is a circular piece of DNA, whereas the genomic DNA of an elephant consists of several pieces of linear DNA. Nevertheless, the focus of molecular biology used to be and, to some degree, still is on generalities, while the focus of evolutionary biology is on variation, with each field of biology holding on to a philosophy that is consistent with its scientific practice.

As outlined in chapter 1, starting in the 1990s, evolutionary developmental biology re-introduced typological thinking to evolutionary biology in the form of what Amundson called "developmental types" (Amundson 2005; Love 2009). This move called for a revision of the philosophical and metaphysical ideas associated with evolutionary biology. Scientific practice, scientific theories, and their metaphysics eventually need to reach a conceptual unity, with each mutually enlightening the other.

Classes and Individuals

Almost all scientific fields initially started out with some kind of classification of objects or processes. Objects are grouped together according to some property that seems relevant for the field of study and we arrive at preliminary generalizations and concepts. For example, one of the first footholds

---

[2] Shortly after my arrival at Yale University in 1991, the Biology Department launched a faculty search in phylogenetics and taxonomy and I served on the search committee. After the advertisement for the position was published, one of my colleagues received an angry phone call from one of the founding fathers of molecular population genetics decrying the stupidity of a faculty search in this area, in particular at a leading academic institution, because "we have waited from a story from taxonomy for decades and it did not arrive." Clearly, there was a very strong opposition against even the most basic support for these fields.

for chemistry was the discovery of two classes of substances, now commonly called acids and bases (Ihde 1964). Each has caustic properties, although they are different, because if one combines an acid with a base, they will neutralize each other and produce a salt. However, if one combines an acid with an acid, they will not neutralize each other.

The properties of each class were then used to define these concepts, even though initially not all of these properties were identified correctly. For example, it was thought that the presence of oxygen was essential for acids, which gave oxygen its name: the thing that causes acids to be sour (in German the connection is more obvious, as oxygen is called "Sauerstoff," which roughly means "sour substance"). This was a reasonable generalization because many paradigmatic acids do have oxygen: sulfuric acid (H_2SO_4), nitric acid (HNO_3), vinegar (CH_3COOH); although one important acid, hydrochloric acid (HCl) does not contain "oxy"-gen. Nevertheless, the correct generalization was eventually obtained and it is now understood that an acid is a substance that, in water, donates protons, H^+, while a base is a substance that donates hydroxyl ions, OH^-.

The reason why I explained the history of the concepts of acids and bases in somewhat more detail than necessary to introduce the idea of a class is that it illustrates another important point about concept development. It is important to not jump from preliminary classifications to "definitions" too early in the development of a field. If chemists had stuck to the original "definition" of an acid, HCl would not, by definition, be an acid and the entire enterprise of chemistry would have been compromised. The development of a proper scientific concept requires flexibility and the willingness to revise "definitions" to accommodate new empirical findings, like HCl lacks oxygen.

In the same way, biology cannot make progress by sticking to any particular definition of homology and treat each finding that does not fit the definition either as evidence that the concept is meaningless or that the homology concept does not apply. In the case of acids it turned out that the concept of an acid was and is meaningful, even though the original definition needed to be revised. Similarly, the definition of homology as a character derived from the same character in a common ancestor leads to difficulties when considering parallelisms, developmental variation, and atavisms. Rather than disposing of the homology concept, we need to be willing to revise definitions (Griffith 2007), or to even abandon the quest for formal definitions all together, at least until a deeper understanding of the development and evolution of characters has been achieved. This point will be discussed later in this chapter.

The approach of defining classes of objects based on common, so-called "defining" properties has served science very well over the centuries. Initially it seemed that it also worked well in biology. A species was defined as a class of individuals that shared species-defining characters, like the pileated woodpecker is a woodpecker with a flaming red crest that, in the male, extends to the forehead and a red stripe on the cheek, while the similar ivory-billed woodpecker has a red crest that does not extend to the forehead in the male. Similarly, higher taxonomic groupings were conceptualized as classes with defining properties, like mammals are defined as that class of animals with hair and mammary glands and a secondary jaw joint (and many more characters), while birds are defined by feathers and a modified tail, the pygostyle, and other characters. Much of taxonomic practice in defining taxa by their shared characteristics still reflects the idea that taxa are classes, even though class-property metaphysics has been rejected within evolutionary biology.

The problem with the idea that biological taxa are classes is that biological objects are inherently variable, even in their "defining" characteristics. In contrast, all oxygen atoms have eight protons that, indirectly, are the reason for oxygen's chemical properties. Of course, chemical kinds are also variable, namely in the number of neutrons, which results in atoms of different mass called isotopes. But variations in the numbers of neutrons do not affect (most) of their chemical properties (except atomic weight and subtle differences in reaction rates).

In contrast, biological species do not have essential characteristics, at least none that both define a species and do not vary between individuals and during evolution. This elementary fact makes classes with defining properties, which work so well in chemistry and other inorganic sciences, inappropriate for evolutionary biology. To resolve this problem, Michael Ghiselin (Ghiselin 1966; Ghiselin 1974)[3] proposed changing the metaphysics of taxa from one based on "object-property-class" to one of "individuals" (see also Hull 1978). His claim was that species, lineages, and clades were names for individuals, and not for classes.

According to Ghiselin (Ghiselin 1997, pp. 38–45), an individual is an entity that is

- Spatiotemporally restricted, while classes are unrestricted
- Is not instantiated, but has parts
- Has no defining properties, and
- Is concrete as opposed to abstract

Spatiotemporal restriction of individuals means that an individual has a definite historical beginning and, ultimately, an end, and that it is limited in spatial extent. In contrast, abstract classes do not have definite beginnings, as these apply to anything that has the defining property, regardless of when and where this entity occurs in time and space. A class does not die out;

---

[3] My account of Ghiselin's philosophy is based on his 1997 book, *Metaphysics and the Origin of Species*. This book is the summary of his thinking in this area and better reflects his views than earlier accounts. Ghiselin, M. T. (1997). *Metaphysics and the Origin of Species*. New York, State University of New York Press.

species do. This property is closely linked to the second characteristic of individuals; namely, it has no instances.

My cat is not an instance of the class of cats; rather, it is a member of a population that is the species *Felis domestica*, and arose from another member of that population. In contrast, the oxygen atom in a particular molecule of water is a member of the class of oxygen atoms and does not have any historical relationship to other oxygen atoms.

By introducing the notion of an individual we have already talked about the fact that biological species are variable, which is also the reason why they are evolvable. As a consequence, so Ghiselin argues, a species cannot have defining properties and is, thus, an individual. Any person, a clear individual, also has no defining properties. A person changes over her/his lifetime due to developmental and aging processes or because of accidents. A person can lose body parts due to injury, but does not cease to be that same person, and may even change "identity" by growing a beard or changing hair color.

Finally, individuals are understood as specific and concrete entities, which do not exist as abstract concepts, like "all atoms that have eight protons" or all persons who are unmarried sexually mature males (a.k.a. bachelors), or the set of all triangles in the Euclidean plane.

Ghiselin also applied his ideas to homology—or, better, homologs—and concluded that homologs were also individuals rather than classes (chapter 13 in Ghiselin 1997 and Ghiselin 2005). Ghiselin first notes, and I agree (see Wagner 1989, and chapter 2 of this book), that homology applies to concrete objects (i.e., body parts), and not to properties or attributes of organisms. Homology is a relationship between body parts of different organisms, including organisms from different species.

Furthermore, body parts are inherited and, thus, form lineages, as species do (see also McKitrick 1994; Butler and Saidel 2000; Geeta 2003; Hall 2003; Wiley 2008), and they evolve. For Ghiselin, the latter characteristic implies that homologs cannot have a defining characteristic, at least none that is intrinsic, like the structure of the atomic nucleus that defines chemical elements, but are extrinsic and historical (i.e., descent from a corresponding part in a common ancestor).

This analysis has a lot to be recommended, but I think that the complexity of homology and homologs renders the distinction between individuals and classes less crisp and less uncompromising than Ghiselin wants us to believe. In the case of homologs, the problems with the individual-class distinction are both formal and biological.

Ghiselin notes (Ghiselin 1997, p. 206) that the homology relation has two formal properties:

- *Symmetry*: if A is homologous to B, then B is homologous to A. This is a consequence of the fact that both A and B are descendent from a common ancestor and neither is more related to the common ancestor than the other.

- *Transitivity*: if A is homologous to B and B is homologous to C, then A is homologous to C. This property differentiates homology from similarity, which is also a relationship between body parts, although similarity is not transitive. If A is similar to B and B is similar to C, does this not mean that A has to be similar to C, because modification can lead to dissimilarity between A and C.

If we add the somewhat trivial property of *reflexivity*, which means that A is homologous to itself, then one obtains all the formal properties of an equivalence relationship "⇔", as defined in elementary set theory (see, for example, Halmos 1972). The property of reflexivity is biologically somewhat silly, although it is conceptually innocuous. The point now is that any equivalence relationship (i.e., any relation that is reflexive, symmetrical, and transitive) implies an equivalence class in the sense of elementary set theory. Hence, at any point in time, the collection of homologous characters is a class, at least from a formal, mathematical point of view.

Thus, there does not seem to be any harm in using the language of classes when we talk about homologs, as long as we are clear on what relationship defines the class. There is no need to invoke an intrinsic defining property to use the notion of a class; equivalence relationships are sufficient. And homology has the mathematical properties of an equivalence relationship, even in Ghiselin's lexicon.

It may even be argued that the notion of intrinsic properties or essences is a post hoc rationalization that empirically is based on equivalence relations, even in the inorganic sciences. What empirically led to the discovery of chemical elements and the associated commonalities in terms of atomic structure was the original realization that different samples of substances could replace each other in chemical reactions (i.e., are chemically equivalent).

The biological complication with the individuality notion, as applied to homologs, is more interesting and, ultimately, more consequential than the formal objection just discussed. One of the centerpieces of Ghiselin's individuality concept is that an evolvable entity cannot have an "essence" (i.e., an invariant defining property). According to Ghiselin, an evolvable essence is a metaphysical impossibility. This might be so, but if applied to biological reality, this argument does not hold much water.

Let us consider as an example the eukaryotes. In contrast to archea and bacteria, eukaryotes are all those organisms that have a cell nucleus. Clearly, eukaryotes are a clade, and they evolve. Nevertheless, they all share a number of intrinsic characteristics and, if so inclined, one can say they share an "essence." The "essence" of being a eukaryote resides in the manner in which their cells are organized and how their genetic material is packaged (i.e., have a nucleus and have chromosomes of linear DNA and chromatin). Hence, having a shared, conserved property has not kept the eukaryotes from evolving in many other respects than those of the defining cell structure characteristics.

In fact, eukaryotes are the lineage that brought about the most complex forms of life, even though they are not the most abundant form of life. The reason that their "defining" properties remain conserved is probably that much, if not all of their physiological and adaptive business is conditional upon their cellular infrastructure and, as such, faces strong selective or structural constraints for maintaining these characters.

One may say that the features shared among all eukaryotes are those that provide the means for the possibility for all the other things these organisms do and, thus, are largely or completely outside the reach of adaptive modification. The idea that certain structural features of organisms remain conserved because of the role they play in enabling other, adaptive physiological or developmental functions goes back to Rupert Riedl (Riedl 1978) and was called "burden" or "generative entrenchment" by William Wimsatt (Schank and Wimsatt 1986). Mechanistically, the notion of a defining property of a collection of species that share a common ancestor is very well compatible with evolvability and may even be an enabling mechanism for adaptive change of other parts of the organism.

The biological basis for Ghiselin's theory of individuals was the evolutionary biology of the mid-twentieth century, prior to the discovery of highly conserved developmental genetic mechanism shared among a wide range of species (see chapters 1 and 3). Thus, it seems to emphasize a theoretical position that has since lost its empirical basis. Or, it never had an empirical basis, as argued by Amundson (Amundson 2005), because it was always based on absence of evidence due to the methodological limitations of transmission genetics.

Thus, we are faced with a situation in which we agree with Ghiselin on many points, as for example that homology applies to body parts and not to attributes and that homologs form lineages, but disagree with some of his positions. In particular, the role of conserved genetic and developmental mechanisms and, thus, the possible existence of biologically meaningful "defining properties," which are excluded from Ghiselin's metaphysics.

Therefore, we have to forge a new metaphysics that accommodates the new empirical findings of evolutionary developmental biology. This new metaphysics is arising from a reconsideration and extension of the notion of "natural kinds" that makes this idea more applicable to biological concepts than the classical notions derived from nineteenth-century chemistry and physics (see, for example, Wagner 1996; LaPorte 2004). In the next section we will discuss these recent developments.

Individuals and Natural Kinds

The discussion about individuals and classes in the last section was in fact hiding an even more important issue, because Ghiselin's notion of classes covers abstract kinds (e.g., triangles or linear equations), artificial kinds (tables and cars), as well as natural kinds (oxygen and gold). In the natural sciences, the question of natural kinds is the most relevant. In Ghiselin's account, individuals stand in sharp contrast to natural kinds, mostly because he refers to a rigid and somewhat outdated notion of natural kinds, as mentioned toward the end of the previous section. But the notion of natural kinds has changed since the early 1990s to accommodate biological kinds like species, higher taxa, and homologs (Boyd 1991; Wagner 1996; Griffiths 1997; Brigandt 2007; Assis and Brigandt 2009). In this section I want to summarize the recent developments in the thinking about natural kinds and what they imply regarding the conceptual status of homology and the biological accounts of homology advocated in this book.

The classical notion of natural kinds was inspired by the late nineteenth-century account of the structure of matter. This established the role of atoms and configurations of atoms (molecules) in making up the kind of matter we encounter in everyday life (as opposed to what is going on in the core of the sun or in the dark recesses of a black hole). Hence, natural kinds were conceptualized as a class of objects that shared an essential intrinsic property, like the number of protons in the nucleus of an atom.

Natural kinds were also thought of as classes of entities that could figure in laws of nature.[4] Clearly, this relatively rigid notion of natural kinds excludes a large number of scientific concepts that nevertheless play an important role in theorizing about nature, not the least of which are notions of genes, populations, species, and many more (see Wilson, Barker et al. 2007). In 1991, Richard Boyd initiated a movement to rethink natural kinds in more flexible terms that are more suited for the kinds of concepts that arise in the biological and social sciences (Boyd 1991). Boyd conceived of natural kinds as "homeostatic property clusters" (HPC) that exist because of some kind of homeostatic mechanism.

The HPC notion of natural kinds is clearly compatible with the classical notion of natural kinds. For example, the quantum laws that govern the configurations of electrons around a nucleus of a certain charge and, thus, determine chemical properties certainly qualify as a homeostatic mechanism. Regardless of how many electrons an atom loses, it will return to the same

---

[4] The fact that early accounts of the notion of natural kinds also often contain superficial discussions of biological species and taxa was misguided, as pointed out correctly by Ghiselin (1966) and Hull. Hull, D. L. (1980). "Individuality and selection." *Annu. Rev. Ecol. Syst.* 11: 311–332, for the reasons summarized above.

state when it recaptures those electrons and it will always behave in the same way when interacting with other atoms to form chemical bonds.

However, the HPC notion of natural kinds is more flexible because it does not require all members or instances of a natural kind to be identical with respect to some defining property. All we need is a cluster of properties that are correlated in their presence or absence because of some kind of homeostatic mechanism. It also allows for evolutionary change of a natural kind as long as the natural kind changes as a unit due to its causal cohesion. For example, sexually reproducing species clearly qualify as a HPC natural kind because of the cohesion of the gene pool caused by interbreeding. Even higher taxa can be seen as natural kinds because of descent from a common ancestor and the effect of the mechanisms of inheritance.

The HPC account of natural kinds even opens a conceptual window that allows reconciling the historical notion of homology with the various "biological" accounts. Below I will follow the most promising account, in my opinion, of natural kinds developed by Ingo Brigandt (Brigandt 2007; Assis and Brigandt 2009): and Wilson, Barker, and Brigandt (Wilson, Barker et al. 2007). What follows is a much abbreviated account of a complex and extensive literature published over the last 20 years.

An important step made by Brigandt and his colleagues was to point out that the metaphysical notions of natural kinds and individuals were actually compatible rather than opposed, as Ghiselin insisted (Ghiselin 1997; Ghiselin 2005). Their key insight was that for populations to be historical individuals in the sense of Ghiselin, the mechanisms that defined a species as a HPC natural kind makes it necessary for them to be individuals (i.e., being able to form cohesive lineages of descent and, thus, be an individual in Ghiselin's sense). Genetic cohesion caused by sexual reproduction made organisms part of a population and, thus, part of a unit of evolutionary change. If we accept this notion, then individuals are a special case of natural kinds. This way of thinking also fits well with the approach to homology advocated here.

In many respects, homologs as conceptualized by Ghiselin (1997), Brigandt (2007), and Wagner (1996) are individuals because one can see them as phenotypic units of evolutionary change. They also form lineages and, thus, are individuals. But their very ability to form lineages of descent depends on their variational and developmental individuality, which is subscribed by a host of developmental genetic mechanisms (see chapter 3). These same mechanisms allow body parts to retain their identity during evolutionary change and are, thus, the proper biological basis of homology. One model to mechanistically account for developmental and variational individuality is that of character identity networks (ChINs) that was detailed previously (Wagner 2007). What is interesting about Brigandt's theory of natural kinds is that it even points to an extension of this notion of homology to include cases not covered by the ChIN model advocated here.

An important liberalization of the natural kind concept is to allow homeo-static mechanisms to be external to the natural kind itself. In the classical no-tion of natural kinds, "essences" are due to intrinsic features and mechanisms, like the structure of the atomic nucleus or the sexual reproduction of a spe-cies. But the HPC notion of natural kinds also allows for external causes—for example, ecological factors that ensure species cohesion (Assis and Brigandt 2009). Following this line of thinking, one can imagine that some cases of homologous body parts can be covered and for which ChINs are unlikely to explain their identity, as is the case with the mechanisms envisioned by the "organizational homology concept" of Gerd Müller (see chapter 2).

For example, a certain segment of the circulatory system of a mammal, the *aorta carotis communis*, is defined as the artery that originates after the branching of the *aorta subclavia* and ends after this branching in the *aorta carotis interna*. All of these parts of the circulatory system are named entities in comparative anatomy books because of their consistent presence, but are unlikely to be individuated by a character-specific gene regulatory network. All are parts of the artery system and probably follow the same developmen-tal program (i.e., are unlikely to have different gene regulatory networks be-stowing them with developmental individuality).

What individuates them is their relative position and function in a larger, conserved system possibly called a body plan or archetype, depending on one's taste (Young 1993). From a mechanistic point of view, what individuates these characters in evolution is their developmental and functional integra-tion into the overall circulatory system ("burden" *sensu* Riedl and "generative entrenchment" *sensu* Wimsatt) and, thus, appears like an extrinsic homeo-static mechanism.

Burden is homeostatic, at least with respect to evolutionary change. While this idea points to a broader notion of "biological homology" than proposed here, we leave the development of this idea to a later time.[5] This seems to be legitimate because it points to an extension rather than an alternative way of accounting for homology and its explanation.

Definitions and Models

In the sciences and elsewhere, it is usually good advice to define a term when it is first used in a text or in a talk. The purpose, of course, is to give precise meaning to the term in order to facilitate communication. There is, however, a more insidious side to this advice that is usually overlooked and which can

---

[5] It is also interesting to note that novel developments in the "metaphysics" of natural kinds paves the way to an extension and modification of scientific concepts, like homology and its mechanistic explanation.

be damaging in a field of research that is still in flux and for which precise meanings are hard to come by, as briefly exemplified in the last section with the history of the concept of an acid.

Chemists had to change their definition to come to a deeper understanding of the nature of acids and bases. Rather than being a nuisance, changing definitions can be a good thing if the facts ask for it. But perhaps we can avoid definitions altogether, and yet still gain precision without them. Before we explore this more radical possibility, let us consider first those situations in which definitions work well.

There are two general situations for which definitions, as a precise and terse statement of the meaning of a term, work well. These are in the formal sciences and in mature empirical sciences. By formal sciences I mean mathematics, logic, and computer science for which definitions are used to construct concepts. "A circle is the set of all points in the Euclidean plane that all have the same distance to a given point." "A semigroup is a set S with an operation $*$, such that for all a, b, and c in S, we have $a*(b*c) = (a*b)*c$." In these examples, the definition, in a way, creates a concept of a "circle" and a "semigroup." All that can be potentially known about these concepts is implicitly contained in the definition. This is the way to go in the formal sciences and the use of definitions is all but unavoidable.

The other situations for which definitions work well are mature empirical sciences in which the theoretical framework and the empirical practices are well established. "Oxygen is the element with the atomic number 8." This is all a chemist needs to know to determine what he/she is talking about when the term oxygen is used. To understand why and how this works, one has to realize that, for a layperson, this "definition" is pretty useless. What does it mean to talk about a "chemical element," or of an "atomic number?" Try to explain this to your grandmother and, unless she already has a PhD in one of the sciences, you will not get very far in your explanation. Even the explanation that the atomic number corresponds to the number of protons in the nucleus will not help much. Besides, how do I know how many protons are in anything so that I find out whether something is an oxygen or not?

The definition of oxygen makes sense only within the broader theoretical context of atomic physics and the experimental practice of chemistry. Trained chemists have a variety of ways by which to tell whether something contains oxygen or not, and all of this makes sense because the experimental practice and the theoretical models match perfectly. This example shows that, in the empirical sciences, definitions work when the science has matured to that point where the theoretical understanding of the subject matter and its empirical practice have converged to a perfect equilibrium (i.e., when we "know what we are talking about"). This, however, is not always the case, in particular in the exciting fields of research where much is still to be discovered. Here is the reason why.

Giving a definition for a term or a concept presumes that we already fully understand our study object. This state of knowledge is summarized in the definition, which then also fixes the meaning of the terms used in expressing this knowledge. But giving a "definition" in a situation where the state of knowledge has not advanced to this point has the unfortunate effect of suggesting precision where there is none. Psychologically, definitions suggest finality and there is a reluctance to change definitions, because it seems immoral.

I think that, in many parts of biology, we are far from a level of understanding for which definitions can be useful rather than detrimental to scientific advancement. Philosophers speak of an "epistemology of the imprecise" and point out that a too rigid definition of a "gene" would have hindered the progress of genetics (Falk 1986; Rheinberger 2000; Stotz and Griffith 2004). This is also the case in evolutionary biology and, within evolutionary biology, nowhere more evident than with respect to homology. In other words, nothing that was said in previous chapters about homology, variational independence, and gene regulatory networks should be construed as formal definitions.

But what can we put into place instead of definitions to clarify meaning to the extent possible? One possibility is to think in terms of models rather than definitions when talking about homology, characters, and similar concepts. Let us first reconsider the classical "homology definitions" in the light of what has been said here.

There are two widely used and accepted "definitions" of homology, as discussed in the previous chapters. One is by Richard Owen that says that two organs are homologous in any shape or function if they are the same organs in two species. As pointed out before, the defect of this definition is that there is no precise meaning attached to the notion of "sameness."

The other widely used definition is that of Lankester (1870) who, following Darwin's idea of descent with modification, defined homology as the relation between two characters that correspond to the same character in a common ancestor. This definition adds the phylogenetic dimension to the term. But, as pointed out above (and in Wagner 1994), this "definition" also relies on the same undefined notion of sameness as that proposed by Owen and, thus, is not much more precise or deeper than that of Owen.

Basically, these two definitions say that characters are homologous if they stand to each other in an equivalence relationship, called sameness; however, we do not know what this equivalence relationship is. Of course at this point in history, we have a much better understanding of what sameness could mean than did either Darwin or Owen. Yet still I would argue that anything we can say at this point in time does not amount to a precise definition, comparable to the definitions of oxygen or gold.

So what is there to do in order to allow discussions and debate within the scientific community to advance knowledge? After all, science is a community effort, and without effective communication scientific progress is hard to come by. I think we can achieve effective communication and sufficiently precise meaning if we remain specific, rather than abstract, when talking about characters, and if we replace abstract definitions with models of what we think is going on.

The inspiration for talking about models to replace definitions derives, in part, from the experience with genetic concepts. One may ask whether there is a precise definition of a "gene," certainly the fundamental concept of genetics. There are various definitions of genes, depending on experimental context—for example, as a unit of segregation or as a piece of DNA encoding for a protein or some other agent, like a ribosomal RNA or a micro-RNA. Yet all these definitions have their obvious defects.

If defined as a unit of segregation, then by definition only sexually reproducing species would have genes, which is obviously a counterproductive stance. There is also essential genetic information coded in non-transcribed parts of the genome, like cis-regulatory elements; so, definitions based on transcription are too short in their reach. In some cases the information of the transcript is not contained in the genome; rather, it arises by modification of primary transcripts by RNA editing.

Without going much further into these issues, it is clear that biological reality is much more complex and our knowledge of them is too preliminary to make abstract definitions useful. Does this mean that geneticists are not effective in communicating about their study objects? Obviously geneticists are very effective, as genetics is one of the most productive areas of biological research and affects all other branches of biology. If precise definitions of the basic unit of study, the gene, which gives genetics its name, are not available, how do geneticists do their job and effectively communicate about it? The answer is interesting and potentially an example for the study of homology.

The way that geneticists achieve clarity of meaning to the degree necessary to conduct research is the use of limited models. In the context of transmission and population genetics, we think about genes as little balls with different colors that follow certain combinatorial and probabilistic rules in transmission between generations, called Mendelian "laws." We all know that genes are not little colored balls, and we know that the Mendelian "laws" can be violated; in fact, this was realized as soon as these laws were re-discovered in 1905, as pointed out by deVries. Nevertheless, while keeping these limitations in mind, the Mendelian model (if we may call it that), rather than calling it a law, allows geneticists to determine whether two mutations that affect a trait are allelic (i.e., variants of the same gene), or are mutations at different genes or loci. This is done by a so-called complementation test.

This is an important step in the genetic dissection of a trait. The Mendelian model implies a set of empirical tests to distinguish between allelic and non-allelic variations. In the same way in bacterial genetics, the operon model provides guidelines for how to understand gene regulation in response to external stimuli. No serious geneticist would doubt that there are cases that do not conform to the operon model. Nevertheless, this model is a great way to dissect and interpret data about gene interaction and gene expression. And the list of models geneticists use to organize their work goes on. They represent the body of ideas and empirical practices that make up genetic knowledge.

Clearly, none of the models used in genetics amounts to an abstract definition of the basic ideas of genetics. But they do provide the conceptual and practical tools to do genetic research and allow the application of genetic principles in medicine, agriculture, and biological engineering. Hence, the absence of precise formal definitions does not do any harm to genetics, and *it may not do any harm to abandon the quest for definitions of homology.*

If one speaks of models, then one implicitly acknowledges the approximate and preliminary nature of what one is talking about. The model is not the reality, as is obvious with the Mendelian model. The model exists to be changed if empirical findings suggest that this should be so. Models are not seen as giving a definite meaning to a term. Eventually, models can lead to precise definitions as a science matures. Note that the paradigm of a precise definition I used above, that of chemical elements, was dependent on a model of how the atom is structured; it is in the context of this model that saying "oxygen is the element with atomic number 8" has any meaning.

Models, once developed to a certain degree, lead to precise definitions. But they do not do this by stipulation, as is the case in the formal sciences; rather, they summarize empirical knowledge. How can this be done in the developmental evolutionary study of body parts?

One approach to avoid definitions as the sole means to specify concepts in the natural sciences was developed in recent years by Brigandt and Love (Love 2009; Brigandt and Love 2010; Brigandt and Love 2012). They note that concepts can have three functions: classify objects, which amount to classical definitions as discussed above; explain phenomena; and structure explanatory agendas. Brigandt and Love argue that in situations like the one we face with homology and innovation, the latter function is the most critical.

A research agenda is a structured set of problems and the ways of how to address them. A research agenda focuses the questions toward a larger set of problems, integrates the contributions from different disciplines, and establishes standards of evidence. Within a given research agenda, different non-congruent classificatory definitions can co-exist without harm because, ultimately, the research accomplished as inspired by the research agenda will drive science forward.

Another possible means to achieve focus on specific models in the study of character evolution and development is to avoid making statements about homology in a biological vacuum. Instead, focus should be on the specifics of the biology of each homology hypothesis. *That is to say, we shall strive to fill in with biological detail what we mean when we say two characters are the same.* What will such a model have to cover? The answer is hard to give because, to my knowledge, this approach has not been used in morphological or developmental evolutionary research. However, based on the ideas presented in the previous chapters, one may sketch what a "character model" might look like.

If a character is conceptualized as a unit of evolutionary phenotypic change, then a character model would have to include hypotheses about: 1) how the character is individualized from the rest of the body; 2) what the variational constraints and biases are on the character; and 3) the phylogenetic context. The first agenda, a model of the individualization of the character, would have to propose the limits of the character (i.e., how much of the body of an organism is included in the character). For example, is the shoulder girdle part of the limb or does it belong to another character? In what organismal context does the character exist, and when during development and how does this unit of phenotypic organization arise?

The second agenda, characterizing the variational properties of the characters, has to provide a list of component parts and a list of the constraints on variation of the characters, which result from the developmental processes that create the character during ontogeny. It also needs to include whether the character comes in different variational modalities, as mentioned in chapter 2 (i.e., character state classes that differ in their variational properties but still are homologous to each other, as probably is the case with feathers and reptile scales). In a character model of feathers, one could say that feathers are a variational modality of the reptilian body scales, which are distinguished from body scales by their developmental derivation from an epidermal tube, and the "cutting-open" through selective cell death. This mode of development leads to a characteristic range of topologically possible morphologies (Prum 1999; Prum and Brush 2002), which differs from that of scales.

There is phylogenetic continuity with scales and, thus, feathers are not a character different from scales; rather, they are a variational modality of scales. This might turn out not to be true as more research is done, but at least we have a clear statement of what we mean when we talk about feathers. We also provide a clear list of empirical questions to address if we want to test this model.

Finally we need to say what we think the phylogenetic extent is of the character, when and how it originated, and how far it was transmitted to descendent species.

This is not the place to give more detailed examples. However, the idea of a character model will guide the chapters in the second part of this book, which is dedicated to in-depth discussions of specific examples of characters and kinds of characters, like cell types, body appendages, and so on, and how the ideas developed in the first part of this book can be incorporated into a specific research agenda. *Any concept is only as good as the research program it inspires.*

PART II

· · · · · · · · ·

Paradigms and Research Programs

The first part of this book was dedicated to the conceptual problems associated with the notions of homology and character identity. The objective was to clarify the concepts to a degree that would allow them to be connected to the mechanisms of development and evolution. In this first part, empirical facts were primarily used to exemplify conceptual ideas. The degree of detail explained in Part I was, by necessity, limited and certainly not sufficient to give a satisfactory account of biological reality.

In the second part of this book, the relationships between conceptual ideas and biological facts are reversed. Each of the following chapters focuses on a specific biological system (e.g., cell types, vertebrate limbs, flowers, and others), which will be discussed in sufficient detail to assess the conceptual ideas developed in Part I. The emphasis will be on "reading" what is known about each of these systems with the concepts explained in the first part of the book in mind. The goal will be to probe whether the ideas regarding homology can be productively integrated into research programs on these specific biological systems. The examples discussed in the following chapters are limited and somewhat idiosyncratic, as they reflect my background and research interests as well as those of my friends whom I learned from in recent years.[1]

Chapter 8 is dedicated to the developmental evolution of cell types. Cell types are the lowest level of biological organization for which questions of identity (i.e., cell identity) play a major role and for which the ideas discussed in Part I should apply. Cells do directly replicate, but specific cell types usually are not transmitted across generations. For example, muscle cells do not usually beget other muscle cells, particularly when we consider muscle cells in different generations. Hence, specific cell types are entities that, unlike genes, do not directly replicate and for which identity is maintained in spite of indirect transmission. Indirect transmission occurs when the thing that is transmitted is generated anew within each generation, yet still maintains a recognizable identity. This area is a very active and expanding branch of

---

[1] In particular the chapters about feathers and flowers grew out of many discussions with my friends and colleagues Michael Donoghue, Vivian Irish, and Richard Prum.

research in developmental and cell biology, as well as increasingly in evolutionary biology in great part due to the efforts of Detlev Arendt at the European Molecular Biology Laboratories in Heidelberg, Germany.

Chapter 9 discusses amniote skin characters (i.e., scales, feathers, and hair). I became interested in these characters thanks to the work of my friend Richard Prum who, in 1999, published a developmental account of the evolutionary origin of feathers. Because I had the privilege to edit his first paper on this subject for the *Journal of Experimental Zoology*, I had a front row seat to witness the controversies regarding the origin of feathers at the time when the first feathered dinosaurs were discovered. All of these characters, scales, feathers, and hair, are essentially tissue level novelties and, thus, the next logical level up from cell types to examine questions about character identity from a developmental and evolutionary perspective.

Chapter 10 deals with fins and limbs, a subject of my scientific interests for a long time. Limbs and fins are among the best understood organ systems in terms of developmental mechanisms and genetics and also for which a large amount of comparative anatomical and paleontological work has been done. Fins and limbs were the paradigms for which Richard Owen developed his ideas about homology and, thus, are an important test case for the mechanistic conception of character identity and homology.

Chapter 11 is a continuation of chapter 10 in that it deals with parts of vertebrate limbs: digits and digit identity. The reason for separating the discussion of digits and digit identity is that it occupies a disproportionate fraction of the controversies about the homology of parts of the vertebrate limb. In particular, there is an extensive literature regarding the identity of the digits in the avian wing that will be reviewed. In my lab, some of the first applications of the biological homology concept were developed in this and allied areas. Another interesting evolutionary problem regarding digit development is that of the urodele hand, which develops unlike any other tetrapod hand. Finally, the question of digit loss and digit re-evolution became an active research area from both the comparative and the developmental point of view.

Chapter 12 is about flowers and flower organ identity. The genetics of flower organ identity is one of the paradigms of organ identity genetics and has sparked active investigations into the origin and genetics of flower organ evolution. The origin of flowers is also a major innovation and, thus, comparable to fin and limb evolution in terms of its value in evaluating a research program on the origin and evolution of character identity. In fact, when studying this subject area I realized that some of the most relevant research on the evolution of character identity has been done in plant biology.

This list of paradigms of developmental evolution has some obvious gaps. Most notable is the absence of discussions about arthropod segments, arthropod limbs and insect wings, and wing patterns like butterfly eyespots.

A large amount of work has been done in these areas and these systems are, thus, also important examples for the development and origin of character identity. By the same token, these areas are highly specialized and this author does not feel prepared to adequately summarize these areas of research. I would hope that others who are more familiar with the details of these fields will eventually assess the biological basis of homology using this vast amount of data.

8

Cell Types and Their Origins

... [T]he differentiated cell-types seem to be islands
in a sea of developmental change.

—Arendt, 2008, p. 879

One of the elementary facts of biology is that higher organisms consist of functionally specialized cells. There are muscle cells that aid the body in locomotion or with moving body fluids or gut contents, and aid in reproduction (e.g., birth in mammals). There are liver cells that detoxify the blood, sensory cells of all kinds that monitor signals from the environment, and many more. These cells have been classified according to their function and their phenotype into cell types, such as striped and smooth muscle cells, neurons and glial cells, to name but a few.

In most cases,[1] all cells and, thus, all cell types of the body arise from a single fertilized or unfertilized egg cell through a process of multiplication, stepwise commitment to a cell fate and, finally, terminal differentiation. During the phase of cell fate commitment, cells receive signals from neighboring cells and from the environment that induce changes in their gene expression status. Eventually, these changes cause the expression of the cell type specific phenotype. Further, the transcriptional status of the cell becomes independent of the external signals. At this point the cell is committed to its fate. Finally, during terminal differentiation cells assume their characteristic phenotypes that are necessary for performing their functions.

During evolution, the number of morphologically recognizable cell types has increased, at least in some lineages, and provides an intuitive measure of organismal complexity. The simple invertebrate metazoan *Trichoplax adhaerens* that has no internal organs, nor even a definite body axis, has been described as consisting of only four or five cell types: dorsal and ventral

---

[1] Of course the exception is propagation by fragmentation in which whole parts of the body form a new individual.

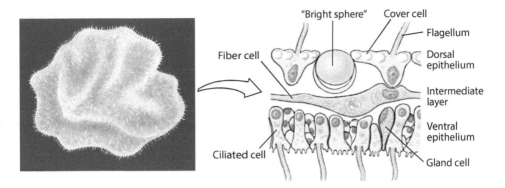

FIGURE 8.1: *Trichoplax adhaerens*, the simplest free living metazoan. (Left) Habitus of *Trichoplax*, as a "hairy spot" as its technical name suggests. (Right) Cross section of an individual showing the principal cell types (after Grell and Benwitz, 1971, *Cytobiologie* 4:216–240).

monociliated epithelial cells, fibrous contractile cells, and gland cells in the ventral epithelium (Syed and Schierwater 2002) (figure 8.1). In contrast, about 400 cell types have been described for the human body (Vickaryous and Hall 2006). Elementary inference suggests that new cell types must have arisen during evolution at some time (Valentine, Collins et al. 1994; figure 8.2).

There are very few molecular correlates with body plan complexity, as measured by the number of cell types. Neither the genome sizes (haploid nucleotide amount[2]) nor gene numbers are strongly correlated with cell type numbers in animals (Pertea and Salzberg 2010; figure 8.3). The only exception is the number of micro-RNAs (Sempere, Cole et al. 2006; see also chapter 6), which seem to be primarily involved in cell differentiation. Thus, it is plausible that the number of unique miRNAs tracks the number of cell types.

Yet in spite of the intuitive appeal of cell types as basic building blocks of multicellular organisms, the concept of cell types raises serious conceptual and practical questions. What exactly makes two cells belong to two different cell types? Are the cell type counts in different organisms, as cited above, really meaningful? How different do cells need to be to belong to different cell types? All these issues can be summarized by one fundamental question: *are cell types real?*

If they are real, in what sense are they real? Of course, this is a paraphrase of the question that motivated this entire book project: are any of

---

[2] It is usual to talk about DNA amount, but in fact the molar amount of DNA is equal to the number of chromosomes (each consisting of one DNA macromolecule) regardless of the size of the chromosome. What we usually call DNA amount is in fact the number of nucleotides, or the mass of DNA in a haplotype.

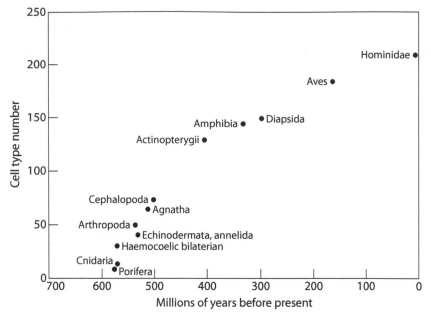

FIGURE 8.2: Relationship between the number of cell types and the age of a taxon. This diagram suggests an increase in cell types during metazoan evolution (from Valentine et al., 1994, *Paleobiology* 20:131–142). Used with permission from the Paleontological Society.

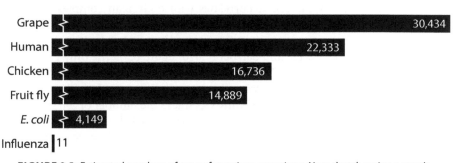

FIGURE 8.3: Estimated numbers of genes for various organisms. Note that there is no correlation between the number of genes and phenotype complexity (gene numbers after Pertea and Salzberg, 2010, *Genome Biology* 11:206).

the phenotypic building blocks of multicellular organisms, cell types, organs, characters, and so on, "real" or are they just imaginations of our regularity-seeking brains?

If we assume that we can answer "yes" to this first question regarding the reality of cell types, then two subsidiary questions arise that are, in fact, more interesting and productive. *What are the molecular mechanisms underlying cell type identity? How did cell type identity originate during evolution?*

Developmental Genetics of Cell Types

One way to address the question about whether the concept of cell types is scientifically meaningful is to ask whether there are identifiable genetic mechanisms that underlie the perceived differences among cell types. Of course, a phenotypic difference like that between a muscle cell and a red blood cell is caused by the expression of different genes that produce cell type–specific proteins, myosin versus hemoglobin, among others. This is a correct and even intuitively appealing answer, although it is misleading.

The problem with this answer is that even the same "kinds" of cells undergo cyclical or environmentally induced changes in their regulatory states. We do not want any two cells with a measurable difference in gene expression to be called different cell types, as this would be useless. *If cell types are real, then they should reflect some stable differences in the reaction norm of their gene expression.* By reaction norm I mean that, given an external stimulus, say an electrical field or mechanical stress, two different cell types will react in a cell type–specific way. A cartilage cell may react with the secretion of collagen II, a nerve cell with membrane depolarization, and a skin cell with keratinization.

The notion of a cell type requires that we allow for some variation in the phenotype of the cell. When the comparisons extend beyond a single organism, and even beyond a single species, the notion of cell types becomes even more abstract (see section on cell type evolution). That is, there are homologous cells that have different functional specializations, just like macroscopic organs can be homologous in spite of different functions.

At this point we have to make a pragmatic decision. Shall we retain a notion of cell types that is completely descriptive, as simply reflecting phenotypic differences related to differences of function, or do we want to link the notion of cell type to that of homology (i.e., a cell type is a "clade" of phylogenetically related cells) in the same sense as the identity of multicellular characters or genes? My preference is to link the notion of cell types to homology as it applies to cells. This is a pragmatic question rather than one of empirical fact. The reason I prefer linking the notion of cell types to that of homology is that it is conceptually richer than any descriptive concept (see also Ereshefsky 2012).

Linking the notion of cell type to that of homology raises the question of whether cell types are distinguished from each other by a cell type–specific regulatory state that individuates them, even before and in a way that is independent of their "typical" phenotype. In other words, I am asking whether cell types are, to some degree, abstract, whether they have an identity that goes beyond the possession of a particular phenotype. The insistence on an "abstract," but still mechanistic notion of cell identity, derives from the very notion of homology, which refers to character identity "in all varieties of form and function," as in Owen's influential definition of homology. Hence, thinking about cell types as potential homologs forces us to look beyond or underneath the phenotypic characteristics that usually are associated with the notion of cell types.

Going back to the last chapter of Part I, chapter 7, the question is whether cell types are natural kinds in the sense that they have a causal homeostatic mechanism that maintains a property cluster. In a few cases, the molecular details have been worked out to such a degree that it is possible to answer this question affirmatively and outline the structure of the gene regulatory network and the molecular mechanisms involved that endow cells with a specific identity. In this section, I will review examples that are particularly well worked out and that reveal common structural features of cell identity networks.

Transcription Factor Antagonism: Hematopoietic Cell Fate Decisions

The first evidence for the critical role of transcription factors in cell fate determination came from the observation that fibroblast cells could be converted into muscle cells through the forced expression of the transcription factor MyoD (Davis, Weintraub et al. 1987). These experiments showed that, in certain cell types, MyoD was sufficient to cause the assumption of a specific cell phenotype. But the question of how a critical transcription factor excludes the expression of alternative cell phenotypes was first worked out in the blood cell lineages. Specifically, the cell fate decisions between monocytes, which act as macrophage precursors (also called the myeloid lineage), and erythroid cells that lead to red blood cells are well understood (Laiosa, Stadtfeld et al. 2006).

The distinction between myeloid and erythroid cell fates is based on the functional antagonism between a pair of transcription factors, PU.1 and GATA1. Myeloid cell fate is associated with PU.1 expression and erythroid cell fate with GATA1 expression. When GATA1 is expressed in myeloid cells, it not only results in the up-regulation of erythroid marker genes, but also results in the down-regulation of myeloid markers. Conversely, forced PU.1 expression in erythroid cells causes the induction of myeloid marker genes *and* the suppression of erythroid genes. What is important in our search for the mechanistic basis of cell type identity is the molecular mechanisms that realize this transcription factor antagonism. The following summary is based on Graf and Enver (2009) and the references therein.

In order to activate the target genes of erythroid cells, GATA1 recruits the histone acetylase CREB-binding protein (CBP). CBP is an activating cofactor of many transcription factor complexes. In the presence of large amounts of PU.1, however, CBP becomes physically displaced from GATA1 by PU.1, which also can directly bind to GATA1. Once bound to GATA1, PU.1 recruits

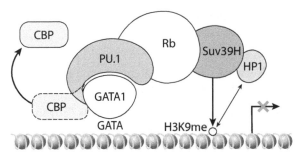

FIGURE 8.4: Mechanism of transcription factor antagonism in cell fate decisions between myeloid and erythroid cells. GATA1 promotes erythroid cell fate and is opposed by PU.1, which promotes myeloid cell fate. To activate erythroid target genes, GATA1 recruits the co-factor CBP. PU.1 displaces CBP from GATA1, associates with GATA1, and then recruits repressive cofactors, like Rb and others, which results in the production of repressive histone marks, H3K9me. (after Graf and Enver, 2009, *Nature* 462:587–594)

two other proteins, Rb and Suv39H, which, like CBP, are also chromatin-modifying enzymes, but with the opposite effect. These proteins result in the methylation of lysine 9 in histone H3. Furthermore, they recruit HP1 that, in conjunction with histone methylation, results in repressing GATA1 target genes (figure 8.4).

In myeloid cells, the cell type–specific factor PU.1 activates target genes through the recruitment of its cofactor c-Jun. The antagonism of GATA1 for PU.1-dependent gene expression again is through a direct protein-protein interaction. GATA1 binds to PU.1 and displaces the activating cofactor c-Jun, which results in a loss of myeloid gene expression patterns.

What is interesting about this mechanism of transcription factor cross-antagonism is that the repression of the alternative set of target genes is not accomplished through alternative, repressive cis-regulatory elements. To the contrary, this repressive activity arises from the same cis-regulatory element as the activating one. The competing transcription factor turns the activating protein complex into a repressive complex that is still bound to the same cis-regulatory sequence. This is a very effective way of causing the repression of an alternative genetic program, because all one needs is a specific protein-protein interaction, as for example the ability of PU.1 to bind to GATA1 to displace the activating GATA1 cofactor, CBP.

This one type of interaction acts on all GATA1 target genes and, thus, causes a unitary switch from one cell fate to another. Also from an evolutionary point of view, the evolution of a novel protein-protein interaction as a means to change the regulatory state of a large number of target genes is very effective. The evolution of a novel protein-protein interaction can lead to repressing all of the target genes of a certain transcription factor. Disrupting a cell type–specific transcription factor complex is also found in another case of cell type specification: the decision between spinal motor neurons and V2 interneurons in mouse and chick (see below).

Core Regulatory Networks and Core Regulatory Transcription Factor Complexes: Embryonic Stem Cells

It might seem odd to include embryonic stem cells (ES) in a discussion of cell identities because, by definition, stem cells have not yet acquired a definite cell fate. Yet in mammals, embryonic stem cells have already made a cell fate decision, which is that between the trophoblast and the embryoblast or inner cell mass. ES play a well-defined role in mammalian development. They play their role according to a genetic program that includes reacting in a cell type–specific manner to external signals. The fact that these reactions are certain downstream cellular fates does not make them any less a cell type, as with a cell that reacts to stimulation with contraction or with the secretion of a hormone. The phenotype of ES is not dedicated to an adult function, but to a developmental function. Thus, it is an "inter-phenotype," as Riedl (Riedl 1978) called developmental homologs, rather than being a terminal, adult phenotypic character.

During mammalian development, the zygote undergoes cell division and soon separates into two cell populations: the trophoblast, which forms the extraembryonic tissues including the placenta, and the inner cell mass, which becomes the embryo proper. It is the inner cell mass from which, ultimately, all the cells of the adult mammalian body derive. This cell mass was first isolated in 1981 and was called embryonic stem cells. ES could be maintained in their pluripotent state in culture for extended periods of time (Smith, Heath et al. 1988).

It was subsequently found that two signaling molecules were essential for maintaining stem cell character. These were leukemia inhibitory factor, LIF, and members of the bone morphogenetic proteins (BMPs). Specifically, the effect of BMPs is through the transcription factor Smad1 to activate a set of genes known as the Id gene family, where Id stands for "inhibitor of differentiation." LIF acts through its proximate transcription factor STAT3. LIF and BMPs, however, do not directly determine stem cell identity; rather, they interact with a group of transcription factors that are expressed by ES: OCT4, SOX2, and NANOG (Boyer, Lee et al. 2005). Each of these has been found to be essential for ES-identity in humans and early embryonic development, and OCT4 and SOX2 are sufficient to reprogram fibroblasts into stem cells.

OCT4 and SOX2 have been found to form a protein complex and jointly regulate NANOG. Furthermore, the OCT4:SOX2 complex cooperates with NANOG to regulate downstream target genes in a feed forward network motif (figure 8.5A). This motif is one of the most frequently identified features in eukaryotic gene regulatory networks (Mangan and Alon 2003). The joint regulation of downstream target genes by SOX2, OCT 4, and NANOG explains why, in knock out or knock down experiments, each of these transcription factors is necessary for stem cell identity. In other words, they are jointly necessary, just as we saw in examples of character identity networks like the compound eye determination network in insects.

What makes the role and the biology of these three transcription factors so interesting is that they not only jointly regulate downstream target genes, but also regulate each other's expression in the form of a positive feedback loop (Boyer, Lee et al. 2005; figure 8.5B). This feature has several important dynamic consequences. First, it ensures that all of the genes that cooperate in regulating target genes are jointly active in a cell that is governed by their expression. It also explains why at least OCT4 and SOX2 are each sufficient to cause stem cell character, because they each activate the expression of the other to jointly cause the ES phenotype.

Another feature of this gene regulatory network is that, because of the joint regulation of these transcription factors, disturbances in the stoichiometric proportions of the transcription factor proteins result in loss of the ES character of these cells. This occurs, for example, in experiments in which one transcription factor gene is overexpressed and, thus, is present in molar concentrations different from those of its interaction partners. Experiments of this sort usually give unclear results rather than revealing the unique functional role of the transcription factor.

This is probably because there is no unique functional role of individual transcription factors. An individual transcription factor does not have a definite functional role because it is not the unit of function. In contrast, molecular analysis suggests that it is the complex of three, and probably more, transcription factor proteins and cofactors that is the actual functional unit for regulating target genes.

In terms of their joint action on target genes, these three transcription factors target a total of at least 354 genes. These can be divided into two classes (Boyer, Lee et al. 2005). One class is activated by these transcription factors and the other class is suppressed. The class of actively transcribed genes includes genes for chromatin modification factors and other transcription factors. Among the latter group is, for example, REST, a transcription factor that suppresses the expression of neuronal differentiation genes. On the other hand, the class of genes that bind to the OCT4, SOX2, NANOG transcription factor complex and are inactive are known players in cell fate determination and differentiation. While the latter is not surprising, given the role of ES, what is important is the shallow hierarchy between the key regulatory genes for ES identity and the effector genes, which seems to be a key feature of a gene regulatory network for cell identity (Mangan and Alon 2003).

OCT4, SOX2, and NANOG are the core set of regulatory genes for ES identity. But there is another question: how do external signals for maintaining ES

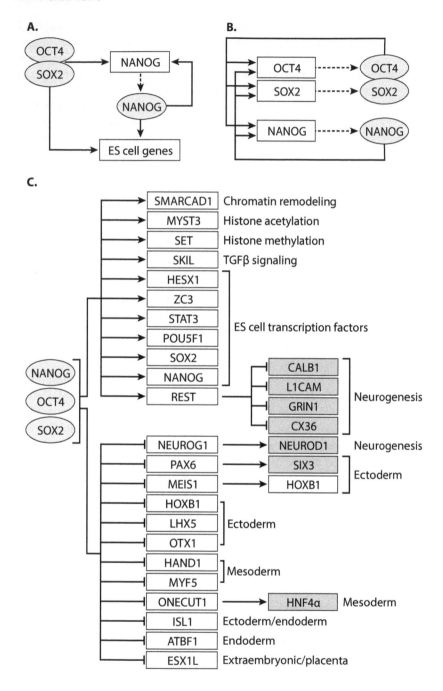

FIGURE 8.5: Model of the core gene regulatory network of embryonic stem cells (ES). A) Regulation of ES genes is caused by a feed forward loop by OCT4, SOX2, and NANOG. Feed forward loops are the most common network motifs in eukaryotic gene regulatory networks. B) OCT4, SOX2, and NANOG also form a positive feedback loop like a ChIN, which maintains the expression of the core regulatory genes. C) OCT4, SOX2, and NANOG jointly up-regulate target genes that promote stem cell phenotype and repress differentiation genes. Note that the regulatory hierarchy is quite shallow, where a large number of target genes is regulated by the same set of core regulators (redrawn after Boyer et al., 2005, *Cell* 122:947–956).

identity interact with the core network? We noted above that LIF and BMP are necessary external signals to maintain ES character. A ChIP[3] study that targeted the effector transcription factor of these signals, Smad1 and STAT3, suggested that these transcription factors were part of the transcription factor complex formed by the core regulators OCT4, SOX2, and NANOG (Chen, Xu et al. 2008). Chen and collaborators identified so-called MTLs, or "Multiple Transcription factor binding site Loci," and found that among the 667 MTLs that contained Smad1, 649 also had a binding signal for NANOG, OCT4, and SOX2. Similarly, 521 of the 718 MTLs that bound to STAT3 also bound to the set of the core regulators.

The MTLs identified in the study by Chen and collaborators (Chen, Xu et al. 2008) had the signatures of binding transcription factor complexes, so-called enhanceosomes (Thanos and Maniatis 1995). First, transcription factor binding was localized to a very compact region of the genome. Second, these regions acted as enhancers if tested in a reporter assay. Third, they had the chromatin modification mark of active enhancers—for example, H3K4me1 (i.e., methylation of lysine 4 of histone 3). Finally, they also recruited the transcriptional co-activator CBP/p300 to their sites of action.

This model explains the maintenance of ES identity, but does not explain how the alternative cell identity, the trophoectoderm, is prevented. It turns out that, similar to hematopoietic cell fate decisions (Graf and Enver 2009), cell fate decision in the early mouse embryo is also based on transcription factor antagonism (Niwa, Toyooka et al. 2005). In this case, the antagonism is between OCT4 and Cdx2, for which Cdx2 favors trophoblast fate by directly interacting with OCT4 to prevent ES fate. Conversely, OCT4 negatively interacts with Cdx2 and a redundant factor Eomeso. Interestingly, this antagonism seems to have evolved in the stem lineage of mammals, as it exists in platypus and eutherians, but not outside the mammals (Niwa, Sekita et al. 2008).

---

[3] ChIP stands for chromatin immunoprecipitation, in which the chromatin of a cell is fragmented and then an antibody against a transcription factor is used to enrich those DNA fragments associated with the transcription factor or other chromatin protein. These DNA fragments are then sequenced and mapped to the genome. Locations in the genome that are occupied by a particular transcription factor are then detected by a high representation of sequences in the DNA enriched by immunoprecipitation.

Detailed molecular analyses of ES and blood cell identity revealed two key features of cell type identity in general. The first was that of a core regulatory network that, in conjunction with external signals, cooperatively regulated target genes. The second was that a critical part of the cooperative regulation of cell type specific genes was mediated through transcription factor complexes and that, in some cases, the suppression of the alternative cell identity was caused by disrupting the core regulatory complex, which demonstrated the essential role of these transcription factor complexes.

A Long Chain of Cell Fate Decisions: Mammalian Motor Neurons

Embryonic stem cells are the result of a simple binary decision: either trophoblast cell or ES. In contrast, the development of a neuronal cell is the result of multiple cell fate decisions that lead to one of the many different cell fates in the central nervous system (CNS). The human brain comprises 50 to 100 billion nerve cells, and each of these cells has to decide from among a long list of alternatives, beginning with the decision between neuronal and non-neuronal (glial) cell fates. Among the neuronal cell types there are broad categories: sensory neurons, interneurons, and motor neurons according to the type of target cells they innervate and from where they acquire their information.

Sensory neurons receive their information from outside the CNS, interneurons communicate with cells within the CNS, and motor neurons contact cells outside the CNS, mostly muscle cells, but also glands. Within each of these categories are many cell types that differ in their precise connections. For example, among the motor neurons of the spinal cord there are neurons that innervate the axial muscles, others that innervate the muscles of the body wall, and yet others that innervate the muscles of the limbs. It is clear that this level of complexity necessitates several layers of regulation to arrive at a specific cell fate. A well-studied example is cell fate determination in the mammalian spinal cord.

Before we consider cell fate determination in the spinal cord, we need to briefly review CNS development in vertebrates (figure 8.6; Rao and Jacobson 2005). The CNS of vertebrates (i.e., the brain and the spinal cord) is derived from the neural tube. The neural tube forms from a fold of the embryonic skin,[4] the primary ectoderm. Hence, the vertebrate CNS is developmentally derived from an epithelial hollow structure inside the body cavity. The cavity of the neural tube becomes the cavities of the brain, the so-called ventricles and the cerebrospinal canal.

---

[4] The exception is the teleost fishes in which the neural tube is not formed as a fold, but as a compact rod of cells that delaminate from the primary ectoderm and then form a tube by dehiscence (i.e., by forming the cavity by cell separation).

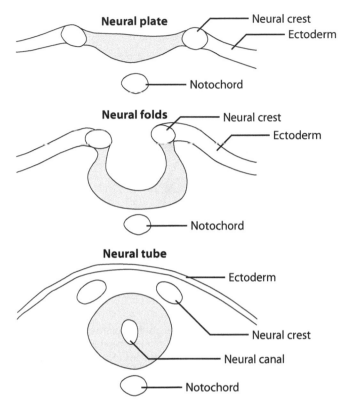

FIGURE 8.6: Embryonic origin of the vertebrate central nervous system. The central nervous system derives from a part of the embryonic skin called the neural plate. The neural plate folds in and finally detaches and forms a tube. At the edge between the neural plate and the definite ectoderm, a population of migratory cells is mobilized and becomes the neural crest.

The CNS has two surfaces. The ventricular surface that faces the cavity of the neural tube corresponds to the apical surface of the original epithelium. The second surface is the pial surface that forms the outer surface of the brain, which corresponds to the basal side of the original epithelium. Cell proliferation primarily occurs in the ventricular layers of the neural tube. From there, neuroblasts migrate to their definite places and then form their cellular processes, the dendrites and axons.

Now we can fill in this general picture with the details that are relevant for spinal cord development (Lee and Pfaff 2001; Shirasaki and Pfaff 2002). Cells in the ventricular layer of the neural tube are divided into two broad domains: the dorsal domain from which spinal sensory cells arise and a ventral one from which the motor neurons arise along with a number of interneurons. Let us now focus on the ventral part of the developing spinal cord.

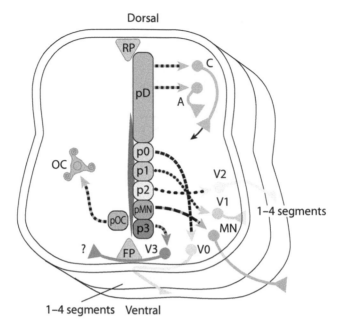

FIGURE 8.7: Neuron differentiation in the spinal cord. Neurons originate from a cell layer close to the spinal canal. This population is divided into dorsal and ventral compartments. Motor neurons (MN) derive from the ventral compartment. In this ventral compartment, five sub-populations are distinguished, with the precursors of motor neurons second from ventral (pMN). The other cells in the ventral compartment form interneurons V0 to V3 derived from precursors p0 to p3. Migration routes of neuroblasts are indicated by dashed lines, solid lines are axons. FP and RP are the cells of the floor plates and roof plates respectively, from which inductive signals emanate, Shh from FP and BMP from RP. OC are oligodendrocytes, a form of glia cell. C are commissural neurons that project to the other side of the spinal cord, A are associative cells that connect to other neurons in the spinal cord, and pD are precursor cells from the dorsal compartment (after Lee and Pfaff, 2001, *Nature Neuroscience* Supplement 4:1183–1191).

The ventral domain is further subdivided dorso-ventrally into five so-called progenitor domains (figure. 8.7). These are called, from dorsal to ventral, p0, p1, p2, pMN, and p3. From the pMN domain, the progenitors of motor neurons arise, while cells from the p0–p3 domains develop into four different interneurons, called v0, v1, v2, and v3, which make connections within the spinal cord.

The cells in the ventral progenitor domain develop under the influence of two morphogenetic signals, Shh and BMP. Shh is first secreted from the notochord that lies underneath (i.e., ventral from) the neural tube, and then by the floor-plate cells, which make up the ventralmost layer in the neural tube. BMP signals come from the roof plate, the cells at the dorsalmost extremity of the neural tube. High Shh concentrations result in p3 progenitors,

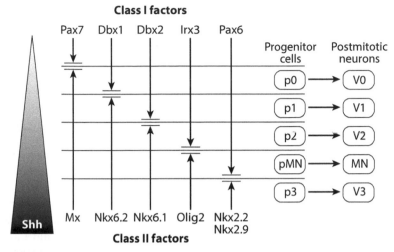

FIGURE 8.8: Transcription factor expression that determines the spatial identity of neuroblasts in the ventral spinal cord. Each limit among spatial identities is determined by the expression of two transcription factor genes that mutually inhibit each other's expression (after Shirasaki and Pfaff, 2002, *Ann Rev. Neurosci.* **25**: 251–281).

while low concentrations of Shh and higher BMP concentrations cause the development of p0 progenitors. The continuous concentration profiles of Shh and BMP are translated into sharp domain boundaries of differential gene expression by a set of 11 to 12 genes,[5] most of which are members of the basic helix-loop-helix class of transcription factors.

These genes fall into two classes. Class I is Shh repressed and BMP induced, whereas Class II genes are induced by Shh. In addition, each Class I gene has a Class II gene counterpart with which it forms a sharp boundary of mutually exclusive gene expression domains. For example, *Pax6*, a Class I gene, is expressed in pMN, p2, p1, and p0, while its counterpart *Nkx2.2/2.9* is expressed only in p3 (figure 8.8). In contrast, *Olig2* is expressed only in pMN and forms a sharp expression boundary along with *Irx3*, which is expressed in p2 to p0. These mutually exclusive expression domains are produced through mutual inhibition among complementary Class I and Class II genes. In this way, each progenitor domain expresses a unique combination of transcription factor genes.

What is important about this system of cell fate determination is that the unique combination of transcription factor genes expressed is not simply a

---

[5] Eleven genes are known but at least one more, a twelfth, is inferred to have to exist based on the overall architecture of the gene regulatory network. See Lee, S. K. and S. L. Pfaff [2001]. "Transcriptional networks regulating neuronal identity in the developing spinal cord." *Nat Neurosci* 4 Suppl: 1183–1191.

passive reaction to the inductive signals Shh and BMP, but is translated into an internal state that is actively maintained through cross-regulatory interactions among the Class I and Class II genes. This fact makes the identity of the progenitor domains, to some degree, independent of the inductive signal. This has been demonstrated in that a loss of Shh signaling in knock out mouse strains does not result in the loss of the progenitor domains. They still form, but somewhat less precisely in their location. Hence, progenitor cell identity is internally represented through the gene regulatory interactions among these genes and reinforced and modulated by external signals.

The next stages in motor neuron development are cell cycle exit, migration to the final location, and cellular differentiation. Motor neurons with different target tissues occupy different locations in the CNS (figure 8.9). For example, in the brachial (i.e., innervating the arms) segment of the spinal cord, motor neurons form three columns with different mediolateral positions. These are called medial and lateral motor columns, MMC and LMC, respectively. The neurons of the MMC innervate the axial musculature, while the LMC neurons innervate the muscles of the limb. The LMC is further subdivided into a medial and lateral sub-column, called LMCm, for the medial sub-column, and LMCl, for the lateral sub-column of the lateral motor column. The LMCm neurons innervate the ventral limb muscles and the LMCl innervate the dorsal limb muscles.

In the chick, the transition from the neuronal precursors to postmitotic neuroblasts is regulated by homeodomain proteins, MNR2 and HB9, which are closely related proteins. In motorneuron precursors, perturbations of this transition often result in a cell fate change to V2 interneurons, which will be considered in greater detail below. The genetic regulation of cell cycle exit

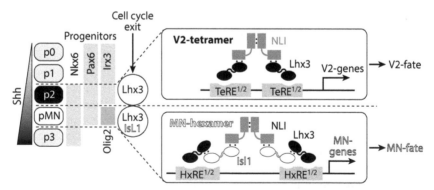

FIGURE 8.9: Regulation of motorneuron (MN) and V2 interneuron identity by different transcription factor complexes. V2 identity is caused by the activity of the tetrameric complex comprising two Lhx3 and two NLI proteins, while MN identity is caused by a hexameric complex that contains the two Isl1 proteins in addition to the four that are contained in the tetrameric complex (after Lee et al., 2008, *Developmental Cell* 14:877–889).

seems to be phylogenetically plastic, as the mouse does not have an MNR2 protein and the HB9 ortholog has another function in mouse spinal cord development than that in chicken spinal cord development.

Final cell identity is regulated by another class of transcription factors, the so-called LIM-HD proteins: Lhx3, Isl1, Isl2, and LIM1 (Lee, Lee et al. 2008). As their name indicates, these proteins have two characteristic domains. These include a homeodomain (HD), which includes the DNA binding domain, and the LIM domain, which is necessary for protein-protein interactions. The LIM domain interacts with another protein called NLI which also has an intrinsic dimerization domain. The result is that complexes of LIM-HD and NLI proteins lead to tetrameric and hexameric complexes like LIM-HD:NLI:NLI:LIM-HD. These complexes are the transcriptionally active molecular species that direct cell type–specific gene expression. LIM-HD proteins seem to be an ancient feature of motor neuron identity, as the homologs of the amniote LIM-HD genes are also involved in motor neuron identity determination in Drosophila (for references, see Arendt 2008).

As briefly mentioned above, the differentiation between MN and V2 interneurons is particularly vulnerable, which might suggest a relatively recent origination of the two cell types. But, to my knowledge, direct evidence for this is not available. The determination of MN identity is closely linked to suppressing the V2 neuron identity, and vice versa.

The expression of the V2-specific and the MN-specific developmental programs is caused by two different transcription factor protein complexes (Lee, Lee et al. 2008; Song, Sun et al. 2009; figure 8.9). The V2 phenotype is caused by the expression of a tetrameric complex (TeC), Lhx3:NLI:NLI:Lhx3,while the MN phenotype is caused by the expression of a hexameric complex (HxC), Lhx3:Isl1:NLI:NLI:Isl1:Lhx3. Hence, one difference between V2 interneurons and MN is Isl1 expression in MN.

The problem, though, is that in both cases, transcription factor complex binding to DNA is done by the same transcription factor, Lhx3. Thus, TeC and HxC can potentially bind to the same binding sites. What then distinguishes the cis-regulatory elements of TeC target genes, the so-called TeC response elements, TeRE, and those for HxC, HxC response elements, HxRE?

To cause specificity of cell fate, each cell type produces ancillary regulatory proteins that eliminate this ambiguity. V2 cells express a protein called Chx10, which specifically binds to HxRE and blocks the binding to TeRE. In MN, the TeRE are blocked by HB9, and HB9 is positively regulated by HxC. In addition, the MN expresses a protein called LMO4, an LIM-only protein that can associate with NLI. The presence of LMO4 in MN disrupts the formation of the TeC. The latter is necessary because the HxC includes the same transcription factors as the TeC, namely Lhx3 and NLI, and the affinity of an Lhx3:Isl1 complex to NLI is about the same as Lhx3. Thus, one would expect the formation of large amounts of TeC in MN simply by the laws of mass action. The role of LMO4 is to prevent the formation of TeC in MNs and favor the formation of HxC.

Hence, the ease with which MN fate can be converted to a V2 cell fate is related to the fact that both cell identities are caused by very similar transcription factor complexes, and special mechanisms are needed to ensure the correct cell identity. Once established, MN cell precursors further segregate into their respective subtypes according to their final position in the spinal cord and their target tissue. These MN subtypes are realized through the expression of a larger set of LIM-HD genes. For example, MMCm neurons express Isl1, Isl2, Lhx3, and Lhx4, while LMC neurons down-regulate Lhx3.

The role of LIM-HD genes in motor neuron specification is phylogenetically old. Homologs of the vertebrate LIM-HD genes also define MN types in Drosophila, and even have similar, if not homologous functions. For example, in vertebrates Lhx3 is essential for MN axons to choose a ventral path or grow toward the limbs. Similarly, the Lhx3 homolog in Drosophila, Lim3, is also essential for distinguishing between dorsal and ventral axon growth patterns (Allan, Park et al. 2005).

It seems that the genes that play roles in the proximate determination of cell types are highly conserved, while the inductive signals and the developmental pathways that generate MN are quite variable, even between chicken and mouse (i.e., the roles of MNR2 and HB9 in cell cycle exit; see above). This is similar to the situation described in chapter 3 for multicellular characters for which a character identity gene network tends to be conserved, but the inductive signals that activate this network vary between species.

Overall, neurons have a more complicated path to their cell fate than do embryonic stem cells and, perhaps, even than that of blood cells. But in the end, the proximate mechanisms of cell identity determination are quite similar. There is the role of transcription factor complexes in regulating target gene transcription (i.e. transcription factor cooperativity), and the cross-antagonism through interference with the function and integrity of the alternative transcription factor complex (TeC versus HxC; see above). Furthermore, motor neurons are also an example of the conservation of the cell identity mechanisms, as homologous transcription factors determine motor neuron cell identity in both amniotes as well as flies.

On the Nature of Cell Types

From the examples summarized above it is clear that cell type identity is subscribed by gene regulatory network states that cause and maintain the cell type–specific phenotype. Although many details differ among cell types, there are a number of similarities that suggest a model of cell type identity. These include:

1. Each cell type expresses a cell type–specific combination of gene regulatory molecules, sometimes called the *core network or cell type*

identity network. Most of the known regulators are transcription factors, but likely include non-coding RNA as well.

2. The core network forms a positive feedback loop and, thus, maintains its own expression, in addition to maintaining cell type identity. In this way, the gene regulatory state may become independent of the cell fate inducing signals.

3. The hierarchy between the "high level" core regulatory genes and the genes that determine the cell phenotype (i.e., the effector genes) is shallow. Often, the core regulators directly regulate all or most of the effector genes.

4. Core regulators do not act alone but form transcription factor complexes that are the actual regulatory units. We may want to call this entity "*Core Regulatory Complex*" (CRC) to emphasize its pivotal role.

5. Cell type identity is also maintained through the active suppression of alternative cell identities both at the level of cross-antagonism between the core regulatory networks and in terms of suppressing the alternative set of effector genes or disrupting the alternative CRC.

Some aspects of this generalization were summarized in a cartoon model by Graf and Enver (2009, their figure 3c; see figure 8.10A). In this model, genes are color labeled according to their role in cell type identity: "light gray" and "dark gray" cell types that are caused by the expression of "light gray" and "dark gray" transcription factor genes. This model assumes two layers of gene regulation: core regulators on the top level and effector genes at the bottom level. The core regulators are auto-regulatory, as indicated by the arrow pointing back at each core regulatory gene, and cross-antagonistic, as indicated by the blunt arrows between the core regulator genes.

Each cell type–specific core regulator activates its set of target genes and inhibits the expression of the target genes of the alternative cell type. This model explains the self-maintenance of cell type identity through the auto-regulatory effect of the core regulator genes. Cross-antagonism is represented by the inhibitory interactions among the core regulators and the inhibitory influences on the effectors of the alternative cell type phenotype. What this model does not represent is the role of the core regulatory complex of transcription factors. To accommodate these findings, the Graf-Enver model is extended in figure 8.10B.

The extended Graf-Enver model explicitly acknowledges that the regulation of target genes (TG) by their core regulatory genes (CR) is mediated by a core regulatory complex (i.e., a complex of transcription factor proteins encoded by the core regulatory genes). This was shown most clearly in the examples summarized above for embryonic stem cells, hematopoietic cells, and motor neurons. In addition, regulation mediated by the CRC implies that the contributions of all core regulatory genes are jointly necessary to obtain the correct output from the effector genes.

A.

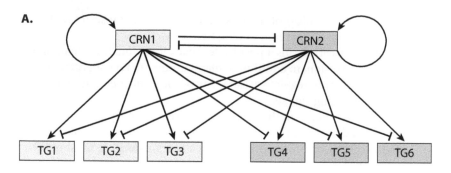

B.

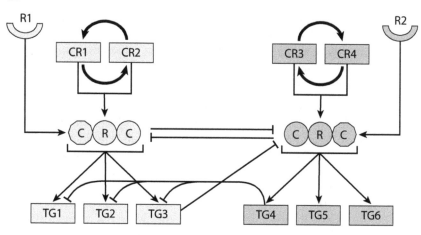

C.

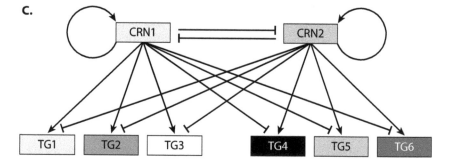

FIGURE 8.10: The Graf-Enver model of cell type identity. A) Basic model structure according to Graf and Enver (2009). Each gene regulatory state defining a cell type is driven by a core regulatory network, CRN1 and CRN2. Each CRN forms a positive feedback loop to maintain its own activity and inhibit the activity of the alternative CRN. Also, each CRN activates the effector genes necessary for the cell phenotype and inhibits, directly or indirectly, the expression of genes for the alternative cell phenotype. B) The extended Graf-Enver model of cell type identity. This model acknowledges that the genes from the core regulatory network (core regulator genes, CR) often form a transcription factor protein complex (core regulatory complex, CRC), and that CRC activity is modulated by signal transduction pathways originating from some receptors (R1 and R2). This model also shows that mutual suppression of alternative cell fates can be realized by different mechanisms: either direct interference between the core regulatory complexes, the expression of a target gene that then interferes with the alternative CRC, or direct suppression of the alternative target genes by a target gene of the other cell type. C) The "abstract" nature of cell type identity as explained by the Graf-Enver model (compare with A). By "abstract," I mean that the identity of a cell is not rigidly tied to a particular phenotype; rather it is tied to the activity of a certain core gene regulatory network. Two cells can be the same cell type but have different phenotypes. This can occur if they express the same core regulatory network but their CRN activates different sets of effector genes, indicated here by different shades as in part A.

The CRC also contains transcription factors that are activated by some receptor (R) through which the cell type responds to signals from other cells or the environment. This was documented above for the roles of STAT3 and Smad1 as mediators of BMP and LIF signaling for maintaining embryonic stem cell identity. The auto-regulatory nature of the core gene regulatory network is symbolized here as a direct cross-activation of two core regulatory genes. However, from the examples summarized above, it is not clear how this regulation works. For example, it could be that the cross-regulation requires the CRC to act on the members of the core regulatory network, although this would make it difficult to explain why some core-regulators can activate other members of the network.

The cross-antagonism between the gene regulatory networks of different cell types is symbolized by blunt arrows between the CRCs of the two cell types. This means that cross-antagonism is, at least in some cases, realized through disrupting the alternative CRC. This is best documented for hematopoietic stem cell fate regulation for which cross antagonism works through the displacement of the essential activating cofactor of the alternative CRC.

Another mode of cross-antagonism is the exclusion of the CRC from the cis-regulatory elements of its target genes, symbolized here as the inhibitory influences of TG4 on TG1, TG2, and TG3. An example of this mode of action is the blocking of the tetrameric response elements by the homeodomain protein HB9 in motor neurons and the role of Chx10 in preventing hexameric transcription factor complex binding in V2 interneurons. Disrupting an alternative CRC can also be accomplished by a target gene of the cell type–specific CRC. This mode of action is symbolized by the inhibitory influence of TG3 on the "dark" CRC. In motor neurons, an example is the role of LMO4 in preventing the formation of the tetrameric complex, which is the CRC for V2 interneurons.

Overall, this model and the experiments upon which it is based show that one can identify a set of cell-internal molecular mechanisms that cause a certain cell identity. The cell identity caused by the activity of a core regulatory network is "abstract," because the activity of the core activity network does not rigidly specify the particular set of target genes it regulates. Naturally, the target genes that are activated by the core regulatory network depend on the existence of cis-regulatory elements in the target genes, which are responsive to the cell type–specific CRC. In principle, different sets of target genes can be activated by the same core cell identity network if different target genes have response elements to the cell type–specific CRC.

Different cell phenotypes can be caused by the same cell identity network if the CRC regulates a different set of target genes (compare figure 8.10A and C). Hence, *cell identity is abstract relative to the cell phenotype.*

The cell identity network's relative independence from its target genes explains how cell function and cell identity/homology can be decoupled. For example, this occurs for retinal ganglion cells and photoreceptors (see below). Ganglion cells have the cell identity of photoreceptors (i.e., are homologous to photoreceptors), even though they do not possess specialized photosensitive structures, but primarily function as interneurons (Arendt 2008).

Based on this interpretation of experimental facts, cell identity is associated with certain phenotypes, but is not rigidly defined by them. Cell identity is linked to the continuity of the core gene regulatory network. This notion requires a change in the structure of the homology concept. Currently, homology is defined through historical continuity of inheritance; hence, many authors resist the association of specific features of characters with the definition of homology. But this historical or genealogical notion of homology, as discussed in chapter 2, begs the question of what it is that has continuity of inheritance in the case of body parts or cell types.

There has to be a mechanistic basis for the individuation of what is inherited and what has historical continuity. The *mechanistic basis for individuality*, in the case of cell types, *is the cell type–specific core regulatory network* that promotes cell type–specific phenotype expression and the suppression of alternative phenotypes. *Continuity of cell type identity (cell homology) is tied to the historical continuity of the core regulatory network.*[6]

---

[6] These core networks are highly conserved and it is tempting to tie cell-type identity to the identity of the core gene regulatory network. However, it is likely that even core networks do undergo some degree of evolutionary modification, in particular after gene and genome duplications or invasions by transposable elements, as was the case with the embryonic stem cell network. See Kunarso, G., N. Y. Chia, et al. (2010). "Transposable elements have rewired the core regulatory network of human embryonic stem cells." *Nat Genet* 42(7): 631–634. Hence, it is important to speak of historical continuity of the core network rather than identity of the core network.

The abstract nature of cell identity is also important for the evolvability of cell phenotypes. There is little, if any, integration among the large number of target genes that are regulated by the core gene regulatory network. Target genes can be removed or added simply by adding or losing a cis-regulatory element that is responsive to the core transcription factor complex. This explains the fact that most morphological changes are due to mutations in cis-regulatory elements, as pointed out by Sean Carroll and his school of evolutionary developmental biology (Carroll 2008).

Hence, the cell phenotype seems to be freely evolvable without the need to change the core gene regulatory network. The core gene regulatory network, in turn, only ensures the possibility of differential gene expression in different cell types. The number of different core regulatory networks determines the maximal number of different cell phenotypes that are simultaneously possible in a given species.

Furthermore, the core gene regulatory network is conserved because a large number of effector genes require the core transcription factors for their regulation. Natural selection that maintains effector gene expression also maintains the core transcription factor network. In addition, the members of the core network are not replaceable by transcription factors with similar DNA binding domains, because their participation in the CRC requires phylogenetically derived protein-protein interactions with other members of the transcription factor complex. Consequently, even though there are only a handful of DNA binding domains, and many transcription factor proteins that share the same DNA binding domain, transcription factors with the same DNA binding domain cannot replace each other during the determination of cell identity because of differences in their ability to interact with other transcription factors[7] and, possibly, non-coding RNAs.

Another striking feature of cell identity determination is that these core regulatory networks share structural features with those known to be involved in the identity determination of multicellular organs discussed in chapter 3. For example, the insect eye developmental pathway has a core gene regulatory network that includes mutually regulating transcription factors and cofactors, *ey*, *so*, and *dach*, which also form a protein complex for regulating target genes (for details, see chapter 3).

Thus, it seems that there is an emerging theme regarding the molecular basis of character identity at different levels of biological complexity starting from the level of cell types up to complex multiple tissue organs, like the eye. Naturally, in cells, the regulatory networks are limited to interactions among transcription factors within the cell, while the identity of multicellular characters includes cell-cell signaling, as for example in the segment polarity

---

[7] This constraint is of course lifted after gene duplications, at least with respect to first order paralog transcription factors (i.e., "sister" transcription factors).

network of insects. In each case, the mechanism for character identity is potentially abstract, as different sets of target genes can be regulated by the same core gene regulatory network in different species.

As based on the continuity of the core gene regulatory network, does the concept of cell type identity lead to a resurrection of "essences" as some authors suggest (Ghiselin 1996, 2005)? The answer is "Yes and No." Yes, in a sense, because this idea relates the abstract notion of character and cell identity to a specific material entity, the core gene regulatory network or the character identity network. The answer also is No, since the core gene regulatory network is not necessarily immutable, as implicit in the notion of an "essence," and can evolve, although probably very slowly.[8]

The Evolutionary Origin of Cell Types

Historically, cell types have been identified on the basis of their functionally specialized phenotypes. Most of the classical cell type categories are named after specialized functions or their tissue of origin; as examples, muscle cells, sensory cells, osteoblasts, and osteoclasts. Given this classification of cell types based on their functional roles, it is tempting to speculate that novel cell types originated to accommodate novel functions. The idea is that when evolution "invented" or "needed" a novel function, it did so by inventing a dedicated cell type. In other words, it seems as if cell type function and cell type identity arose at the same time and, thus, are in a way the same thing.

While this view is intuitively plausible, the description of cell types from basal invertebrate lineages at least suggests an alternative model. What is surprising when studying invertebrate histology is that one encounters many examples of cells that perform multiple functions that, in more familiar organisms like humans, are performed by specialized cells.

An example of multifunctional cells in invertebrates is the epithelial muscle cell.[9] As their name suggests, these are cells that perform both the functions of an epithelial cell as well as those of a muscle cell. They are found in cnidarian larvae, as well as echinoderm tube feet, and participate in forming the apical epithelial surface and have a cilium. They behave like any other epithelial cell. At their basal pole, they are elongated, have an accumulation

---

[8] This situation is similar to the solution of the problem of the nature of chemical elements. It has been shown that the "essence" of a chemical element resides in the atomic number (i.e., the number of protons in the nucleus). But this "essence" is also not immutable, because nuclear fission and fusion can change this "essence," although at energy levels different from that of typical chemical reactions. Thus, these "essences" are invariant with respect to energy levels typical for chemical processes, but not absolutely invariant. In the natural sciences, notions of "essences" are relative to certain types of processes, not absolute, as implied by certain philosophical theories.

[9] The epithelial muscle cells are not the same as the myoepithelial cells from mammary glands. Myoepithelial cells are smooth muscle cells located in the basal layer of glandular epithelia, but do not have epithelial function.

of actin and myosin fibers, and are contractile. These cells are sensitive to mechanosensory stimuli and also have secretory vesicles.

Another surprisingly multifunctional cell is the ocellus that is found in sponge larvae as well as cnidarian larvae. These cells have a motile cilium that performs locomotor (steering) function, have photoreceptive functions, and perform the function of a pigment cell by shading the photosensitive parts of the cell. This is a complete sensory-motor unit realized within a single cell. At the very least, these examples show that functions that, in more derived species are performed by distinct cell types, can be accommodated by single cells. Thus, the existence of these cells raises the possibility that novel functions arose as derived cell phenotypes within existing cell types and that only later, and not in every lineage, segregated into the different cell types. These observations inspired a model of cell type evolution that could be called the *sister cell type model* (Arendt 2008).

The Sister Cell Type Model

The existence of multifunctional cells like the epithelial muscle cell in lower invertebrates suggests that specialized cell types may have arisen through the segregation of functions that had already evolved in the ancestral cell type. According to this model, *novel functional needs were primarily met by phenotype innovations within existing cell types*. For example, epithelial cells acquired the ability to contract at their basal pole in response to some stimuli. Adaptive improvement of this novel function led to a cell phenotype that is exemplified by the epithelial muscle cells of some invertebrate groups. Then, according to this model, came a phase during which some cells in the epithelium emphasized the contraction function, while others emphasized the protective and communicative functions of typical epithelial cells. The adaptive scenario that encourages specialization will be discussed later in this chapter and has been modeled by Rueffler and collaborators (Rueffler, Hermisson et al. 2012, see also chapter 4).

It is during that phase of increased specialization that the evolution of a genetic mechanism occurs that enables and enhances more and more differential gene expression of the cells with different functional specializations. This is when, according to this model, the origin of novel cell types occurs. Eventually, the emerging cell types fully segregate their functions and the differences between the cell types becomes qualitative. This hypothesis was developed by Detlev Arendt and turned into a research program regarding the phylogeny of cell types (i.e., cell typogenesis; Arendt 2008).

This model for the origin of cell types by segregation of function has a number of important implications that were the framework for Detlev Arendt's cladistic research program on cell typogenesis. According to the sister cell type model, cell types arise from existing cell types by a process of gradual specialization and, eventually, "cell type splitting" (figure 8.11). If this is the

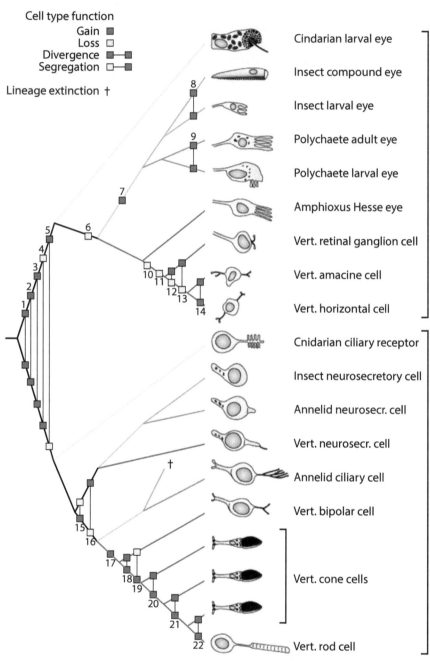

Cell type function
Gain ■
Loss □
Divergence ■—■
Segregation □—■

Lineage extinction †

Cindarian larval eye

Insect compound eye

Insect larval eye

Polychaete adult eye

Polychaete larval eye

Amphioxus Hesse eye

Vert. retinal ganglion cell

Vert. amacine cell

Vert. horizontal cell

Cnidarian ciliary receptor

Insect neurosecretory cell

Annelid neurosecr. cell

Vert. neurosecr. cell

Annelid ciliary cell

Vert. bipolar cell

Vert. cone cells

Vert. rod cell

Rhabdomeric photoreceptor cells

Ciliary photoreceptor cells

FIGURE 8.11: Cell type evolution by cell lineage splitting (sister cell type model after Arendt, 2008) as illustrated by the evolution of photoreceptor cells (see text for explanation).

case, then each new cell type comes into existence together with a sister cell type. The sister cell types divide among themselves the functions that were performed by their ancestral cell type. The result is that, after typogenesis, the two cell types form independent lineages of descent and modification.

Thus, at a formal level, cell types arose by a process of lineage splitting, just like species tended to originate by splitting of populations or novel genes arose by gene duplication. This implies that the phylogeny of cell types forms a tree of descent, similar to the tree of descent for species or for genes that originate by gene duplication. Hence, it should be possible to reconstruct a tree of "cell typogenesis" for all cell types in multicellular organisms. How this can be done is discussed below.

Another important implication of the sister cell type model is that sister cells would be predicted to be very similar in their ontogenetic development and also in their gene expression profiles. According to this model, cell types arose during evolution from functional specialization of terminally differentiated cells. The modifications of cell development that were responsible for the origin of novel cell types were changes in the terminal differentiation cascade, and not due to changes in the early developmental decisions of cell ontogeny.

Phylogenetically young sister cell types should at least share a developmental pathway up to the point of terminal differentiation. Also, sister cell types should be very similar in terms of their gene expression patterns. Differences in gene expression should reflect only the functional specializations and the need to devise a regulatory network that allows for differential expression. Thus, on a statistical level, it is expected that sister cell types have more similar transcriptomes than each of them compared to other cell types in the same species.

This hypothesis suggests that transcriptome data can be used to identify the cell-typogenetic relationships among cells in the same species as well as across closely related species. In the next section we will discuss a few examples for which this approach was used to identify hypothetical relationships among cell types. Here I want to reflect a little more on the implications and the plausibility of the sister cell type model.

The sister cell type model explains the origin of novel cell types with the implication that, during evolution, novel cell types arose as pairs and, thus, initially shared much of their developmental pathways. This model is silent regarding the likelihood of subsequent modifications of the developmental pathway. It has to be expected that, once established, the developmental pathway of the cells can become modified, as is the case with the developmental pathways of all characters. Perhaps one of the most dramatic changes in cell type development was the invention of the neural crest.[10]

---

[10] The modification of development in holometabolic insects is another example (see chapter 3).

The neural crest is a derived developmental character of vertebrates that added considerable evolvability to the vertebrate body plan (Gans and Northcutt 1983). The neural crest is a population of cells that originates at the time of neural tube formation from the border region between the neural plate and the general body ectoderm. These cells become migratory and spread throughout the embryo to form a large number of cell types, ranging from cartilage and bone cells to pigment cells, and glandular cells and neural cells of the peripheral nervous system. It seems that the neural crest is essentially a shortcut toward a large number of different cell types by not going through the historically dictated tree of cell fate decisions (i.e., without recapitulating the phylogenetic history of cell typogenesis). In fact, any broad taxonomic comparison of cell type development shows that the developmental pathways of homologous cell types can be very different and, thus, derived in one or the other way.

A comparison of the body plans of bilaterian animals suggests that the most recent common ancestor of humans and fruit flies already had a nervous system that directed the activity of muscle cells. Hence, their ancestor must have had motor neurons. The fact that motor neuron determination is accomplished by homologous sets of transcription factor genes (see above) further supports the inference that human and fly motor neurons may be homologous. Yet the developmental pathways of motor neuron development cannot be more different in these two species. Human motor neurons arise within the epithelium of the neural tube. In contrast, Drosophila motor neurons arise from mesenchymal cell populations.

Even less far-fetched comparisons show important differences in cell ontogeny. For example, among different species of nematodes, the development of homologous vulva cells is different (Sommer and Sternberg 1994), and the induction of the notochord in the ascidian larva and during vertebrate development is accomplished by different signaling pathways (Satoh 2003). Examples of this type are legion and have inspired the epigraph of this chapter: "*The differentiated cell-types seem to be islands in a sea of developmental change*" (Arendt 2008, p. 879). This means that the homologous cell types maintain shared determination mechanisms, but evolve different developmental pathways to reach the point of terminal differentiation, as discussed in chapter 3 regarding the developmental genetics of homology among multicellular organs.

Functionally defined cell type categories, like smooth muscle cell, do not necessarily define a biologically homogeneous and homologous collection of cells. It is plausible that contractility can evolve in different cell types, which do not need to be closely related. This was likely the case for mammalian *myo-epithelial cells*, which are not homologous to the epithelial muscle cells mentioned above. Rather, they are smooth contractile cells that originate in the glandular epithelium of mammalian glands, as for example sweat glands, salivary glands, and mammary glands.

These cells are distinguished from other smooth muscle cells in that they express marker genes that are typical for epithelial cells (e.g., pre-keratin proteins; Franke, Schmid et al. 1980), rather than the markers for myofibroblasts, which are contractile cells derived from fibroblasts. Another epithelial identity marker is P63, which is typical for basal epithelial cells but is absent in myofibroblasts (i.e., muscle cells outside the glandular epithelium, but present in myo-epithelial cells; Barbareschi, Pecciarini et al. 2001). Hence, it is clear that non-equivalent cells can assume the phenotype of a smooth muscle cell and perform the same function that other muscle cells do.

Functional cell type classifications can be misleading for recognizing cell identity. One has to look at more characters than just the functional phenotype to reliably recognize the identity of a cell. These other characters should preferably be gene products that are not directly related to cell function, but indicative of cell types regardless of cell function, such as the intermediate filaments and transcription factors.

The discussion of the neural crest above raises an issue that has so far been ignored in the discussion of the sister cell type concept (but see section above on embryonic stem cells). This is the question regarding the existence and identity of developmental cell types. These are cell types that are not part of the adult organism but evolved either to aid in development or are intermediate cell types during development. The neural crest is likely to be one of these, as it is clearly derived in vertebrates. The mammalian embryonic stem cell is another.

In all non-mammalian amniotes (i.e., those vertebrates that share with mammals the presence of fetal membranes), early development begins with cleavage and gastrulation and the formation of the basic body organization. Only then do the extraembryonic tissues form (i.e., the allantois, amnion, etc.). In contrast, in mammals the difference between the embryo proper and the extraembryonic tissues is the first cell fate decision leading to the trophoblast and the inner cell mass or embryoblast, which consists of the embryonic stem cells. The latter eventually forms the embryo proper.

Thus, it is tempting to speculate that the embryonic stem cell is a derived feature of mammals and that it does not have a homolog in reptiles, including birds (Niwa, Sekita et al. 2008). Nevertheless, the embryonic stem cell has the molecular machinery that allows it to be a well defined cell type (see above), although it is a cell type that has no functional role other than to provide progenitor cells for the formation of the embryo. Hence, the number of cell types of a species should not only include the cells that can be identified in the histology of the adult organism, but should include those that play specific roles during development. These could be called developmental cell types.

These cells are difficult to recognize as being distinct from just transient stages in cell differentiation, like the neuroblasts that travel from their birthplace to their site of terminal differentiation. The existence of developmental cell types also challenges the generality of the sister cell type model, as pointed out by Koryu Kin.[11] Thus, we may assume that the sister cell type model applies only to terminally differentiated cells.

In some invertebrate clades, homologous cell types have been recognized among the earliest cells formed: the blastomeres of the spiralian embryo. Spiralian animals are part of the lochotrophozoa clade that have a highly stereotyped pattern of cleavage divisions. This allows tracing cells through a number of divisions simply by their positions relative to each other and by their size. Corresponding cells tend to give rise to the same body tissues, like the germline or the mesoderm. Hence, the phylogenetic history of cell types has to include the history of developmental cell types.

Returning to the sister cell type model, we have to recognize that this model, of course, is a hypothesis and neither a logical necessity nor an empirical fact—at least not yet. For example, it is conceivable that a novel cell type arose from a cell lineage that, during development, diverged earlier from its sister cell type lineage than at the last stage of terminal differentiation, as assumed by the sister cell type model (Koryu Kin, pers. comm.). On the other hand, it is not yet entirely clear whether novel cell types always originated from functional segregation from an ancestral cell type that was multifunctional. Hence, it is not clear whether the sister cell type model is general, or if it only applies to a subset of cell typogenetic events. Nevertheless, the sister cell type model provides a rationale and a set of well-defined expectations to structure a research agenda (*sensu* Love 2009; Brigandt and Love 2012) for the evolution of novel cell types.

One might raise the question of whether and how this model can be tested, given that we do not have independent information on how cell types evolved. All we have are descriptions of cell types among species, a species phylogeny, and data for the similarity of cell types. The sister cell type model provides an interpretation for these data, which results in a hypothesis regarding the evolution of cell types. But how can these data then be used to test the validity of this model?

I think that this situation is the same as in any other case of fundamental scientific inference, as for example in phylogenetic inference or, in the first half of the nineteenth century, inferences on the nature of chemical elements and their fundamental properties of atomic weight and valence (Ihde 1964). The ultimate criterion for the acceptance of any scientific model, like the sister cell type model, derives from its ability to accommodate a large amount of data.

If the model leads to conflicting, non-parsimonious interpretations of the basic data, in this case species phylogenies, and similarity of cells and gene

---

[11] Koryu Kin was a graduate student in my lab working on the origin of the endometrial stromal cells when he raised this important point during a seminar taught in the spring of 2011.

expression patterns, the model will have to be modified or even abandoned. If, however, the sister cell type model mostly leads to parsimonious, coherent data interpretations, the likelihood is high that it is correct, or at least useful. Hence, the question of whether the sister cell type model shall be accepted can only be answered by applying it as an interpretative framework to as many case studies as possible. In the next section we will turn our attention to such case studies.

The Functional Benefit of Cell Type Segregation

If a single cell can perform multiple functions, then what is the benefit of having evolved multiple specialized cell types? This question has received surprisingly little attention from both theoretical and experimental biologists. Most work on this question has been done in the context of the evolution of multicellularity and the associated evolution of soma-germline segregation. In single celled organisms, the cell performs all those functions that are necessary for survival and reproduction. In many multicellular organisms, survival and reproduction are performed by different cell lineages. The one that forms the "body" of the organism is mostly dedicated to survival functions, and the germline focuses on reproductive functions.

The evolution of the germline-soma differentiation is nicely illustrated by volvocine algae, a group of photosynthetic flagellated eukaryotes (Kirk 1998). Within the volvocine algae one finds all cases from single celled organisms (e.g., *Chlamydomonas*) to multicellular forms with a fully differentiated germline (e.g., *Volvox*), including transitional forms like small multicellular forms without soma-germline differentation (e.g., *Gonium* and *Eudorina*) and forms that have late germ line differentiation (e.g., *Pleoeudorina*). Obviously, all of these forms are viable and successful in the sense that they still survive today. What then drove the evolution of germline separation? The most likely answer is a ("convex") tradeoff between the reproductive and survival functions (Michod 2006; see also chapter 4 and Rueffler, Hermisson et al. 2012 for a generalized model).

A tradeoff between different functions exists when the performance of one function interferes with the performance of another function. In the case of volvocine algae, and probably in the choanoflagellate ancestors of animals as well, a dividing cell has to give up its flagellum and, thus, cannot perform an important survival function while engaged in reproduction (for a discussion, see Buss 1987). The more time that a cell is engaged in reproduction, the lower will be its performance in survival functions, and vice versa.

In the case of a group of cells, the fitness of a group is some cumulative result of the reproductive and survival efforts of all its cells. Now the question is, under what circumstances is it better for the group of cells (i.e., the multicellular individual) to be composed of specialized cells that either only do reproduction (germline) or survival functions (soma) rather than having each cell participate in both? The answer is that soma-germline specialization is favored by natural selection if there is a "convex" tradeoff between survival and reproductive functions.

In the case of a linear tradeoff, a cell that increases in, say, reproductive performance loses an equivalent amount of survival effort. Hence, it does not matter how specialized each cell is, as the overall performance of all cells together is the same. If the tradeoff, however, is convex, then increasing performance in one function results in a less than equivalent loss of performance of the other function. In other words, in the case of a convex tradeoff, a group of specialized cells in which each cell either does survival functions or reproduction, but not both, is better off overall than a group of cells in which each cell contributes to both functions. In contrast, if the tradeoff function is concave, a group of cells does best if all cells contribute to both functions.

The logic of this model can be generalized for all kinds of characters or cellular functions (see chapter 4 and Rueffler, Hermisson et al. 2012) and provides a general model of the functional benefits of cell specialization and cell type origins. What remains unclear is how the gene regulatory network arises that allows the segregation of functions among a pair of sister cell types.

Case Studies of Cell Typogenesis

Heterocysts in Cyanobacteria

Cyanobacteria, or bluegreen algae, are the clade of bacteria that evolved photosynthesis and are also the ancestors of plant chloroplasts, which are endosymbiotic cyanobacteria. Cyanobacteria exist in single celled and multicellular forms, and even differentiate into a limited number of cell types. The most common forms of cell types are the vegetative, photosynthetic cells and the so-called heterocysts, which are specialized for N_2 fixation (Wolk 1996). Heterocysts are terminally differentiated cells that lost their ability to divide. The evolution of heterocysts has similarities to the soma-germline differentiation discussed above.

The difference from the soma-germline differentiation, however, is that in the case of vegetative cells of cyanobacteria, the proliferative cells also perform important survival functions (photosynthesis) and, thus, are not completely specialized for reproductive functions, as are the gametes of animals. In the case of the cyanobacteria, some of the survival functions (i.e., photosynthesis) are compatible with the reproductive function and, thus, the driving force for cell differentiation is different from that in germline-soma differentiation in animals and volvocine algae. In this case, the driving force is a conflict between two biochemical processes.

Photosynthesis and N_2 fixation are biochemically incompatible and, thus, cannot occur in the same cell at the same time. N_2 fixation, however, is not limited to cyanobacteria with differentiated heterocysts, as it also occurs in unicellular and undifferentiated multicellular cyanobacteria. In these organisms, N_2 fixation occurs during the night when photosynthesis is shut off. A phylogenetic analysis of cyanobacteria showed that the evolution of heterocysts likely occurred only once from undifferentiated multicellular cells that performed N_2 fixation during the night (Rossetti, Schirrmeister et al. 2010). A mathematical model of heterocyst evolution suggests that heterocyst-forming bacteria are favored in those environments with short daylight periods and, thus, a limited opportunity for photosynthesis (Rossetti, Schirr-meister et al. 2010).

The evolution of heterocysts in cyanobacteria illustrates the principle that functional innovations, in this case N_2 fixation, occurred prior to and independent of the origin of a functionally specialized cell type. The novel function was initially part of the reaction-norm of the ancestral cell type (i.e., N_2 fixation occurs in the absence of light when photosynthesis cannot occur and the conflict between photosynthesis and N_2 fixation is avoided). Only under certain ecological conditions, probably short daylight periods, was the evolution of a dedicated and terminally differentiated cell type favored. *Functional innovation and the evolution of a novel cell type were two different evolutionary events.*

The Evolution of Photoreceptor Cells

No other cell type has attracted so much debate about its origin and evolutionary diversification as that of the photoreceptor cell type family (see Salvini-Plaven and Mayr 1977; Land and Fernald 1992; Gehring and Ikeo 1999; Arendt 2003; and many more). Fortunately, this debate has also motivated a large amount of empirical work, so that photoreceptors are perhaps the best understood cell type in terms of their comparative biology. All photoreceptors expand the amount of cell membrane that is dedicated to the deployment of photosensitive pigments and the associated signaling molecules. It has long been recognized that there are two principal ways in which cells expand their cell membrane: the ciliary type and the rhabdomeric type.

The ciliary type is best known from vertebrate rods and cones. In this kind of receptor cell, the additional membrane is derived form the cilium, as the name suggests. Rhabdomeric photoreceptors are mostly known from invertebrates, like the compound eye of Drosophila. In these cells, the membrane to accommodate the photosensitive pigments derived from microvilli. From an evolutionary point of view, the question is whether all photoreceptors are broadly homologous or independently derived. Similarly, one needs to ask whether the ciliary and rhabdomeric photoreceptor cells are natural units (are all homologous) and what the phylogenetic relationships among all these cells are. The discussion below is based on (Arendt 2003; Arendt, Tessmar-Raible et al. 2004; Arendt 2005; Koyanagi, Kubokawa et al. 2005; Arendt, Hausen et al. 2009; Plachetzki, Fong et al. 2010).

All photoreceptors are broadly homologous[12]: the similarities among the photoreceptor cells from different animals go far beyond the defining functional role of a photosensitive cell. For photoreception, all animal photoreceptor cells use a carotenoid pigment associated with an apo-protein called opsin. Phylogenetic analysis of all known opsins shows that they are homologous. In all photoreceptor cells, the activated opsin interacts with an alpha subunit of a G protein to initiate intracellular signal transduction. In addition, in all photoreceptor cells, activated opsin is quenched by homologous molecules: a rhodopsin kinase that phosophorylates activated opsins and arrestin that competes with the opsin for association with the $G\alpha$ protein.

It is unlikely that independently derived photoreceptors would have converged on the same machinery for photoreception and quenching. This conclusion is also supported by the similarity of the transcription factor repertoire involved in the differentiation of photoreceptors across the animal kingdom. Among these, the most notable is *Pax6/ey*, already discussed in chapter 3. Other transcription factors with very broadly distributed expression among photoreceptors are *otx*, genes homologous to the Drosophila *orthodenticle* gene, and *Six* genes related to the Drosophila gene *sine oculis*.

Ciliary and rhabdomeric photoreceptors are sister cell types. However, even though all photoreceptor cells are homologous, several questions remain to be answered. First is whether the distinction between ciliary and rhabdomeric receptors cells is natural in the sense of demarcating phylogenetically unitary cell types. As mentioned above, the primary defining feature of ciliary and rhabdomeric cells concerns the morphology of the photosensitive membrane. This character could have originated multiple times. However, a molecular comparison of ciliary and rhabdomeric receptors reveals that they likely represent natural phylogenetic units.

For one, the opsins expressed in ciliary receptor cells form a clade, the c-opsins, and those expressed in rhabdomeric cells form another clade, the r-opsins. Their interaction with a G-protein is also characteristically different in terms of the alpha subunit subtype of the particular G-protein complex. In rhabdomeric cells, opsin interacts with a Gq_α protein, while in ciliary cells opsin interacts with a Gi_α subunit. The intercellular signaling cascade is also different between these two cell types. Ciliary cells use the phosphodiesterase pathway, whereas rhabdomeric cells use the phosopholipase-C pathway. The homeodomain transcription factor rx (retinal homeobox) is specific for

---

[12] The term "broadly homologous" means derived from a common ancestor regardless of whether the cells are directly homologous or paramorph in the sense defined in chapter 2.

ciliary photoreceptors, but is not expressed in rhabdomeric photoreceptors that continue to express *Pax6*.

The next question we need to answer is whether ciliary and rhabdomeric photoreceptors are strictly homologous (i.e., "orthomorph"), or whether they arose by cell type duplication and, thus, are "paramorph" (see chapter 2 for a discussion of these concepts). If, for example, the ciliary cell types only existed in vertebrates, in which they were first described, and all other animals only had rhabdomeric cells, then the difference between ciliary and rhabdomeric cell types, in spite of all their differences, could be just two different character states of a strictly homologous cell type. On the other hand, it is possible that these cell types arose through specialization and, thus, would be two duplicated cell types (i.e., sister cell types).

The most convincing way to distinguish between these two scenarios is to find animals in which both cell types co-exist. This is the case in the annelid *Platynereis dumerilii*, a marine ragworm (polychaete). Based on ultrastructural and morphological criteria, it has been known for a long time that ciliary and rhabdomeric cells exist in the brain of polychaetes (Dhainaut-Courtois 1965; Whittle and Golding 1974). With molecular techniques it was confirmed that these cells, in fact, were ciliary and rhabdomeric photoreceptors (Arendt, Tessmar-Raible et al. 2004). There is also evidence that ciliary receptors are present in some gastropod veliger larvae, while the adult eye contains rhabdomeric cells (see Blumer 1995 and references therein). Hence, it is clear that these two cell types have to be paramorphs (i.e., individualized copies of an ancestral photoreceptor type), because they can co-exist in the same animal. It turns out that this co-occurrence is not limited to ragworms, but that there are even true homologs ("orthomorphs") of rhabdomeric photoreceptors in vertebrate retinas, including the human retina (see below).

The paramorph character of ciliary and rhabdomeric photoreceptor cells is paralleled by the paralogous nature of some of the genes characteristic of these two cell types. As mentioned above, the c- and the r-opsins are paralogs and probably arose at about the same time as the cell types specialized. In addition, the Gq_α and Gi_α subunits that specifically interact with the r- and c-opsins, respectively, are paralogs. Hence, the balance of evidence clearly shows that the rhabdomeric and ciliary photoreceptor cells are sister cell types, or at least cell types that arose by differentiation from an ancestral cell type (i.e., are "cousin cell types" to some degree).

Given the paramorph character of ciliary and rhabdomeric photoreceptor cells, it is clear that some ancestor of vertebrates had to have both cell types. Then what happened to the rhabdomeric photoreceptors in vertebrates? The well-known rods and cones of the vertebrate retina are both ciliary cells. It turns out that three interneurons and projecting neurons of the vertebrate retina are, in fact, derived from rhabdomeric photoreceptors, even though they do not display a highly elaborate cell membrane characteristic of typical photoreceptor cell.

These cells are the retinal ganglion cell, the amacrine cells, and the so-called horizontal cells (figure 8.11). Even though they do not have the typical photoreceptor morphology, they nevertheless are somewhat photosentitive and express a form of opsin, called melanopsin. A phylogenetic analysis of melanopsin sequences identified melanopsin as belonging to the clade of r-opsins (figure 8.12; Plachetzki, Fong et al. 2010). Furthermore, they express Pax6, typical for rhabdomeric cells, rather than the transcription factor rx, typical for ciliary cells. Another transcription factor that is characteristic of rhabdomeric cells is *atonal*, which is found in insect and annelid rhadbomeric precursor cells.

In the mouse, the retinal ganglion cells and the amacrine cells are also derived from cells that expressed *atonal* homologs, *Math1* and *Math3*. Hence, it is very likely that these three retinal neuron types derived from rhabdomeric photoreceptors and, thus, are properly members of the family of rhabdomeric photoreceptor cell types, even though they no longer are specialized for photoreception.

Given that retinal ganglion cells, amacrine cells, and horizontal cells are clearly distinct cell types in the vertebrate retina and that they derived from an ancestral rhabdomeric photoreceptor cell, they also have to be considered sister/cousin cell types. Among these three cell types, the retinal ganglion cell and the amacrine cell are perhaps more closely related to each other than each is to the horizontal cell (Arendt 2003). This is indicated in that they express *atonal* homologs, whereas these genes are not expressed in horizontal cells. They also share AcCh and, perhaps, even dopamine as a transmitter.

Overall, the evidence from photoreceptor diversity and phylogenetic distribution supports the sister cell type model of cell type evolution (see above). Sometime during early metazoan evolution, a still unknown cell type specialized into two cell types, the ciliary and rhabdomeric cell types, and both certainly co-existed in the most recent common ancestor of bilaterian animals. These cell types or their derivatives still co-exist in many animals, including humans. Further specialization of multiple eye types, like larval eyes and adult eyes, led to further sub-specialization and to the origin of more photoreceptor cell types, including cells that no longer function as photoreceptors (figure 8.11).

What then was the ancestral photoreceptor, a ciliary or a rhabdomeric photoreceptor? Given that the morphological characteristics of ciliary and rhabdomeric cell are their elaboration of a cilium or microvilli, respectively, a plausible candidate for an ancestral cell type would be a cell that had both microvilli as well as a cilium. In fact, such cells are well known: choanocytes. Choanocytes are best known from the inside of sponges, and are highly reminiscent of the morphology of the closest relatives of animalia, the choanozoa,

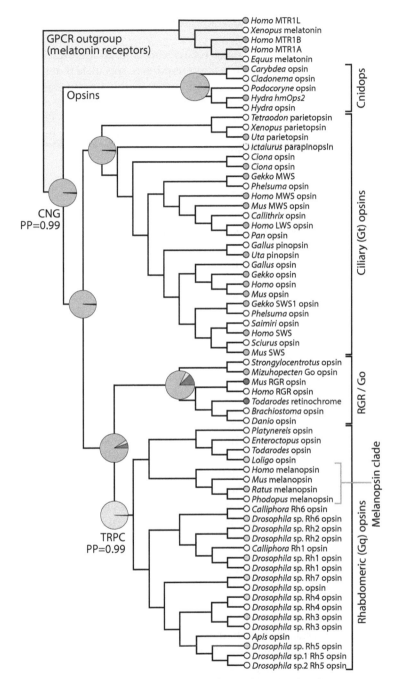

FIGURE 8.12: Phylogeny of opsin genes. Note that melanopsin, the photopigment in vertebrate retinal interneurons, is nested in the clade of rhabdomeric opsins. This suggests that these interneurons, even though they do not function as specialized photoreceptors, are derived from the "invertebrate type" photoreceptors (redrawn after Plachetzki et al., 2010, *Proc. R. Soc. B* 277:1963–1969).

a small group of protists with a characteristic arrangement of a cilium and a collar of microvilli.

Choanozoans are also the closest relatives to animals among all the remaining eukaryotic lineages. Hence, it is tempting to speculate that there was a choanocyte that acted as a photoreceptor and in which opsin was in both the cilium and the collar of microvili. In fact, such a cell has been described in the eye of an arrow worm, *Sagitta scrippsae*, a chaetognath (Eakin and Westfall 1964). However, it was not clear from that paper whether the microvilli had a photoreceptor function.

The Micromere Lineage and Larval Skeleton of Sea Urchins

Among all the model systems of evolutionary developmental biology, the early development of sea urchins occupies a special place. Thanks largely to the decades long focused work in Eric Davidson's lab at the California Institute of Technology, we know more about the structure of the gene regulatory networks underlying early lineage decisions and their variability among species than about any other system. The work on sea urchin development could have been cited in any of the previous sections, but I describe it here at the end of this chapter because it is a culmination point of dedicated work into the structure, function, and evolution of developmental gene regulatory networks underlying cell fate decisions.

Two cell populations have received particular attention, the determination of the larval endoderm (gut) and the micromere lineage, which gives rise to the larval skeleton of the eu-echinoids. Of these, the developmental biology of the endoderm is a paradigm for an ancient character with a highly conserved core gene regulatory network (Hinman, Nguyen et al. 2003). The other, the micromere lineage, is a paradigm for the origin of a novel cell type, as micromeres and the larval skeleton are derived features of sea urchins.[13] For this reason, we will focus on the micromere cell lineage. Unless indicated otherwise, the discussion of micromere development and evolution is based on (Hinman and Davidson 2007; Gao and Davidson 2008; Oliveri, Tu et al. 2008).

Micromeres and Larval Skeletons

Micromeres arise early during the development of the "typical" sea urchin embryo through the third and fourth cleavage divisions. The plane of the third cleavage division is horizontal and is the first unequal division at the vegetal pole of the embryo. It creates a set of four small micromeres and four so-called macromeres. The fourth division of the micromeres is also unequal, generating a set of even smaller cells, the four "small micromeres" and the

---

[13] Whether the larval skeleton is an echinoid apomorphic character an apomorphic character or for a clade including the brittle stars is not entirely clear; see below for more discussion.

four "large micromeres." The small micromeres do not participate further in early development and will not be considered here. The large micromeres, which will be just called micromeres in the following, send signals to their surrounding cells, and then ingress just prior to gastrulation and give rise to the larval skeleton. The macromeres will develop into other mesodermal cell lines, like pigment and adult skeletogenic tissues and the gut endoderm, partly under the influence of the micromeres.

Before ingression, the micromeres sequentially send three signals. First is a Wnt8 paracrine and autocrine signal. Second is a signal called the "early signal" (ES), which has not been identified at the molecular level, but which is necessary and sufficient for endo-mesoderm specification. Finally, the micromeres express the Notch ligand Delta-N, which signals to the cells immediately neighboring the micromeres. The Delta-N signal is necessary for mesoderm specification. The micromeres then ingress into the blastocoele and form the larval skeleton. During skeletogenesis, micromeres largely deploy the same biomineralization genes as during the development of the adult skeleton.

In a phylogenetic context, one has to distinguish between the origin of the larval skeleton and the micromeres. Among sea urchins, only the euechinoids have micromeres, while the cidaroids do not. The cidaroids are the most basal lineage of living sea urchins, which retain many plesiomorphic features (Smith 1992). They do possess a larval skeleton, although it develops much later during gastrulation (Bennett 2009). The closest outgroup to sea urchins is the sea cucumbers, or holoturians. Holoturian larvae do not have a larval skeleton. What the next outgroup of echinoids may be is not entirely clear. It could be either the sea stars, which do not have a larval skeleton, or the brittle stars, the ophiuroids. The latter do have a larval type that is very similar to the sea urchin larva, called the ophiopluteus (the sea urchin larva is called pluteus), which does have a larval skeleton in many ways similar to that of sea urchins. Hence, from that data it is unclear whether the skeleton of the pluteus larva evolved in the stem lineage of sea urchins or in the stem lineage of the clade (ophioids, holoturians, echinoids).

The uncertainty derives from the relative phylogenetic position of sea stars and brittle stars, as mentioned above. If sea stars are closer to sea cucumbers and sea urchins, then independent derivation of the ophioid and echinoid larval skeleton would be more parsimonious, even though only by one step. If the brittle stars are closer to the sea urchins than the sea stars, then the reconstruction is ambivalent. The origin of a larval skeleton in the stem of brittle stars and sea urchins and a loss in sea cucumbers is equally parsimonious as independent derivation of the brittle star and sea urchin larval skeletons, assuming equal weight for the gain of the character as its loss. If the loss of a character is considered more likely than its gain, then homology of the brittle star and sea urchin larval skeletons is more parsimonious.

In any case, the larval skeleton is older than the micromeres because the basal sea urchins, the cidaroids, have a larval skeleton, but no micromeres. Thus, it seems that the micromere lineage reflects a change in the embryological origin of the skeletogenic cells that form the larval skeleton. In cidaroids, these cells ingress together with other mesodermal cells during gastrulation. In contrast, micromeres are distinguished by ingressing before gastrulation. In cidaroids, it would be interesting to know whether there are cells that perform the signaling functions of micromeres. At least the signaling function of micromeres is not likely to be plesiomorphic for sea urchins because it is not found in sea stars (Hinman and Davidson 2007).

The Gene Regulatory Network of Micromere Specification
(Oliveri, Tu et al. 2008)

The structure of the gene regulatory network of micromeres is best understood in terms of modular sub-networks dedicated to a number of tasks. These tasks are:

- Interpretation of maternal positional information
- Expression of intercellular signals
- Progression and stabilization of the regulatory state
- Exclusion of alternative cell fates
- Execution of the skeletogenic differentiation function

The micromere lineage becomes specified through cytoplasmatically localized maternal factors at the vegetal pole of the blastula. These include maternal beta-catenin ($c\beta$), Ets, Otx, and Dsh. The localized presence of Dsh is critical, as it causes the nuclear localization of maternal beta-catenin. This is the first molecular sign of progression toward micromere fate commitment. In the nucleus, $c\beta$ forms a complex with the transcription factor Tcf and activates a key zygotic regulator, Pmar1. Pmar1, in turn, suppresses a constitutively active inhibitor, HesC, which is a general suppressor of micromere specific gene activity. Upon HesC suppression by Pmar1, the door to micromere specification is opened.

This is a double negative gate, as Davidson likes to call this mechanism. By activating a suppressor, Pmar1, that in turn suppresses another suppressor, HesC, activating the gene regulatory network is achieved through ubiquitous activating factors. All three of the micromere signals, Wnt8, ES, and Delta, are under the control of this double negative gate, as well as the larval skeletogenic pathway. Here we will continue with the skeletogenic gene regulatory network.

The skeletogenic regulatory state is achieved in two steps. First, the double negative gate of Pmar1 and HesC de-represses three transcription factor genes, *Alx1*, *Ets1*, and *TBr*, which in turn activate another set of skeletogenic regulators: *Erg*, *Hex*, and *Tgif*. Second, these three latter transcription factor genes engage in what Davidson calls a "cross-regulatory embrace" that locks

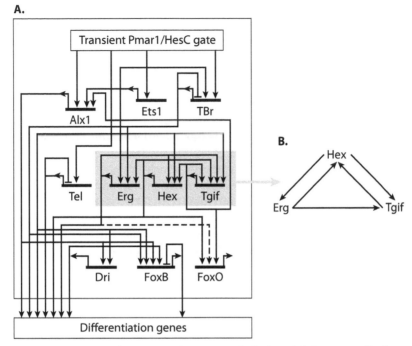

FIGURE 8.13: Structure of the core regulatory network of echinoid skeletogenic cells of sea urchins, in particular larval skeletogenic cells. A) Outline of that part of the network that results in maintaining the gene regulatory network state of skeletogenic cell. B) Abstracted core network of skeletogenic cells. This core network is shared between larval and adult skeletogenic cells. The larval cells differ from the adult skeletogenic cells by the participation of *TBr, Tel, FoxB*, and *FoxO* (redrawn after Oliveri et al., 2008, *PNAS* 105:5955–5962).

the cell in a self-maintaining regulatory state. This regulatory embrace results from the mutual activation and self-activation (in the case of *Tgif*). The structure of this cross-regulatory network is sketched out in figure 8.13.

This is the best candidate for the core regulatory network of skeletogenic micromeres. The genes from this core network together with their immediate upstream regulators jointly activate the genes for the biomineralizing matrix and, thus, the genes that form the actual skeleton.

This summary is a great simplification. For more details, see Oliveri et al. (2008) and the other papers upon which this model is based. However, it summarizes the main causal narrative that emanated from the work on this gene regulatory network and illustrates the overall logic of the network: interpretation of positional signals; the translation of a positional signal into de-repressing the cell specific functions; and the assumption of a self-stabilized regulatory state, which finally activates the genes necessary to produce the skeletal matrix.

We will now turn to a comparison of this network model with other gene networks. Two comparisons are particularly enlightening: a comparison with the genes responsible for the development of the adult skeleton in the sea urchins (Gao and Davidson 2008); and a comparison with the mesoderm specification in sea stars (Hinman and Davidson 2007), which plesiomorphically lack a larval skeleton.

Gao and Davidson (2008) compared the micromere specification network with the gene expression inventory of the adult skeleton of the sea urchin *Strongylocentrotus purpuratus* and a sea star, *Asterina minima*. The result was that not only the same set of skeletogenic matrix genes were expressed, but all the upstream regulators were expressed, which in the pluteus larva were downsteam of the *Pmar1*/*HesC* double negative gate. In addition, there were four genes expressed downstream of *Pmar1*/*HesC* that were not involved in adult skeletogenesis. These were *TBr*, *Tel*, *FoxB*, and *FoxO*. Also, micromere-specific signaling seemed not to be present in sea stars. In contrast, Delta-Notch signaling was present, but with an opposite purpose. In sea urchins, Delta-N is necessary for mesodermal identity, while in sea stars the Delta-N signal suppresses the mesodermal fate (Hinman and Davidson 2007).

Interpretation of the Data

When considered from the standpoint of cell type identity and cell type evolution, it is important to separate the issue of larval skeleton evolution from that of micromere origins. Larval skeletons are clearly older than micromeres, as shown by the presence of a pluteus larva with internal skeletons in the cidaroid sea urchins. From this scenario it also follows that larval skeletogenic cells are most likely a sister cell type of the adult skeletogenic cells. They both derive from late gastrulation mesoderm and deploy a very similar gene regulatory network when it comes to their skeletogenic function.

The data from Oliveri et al. (2008) as well as Gao and Davidson (2008) further show that the two types of skeletogenic cells also share much of what can be called the skeletogenic core-regulatory network (i.e., that network of genes that lock in the regulatory state through a dynamic feedback among them and which jointly regulate the genes for biomineralization and morphogenesis: *Erg*, *Hex*, and *Tgif*). The fact that this part of the network is shared between larval and adult skeletogenic cells makes the conclusion nearly inevitable that these are sister cell types.

While the core gene regulatory network is shared among these cell types, the regulatory network of larval skeletogenic cells is clearly modified relative to that of the adult skeletogenic cells. As mentioned above, there are four transcription factor genes that are expressed and which have essential functions in larval skeletogenesis, but not in the juvenile or the adult: *TBr*, *Tel*, *FoxB*, and *FoxO*. Of these, the function of *TBr* is most puzzling because, in other deuterostomes, it is necessary for endomesoderm determination. In sea urchin micromeres, this gene is necessary together with the genes of the core network for skeletogenesis. On the other hand, based on known regulatory links, *TBr* does not seem to partake in the dynamic stabilization of the regulatory state and, thus, is not in the narrow sense part of the core network. *TBr* has activating regulatory input on all three core network genes, but there seems to be no feedback from these genes to *TBr*.

These observations about the role of *TBr* in relation to the core network of larval skeletogenic cells suggests a mode of evolution of cell type identity networks. In the micromeres, the core self-stabilizing interactions are the same as in the adult skeletogenic cells, but other regulatory genes partake in regulating the downstream executor genes, most notably *TBr*. Of course micromeres do form a type of skeleton other than the adult skeleton and there have to be differences in the regulatory inputs and the execution of the developmental program.

It is likely that the derived genes in the skeletogenic network were acquired for that purpose, as with the upstream regulatory machine, but not necessarily integrated in the core network that stabilizes the regulatory state. Hence, sister cell types do not necessarily differ in the core network per se; rather, they differ in the context in which the core network is deployed, both in terms of the genes that activate the core network and the set of other genes that contribute to regulating the developmental program. A similar conclusion can be reached from comparing different eye determination networks in insects for larval, compound eyes, and ocelli, as discussed in chapter 3.

Now we need to discuss the evolutionary/developmental significance of micromeres. As mentioned above, micromeres are phylogenetically younger than the larval skeleton. In terms of larval skeleton development, one can consider micromeres as a means to deploy larval skeletogenic cells earlier than in ancestral species, essentially a heterochrony (Gao and Davidson 2008). Micromeres respond to maternal factors with the activation of signaling and, eventually, the skeletogenetic pathway through the activation of the *Pmar/HesC* double negative gate. The gene regulatory network that mediates between the maternal factors and any of its cellular functions is novel. No trace of this has been found in sea stars (Hinman and Davidson 2007).

Interestingly, there is a small auto-regulatory network included in the micromere-specific gene regulatory network. It includes Wnt8 regulation by Blimp1 and the fact that Wnt8, in turn, feeds back onto Blimp1 expression through an autocrine mechanism (Oliveri, Tu et al. 2008). Hence, there is a small network that stabilizes the micromere regulatory state before the core network for the skeletogenic regulatory state kicks in. Thus, micromeres can be considered a separate, transient cell type from the larval skeletogenic cells.

Gao and Davidson (2008) speculated that micromeres originally evolved as a signaling center at the vegetal pole of the blastula. Consistent with this idea is the fact that most of the signaling functions of micromeres are directly regulated by the micromere-specific gene regulatory network rather than the core network of skeletogenic cells. Also consistent with this idea is that even sea stars have a vegetal pole signaling center, although with a different gene regulatory network than that of micromeres (Hinman and Davidson 2007). Hence, a plausible scenario for the evolution of eu-echinoid pluteus development is that there were two more or less independent evolutionary events. One of these led to the development of a larval skeleton and, thus, a larval skeletogenic cell type, which originally developed during gastrulation, like in the cidaroid sea urchins. The second event was the origin of micromeres as a signaling center. Then micromeres acquired the capability to progress to the gene regulatory network state of the larval skeletogenic cells and, thus, to differentiate early. Thus, this is a shortcut to the larval skeletogenic fate in the way that the neural crest in vertebrates is a shortcut to a variety of different cell fates.

Research on the development and evolution of the micromere lineage illustrates a number of issues that can also be pulled together from other examples in this chapter. These are the existence and importance of a core gene regulatory network that causes the stabilization of a gene regulatory state and, thus, explains cell type identity. They also provide an exquisitely detailed picture into how plesiomorphic gene regulatory networks become integrated and modified in a new context.

What the current results do not address is the way in which the different inputs of the core regulatory genes jointly regulate the target genes. In particular, the role of protein-protein interactions among transcription factor proteins in this process is not addressed, a feature that is better known from other systems like human hematopoietic cells or neuroblasts. The fact, however, that knockdown of a single transcription factor gene, for example *TBr*, can severely disrupt skeletogenesis suggests that there is also at least some degree of cooperativity among the participating transcription factors in these cells. It also seems that micromeres and larval skeletogenic cells are different cell identities. In addition, the larval skeletogenic cells are probably just a modification of the general skeletogenic cell fate, because the adult and the larval skeletogenic cells share the same core network, although with some different necessary modulating inputs to account for the different skeletal morphologies.

Concluding Reflections

In many ways the evolution of cell types is a critical proving ground for any theory of character identity and homology. The evolution of cell types is the lowest level of biological complexity for which questions of character identity can be investigated and which is neither directly a gene, and a direct replicate of ancestral entities, nor more or less direct copies of a gene, like RNAs and proteins (*modulo* RNA editing and post-translational modification). Hence, the question of whether the notion of a cell type, or more generally the notion of character identity, is biologically meaningful has first to be answered at the level of cell types.

Cells are also the entity for which a complete understanding of the mechanistic basis of cell/character identity is likely to be possible in the near future. Through the boost provided by stem cell research in recent years, a detailed understanding of the mechanisms of cell fate decisions is emerging (Graf and Enver 2009). This research shows that cell fate and, thus, cell identity is usually determined by a small core network of regulatory genes that controls both the expression of cell type characteristic genes and suppresses genes that belong to alternative cell identities. These results support the hypothesis that character identity is subscribed by a core gene regulatory network, which I earlier called a character identity network (ChIN) that directs the development of the character typical phenotype.

The structure of core gene regulatory networks for cell type identity also matches well with the *abstract nature of the homology concept*, which denotes character identity "regardless of shape or function" (i.e., regardless of specific phenotype). The core cell type identity networks are abstract in the sense that they can direct the expression of any set of target genes. *The nature of the target genes is not prescribed by the core gene regulatory network* and its components, mostly transcription factors; rather, it is prescribed by the presence or absence of cis-regulatory elements that are responsive to the transcription factors characteristic of the cell type. This fact was confirmed recently by a comparison of the gene regulatory networks of mouse and human embryonic stem cells (Kunarso, Chia et al. 2010).

The coherence of cell types in development and evolution seems to be enhanced by the fact that, in many cases, the target genes of the core regulators are regulated by an obligatory protein (and, perhaps, RNA) complex rather than by individual transcription factors. The causal cohesion of a cell type is supported by the functional and phylogenetic cohesion of trans-regulators, called Core Regulator Complexes here, which act cooperatively at the molecular level to achieve a typical phenotype.

The evolution of cell types seems to follow a pattern of diversification by lineage splitting, called the sister cell type model (Arendt 2008). This principle is perhaps best illustrated in the case of photoreceptor cell types in which the major forms of photoreceptors are clearly paramorph cell types based on the phylogeny of the genes expressed in these cells, as well as the fact that they can co-exist in the same animal (Patterson's criterion of non-homology). The sister cell type model is best supported in the case of terminally differentiated cells like photoreceptors. However, it is not yet clear whether and how this model can also account for the evolution of developmental cell types, like the embryonic stem cell that might be a mammalian innovation, or the neural crest cells that are a vertebrate innovation, or the micromeres of eu-echinoids.

9

Skin and a Few of Its Derivatives

Although less conspicuous than the endoskeleton or the brain and eyes, one of the key novelties of vertebrates is their skin. Vertebrate skin is unique among metazoans in at least two respects. For one, vertebrates are the only phylum for which the body is completely covered by a multilayered epidermal cover (figure 9.1).[1] This facilitated the evolution of many functional specializations, including the fully keratinized skin of amniotes (reptiles and mammals), also a key innovation for terrestrial life.

The other peculiarity of the vertebrate skin is that it is a composite structure comprising the epidermis, which corresponds to the embryonic ectoderm, and a mesodermal layer that is underneath the epidermis and is called the dermis. In fact, many of the typical skin characteristics, like scales, feathers, and hair, are the products of an interaction between epidermis and dermis, a developmental theme in vertebrate development called epithelial-mesenchymal interaction. This type of interaction underlies many, if not most, cases of vertebrate organogenesis (Gilbert 2010), and is also critical for the development and elaboration of the integument.

The microanatomy of the vertebrate integument is as varied as are the main clades of vertebrates. Fish skin varies in depth between two cell layers in the larva and more than ten layers in the adult (Whitear 1986). The cells in all layers are metabolically active and are knit together by desmosomal junctions and filaments. Fish show a variety of novel cells in the skin, mostly single-celled glands that provide protection for the skin and mucous secretions. While a major difference between fish and land vertebrates is keratinization of the skin in land vertebrates, keratinization is not limited to them. Even fish skin cells can keratinize, although only at fairly localized spots. For example, the "nuptial pearls" of trout are beads of keratinized cells that appear on the skin of males during the mating season.

---

[1] To my knowledge the only exception to this rule is a small part of the skin of Chaetognathes, which also seems to have a multilayered epidermis (Salvini-Plawen, personal communication).

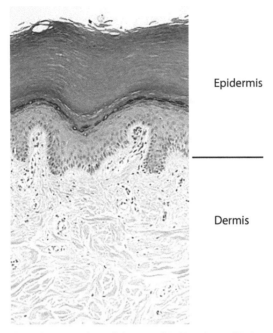

Epidermis

Dermis

FIGURE 9.1: Histological section through human skin. Vertebrate skin is a composite of the epidermis, which is of ectodermal origin, and the dermis, which is of mesodermal origin. Note that the epidermis comprises multiple layers of skin, unlike the skin of invertebrates. This multilayered architecture allowed for the evolution of specialized cell types and glands in the skin (based on http://www.intechopen.com).

The main difference between amphibian and fish skin is the dominance of multicellular glands in amphibians as opposed to single-celled glands in fish. Keratinization also plays an increasing role in amphibians, although not as dominant as in amniotes—for example, the claws of *Xenopus*, which is also called the African *clawed* frog.

The major transition in skin biology, though, was the origin of the fully keratinized integument of amniote animals (figure 9.2). The keratinized skin of amniotes consists of a basal layer in which cell proliferation occurs and various layers above that in which cell death and transformation occur resulting in various layers of keratin. This dominance of keratinization also resulted in a large number of specialized structures that essentially are highly structured forms of keratin, including large structures like the beaks of birds, turtles, and the platypus, claws, nails, and hoofs, as well as smaller, more numerous structures, likes scales, hair, and feathers. Hair and feathers were key innovations associated with the origin of major amniote clades, namely mammals and birds, respectively. Hence, the amniote skin was a

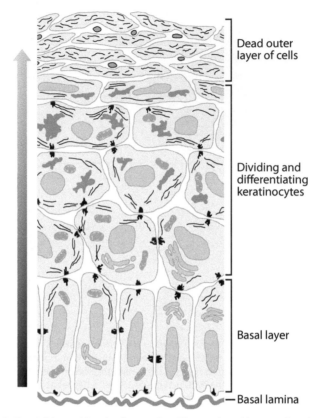

FIGURE 9.2: Keratinizing epidermis of an amniote. The amniote skin comprises three principal layers. At the bottom is the basal layer of living and dividing cells, the *lamina basalis*. The *lamina basalis* cells undergo cell division and produce new cells that are pushed up and start dying and keratinizing. The outer layer is the *stratum corneum* comprising dead cells that form the keratin layer of the skin.

rich source of evolutionary innovations important for many aspects of bird and mammal life.

Developmental Evolution of Amniote Skin and Skin Appendages

Even though amniote skin is just a small fraction of vertebrate skin diversity, it is the only part about which a substantial amount of experimental data are available for attempting a developmental biology of skin evolution. As noted above, amniote skin was also the site of origin for two of the most dramatic innovations that were associated with two major vertebrate radiations: hair

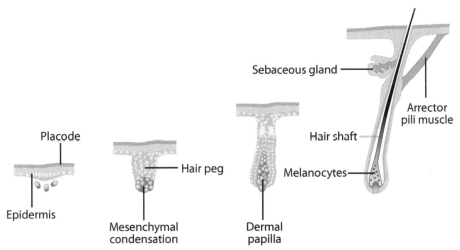

FIGURE 9.3: Development of a skin appendage as exemplified by the development of hair. The very first stage of development is the formation of a placode, which is a localized thickening of the epidermal layer. Underneath the placode, mesodermal cells aggregate and form a condensation. In the case of hair, the epidermal cells then grow into the dermis and form a hair peg. At the tip of the hair peg, the mesenchymal cells enter the hair peg and form the dermal papilla. The developed hair is a composite organ comprising the hair shaft, the follicle, the sebaceous gland, and an associated muscle, the musculus arrector pili (after Fuchs, 2007, *Nature* 445:834–842).

and feathers associated with mammals and birds. Thus, amniote skin biology provides a set of empirical facts that we can use to test ideas about the developmental evolution of character identity and origination.

Developmentally and anatomically, the amniote skin, and the vertebrate skin in general, is a composite of two major contributions: the epidermis, originating from the embryonic ectoderm, and the dermis, derived from the mesodermal germ layer (figure 9.3). Even though the dermal and epidermal components of the skin are closely integrated and interdependent, they remain physically separated by the basal membrane of the epidermis. The development of skin and its appendages results from a developmental conversation between these two cell populations for which regional differences in the skin phenotype depend, to varying degrees, on regional differences between the dermal and the epidermal components. The summary provided below is based on reviews by: Chuong, Widelitz et al. 1996; Chuong and Homberger 2003; Widelitz, Jiang et al. 2003; Wu, Hou et al. 2004; Fuchs 2007; Fuchs and Horsley 2008; Dhouailly 2009.

The first morphological change in the general body skin common to mammals and birds is the formation of a dense dermal layer underneath the epidermis, which is necessary for later appendage development (Dhouailly

2009). The formation of this dermal layer is initiated by an epidermal signal and depends on the transcription factor TWIST2/Dermo1 (Olivera-Martinez, Thelu et al. 2004). The dermis then signals back to the epidermis to form a placode. This signal involves at least one member of the Wnt family of signaling molecules and results in the activation and nuclear localization of beta-catenin.

The placode then signals back to the mesoderm to form a dermal condensation that, in turn, signals back to the placode to initiate the appendage-specific developmental program. The early stages of this conversation between the epidermis and the dermis seem to be conserved among amniotes and among different appendage types. In particular, the need for Wnt/beta-catenin signaling during placode formation is evident, because inhibiting this system affects hair, sweat gland, mammary gland, and tooth development, as well as feather development in chickens, which are all organs that derive from placode or placode-like precursors. Nevertheless, modulating Wnt/beta-catenin signaling has consequences for the development of alternative skin organs, like feather versus scutate scale or gland versus hair development (Widelitz, Jiang et al. 2003; Wu, Hou et al. 2004; Dhouailly 2009).

Besides being the boundary between the organism and the environment, one of the main adaptive roles of the skin was the evolution of skin differentiations and skin appendages that could assume specialized functional roles. A large fraction of vertebrate characters are skin appendages or skin derivatives, from the lenses of the eyes, the nose, and all other deeply entrenched vertebrate body plan characters (Schlosser 2005) to teeth, dermal scales, and taxonomically more restricted appendages like hair, feathers, epidermal scales, as well as horns, hoofs, claws, and many more. A major challenge for the evolutionary biology of the vertebrate skin is clarifying the relationships among different skin derivatives within the same species (are the epidermal scales on bird feet simplified feathers?) and relationships to the skin derivatives of other species (are teeth modified dermal scales or are shark scales heterotopic teeth?; Johanson and Smith 2005). In fact, much of the theoretical and empirical discussions about skin development and evolution center on these two sets of questions. Here I will focus on these questions as they are articulated with respect to amniote skin development and evolution, in particular those related to the origin and evolution of hairs and feathers.

With respect to feather evolution, the most intensely debated questions are the following. 1) Are feathers derived from epidermal archosaur scales or did they originate separately? 2) What distinguishes a feather from a scale (i.e., what are the essential developmental innovations that brought about feathers)? 3) Are the scutate scales on the tarsus-metatarsus of birds homologous to those of the non-feathered reptiles? 4) Or, are the scutate scales simplified feathers (Davies 1889; Dhouailly 2009)?

There is a similar set of unresolved questions associated with the origin of hair in mammals. 1) Are hairs derived from "reptilian" scales? 2) Is hair only an accessory character of skin glands? 3) Is hair derived from tongue scleratizations, so-called filiform papillae (Dhouailly and Sun 1989)? 4) Are mammary glands derived from hair?

All these questions are difficult to answer because of a lack of extant primitive or transitional character states and, until recently, a lack of a fossil record. A further complication derives from the molecular genetics of hair and feather development that has revealed very similar or even identical molecular mechanisms underlying the development of both hair and feathers and, to make things worse, even the development of teeth. Hence, much of currently available molecular developmental data are not immediately helpful for tracing the evolutionary history of skin derivatives, in particular hair and feathers, or even to understand their development.

The problem with the experimental data was brilliantly synthesized by Danielle Dhouailly in a seminal review in 2009 (Dhouailly 2009). Here, I will use her summary as a point of departure for proposing a model of amniote skin developmental evolution that may help sort out the facts into a scheme that, if anything, may help us to focus on the most pertinent experimentally resolvable questions.

As mentioned above, many if not all (see below) amniote skin appendages initially develop through a set of common embryological stages. Most notable is a "placode" (i.e., a local thickening of the embryonic epidermis; ectoderm) and a condensation of mesenchymal cells underneath the placode. These stages are well documented for hair and feather development, and even avian scutate scales, but have not been described for lizard and alligator scales (Alibardi and Thompson 2000; Dhouailly 2009). Similarly, the so-called reticulate scales on the plantar surface of bird feet lack morphologically discernible placodes and dermal condensations. Rather, they develop from undulations of the epidermis caused by localized connective tissue strands that pull the inter-scale epidermis down (Sawyer 1972).

Whether these morphological differences do, in fact, signify a fundamental developmental difference between feathers and hair on the one hand and "reptile" scales and reticulate scales on the other is unclear. This issue can only be resolved by probing developing reptile scales for their expression of key developmental genes like Wnt/beta-catenin, BMP, Shh, and others. Very little is currently known on this point, with the exception of Shh and BMP in alligators (Harris, Fallon et al. 2002).

This mode of development, the formation of a placode and a mesenchymal condensation associated with the placode, is very widespread among vertebrates and is not limited to the development of skin appendages (figures 9.3 and 9.4). Most notably, a large number of fundamental vertebrate body plan features, such as the lenses of eyes, the olfactory epithelium of

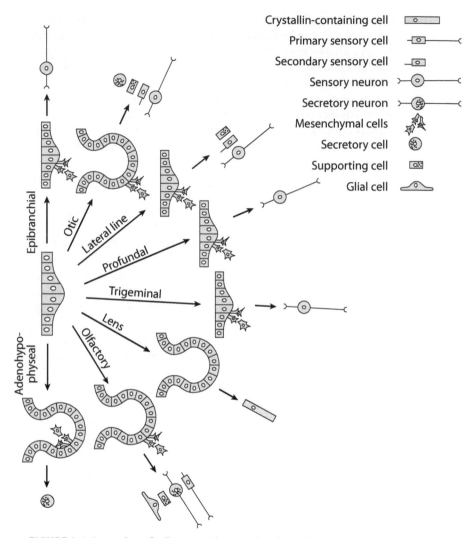

FIGURE 9.4: A sampling of cell types and organs that derive from head placodes during vertebrate development. Head placodes are also known as neurogenic placodes, but not all head placodes form neurons. The counter example is the lens placode of the eye (drawing based on Schlosser 2005 *J. Exp. Zool. B. Mol. Dev. Evol.* 304B:347–399).

the nose, the anterior part of the pituitary, the inner ear, as well as fish and amphibian lateral lines and many more, are all derived from an epithelial-mesenchymal interaction that includes the transient formation of epidermal thickenings (i.e., placodes). Whether there is more than a superficial similarity between these "body plan placodes" (also called neurogenic placodes, although a slight misnomer because they include the lens placode, which is not neurogenic) and those that occur during skin development are unclear. All body plan placodes are characterized by activation of a *Six1, Six4,* and *Eya* network (Schlosser 2005), although not all of these are expressed, as for example *Eya* in hair placode cells (e.g., Rhee, Polak et al. 2006). But I am not aware of a systematic attempt to elucidate the relationships between body plan and skin placodes.

Be that as it may, the question remains of what we can make of the molecular similarities between hair and feather development even though they are clearly not homologous, which means feathers did not arise from hair or the other way around. It is also unclear whether they have the same precursor. There are a number of experimental facts that may be helpful for getting a fuller picture of the developmental evolution of the amniote skin. The most intriguing observations are those related to the development of different regions of the adult and embryonic skin.

In mammals and birds, most of the external body surface[2] is covered by what may be called the General Body Epidermis (GBE) that, in amniotes, is cornified to varying degrees and decorated with clade-specific appendages and glands (hair and sweat glands in mammals and feathers in birds). In addition, there are tissues that are in continuity with the GBE, although they are different and these differences are ancestral for amniotes. As examples, these include the cornea, the epithelium that covers the eye, the plantar skin of the hands and feet of many terrestrial amniotes, which usually does not have hair or feathers (with the exception of rabbit feet, for example), and, in embryos, the amnion that is the extraembryonic cover, which protects the embryo.

The amnion is in continuity with the GBE, as is the cornea. Intriguingly, both the amnion and the cornea can form lineage-specific appendages, hair or feathers, when combined with dermis that is normally associated with the GBE (Dhouailly 2009). The plantar epidermis is less easily transformed to display the phenotype of the GBE. In all cases for which the GBE derivatives, hair or feather, are induced, the inducing agent is activation of the Wnt/beta-catenin pathway in the epidermis, whether in cornea cells or amnion cells. During normal development of the cornea, the activity of this pathway is inhibited by locally expressed transcription factors (Dhouailly 2009).

In the cornea, the Wnt/beta-catenin pathway is inhibited by *Pax6* and in the plantar epidermis this pathway is inhibited by *En1*, a gene specific for ventral limb surface identity.[3] One can say that the GBE phenotype is the default state that is suppressed in specialized skin regions like the cornea

[2] Topologically the covers of the mouth, the gut, and the gut appendages, like the lung and the liver, are also part of the body surface and, hence, make for an "internal" body surface as opposed to the external that is directly exposed to the extra-organismal environment.

[3] The plantar surface is not uniquely characterized by *En1* expression, but by a combination of *En1* and *HoxA13* and *HoxD13*. So the combination of these and perhaps more factors is likely responsible for the differentiation of the plantar skin from the GBE. This inference is also supported by the fact that the whole ventral body surface expresses *En1*, even though it is covered by GBE.

and the plantar epidermis. The question for us is what these developmental facts mean in terms of skin evolution. Cornea, amnion, and plantar skin suppress the execution of derived developmental programs (i.e. those of feather or hair). Does this mean that the phenotypes of the cornea and the plantar skin are evolutionarily derived relative to feathered or hairy skin? This would be suggested by a strict recapitulative view of developmental evolution (Dhouailly 2009).

However, this scenario does not make phylogenetic sense because eyes are phylogenetically at least as old as crown group vertebrates, as is the cornea, and tetrapods had plantar skin differentiations long before the origin of amniotes. And, of course, all amniotes have an amnion. The fact that a phenotype is developmentally secondary (results from suppressing another developmental pathway) does not mean that it was phylogenetically derived relative to the character that is suppressed. To resolve this apparent contradiction between developmental and phylogenetic facts, it is useful to return to the distinction between character identity and character states.

If we think of the GBE and the other parts of the external body surface, cornea and plantar epidermis, as different character identities of the skin (i.e., different developmentally and variationally individualized parts of the body cover), then it is easy to see that these specialized parts of the body surface, cornea and plantar epidermis, evolved by suppressing the default identity, namely GBE. This is because they are phylogenetically younger than the GBE.[4] Note that the whole point of the distinction between character identity and character state is that character identity does not specify character state. That means that the fact that, in the eye, the GBE identity is suppressed is independent of the GBE phenotype or GBE character state.

The fact that *Pax6* suppresses hair development in mice is only a consequence of suppressing the GBE character identity because it happens to be that, in mammals, the GBE grows hair. In the chicken, *Pax6* also suppresses GBE identity, as it likely was in the common ancestor of mice and chickens. Again the fact that *Pax6* suppresses feather development in the chicken eye is also a consequence of GBE character identity suppression.

Hence, *Pax6* expression in the cornea and *En1* expression in the plantar skin are not primarily to oppose the execution of the hair or feather program; rather, they determine non-GBE character identity. Hair (or feathers) is a derived *character state* of the GBE. This situation is very similar to the one regarding the role of *Ubx* in determining haltere and wing phenotypes discussed in chapter 3. It turns out that *Ubx* does not primarily oppose wing development, as it is also active in four-winged insects. But it does determine

[4] This assumes that the GBE of a tunicate is homologous to that of a vertebrate, and there is some molecular evidence to support this. Then it is clear that the cornea is a vertebrate novelty, as is the plantar epidermis and the amnion.

hind wing identity, regardless of whether the hind wing looks like a wing or a haltere. Similarly, one can understand the role of *Pax6* in the cornea as opposing *GBE character identity* rather than hair or feather identity. In fact, hair development suppression is only a downstream consequence of GBE character identity suppression.

The difficulty in interpreting experimental results is that experimental results for any one species cannot distinguish between genetic effects on character identity and character states. Of course, any experimentally induced change in character identity has consequences for the observed character state. However, the molecular mechanism that is affected by the experiment may not be specific for the character state. Rather it may be specific for character identity.

Let us summarize these ideas with a scenario for the developmental evolution of the amniote GBE and, thus, for the evolution and development of hair and feathers. It is clear that the body surface of the very first crown group vertebrates had at least two domains, the general body cover and the cornea, because the vertebrate crown group ancestrally had eyes. Because of their different functions, the phenotypes of cornea and general body cover have to be different and, thus, execute different developmental and cell differentiation programs. Hence, the GBE and the cornea have different character identities.

The differentiation of the body surface into more identities continued with the further diversification of the vertebrates as a group and with further functional specialization of different parts of the body surface. One was the evolution of a specialized surface of the hands and feet of terrestrial tetrapods, derivatives of the ancestral dorsoventral differentiation of the paired appendages, to more specialized skin regions in birds (see below), to minute specializations in the skin covering the fin rays of some benthic fish groups, like the "cuticularized" skin of blennies (see Brandstätter, Misof et al. 1990). For the amniote ancestor, we certainly can assume that there were at least three developmentally individualized skin regions: the general body epidermis; the cornea; and the plantar skin (and probably more). These can be seen as three skin character identities.

Most likely already at that stage of evolution, the GBE identity was the default epidermal character identity. The execution of the GBE developmental program was actively opposed by the ChIN of the cornea and the plantar epidermis, regardless of what the actual phenotype of the GBE was at that time. The evolution of derived skin appendages then proceeded within the framework of the three ancestral epidermal character identities. In mammals, the GBE evolved the hair program, and, in a highly derived clade of sauropsids, the feather development program evolved as a derived character state of the GBE.

In this scenario, it is clear that cornea and plantar identity genes, *Pax6* and *En1* (probably in conjunction with *HoxA13* and *HoxD13* to limit the phenotype to the ventral autopodium rather than the whole ventral limb) would oppose the development of hair or feathers. However, this would have been secondarily as a consequence of suppressing the GBE character identity, and not as specific inhibitors of feathers and hair development. Given this general framework, we can now return to the question of how hair and feathers arose within the GBE.

Mammalian Skin Derivatives: Hairs and Breasts

Mammalian hair that is visible on the surface of the skin is only a minor component of a complex microanatomical organ that has several components (see figure 9.3). It consists of the hair proper along with its follicle, which allows the hair shaft to elongate, a repository for hair stem cells (a.k.a. "the bulge"), one or two types of glands, and a muscle that can erect the hair, the *musculus arrector pili*. The glands can be those that produce a sebaceous lipid-rich secretion or eccrine glands that produce an electrolyte-rich secretion (a.k.a. "sweat"). If we talk about the origin of hair, we mean not only the origin of the keratinized bristle or hair shaft, but the origin of what has been called the *pilo-sebaceous unit* (PSU) or hair–sebaceous gland unit.

This complex of microscopic organs can vary in composition, with some PSUs having only one gland, or having glands but no hair, as is the case with mammary glands of therian mammals. Of course, sweat glands, like those on the hands and feet, can also develop without associated hair. Understanding hair development and mammalian skin evolution thus requires addressing the relationships between glands, hair, muscle, and the various associations that exist between these elements. That various glands, teeth, and hair are developmentally related is also indicated by the commonality of the molecular mechanisms that play a role in their development. For example, defects in the ectodyspasine signaling cascade affect hair, sweat gland, and tooth development (Mikkola and Thesleff 2003). Retinoic acid treatment in the mouse results in glomerular glands that replace vibrissa hair on the face, as well as hyperplasia of sebaceous glands.

As mentioned before, and this is an unavoidable theme in skin biology, the first steps of hair development are shared with other skin appendages: the formation of a placode and a dermal condensation (figure 9.3). Then a column of epidermal cells grows down into the dermis to form what is called the "hair peg." At the base of this hair peg, a dermal condensation forms that eventually will become the dermal papilla housed in a small cavity at the base of the hair root. At the root, cells will remain proliferative and produce cells that both push the hair shaft up and out and contribute to this by apoptosis and keratinization. As the hair shaft becomes individualized and grows outward, the continuation of the epidermis into the hair root is called the outer hair sheath. From this sheath, the glands develop by forming a bulge, and the lumen of the gland develops secondarily.

There are essentially three types of models that have been proposed for the evolutionary origin of mammalian hair. They all agree that the thermal insulating function of the modern mammalian pelage is secondary and is not the driving force for the origination of hair. They differ with respect to the primary function that proto-hairs had and the precursor structures from which hair arose. The first and oldest model assumes that proto-hairs initially evolved for mechanoreceptive functions similar to those of spikes and bristles found in modern amphibians and squamates. The earliest reference to this theory is Maurer (Maurer 1893), was further developed by Elias and Bortner (Elias and Bortner 1957), and in its modern incarnation by Maderson (Maderson 1972; Maderson 2003).

In Maderson's formulation, hair originated as mechanosensory bristles in the inter-scale epidermis. According to this model, the proto-hair was a structure similar to inter-scalar sensory bristles in geckos and other squamates, even though Maderson did not draw a direct homology between these structures and mammalian hair as earlier authors did. This theory is intuitive, given what is observed on mice and rat tails and what is observed on other extant mammals for which what looks like scales and hair co-exist (e.g. armadillos) with groups of three hairs placed between scales.

Another theory drew a connection between filiform papillae on the tongue of mammals and hair by proposing a transformational scheme by which hair derived from tongue papillae (Dhouailly and Sun 1989). Finally, Stenn and colleagues proposed a sebogenic hypothesis in which hair evolved as a derivative of sebaceous glands (Stenn, Zheng et al. 2008; Dhouailly 2009). To give a full assessment of these theories goes beyond the scope of this chapter. Below I will focus on the sebogenic hypothesis because it relates well to the developmental and paleontological data.

At issue is the assumed state of the general body skin in the ancestral mammalian lineage. The split between the "reptilian" and mammalian lineage is the deepest in the crown amniote clade (Gauthier, Kluge et al. 1988). Thus, the skin biology of reptiles is not directly informative regarding the ancestral state of the mammalian skin. Furthermore, skin does not readily fossilize and, thus, the fossil record is quite limited. The only useful paleontological evidence derives from skin impressions from a therapsid fossil from the upper Permian, *Estemmenosuchus* (figure 9.5).

These well-preserved skin impressions show a hairless skin with impressions that have been interpreted as glandular (Chudinov 1965). Hence, this fossil is consistent with the sebogenic hypothesis, which assumes that the ancestral condition in the stem lineage of mammals was a glandular epidermis. The assumption is that at least some of the glands were sebaceous and contributed to the skin-water barrier. The functional hypothesis suggests that

FIGURE 9.5: *Estemmenosuchus mirabilis*, a stem mammalian species from the Permian of the Ural region. This is the second oldest therapsid fossil and, thus, a representative of an early stem mammal. Well-preserved skin impressions suggest a hairless, glandular skin. These skin impressions support the sebogenic theory of the origin of hair, which suggests that hair arose as an accessory organ to a sebaceous skin gland. The functional role of the hair might have been to draw out the sebaceous secretions like a wick of a candle (modified after a drawing by Dmitry Bogdanov).

hair first arose as a "wick" that served to draw the oily secretion out from the gland and onto the external skin surface. The main piece of evidence for the sebogenic hypothesis of hair origin is developmental.

As mentioned above, sebaceous glands and hair tend to develop together, which gave rise to the concept of the pilo-sebaceous unit. However, the interdependency between sebaceous glands and hair follicles is not absolute. There are sebaceous glands without a hair follicle, as for example in the eyelid, the lips, the prepuce, and the labia minora. But hair follicles do not come without a sebaceous gland. Stenn and colleagues (Stenn, Zheng et al. 2008) concluded from this that the dependency between sebaceous gland and hair is asymmetrical, with hair being dependent (at least statistically) on the presence of sebaceous glands, but not the other way around. This form of developmental dependency suggests that the gland is phylogenetically older than the hair, with the hair constructed in a way that requires the presence of a sebaceous gland. If this inference is correct, then it would predict that hairy skin derived from glandular skin, as supported by the skin impressions of *Estemmenosuchus*.

The asymmetrical developmental dependency between gland and hair is mirrored by a dose-response relationship that determines the presence of glands and hair. Multiple lines of evidence suggest that weak inductive signals for "hair" development result in sebaceous glands, regardless of whether Wnt or Shh signaling is involved (Niemann, Unden et al. 2003). Only at higher signal intensities, in particular stronger Wnt/beta-catenin signals, will a hair follicle be formed in addition to the sebaceous gland. Developmentally, the logic of this relationship explains why hair always comes with glands, but glands can develop without hair.

Stepping back from the details of this scenario, the evidence suggests that hair is a novel structure, but that it probably arose as an ancillary structure within an existing skin differentiation, the sebaceous gland. Considered as a structure in itself, hair is a Type I novelty, although when considered as a part of the pilo-sebaceous unit, it contributes to a character state (i.e., a Type II novelty), as a transformation of or an addition to a pre-existing structure.

Origin of Breasts

One of the signature innovations of mammals is the presence of milk (mammary) glands that gave the clade its name. There are many theories about the evolutionary origin of mammary glands that we will not discuss in detail here (see Blackburn 1991). However, the comparative anatomical and developmental evidence is quite clear—namely, the mammary gland is a modified hair or, to be more precise, a modified pilo-sebaceous unit.

All extant mammals provide for their young with a nutritious secretion (a.k.a. milk), including the monotremes, *Platypus* and *Echidna* (Lombardi 1998). In the monotremes, the milk gland is permanently associated with hair, and milk is secreted diffusely into a patch of hair from which the young lick up the milk (figure 9.6). The initial functional role of this "milk patch" may have been as a gland to impregnate the eggshell that was in a shallow pouch of the belly (Oftendal 2002). Nipples are limited to therians—that is, marsupials and eutherians (placentals).

In marsupials, the developing milk glands are transiently associated with hair (Oftendal 2002), but have no hair in the mature mammary gland. In

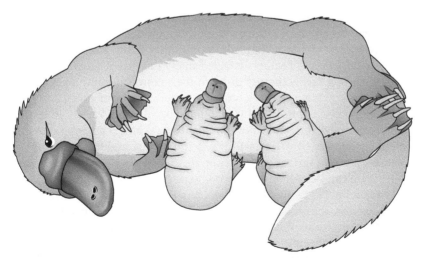

FIGURE 9.6: Nursing female platypus. Although the platypus does not have nipples, it does have milk glands in a diffuse field on its belly. The young lick the milk off the mother's fur as it seeps out of the skin.

eutherians, milk glands are never associated with hair, and strong BMP expression in and around the nipple prevents hair growth in the areola. In addition, the mammary gland is developmentally derived from a single placode origin with the development of an extensive branched milk gland.

This comparative evidence together with the fact that hair and sebaceous glands are developmentally coupled (as discussed above) suggests that mammary glands were, in fact, an innovation through the individualization and transformation of some pilo-sebaceous units in the general body epidermis. The evolution of the derived mammary glands of eutherians was a multi-step process (Blackburn 1991). First was the specialization of the sebaceous gland to produce milk, a grade of organization still found in monotremes. One can think of this as a Type I innovation through the individualization of a repeated anatomical unit, the PSU. Then a number of these milk-producing pilo-sebaceous units were integrated into a unitary organ, which had a common secretory duct and a nipple, and developed from a single developmental precursor.

The synorganization of the development of multiple pilo-sebaceous units into a developmental pathway starting from a single placode may have been the most interesting and possibly most difficult transition. One possibility is that one pilo-sebaceous unit evolved the ability for branching morphogenesis that resulted in an extensive system of milk-producing glandular units, rather than the fusion of multiple pilo-sebaceous units into one organ. The answer to this question must be relegated to future research.

Devo-Evo of Bird Skin: Scales into Feathers

One of the most consequential novelties in the history of vertebrates was the origin of feathers (figure 9.7). Among living animals, the presence of feathers is a diagnostic character of birds, as no other living animal group has this skin specialization. Feathers enabled many important functional innovations, including homeothermy through their contribution to heat insulation, flight by providing a strong but light material to form wings and tails, and behavioral communication by enabling a bewildering variety of feather shapes, structural colors, and coloration patterns. Feathers are even involved in sound production (Bostwick and Prum 2003; Clark and Feo 2008).

Feathers are also unique among skin appendages because of their complex structures and the different morphologies that they can assume (Lucas and Stettenheim 1972). By comparison, the skin appendages of the living relatives of birds, the crocodylians and squamate reptiles, are comparatively uniform, as they are all more or less differently shaped scales. Thus, a deeper understanding of the nature of character identity, homology, and novelties has to confront the origin and identity of feathers as one of the paradigms of a game-changing evolutionary novelty. As will become clear from confronting the details of feather development and variation, the case of feather

FIGURE 9.7: Diversity of feather shapes found on one bird's body, including down feathers, contour feathers, and flight feathers (image taken from http://www.cabrillo.edu/~jcarothers /lab/notes/reptiles/FRAMES/MainFrame.html).

origins and evolution challenges the notions of character identity and evolutionary novelties in interesting ways.

The Anatomy and Variation of Feathers

The anatomical nature of feathers can't be defined without considerable abstraction and knowledge of the morphological variety of which feathers are capable. Thus, it is better to explain the anatomy of feathers by starting at what one could call an archetypical feather and then explain the variations on this theme.

The "archetypical" feather consists of a shaft that is divided into two segments: the proximal calamus and the rachis (figure 9.8; Lucas and Stettenheim 1972). The calamus is partially embedded in the skin of the bird and ends at that point where side branches emanate from the shaft. From this point onwards, the shaft is called a rachis and the side branches are the barbs. Collectively, the barbs form the feather vane on either side of the rachis. From the barbs, secondary branches arise that are called barbules or radii. These barbules can have small hooklets that can connect to the barbules of the next barb and give the vanes of certain feathers some degree of mechanical stability.

There are numerous variations of this "typical" feather. Down feathers lack a rachis and the barbs emanate from the distal end of the calamus. Filoplumes are simple hair-like appendages that occur in association with down feathers. Barbules and hooklets can be absent, which results in more fluffy feathers than the rigid flight feathers. The sizes of the barbs on either side

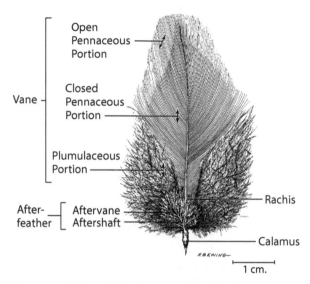

Vane

Open Pennaceous Portion

Closed Pennaceous Portion

Plumulaceous Portion

After-feather { Aftervane Aftershaft

Rachis

Calamus

1 cm.

FIGURE 9.8: Structure of a "typical" feather. The stem of the feather is called the calamus, which continues into the mid rib of the feather called the rachis. The primary side branches are called barbs and the secondary branches are the barbules. Barbules can have hooklets so that barbules of consecutive barbs can interlock. Redrawn by permission from Prum 1999, *J. Exp. Zool.* (*Mol. Dev. Evol.*) 285:291–306. Copyright © 1999 Wiley-Liss, Inc.

of the rachis can differ, which results in aerodynamically adapted flight and tail feathers. A second smaller rachis can arise at the opposite side of the calamus, resulting in feathers with a main rachis and a secondary rachis. The proximal barbs can be downy and the distal part can be vaned, like the feathers of the general body cover that provide heat insulation at the basal part and protective functions on the apical part of the feather.

A comprehensive summary of the morphological variations of feathers can be found in Lucas and Stettenheim (1972). The very brief summary provided here may suffice to illustrate the point that the basic organization of feathers allows for a great variety of morphologies and functions central to the life of birds.

Feather Development

The first morphological stage, again (!), of feather development is the formation of an epidermal placode in association with a dermal condensation (figure 9.9). The epidermal placode is a group of elongated epidermal cells that form a characteristic bump on the surface of the embryonic skin. The dermal condensation is an aggregation of mesodermal cells closely associated with the placode. Reciprocal signaling between the placode and the

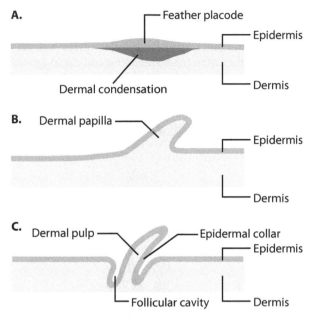

FIGURE 9.9: Early stages of feather development. A) The first morphological sign of feather development is the formation of a placode, which is a thickening of the epidermis and a condensation of dermal cells underneath. B) The critical first step when feather development deviates from that of scales is the formation of a dermal papilla and a tube-like protrusion. C) Finally, the base of the epidermal tube grows inward to form a follicle, which will be the source of feather stem cells and the continuous growth of new feathers during the molting cycle (redrawn after Prum and Brush 2002, *Quart. Rev. Biol.* 77:261–295).

dermal condensation establishes a developmental module. This stage of feather development is still similar to that of many scales.

Recombination experiments that compared epidermal and mesodermal cells established that the identity of the epidermal differentiation that emerged from the placode was determined by the identity of the mesodermal condensation (Sengel 1976). Prospective scale epidermis combined with feather mesoderm forms feathers, and prospective feather epidermis recombined with scale dermis forms scales.

The next stages of feather germ development are formation of a tube-like extension of the skin and formation of a follicle at the base of the feather germ (Lucas and Settenheim 1972; Widelitz, Jiang et al. 2003; Figure 9.9). The follicle results from the epidermis at the base of the feather germ that sinks into the dermal layer and assumes a function as the principal locus of cell proliferation. Feather differentiation starts at the tip of the feather germ and proceeds proximally with new cells added primarily at the base (i.e., at the follicle). Hence, a feather has a mode of development that is quite different from other branched structures, like trees, in that the distal parts of the developing feather are older than the proximal ones and the branches of the feather, barbs, and barbules arise from differentiation of an epithelium (see next paragraph) rather than from sprouting and distal growth.

The characteristic structures of the distal feather, barbs, barbules, and rachis, arise within the ectodermal epithelium of the feather germ, which at this stage is a tube of epidermis filled with a dermal pulp (figure 9.10). The outer layer of the epidermal feather germ forms a decidual layer of keratinized cells, called the sheath, which will fall away as the mature feather

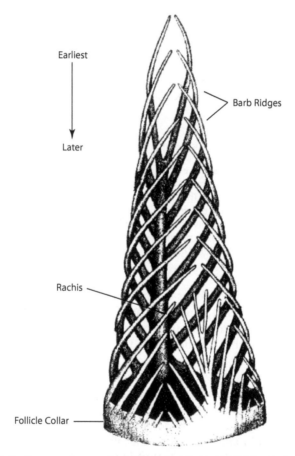

Earliest

Later

Barb Ridges

Rachis

Follicle Collar

FIGURE 9.10: Feather morphogenesis: the branched shape of a feather is developmentally derived from a differentiation of the epidermal tube that constitutes one of the earliest stages of feather development (see figure 9.9C). Essentially, the epidermal tube of the feather germ is "cut open" by the formation of keratinized barb ridges and the falling away of the material in between. Spiral displacement of the location of barb-ridge formation results in the characteristic feather shape and the formation of the rachis. The rachis results from the fusion of barb ridges. Reproduced by permission from Prum 1999, *J. Exp. Zool. (Mol. Dev. Evol.)* 285:291–306. Copyright © 1999 Wiley-Liss, Inc.

unfolds. Underneath the developing sheath, the epidermis forms ridges, so-called barb ridges, by forming deep folds at their basal sides. Each ridge will form one barb and the associated barbules.

Barb ridges have an outer layer that corresponds to the basal layer of the general epidermis by directly sitting on top of the basal membrane of the epidermis. This cell layer is called a marginal plate, in particular those parts that lay opposite neighboring barb ridges. The inside of the barb ridge has large cells arranged in two parallel layers, called barbule plates, which eventually will differentiate into barbules. At the bottom (or tip) of the barb ridge where the barbule plates meet, there are similar cells that will differentiate into the barb. Degeneration and decidualizing of the marginal plates essentially result in the physical separation of neighboring barbs. This "cuts open" the continuous epidermal tube of the early feather germ.

In pennaceous feathers, the barb ridges become radially displaced as their differentiation proceeds from distal to proximal such that, eventually, barb ridges fuse, mostly at the anterior pole of the feather germ, and form the rachis. Most of the morphological variation in feathers can be explained by variations in the rate of differentiation and radial displacement of barb ridge differentiation during distal to proximal differentiation of the feather germ.

The critical feather-typical events in the development of a feather are the formation of an epidermal tube, the formation of the follicle, and the differentiation of barb ridges. Some of the molecular signaling systems that are operative in these events are understood (Harris, Fallon et al. 2002; Yu, Wu et al. 2002; Widelitz, Jiang et al. 2003; Wu, Hou et al. 2004; Dhouailly 2009). Interestingly, the initial polarity of feather and scale placodes is caused by the Shh/BMP module, and this module is later re-deployed during the differentiation of the barb ridges when these signals are expressed in the marginal plate cells.

Origin of Feathers

All outgroup taxa to birds are "reptiles"[5] (*sensu lato*) and are covered with scales. This includes all other amniote taxa, except for mammals, crocodylians, squamates, and turtles. Hence, it is safe to infer that the common ancestor of birds and crocodylians also had scales as the principal skin ornament. Of course we are talking about epidermal scales, and not bony, dermal scales like fish scales. Epidermal scales essentially are localized areas of intense keratinization formed by tough beta-keratin, and separated by more flexible skin with alpha-keratin.

Reptilian scales can assume a variety of morphologies, ranging from simple plaques of keratinization, like the reticulate scales of the palmar surface

[5] Of course birds are technically also reptiles as they are nested within the clade Reptilia.

of the hand and foot, to overlapping scales typical of the squamate body, and to protruding structures like frills, thorns, and horns. Scales also exist on the body of birds, primarily on the palmar surface of the foot called reticulate scales, and the cover of the toes and the dorsal tarsus-metatarsus, called scutate scales.

Given that the ancestors of birds certainly were scaled reptiles, a discussion of the origin of feathers has to address the question of what the relationship is between scales and feathers. At a descriptive level, one may ask what are the similarities and differences between scales and feathers. At a more conceptual level, the question is whether feathers arose as modifications of scales (i.e., whether feathers essentially are a set of character states of scales; feathers as glorified scales). That feathers are, in fact, derived from scales is currently the consensus view. However, the answer to this question is not straightforward and we need to be clear on what the evidence is that supports the connection between scales and feathers.

Assuming that this consensus is correct, that feathers derived from scales, then the downstream question is how the transformation from scales to feathers occurred. Great progress has been made in recent years both in the paleontology of feathers and in the developmental biology of feathers so that these questions are an excellent test bed for ideas about character identity and innovation.

What Is New about Feathers? And What Is Ancestral?

Feathers are distinct from other epidermal appendages at many levels, including the type of keratin used, their histology, their morphology, and their morphogenesis. A once popular theory asserted that the origin of feathers might have been related to the evolution of a novel type of beta-keratin, called feather-keratin or ϕ-keratin (Brush and Wyld 1982; Brush 1996).

Keratins are a large family of intermediate filament proteins found in all animals (Rugg and Leigh 2004; Toni, Valle et al. 2007; Alibardi, Dalla Valle et al. 2009). Two large classes are distinguished: α- and β-keratins. This classification reflects their predominant secondary structure features, with α-keratins having a preponderance of α-helices and β-keratins having a core of β-pleated folds. Sequence similarity between α- and β-keratins is low and provides no evidence for a common ancestral protein from which these two keratin classes could have derived (Dalla Valle, Nardi et al. 2009). In addition, because the structures of these proteins are also very different, it is safe to assume that these two protein families do not share a common ancestor among the keratins.

Taxonomically, α-keratins are found in all amniote animals, representing the only keratin in mammalian skin, but are also found in reptiles *sensu lato*. In contrast, β-keratins are found only in reptiles, including birds, and are a main component of the mechanically and chemically tough substance

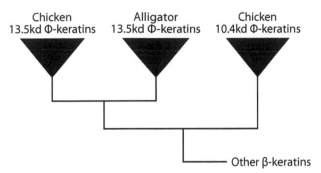

Chicken
13.5kd Φ-keratins

Alligator
13.5kd Φ-keratins

Chicken
10.4kd Φ-keratins

Other β-keratins

FIGURE 9.11: Simplified gene tree of feather type β-keratins (a.k.a. φ–keratins; Dalla Valle et al., 2009). There are two size classes of φ-keratins, the large 13.5 kd and the smaller 10.4 kd type, which is unique to feathers. Alligators also have 13.5kd φ-keratin, which implies that the bird 10.4 kd keratin originated prior to the most recent common ancestor of birds and alligators and, thus, the "feather-typical" keratin gene family is older than feathers.

of scales, claws, and beaks. In feathers, a subfamily of β-keratins is found, called ϕ-keratins (Brush and Wyld 1982; Arai, Takahashi et al. 1986). The ϕ-keratin family was found to consist of two subfamilies, one unique to feathers consisting of 10.4 kd proteins and one larger one of 13.5 kd found in claws, scutate scales, and beak keratin. Members of the ϕ-keratin subfamily were also found in alligator claws (Sawyer, Glenn et al. 2000).

This distribution led to the idea that the feather-specific forms of ϕ-keratins arose coincidentally with feathers (Brush 1996). However, recent phylogenetic studies of ϕ-keratins reveals that the lineage of the 10.4 kd ϕ-keratins is older than that of the larger ϕ-keratins and is shared between birds and alligators (figure 9.11; Dalla Valle, Nardi et al. 2009; Ye, Wu et al. 2010).[6]

These results show that, far from being a feather-specific novelty, so-called feather-specific ϕ-keratins (i.e., 10.4 kd ϕ-keratins) are a protein family older than the most recent common ancestor of alligators and birds. A similar story has emerged for the so-called hair keratins, which have also been found to be much older than the origin of mammals (Eckhart, Valle et al. 2008). Hence, we have to conclude that the structural proteins specific for derived epidermal structures (feathers and hair) can be much older than the structure itself, and that the origin of these structural molecules did not play a part

---

[6] A divergent result by Greenwold and Sawyer [Greenwold, M. J. and R. H. Sawyer (2010). "Genomic organization and molecular phylogenies of the beta (beta) keratin multigene family in the chicken (Gallus gallus) and zebra finch (Taeniopygia guttata): implications for feather evolution." *BMC Evol Biol* **10**: 148] is due to the lack of appropriate outgroups; the authors used crocodile sequences as the outgroup and, thus, forced the root to the archosaur ancestor and created a phylogenetic hypothesis suggesting a derived status of low molecular weight feather keratins.

in the evolutionary origin of the structures themselves. It is the origin of the developmental organization of a structure rather than its molecular constituents that explains evolutionarily novel structures, such as hair and feathers.

At the histological level, feather germs form a complex multilayered tissue with layers of different molecular compositions. Specifically, from the outside to the inside one can distinguish six layers (see table 9.1; from Sawyer, Salvatore et al. 2003). A corresponding number of layers are found in chicken scutate scales, which correspond to those in the feather and also have similarities in their molecular compositions. The difficulty in comparing scale and feather histology is that many of the layers in scales are present only in embryos and are lost upon hatching. Hence, the comparison has to include embryonic stages to reveal corresponding histological units of organization between feathers and reptile scales.

From this comparison, it has been hypothesized that the feather sheath is homologous to the secondary periderm, and that barb ridges are homologous to the subperiderm of scales (Sawyer, Salvatore et al. 2003; Sawyer, Rogers et al. 2005). Hence, it is likely that the histological complexity of the embryonic archosaur scale was the substrate from which was built the morphogenetic complexity of the feather.

Developmentally, both feathers and scales form from localized areas of mesenchymal-epithelial interactions just like many other vertebrate organs (Sengel 1976). Thus, feathers and scales share many inductive signals in common and with other placode-derived structures in a wide variety of vertebrates. Feather anlagen and alligator scale anlagen also share early expression patterns of the Shh-BMP signaling module in which in both the anterior end of the anlage expresses BMP2 and the posterior part expresses Shh. The development of feathers starts to deviate from that of scales when BMP and Shh in the posterior placode compartment are co-expressed coincidentally with the distal outgrowth of the feather germ (Harris, Fallon et al. 2002).

Table 9.1. Comparison of the histological layers of feather germs and embryonic scales (after Sawyer et al. 2003 and 2005). These data suggest that the basic histological organization of the feather germ is homologous to that of the embryonic archosaur scale.

Bird Feather	Bird Scale	Alligator Scale	ϕ-keratin
Periderm	Primary Periderm	Primary Periderm	Yes
Sheath	Secondary Periderm	Secondary Periderm	No
Barb Ridges	Subperiderm	Subperiderm	Yes
Axial Plate	Alpha-Stratum	?	No
Marginal Plate	Beta-Stratum	Beta-stratum	No
Stratum Basale	Stratum Basale	Stratum Basale	No

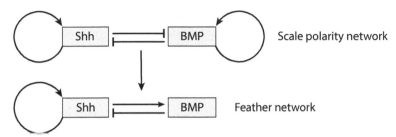

FIGURE 9.12: Model of the gene regulatory network underlying feather morphogenesis. Both alligator scale placodes and feather placodes display a typical spatially exclusive expression pattern of Shh and BMP, probably produced by a gene regulatory network of the type at the top of the figure. When the feather germ develops, the expression pattern changes to one of overlap, which shows signs of a typical activator-inhibitor network, symbolized at the bottom of the figure (Harris et al., 2005).

This is the earliest known molecular difference between feather and scale development and coincides with the origin of the feather-specific tubular organization of the feather germ. Later stages of feather germ development include re-patterning of BMP-Shh expression in the context of barb ridge formation. In fact, experimental evidence shows that these two signaling molecules are causally involved in the development of barb ridges (Harris, Fallon et al. 2002; Yu, Wu et al. 2002). The mode of action is likely a classical activator-inhibitory dynamic, in which Shh plays the role of an activator and BMP is the inhibitor (Harris, Williamson et al. 2005). The transition from the ancestral mode of expression to the derived feather-specific mode was one in which first BMP and Shh were mutually exclusive to one in which Shh induced BMP (figure 9.12).

Two conclusions are supported by the aggregate of the molecular, histological, and developmental data. First, the histological organization and the early stages of development suggest that feathers, in fact, were derived from archosaur scales (i.e., were a derived set of character states of epidermal scales) and, thus, can be considered a Type II innovation. The second conclusion is that the derived characteristics of feathers were caused by derived morphogenetic patterning rather than by a novel kind of structural protein (e.g., low molecular weight ϕ-keratin) or a novel histological organization (Prum 1999; Brush 2000).

However, Prum and Brush have argued that feather origination also created many novel structures that had no antecedents in scales (Prum and Brush 2002). Examples are the barbs and the rachis that can be considered true Type I innovations (i.e., the origination of novel structural parts). Hence, Type II innovations are, in part, realized by Type I innovations nested within an ancestral structural unit.

It seems that there is, perhaps, a more general pattern to be recognized—namely, that some of the more radical transformative novelties (Type II) are realized by novel structures (i.e., Type I innovations) that are nested within the Type II novelty. In this case, the origin of the follicle, barbs, and rachis are Type I novelties, which are nested within the derivation of feathers from epidermal scales. A similar pattern is possibly involved in another major Type II novelty, the transformation of fins to limbs, as will be discussed in chapter 10. In this case, limbs were clearly derived from paired fins (Type II novelty), but included the derived structures of digits, mesopodial elements, and probably the autopodial field (Type I novelties).

Prum's Hierarchical Model of Feather Origin

As feather development deviated from the developmental stages shared with scales, the placodes, a series of morphogenetic events gave rise to the highly complex structure of mature feathers (see above). Each event was hierarchically dependent on previous events (figure 9.13; Prum 1999). For example, distal outgrowth was necessary to transform the flat patch of epidermis into a tubular structure, and the development of a follicle was necessary to allow continuous growth, even as the distal part of the tube started to differentiate. The tubular structure was also necessary prior to the differentiation of barb ridges, and these were necessary for barbules to form. The rachis formed through the confluence of barb ridges and, thus, depended on their prior formation. These hierarchical dependencies would predict a corresponding series of evolutionary steps, assuming that all of these developmental mechanisms could not have originated simultaneously.

According to Prum's model (1999), the initial stages of feather evolution are as depicted in figure 9.13. Three stages are predicted to have led up to the appearance of the pennaceous feather. Stage I consisted of the origin of a hollow tubular outgrowth from the feather placode and involved the origin of the follicle. The follicle originated from an in-folding of the epidermis

FIGURE 9.13: Prum's hierarchical model of feather evolution (Prum 1999). Stage I corresponds to the feather germ stage in development during which the protofeather is only a tubular keratinized skin appendix. Stage II represents the origin of feather barbs from longitudinal differentiation of the feather germ. Stage IIIa represents the origin of the rachis by spiral displacement of barb differentiation and the fusion of barb ridges at the anterior end of the feather germ. Alternatively (stage IIIb), the next step could have been the origin of barbules, the second order branches of feathers. The reason why this model does not predict a particular order for these events is that the origin of the rachis and the barbules do not represent a developmental interdependence like the origin of barbs depends on the presence of the feather tube. Stage IIIa+b represents a case for which both rachis and barbules occur. Stage IV represents the origin of hooklets that interconnect neighboring barbules to form a connected surface. Stage Va is the evolultion of asymmetrical feather shape typical for flight feathers and Stage Vb the origin of an afterfeather. Redrawn by permission from Prum 1999, *J. Exp. Zool. (Mol. Dev. Evol.)* 285:291–306. Copyright © 1999 Wiley-Liss, Inc.

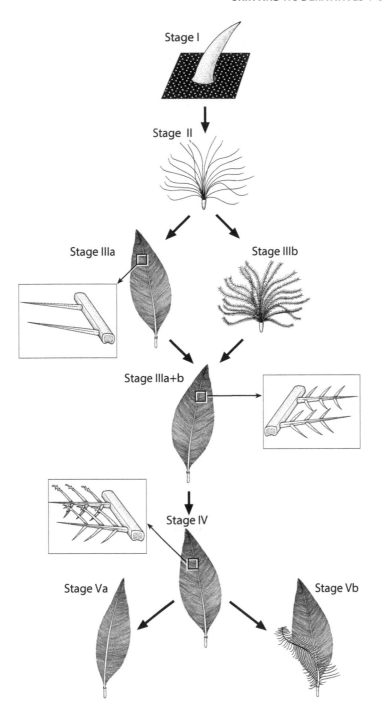

Stage I

Stage II

Stage IIIa

Stage IIIb

Stage IIIa+b

Stage IV

Stage Va

Stage Vb

around the base of the epidermal tube and the establishment of an intimate contact between the rim of this in-fold and the mesoderm. Stage II was the origin of barbs that arose as a longitudinal differentiation of the epidermal tube that, in turn, led to the separation of individual strands of keratinized epidermis. This resulted in a tuft of keratinous filaments similar to that of a down feather.

After the origin of barbs, the next two structures that could arise were the rachis and the barbules. Barbules can exist in feathers with or without a rachis, and vice versa. There is no hierarchical dependency between these two structures and, thus, no prediction can be made as to the order of appearance of these structures during evolution. Thus, Prum's model recognizes two possible stages in parallel: Stage IIIa, which is a feather with a rachis but without barbules, and Stage IIIb, a feather that has barbules but no rachis. Combining both barbules and rachis leads to what Prum calls Stage IIIa+b, essentially a symmetrical pennaceous feather. Stage IV is predicted to be one in which barbules interlock and form the coherent vanes of flight feathers.

Alternative models for the origin of feathers are mostly based on the idea that feathers can be understood as elongated flattened and branched reptile scales (Maderson 1971; Maderson and Alibardi 2000). We will not discuss these models further. They were discredited because they could not account for the fundamental topological facts of feather development. Namely, the outer and the inner surfaces of a feather vane correspond to the apical and the basal aspects of the epidermis, because barbs are created from "cutting open" the epidermal tube of the feather germ. In the "elongated scale" theory, however, the outer and inner surfaces of the feather would correspond to the outer and the inner surfaces of a squamate scale. Thus, this theory is incompatible with the mode of feather development. Further discussion of these theories is not enlightening regarding the biology of feather origination.

Before we move on to the paleontological evidence regarding feather origins, perhaps it is worth reflecting on the nature of Prum's model. This model predicts a parallelism between the ontogenetic development of feathers and their evolutionary history. Thus, it falls within the realm of "recapitulation" theories, similar to that of Haeckel. Haeckel's theory of ontogenetic recapitulation has been discredited because of numerous counter-examples to a literal reading of the "law of recapitulation." Prum's model, however, is different. The reason why Prum's recapitulation theory of feather evolution has more punch than recapitulation in general is that it is based on an understanding of the developmental dependencies among stages of development.

In contrast, the uncritical use of the recapitulation idea is a simple generalization and, therefore, vulnerable to falsification by many counter-examples. However, it has long been recognized that cases of recapitulation can be mechanistically explained if there are functional or developmental interdependencies that force a parallelism between ontogenetic and phylogenetic transformations (Riedl 1978). Hence, one can consider Prum's model of feather evolution as a *mechanistically enlightened recapitulation* model.

Paleontological Evidence of Feather Origination

The first evidence of feathers outside the extant birds was confusing. Associated with the skeleton of *Archeopteryx*, a fossil that displays a mosaic of ancestral dinosaur and derived bird characters, were found impressions of fully formed modern feathers. Hence, at that time, the closest known outgroup of birds was already a flying animal with modern shaped feathers and, thus, did not provide any evidence regarding the initial stages of feather evolution. This situation changed dramatically during the 1990s when fossils of integumental structures were recovered from the Jehol fossil fauna of China (for a review, see Norell and Xu 2005).

Within the stem lineage of birds (i.e., the theropod dinosaurs), integumental appendages have by now been identified in most lineages crownwards of *Compsognathus* (figure 9.14). Specifically, *Sinosauropteryx prima*, a member of the clade of Compsognatha, has been described as having filamentous integumentary appendages on the head, along the spine, and also along the belly and the limb. These structures are clearly separated from the skeleton, up to 30 mm long, and likely hollow (Currie and Chen 2001). Some of these appendages are branched distally, but there is no strong evidence for barbules (Norell and Xu 2005). These structures are easily identifiable as corresponding to Stages I and II of Prum's model of feather evolution.

Similarly primitive integumental filaments have also been described in the basal maniraptorian dinosaurs *Beipiaosaurus* and *Shuvuuia* (Norell and Xu 2005). The latter fossils have even been shown to contain residues of β-keratin (Schweitzer, Watt et al. 1999), which removed all doubts that these were, in fact, epidermal structures rather than collagen or other fibers that might have been exposed during body decomposition.

More advanced feather morphologies, corresponding to Stage III of Prum's model, are also found in more derived lineages of theropod dinosaurs. The most basal lineage among them is the tyrannosaurid *Dilong paradoxicus* (Xu, Norell et al. 2004). Feathers of essentially modern aspect have been found in a number of dromaeosaurids, which belong to the sister taxon of Aveales (the clade uniting all descendents of the most recent common ancestor of modern birds and *Archeopteryx*).

The phylogenetic distribution of morphological feather types thus supports the recapitulation model of Prum. More primitive types of feathers, Stages I and II, are found among more basal lineages and more derived types, Stages IV and V, are restricted to more derived lineages like the dromaeosaurids. Hence, the developmentally based predictions of stages of feather

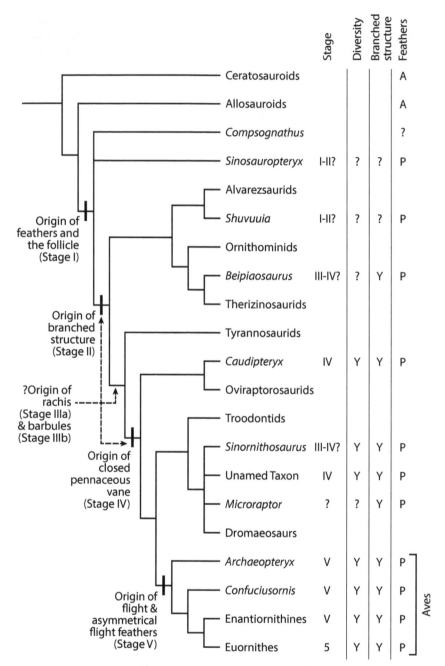

FIGURE 9.14: Phylogenetic hypothesis for the evolution of feathers. The tendency is that more basal lineages show evidence of earlier stages of feather evolution, stages I or II. In contrast, more advanced feathers are found in more derived lineages, as predicted by Prum's hierarchical model of feather evolution (figure 9.13). (Redrawn after Prum and Brush 2002, *Quart. Rev. Biol.* 77:261–295.)

evolution correctly predicted the sequence of morphologies as they appear in the fossil record.

One other observation about the most basal theropod dinosaur with epidermal appendages, *Sinosauropteryx*, is that the appendages are found on all parts of the body, from the head, the back, the belly, as well as the limbs. Thus, it is possible that the first stages of transformation from scale to feather occurred all over the body simultaneously. If this is in fact the case, then the origin of feathers would be a good Type II innovation and unrelated to the differentiation of different parts of the body surface. It would also support an old, but a long discredited idea by H. R. Davies who was the first to recognize the true nature of feathers in his dissertation completed under the supervision of Carl Gegenbaur in Heidelberg (Davies 1889).

Davies argued that the scutellar scales on the tarsus-metatarsus and the digits of bird feet are not homologous to archosaur scales; rather, they evolved in conjunction with the reduction of feathers on the extremities. Davies supported this idea with observations of pigeons with feathered feet and the transition between scaled and feathered parts of the skin in normal pigeons, which shows a transformational series with intermediate morphologies. This scenario would also suggest that feather individuality, as discussed in the next section, is secondary to the origin of feathers per se.

Feather Individuality and Transgressive Variation

Even a cursory look at the variety of morphological feather types of any set of birds clearly shows that feathers on different parts of the body are individualized, at least to some degree (Lucas and Stettenheim 1972; figure 9.7). The primary flight feathers (remiges) of a flying bird are long, asymmetrical, and partly covered by smaller less asymmetrical feathers called coverts. The tail often has a gradation from asymmetrical to symmetrical feathers (rectrices), and the body is covered with smaller feathers that are downy at the base and vaned at the tip. Subsets of feathers can assume extreme phenotypes as exemplified by the peacock's train, which are the coverts of the tail feathers.

The plumage of birds consists of sets of feathers that have assumed a degree of individuality sufficient to support the functional adaptation of parts of the plumage according to their functional roles in flight, thermal insulation, and communication. Hence, differential modification of subsets of feathers due to natural and sexual selection has clearly occurred. Consequently, different sets of feathers also have to have sufficient variational individuality (i.e., their genetic correlation cannot be too high).

Yet in spite of the obvious variational individuality of feathers, it is also clear that all the different feathers on a bird realize variations of the fundamentally identical developmental program. Variations among feather "types" can be understood to be the result of variations in the settings of a few parameters in the feather differentiation program—for example, the rate of displacement of the barb ridge towards the rachis or the angle at which the barbs eventually open after they are released from the epidermal tube (Prum and Dyck 2003, Teresa Feo in preparation). All feathers seem to share the same developmental potential and, thus, do not seem to express alternative mutually exclusive developmental programs, as deeply individualized characters do (e.g., eye development versus general body epidermis in a fly). The relationships among the different feather types are not limited to the expression of the same basic program, almost by definition. In addition, there is also the transfer of novel phenotypes from one feather group to others, a phenomenon that could be called "transgressive variation."[7]

One innovative feather characteristic is the modified tip of the feathers of waxwing birds, genus *Bombycilla*. Mature individuals of all three species of waxwings exhibit flattened red tips on some of their secondary flight feathers (figure 9.15), which are more frequent in the cedar and the Bohemian waxwing than in the Japanese species. These structures probably arose by an incomplete separation of barbs that remained fused to the rachis and, thus, produced a drop-like tip (Brush 1967; Olson 1970; Mountjoy and Robertson 1988). The cortex of these structures contains the carotenoid astaxanthin that gives it its red color (Brush and Allen 1963). Modified tips have been described in widely divergent clades of birds (e.g., rails, the scaled cuckoo, and the Araçari; Brush 1965). The structure and pigmentation of these structures also vary.

In the waxwings, the number of these tips varies and tips are found in both sexes. They play a role in mating, as there is assortative mating with respect to the number of tips.[8] In rare cases, these tip structures are also found in other parts of the body, most often in the tail feathers but also on the primary flight feathers, the great coverts (i.e., the coverts of the primary flight feathers; Yunick 1970; Stedman and Stedman 1989). This is perhaps the nicest example of transgressive variation, because it is a unique feature that arose in one specific group of feathers, the secondary flight feathers, and is occasionally transferred to other groups of feathers, the rectices and the great coverts (at a frequency of about 1% of individuals or less).

A reverse case of character transfer has also been observed. The tail feathers of the cedar waxwing have a yellow or orange pigmentation at their distal ends. Stedman and Stedman (1989) reported a case of one hatchling with distal yellow spots at some of the primaries (i.e., a tail feather character transferred to the wing feathers).

Plumage variation in waxwings clearly demonstrates that a character state that evolved in one group of feathers, the secondary flight feathers, could

---

[7] The term transgressive is used here to describe variation when differentiated serially homologous parts share the same derived phenotype.

[8] This inference assumes that assortative mating with respect to red wing tip number is not due to a correlation with another character for which mating is assortative.

FIGURE 9.15: Feather tips of the cedar waxwing coverts. This structure is formed due to delayed differentiation and separation of barbs and the accumulation of a carotenoid, which gives it the red coloration. This is a novelty among waxwings' wing feathers, but can also be found on other feathers, like the tail feathers at low frequency (about 1%). This is an example of the transfer of a novel phenotype from one part of the plumage to another, here called "transgressive variation" because this variation transgresses the limits of individual feather types on the body of the bird. This phenomenon is a sign of limited developmental individuality of feathers, even though they are sufficiently individualized so that they can adapt to different functional demands. Reprinted by permission of the Yale Peabody Museum.

easily and without loss of cohesion be transferred to other feather groups, primaries, tail feathers, and coverts. This form of variation can be called transgressive because it transfers information from one group of individualized feathers to other groups.

It is likely that the ease of information transfer is aided by the peculiar developmental organization of feathers for which the information for regional identity of feathers and that for the developmental competence to form feather phenotypes is distributed between two very distinct cell populations. As mentioned in the section on feather development, classical experiments show that regional identity of feathers is determined by the identity of the dermal component of the feather follicle, while the competence for feather morphogenesis resides in the epidermis. Specific feather phenotypes are then the result of an interaction between the dermis and the epidermis and, thus, phenotypic characteristics that evolved in one group of feathers were variationally accessible to other feather groups because these feather shapes were within the developmental competence of the epidermis of all feather germs.

The developmental predisposition towards transgressive variation in feathers also limits the degree of individuality expressed by different feather kinds on a bird. These biological facts are not neatly captured by the strict dichotomy of character identity vs. character state. It is a situation that stands somewhere in between well-established individualized characters and simple copies of identical body parts. This situation could be called a *character swarm*, which is

> a group of serially homologous body parts that are variationally individualized, but which still partake in the same character modality, and thus remain phenotypically similar and subject to transgressive variation.

Characters in a character swarm lack definite, stable distinguishing characteristics, even though they clearly can respond differentially to natural and sexual selection. Thus, they are individualized and have their own evolutionary adaptive histories. The example of bird feathers also suggests that character swarms are a distinct form of developmental/variational organization of body parts rather than a transient stage during the evolutionary progression toward full individuality. The reason is that the "incomplete" individuality of feathers is at least as old as birds and certainly much older after including considerable amounts of theropod diversity. One can suspect that this organizational principle also applies to other groups of serially homologous characters, such as teeth, hair, annelid and arthropod segments, and many more.

Transgressive variation among serially homologous parts also shows that the fact of serial homology has variational and evolutionary consequences. Serial homology is a developmentally and evolutionarily relevant fact and, therefore, this concept cannot be eliminated without a loss of biological insight.

10

Fins and Limbs

The origin and evolution of tetrapod limbs hold a particular fascination for us. Some of the most momentous periods in the history of the human lineage involved evolutionary changes to the paired appendages. Modifications of the hind limb and foot were key during the evolution of bipedal locomotion and the erect posture that is characteristic of humans by freeing our hands to perform more sophisticated functions than locomotion, like making and using tools, or playing the piano. The hand was predisposed to these new functional challenges because of previous adaptations for climbing in trees. This led to the evolution of an opposable thumb, which I am now using to hold the pen I use to write this page. The structures on which these advances were based originated during the transition of aquatic vertebrates to land (i.e., during the transition from fins to limbs). Thus, it is not surprising that evolutionary biologists have spent a great deal of time and effort to understand both the origin of paired fins and the transformation of fins to tetrapod limbs.

Paired Fins

The origin of paired fins was the subject of one of the most notoriously unresolved controversies in the history of biology, the controversy between the proponents of the lateral fold theory like Thatcher (Thatcher 1877), Balfour (Balfour 1881), and Mivart (Mivart 1879) and Gegenbaur's archipterygium theory (Gegenbaur 1876). According to Lynn Nyhart, the inability to resolve this controversy was instrumental in the decline of evolutionary morphology toward the end of the nineteenth century (Nyhart 1995). This is an exceptionally difficult problem because it reaches so far back into the history of vertebrates and, correspondingly, relevant evidence is difficult to come by.

Gegenbaur proposed that the paired appendages of vertebrates derived from gill arches in which one of the gill arches became the shoulder girdle and the gill rays developed into a free extremity, which led to what he called

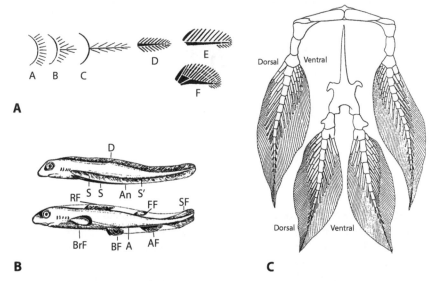

FIGURE 10.1: The two principal theories explaining the origin of paired fins. A) The archipterygium theory of Carl Gegenbaur (1876) states that the pectoral fin originated from a gill ray in which the gill arch became the shoulder girdle. B) The fin fold theory states that paired fins originated from a paired fin fold in continuation of a medial fin fold (Thatcher 1877). C) The fin skeleton of the Australian lungfish, *Neoceratodus fosteri*. Gegenbaur considered the lungfish paired fin a realization of his archipterygium.

the *archipterygium* or original fin (figure 10.1A) (Gegenbaur 1876). The archipterygium was assumed to have comprised a central axis with endoskeletal rays extending at the anterior and the posterior edge of the archipterygium. Gegenbaur saw the fin skeleton of the Australian lungfish, *Neoceratodus fosteri*, as an extant exemplar of his archipterygium (figure 10.1C). This in fact has the structure conceived for the archipterygium, or more likely inspired the archipterygial model.

To a modern evolutionary biologist, invoking the lungfish as representative of an ancestral vertebrate trait is puzzling because the phylogenetic position of the lungfish does not support its role as an exemplar for an ancestral character state. The Gegenbaur school did acknowledge that the lungfish was probably the closest living relative of amphibians and, thus, was a very derived relative to the stem gnathostomes in which paired fins must have originated. It seems likely that the conceptual attractiveness of the lungfish fin as a model for ancestral fin morphology derives from the fact that it can be viewed as a generalized structure from which is it easy to derive fins, which either have anterior radials or any proportion of anterior and posterior radials, as well as limbs.

Another theoretical desideratum that Gegenbaur's theory fulfilled was the idea that any derived structure had to arise from pre-existing structures. This is one of the few points on which Gegenbaur agreed with Anton Dohrn, who opposed Gegenbaur on almost every other issue (Ghiselin 2003). Thus, deriving fins from gills is satisfying if one believes that every novel structure has to have an antecedent structure. After the demise of the Gegenbaur school of evolutionary morphology, the archipterygium theory had few, if any defenders, and it only recently received some attention from paleontologists and developmental biologists because of the similarities between limb and gill ray development (see below).

The major competitor, and by many people's account the successful theory of paired fin origins, is the lateral fin fold theory (LFF) proposed by Thatcher (Thatcher 1877), Balfour (Balfour 1881), and Mivart (Mivart 1879) (figure 10.1B). The idea is that medial fins arose first from the medial fin fold of a basal chordate, similar to that found in *Amphioxus*. The medial fins are also individualized into a number of dorsal fins, a caudal and an anal fin, which must have arisen from portioning the ancestral continuous fin fold. Paired fins may have arisen similarly from the partitioning of a lateral fin fold.

The proponents of the LFF theory point to the similarity between medial fins and paired fins, a similarity that the archipterygium model cannot explain. The LFF theory received a major boost when, in 1876, Balfour described a fold that transiently connected the pectoral and the pelvic fin bud in a shark embryo. This evidence is still cited today, even though more recent examinations of shark embryos failed to find any such structure (Tanaka, Munsterberg et al. 2002). In any case, the LFF theory predicted that the pectoral and pelvic fins arose simultaneously from the partitioning of the LFF. It is this prediction that causes the greatest difficulties for this theory with respect to the fossil record of early vertebrates (see next section).

Fossil Evidence on the Origin of Paired Fins

Although the fossil record regarding the early history of vertebrates is naturally sparse, it nevertheless sets some parameters that constrain the range of possible explanations of the origin of paired fins. Figure 10.2 provides an overview of the likely phylogenetic relationships among the major vertebrate groups. The crown group, gnathostomes, comprises the ray-finned and the sarcopterygian fishes, the extinct acanthodians and the cartilaginous fishes. Ancestrally, this group had two pairs of fins: the pectoral and the pelvic fins. The gnathostomes share this characteristic with their next outgroup, the placoderms, which are also jawed.

Next down the stem of gnathostomes come jawless groups with a polytomy among three groups next to the jawed vertebrates: Osteostracans, Pituriaspis, and Galeaspids. Of these, the Galeaspids had no paired appendages

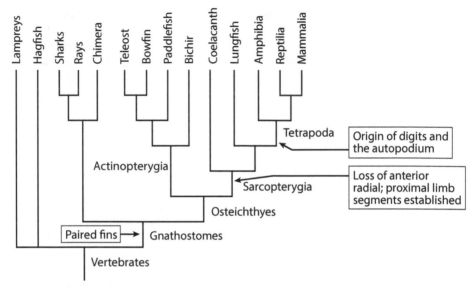

FIGURE 10.2: Overview of the major groups of vertebrates. The two most ancient lineages outside of the jawed vertebrates are the lampreys and the hagfish. There has been a long-running controversy regarding the phylogenetic relationship between them and the rest of the vertebrates as indicated by the polytomy at the base of the vertebrate clade. The next clade is the jawed vertebrates, Gnathostomes, which include all animals from sharks to humans. Osteichthyes are the bony fish that include all descendents of the most recent common ancestor of zebrafish and human (technically, humans are bony fish). Sarcopterygia are the fleshy finned fish, which includes the lungfish, and the tetrapods. Finally, the tetrapoda are the four-legged animals, whether they have legs or not (e.g., snakes), including amphibians, mammals, and reptiles, which include the birds. The origin of paired fins occurred in the stem lineage of gnathostomes, the origin of the proximal limb elements, stylo- and zeugopods, occurred in the stem lineage of sarcopterygia, and finally the so-called fin-limb transition occurred in the stem of tetrapods. Hence, the origin of limbs was a slow process over long periods of phylogenetic time.

and Pituriaspis are only known from their head shields (Coates 2003). Most interesting in this group are the Osteostracans, which are the most basal group with paired fins that contain endoskeletal elements and a pectoral girdle (figure 10.3). Most importantly, these jawless fish only had one pair of fins immediately after the head shield. All the other forms more basal than the Osteostracans either had no paired appendages or had some keeled dermal scales like the Heterostracans or the Anapsid-like fishes (Coates 2003).

From the evidence in the fossil record it seems unlikely that a lateral fold gave rise to the pectoral and the pelvic fins by subdivision, because lateral keels and folds are not widely found among basal jawless fossil groups and the first instance of paired fins with endoskeletal elements only has anterior, pectoral fins. Thus, it is more likely that paired fins first arose in association with the head and were then duplicated to give a second set of paired fins

FIGURE 10.3: *Hemicyclaspis*, a Devonian jawless fish belongs to the osteostracan group, which is the oldest group known to have paired appendages. Note that the oldest fossil with paired fins only has pectoral fins, which contradicts the fin fold theory of the origin of paired fins. The fin fold theory implies that pectoral and pelvic fins originated simultaneously by subdividing a lateral fin fold. (After: http://www.rareresource.com/.)

characteristic of jawed vertebrates, the placoderms and the crown group gnathostomes. According to this reading of the fossil evidence, paired fins are serial homologs that originated from duplication of a pectoral fin.

In other words, the pectoral fin developmental program was redeployed. This scenario is also consistent with the profound similarity of the developmental program of pectoral and pelvic fins as well as the forelimbs and hind limbs. In both cases, the axes of the fin/limb bud are determined by the same set of signaling centers and very similar signaling molecules (Grandel 2003; Mercader 2007).

Recent Developmental Evidence

For most of the twentieth century, the archipterygial theory of paired fin origins was considered untenable. However, this theory recently obtained a new lease on life from molecular evidence. Gillis and collaborators (Gillis, Dahn et al. 2009) showed that gill rays utilize a signaling system with surprising similarities to that deployed in early limb bud development. Gillis and collaborators examined the expression and function of two key limb signaling pathways in the gills of the little skate *Leucoraja erinacea*. These were *Fgf8*, which is expressed in the apical ecdodermal ridge of the fin and limb bud, and *Shh*, which is the signaling molecule from the zone of polarizing activity in the limb and fin bud (see below for a summary of limb development).

They found that these signals were expressed in a manner similar to that in the limb bud in which *Fgf8* was expressed from an apical pseudostratified epithelium. Unlike the limb bud, *Shh* was expressed in the gill epithelium, but its signal was received by mesenchymal cells, like those in the limb as determined by the expression of the Shh receptor Ptc. In addition, as in the limb bud, *Shh* expression could be ectopically induced in gill rays by retinoic acid. In addition, as in the limb bud, *Fgf8* and *Shh* form a positive feedback loop in which suppressing Shh signaling resulted in a loss of Fgf8 and ectopic application of Shh protein resulted in inducing Fgf8 expression.

To my knowledge, this was the first empirical evidence in support of the archipterygium model of paired fin origin since the decline of the Gegenbaur school of evolutionary morphology. However, a thorough evaluation of this new evidence is still lacking. For example, one may question in how many parts of the body has the Fgf-Shh network been found and what is their evolutionary history? What is the relationship between medial and paired fin development? Besides the use of these signaling molecules, what individuates pectoral fin rays from other body appendages that also use Fgf and Shh?

Novel developmental evidence has also been published in support of the LFF theory (Tanaka, Munsterberg et al. 2002; Freitas, Zhang et al. 2006; Yonei-Tamura, Abe et al. 2008; Shearman and Burke 2009). As noted above, the LFF theory states that the fin developmental program first evolved in the medial fin fold after which a lateral fin fold evolved, which gave rise to separate pelvic and pectoral fins. The similarity of medial and lateral fins has been confirmed by gene expression studies of cat shark and lamprey medial fin development (Tanaka, Munsterberg et al. 2002).

It was found that *dHand* and collinear expression of 5' HoxD genes occurred at the posterior edge of the dorsal fin. These expression patterns are also found in the early stages of limb and paired fin development, which suggests that these gene regulatory networks are older than paired fins and were probably co-opted from a medial fin developmental program. Shh is also expressed in the dorsal fins of sharks and skates (Dahn, Davis et al. 2007), although these expression patterns are puzzling because the shark dorsal fin has the expected posterior expression domain, while the skate has both an anterior and a posterior expression domain (figure 1 in Dahn, Davis et al. 2007).

Another major piece of supportive evidence for the LFF theory was the discovery of conserved lines of competence for fin/limb development. In chickens it was first discovered that the inter-limb region (i.e., the region between the forelimb and hind limb buds) can be induced to form additional limbs (Balinsky 1974; Cohn, Izpisua-Belmonte et al. 1995). Both forelimbs and hind limbs originate at the boundary between a dorsal and a ventral compartment of the embryonic skin, which is marked by expression of the *engrailed-1* gene in the ventral compartment (Michaud, Lapointe et al. 1997; Tanaka, Tamura et al. 1997). This boundary marks the future apical edge of the fin/limb bud, and also continues between the fin buds.

There are also dorsal and ventral lines of competence from which medial fins arise during development. These lines of competence are very old and can be found in fish and sharks (Mabee 2000; Mabee and Noordsy 2004; Yonei-Tamura, Abe et al. 2008; Shearman and Burke 2009). Hence, the reasoning is that these extended lines of appendage-forming competence are the developmental remnants of the extensive lateral fin fold that preceded the individualization of pectoral and pelvic fins. However, it may be questioned whether the dorsal and ventral compartments of the embryonic fish ectoderm are really the traces of an ancestral lateral fin fold.

It seems more likely that these compartments are ancient features of the vertebrate body plan rather than having evolved to allow for the expression of a lateral fin fold. In general, compartment boundaries are privileged places for the development of signaling centers and, thus, are not necessarily traces of the LFF. At this point one can conclude that the LFF theory is plausible on developmental grounds, even though the fossil evidence is at least ambiguous, if not wholly contradictory (see above).

Conclusions on Paired Fin Origin

To this point, research on the origin of paired appendages in vertebrates has not provided a coherent explanation. It does, however, illustrate an important point for the developmental evolution of appendage structures—namely, the extensive commonality of developmental mechanisms that are shared among medial and pectoral fins as well as branchial rays. These data are, of course, biased because all relevant studies have focused on known limb developmental genes rather than determining what is specific and different among these structures at the genome-wide level. With these limited available (and even possible) comparative data on the early stages of vertebrate evolution, it seems impossible to determine what the sequence of genetic changes was that led to derived structures like paired fins. In the future it will be important to investigate entire transcriptomes and other genome-wide data from basal and derived species to more fully understand the genetic changes that caused these major transitions in vertebrate phylogeny.

From Fins to Limbs

The fin-limb transition was, and still is, the subject of intense and increasingly interdisciplinary research. It is perhaps that place where paleontologists and molecular biologists work most closely together to answer one of the big questions of evolutionary biology.

Since the objective is to explain the origin of the tetrapod limb, it is important to start with a characterization of the derived character: the tetrapod limb. This does not mean that I believe that evolution was or is a goal-directed progression toward a particular "higher form of being"—quite the contrary. Evolution proceeded along various lines of descent leading to a large variety of "forms of perfection" from sharks, as the perfect aquatic predator, to the highly diverse and disparate teleosts, and to tetrapods, including humans.

However, because evolution is essentially undirected, it is also important to have a standard against which we can evaluate our fragmentary knowledge of the past as to its relevance for explaining the origin of the tetrapod limb. As will be clear from the discussion below, we still struggle with some of the most fundamental questions regarding the fin-limb transition and, thus, we still have to evaluate the paleontological and developmental evidence as to its relevance to the question at hand. To do this, we need to have a clear picture in what way limbs differ from fins in terms of their morphology as well as in terms of their development.

Comparative Morphology and Embryology of Tetrapod Limbs

When we talk about the morphology of the tetrapod limbs, most often we talk about only the skeletal parts of the limb. Of course, real limbs are a combination of bones, connective tissues, muscle, blood vessels, nerves, and skin. Of these, bones are the easiest to study both in extant and in fossil forms. Even during development, skeletal elements are relatively easy to visualize and have, thus, become a token for "the limb" itself. But there are also more defensible reasons to focus on the skeleton. The skeleton determines the spatial and mechanical units of the limb and, thus, is a good token for the overall organization of the limb, which after all has primarily a biomechanical function unlike other body parts, such as the brain or the pancreas.

The skeleton of the tetrapod limb is an endoskeleton, which means that the majority of it consists of elements that arise from aggregations of cells (called mesenchymal condensations) that express cartilaginous extracellular matrix (ECM) components, like collagen II and chondroitin sulfate. Ultimately, the cartilage is replaced by bone substance.[1] This mode of development is in contrast to the most visible part of the skeleton of a typical teleost fin (i.e., the fin rays), which are dermal skeleton. Fin rays are similar to fish scales in that they develop at the interface between the epithelial part of the skin and a mesenchymal condensation that aggregates at the base of the epidermis (Grandel and Schulte-Merker 1998). There is broad agreement that the tetrapod limb skeleton has no part that corresponds to dermal fin rays. Any comparison between the tetrapod limb and fin skeleton has to be limited to the endoskeletal parts of the fin.

---

[1] The only exception from this mode of development is sesamoid bones that develop within tendons and other parts of the connective tissue, but in most animals do not contribute major parts of the skeleton. Examples for which sesamoids contribute bigger elements are the moles and the "panda's thumb."

A.

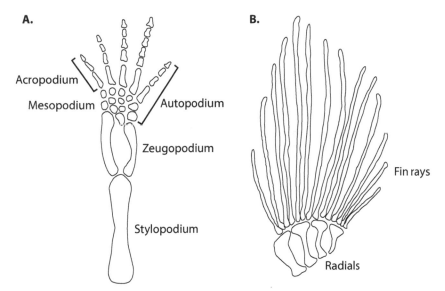

Acropodium

Mesopodium

Autopodium

Zeugopodium

Stylopodium

B.

Fin rays

Radials

FIGURE 10.4: Comparison of the basic structures of the skeletal elements of (A) the tetrapod limb and (B) the teleost pectoral fin. The tetrapod limb has one proximal element, two middle elements, and many distal elements. In contrast, the pectoral fin of teleosts always has four proximal radials and a variable number of fin rays, which are absent in the tetrapod limb (from Wagner and Chiu, 2001, *J. Exp. Zool. B. Mol. Dev. Evol.* 291:226–240).

All tetrapod limbs articulate with a single bone to the body and have three clearly delimited segments along the anterior-posterior axis (figure 10.4). The most proximal segment articulating with the body is called the stylopod: *humerus* and *femur* in the forelimb and hind limb, respectively. The next segment originally consists of two elements in anterior-posterior arrangement called *radius* and *ulna* in the forelimb and *tibia* and *fibula* in the hind limb. Then there is the autopod, which is the most complex and variable of the limb segments: the hand (*manus*) and the foot (*pes*). Among crown amniotes, the autopod ancestrally had five digits (pentadactyl). The situation at the base of the crown group tetrapods is less clear (see below).

Ancestrally, the autopod comprised two parts: a hinge region consisting of a variable number of nodal elements called the mesopodium (i.e., the wrist and the ankle, respectively) and a distal region consisting of the metapodials (i.e., metacarpals and metatarsals) and the digits. The latter segment, the metapodials and digits, has been call acropodium. I will use the terms stylo-, zeugo-, auto-, meso-, and acropodium whenever I do not make a distinction between forelimbs and hind limbs, which seems legitimate because of the fundamental similarity of their organization (figure 10.4).

During development, the three segments of the tetrapod limb arise in a proximal-to-distal sequence (figure 10.5; Shubin and Alberch 1986). As

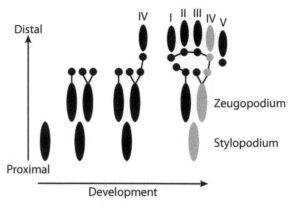

FIGURE 10.5: Simplified representation of the developmental sequence of the skeleton of a typical tetrapod limb. Development starts proximally with the condensation for the humerus/femur (stylopod) and proceeds distally with radius/tibia and ulna/fibula (zeugopod). Most autopodial elements derive from the posterior zeugopodial element with the intermedium and the proximal mesopodial and then digit IV. Digits III, II, and I are added in a posterior to anterior direction, while digit V develops largely independently of the other digits. The metapterygial axis is indicated in gray.

mentioned above, the limb skeleton develops from pre-chondrogenic condensations of mesenchymal cells starting proximally with a single condensation that gives rise to the humerus/femur. Later and more distally, the condensation splits into two condensations, which give rise to the two zeugopodial elements: radius/tibia and ulna/fibula. Later on during development, the anterior and the posterior condensations differ greatly.

The anterior elements, the radial and tibial condensations, continue to add elements by segmentation distally. In particular, they add a proximal mesopodial element, the radiale and the tibiale, and sometimes a digit-like structure, the pre-pollex and the pre-hallux. In contrast, the posterior condensations continue with a bifurcating mode of development during which they contribute elements in direct continuation of their own direction of development and split off elements to the anterior. The series of elements in continuation of the posterior zeugopodial element (ulna or fibula) is sometimes called the "primary axis" or the metapterygial axis for reasons explained below.

Then, in line with the primary axis, another proximal mesopodial element arises, the ulnare or the fibulare, distal mesopodial 4, and metapodial 4 and digit 4. Splitting off from the primary axis first is the intermedium, which is the third proximal mesopodial that comes to lie between the ulnare and the radiale in the forelimb and the fibulare and the tibiale in the leg.

Most digits of a typical pentadactyl limb develop anterior to the primary axis, primarily in a posterior to anterior sequence. Digit 5 seems to develop independently and posterior of the primary axis. In between the proximal and distal mesopodial elements, additional mesopodial bones can occur that are both variable in their occurrence among species as well as in their mode of development.

Digit development is more variable than that of the proximal elements. There are developmental variations that reflect the variability of the adult hand skeletons—for example, in the context of digit reduction, as occurs in ungulates (Sears, Bormet et al. 2011) or in skinks (Young, Caputo et al. 2009), or in the context of the evolution of the bird hand (Young, Bever et al. 2011). However, a major deviation from the schema outlined above also occurs that is not obviously associated with a major transformation of the adult hand, which is the development of the urodele limb.

In urodeles (salamanders and newts), the first digit to develop is not digit 4; rather, digits 2 and 1 develop first. The remainder of the digits develop in an order inverse to that of other tetrapods. This difference was once used to argue that tetrapods were not monophyletic, but rather that anurans and amniotes ("eu-tetrapods") derived independently from urodele amphibians. However, molecular phylogenetic evidence does not support this scenario (Zardoya and Meyer 2001). Therefore, the mode of hand development observed in urodeles is now viewed as derived, in particular an autapomorphy of the Urodela (Wake and Shubin 1994; Wagner, Khan et al. 1999; Frobisch and Shubin 2011). This problem will be discussed further in chapter 11.

What is the phylogenetic origin of the tetrapod limb? The endoskeleton of a zebrafish or any other teleost fish, by far the biggest group of living vertebrates, has no resemblance to that of a limb (figure 10.1A and B). However, a quick look at the gnathostome phylogeny shows that teleosts are a highly derived clade within the jawed vertebrates (figure 10.2). The evidence for the ancestral conditions is more likely to be found in fish that branched off closer to the base of the bony fish clade or even the cartilaginous fish, sharks, rays, and chimeras. There are only two living relatives (actually small clades) to tetrapods: the lungfish and the coelacanths. Coelacanths are not accessible for developmental studies, and lungfish have highly derived fin skeletons (but see below). In recent years, most developmental work was done on either the shark *Scyliorhinus canicula* or the skate *Leucoraja erinacea* or the basal ray-finned fish *Polyodon spathula*, the paddlefish (Metscher and Ahlberg 1999). The endoskeleton of the paired fins of these fishes suggests an ancestral character state from which both the teleost fin skeleton and that of sarcopterygian fins and the tetrapod limbs could have derived.

In both the shark, *Scyliorhinus canicula* (figure 10.6), and the skate, *Leucoraja erinacea*, the pectoral fin skeleton articulates with the girdle with three cartilaginous plates. This condition is called the "tribasal" fin. From anterior to posterior, these plates are called pro-, meso-, and metapterygium. Distal of these plates one finds a larger number of smaller elements: so-called radials.

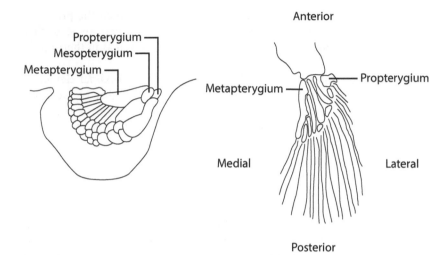

FIGURE 10.6: The tribasal paired fin of a shark (left) and a basal bony fish, right (paddlefish, *Polyodon spathula*). The tribasal fin has three basal elements, the pro-, meso-, and metapterygium. The metapterygium is the direct homolog of the limb skeleton of tetrapods. The other parts of the fin have been lost in tetrapods. In contrast, in the teleost lineage the meso- and metapterygium have been lost. Consequently, there are no homologous skeletal elements shared between mouse limbs and zebrafish pectoral fins (see Figure 10.4).

This tribasal pattern is likely to be ancestral for the paired fins of crown gnathostomes (evidence summarized in Coates 2003), and is also found in the osteichthyan stem group of acanthodians. A similar situation is also found in the recent basal actinopterygian *Polyodon* that has four radials attached to the shoulder girdle (figure 10.6).

The derived fins of teleosts and those of sarcopterygians, including tetrapods, can be understood as derived from the tribasal condition by opposite trends in endoskeletal reduction (Grandel and Schulte-Merker 1998; Mabee 2000; Mabee and Noordsy 2004). The four radials of the teleost pectoral fin (figure 10.1B) are thought to represent the pro- and the mesopterygium, and thus arose through the loss of the metapterygium. The sarcopterygian fin can be understood as homologous to the metapterygium. The sarcopterygian fin skeleton and, thus, that of the tetrapod limb is understood as homologous to the metapterygium and arose through the loss of the pro- and mesopterygium.

The pectoral fin of the paddlefish, *Polyodon spathula*, is interesting. It is similar to the ancestral condition and easily accessible for developmental studies because it is bred as a game fish in the U.S. The pectoral fin skeleton of the paddlefish comprises a propterygium, two middle radials, and a metapterygium (figure 10.6). This somewhat specialized situation arises in development from a typical tribasal stage (figure 4A in Mabee and Noordsy 2004) in which the anteriormost cartilage gives rise to the propterygium and radial 1 (figure 4B in Mabee and Noordsy 2004).

For the evolution of the tetrapod limb, the most interesting feature of the primitive fin is the metapterygium. In many sharks the metapterygium branches off radials at its anterior edge (figure 4 in Coates 2003) as in the primary axis in limb development (see above). There are exceptions. For example, in the dog fish (paradoxically, a.k.a. cat shark, *Scyliorhinus canicula*), the radials are found distally, and in the fossil *Orthacanthus*, a xenacanth elasmobranch, radials are found on both the anterior and the posterior edge of the metapterygial axis (Braus 1906). In spite of these variations, the predominant pattern is one in which secondary skeletal elements branch off at the anterior edge of the metapterygium.

This trend is maintained in the tetrapod stem groups like Eustenopteron. Even in the highly derived biserial fin of the Australian lungfish *Neoceratodus* (figure 10.1C), which comprises a long row of so-called metameres and anterior and posterior radials in a pinnate pattern, a remnant of the anterior branching metapterygial pattern is seen during development (Joss and Longhurst 2001). Hence, there is a broad consensus that the proximal two segments as well as the proximal mesopodials correspond to the metapterygium. The ancestral gnathostome fin and the tetrapod limb share a deeply engrained early pattern of morphogenesis: the anterior branching of the metapterygual axis (figure 10.5; Shubin and Alberch 1986).

Evolution of the Autopodium

Although there is a broad consensus regarding the derivation of the proximal and middle segments of the tetrapod limb, the main difficulty is understanding the origin of the distal segment of the limb: the autopod. The distal parts of the pectoral fin skeletons of stem group tetrapods, those that do not have a clearly recognizable hand, are extremely variable (figure 10.7). Distal elements can include a large number of radials arranged in an anterior-posterior series, as in *Sauripterus* (figure 10.7B), or consist of a continuation of the metapterygial axis with continued branching off of anterior radials, as in *Eusthenopteron* (figure 10.7C), *Panderichthys* (figure 10.7D), and *Tiktaalik* (figure 10.7E). With regard to *Panderichthys*, which has been described as lacking distal elements, it should be noted that a CT scan of a specimen partially embedded in matrix revealed distal elements very similar to those of *Tiktaalik* (Boisvert, Mark-Kurik et al. 2008).

In all cases, the radius extends farther distal than the ulna, and does not participate in a zeugo-autopodial joint typical for tetrapods. The recently discovered *Tiktaalik* (Daeschler, Shubin et al. 2006) is somewhat special in that the distal radials seem to line up as if to allow for two or three levels of flexion,

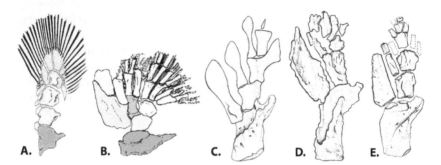

FIGURE 10.7: Variations of pectoral fin skeletons in stem tetrapod and non-tetrapod sarcopterygians. A) Coelacanth *Latimeria*, B) *Sauripterus*, C) *Eusthenopteron*, D) *Panderichthys*, E) *Tiktaalik*. Anterior is to the left and distal is up. (A and B reproduced with permission from Coates, 2002. © 2002 Blackwell Publishing Inc.; C, D, and E reprinted by permission from Macmillan Publishers Ltd: *Nature*, Boisvert et al., 2008, copyright 2008.)

functionally similar to a tetrapod limb mesopodial-metapodial joint and the metapodial-digital joints. For this reason, the authors argued that *Tiktaalik*, and perhaps *Panderichthys*, already had an autopodium, even though the similarity with even slightly more crownward taxa is quite limited.

Three fossil taxa represent stem tetrapods with clearly recognizable hands and feet. These are *Acanthostega*, *Ichthyostega*, and *Tulerpeton* (figure 10.8). The transition to a clearly limb-like morphology was stepwise during which the shape of the stylopod led the trend. A limb-like humerus was first described in the still fish-like *Panderichthys* and was retained in *Tiktaalik* and more advanced forms. The mid-section remained the most conserved, while the next transition was the origin of digits. Digits have been related to distal radials.

Digits are a series of bones that are smaller segmented superficially, similar to the distal radials found in *Sauripterus* and *Tiktaalik*. The distinguishing characteristics of digits, however, are their arrangement along the anterior-posterior axis and the complete absence of branching patterns distal to the mesopodium. Digits first appeared in numbers in excess of the canonical five (of crown tetrapods), with eight in *Acanthostega* and in *Ichthyostega* and six in *Tulerpeton*. The mesopodium was the last part of the autopod to be completed. It was not fossilized in *Acanthostega*, probably because of absent or late ossification, as in many extant amphibians. The hind limb of *Ichthyostega*, however, showed an early stage of mesopodial elaboration. The posterior two to three digits directly articulated with the large fibulare, and then there were three distal tarsals, whereas the situation at the anterior end was unclear. Central tarsals were completely absent.

In summary, the autopod arose by suppressing distal branching of elements, establishing a segmented series of smaller bones called digits, and

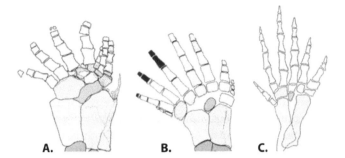

FIGURE 10.8: Auto- and zeugopods of stem tetrapods that have recognizable autopods. A) *Ichthyostega* hind limb, B) *Acanthostega*, C) *Tulerpeton*. Note the "supra-penta-dactylous" condition of first tetrapods (i.e., the early tetrapods had more than five digits). Reproduced by permission from Coates 2002. © 2002 Blackwell Publishing Inc.

gradually elaborating the mesopodium to support a hinge between the zeugo and the acropodium, which is essential for the tetrapod limb's function in body support. What is not clear is whether these changes heralded the origin of a novel body part, what anatomists call the autopod (i.e., a Type I innovation), or whether it represented a character state transformation (i.e., a Type II innovation). In the latter case, it could be that an autopodial equivalent was already present in the fin before the morphological characteristics of a morphologically recognizable autopod arose. To address this question, we first have to have a look at the molecular mechanisms of fins and limb development.

Development of Fins and Limbs

Fins and limbs develop from out-pockets of the body wall, called fin and limb buds, respectively. These consist of an epidermal cover and a mesenchymal filling derived from the lateral plate mesoderm. Along each side of the body, there is a boundary in the epidermis that separates the dorsal from the ventral compartments. These compartments are clonally restricted, which means that cells that originate in one compartment do not mingle with cells from the other compartment. It is at this border that signals from the mesoderm initiate the outgrowth.[2]

A number of fibroblast growth factors (Fgfs) are sufficient to initiate limb bud development, although Fgf10 is the most likely native signal. The mesenchymal Fgf10 signal induces Fgf8 expression in the epidermis, which then

---

[2]The basic outline of limb development is now textbook knowledge and, thus, detailed references will not be provided for this section. Consult for example Gilbert, S. F. (2010). *Developmental Biology.* Sunderland, MA, Sinauer Associates.

forms a self-sustaining feedback loop with the mesenchyme that allows the limb bud to continue to develop in the absence of external signals. There is also evidence that Wnt signals participate in this loop.

Fgf expression in the epidermis occurs in a specialized group of cells that form a strip of elongated cells along the leading edge of the outgrowing limb bud. These cells form the apical epidermal ridge (AER), one of the two key signaling centers in fin and limb development. The zebrafish fin bud is also initiated and maintained by Fgf, although with slightly different Wnt family members of signaling molecules. Later, the AER continues to grow out and away from the mesenchymal pocket to form the fin fold in which the bony fin rays develop.

The spatial pattern of limb development is controlled by signals that determine the three principal axes of the limb bud: the proximodistal; the dorsoventral; and the anterior-posterior axis (figure 10.9). Each axis is established by its own system of signaling molecules and all of them are integrated and interdependent.

The proximodistal pattern depends on the AER, which signals to the underlying mesenchyme to remain undifferentiated, to continue proliferating, and forming what is called the progress zone (PZ). The exact mode of proximodistal patterning is currently unclear. Two phenomenological models can explain most experimental data (Mariani and Martin 2003). The classical model is the progress zone model, which states that the developmental fate of the cells in the PZ is determined by the amount of time they spend in the PZ. When these cells leave the PZ and because they are separated from the AER, they begin to differentiate into structures according to the number of cell divisions they underwent within the PZ.

This model is mostly based on early surgical experiments with chicken limb buds that showed: 1) the signal of the AER is permissive rather than

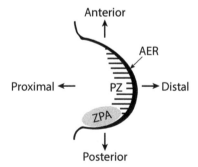

FIGURE 10.9: Major signaling centers of the limb bud. ZPA is the Zone of Polarizing Activity that produces Sonic hedgehog (Shh) as a signal; AER is the Apical Ectodermal Ridge that signals with Fgf's to the PZ, the Progress Zone, which is the proliferative zone in the limb bud.

instructive (i.e., the age of the AER does not determine the structures formed by the cells receiving the signal); and 2) transplanting PZ cells from limb buds of different ages shows that the identity of the proximal cells of the host limb bud does not influence the developmental fate of the grafted PZ cells. This implies that the cells proximal to the PZ are determined in their developmental fate.

In recent years the PZ model has been challenged because one of the key experiments that supported this model was found to be flawed. This was an experiment in which the AER was removed from limb buds at different stages of development. Removing the AER during early stages resulted in proximal structures only, whereas more distal structures could develop if the AER was removed later. This effect could be explained by the formation of a necrotic degeneration zone as a consequence of AER removal. This necrotic degeneration zone is of constant width and affects a larger fraction of the cells in a young limb bud than in an old one. Thus, these experiments did not necessarily support the PZ model.

In response to this, various forms of pre-specification models have been proposed for which the proximodistal differences are established in the early limb bud and the segments expand in size only later during development. One model suggests that the proximodistal axis is established by two gradients. One gradient is the Fgf signal from the AER, and the second is a gradient of retinoic acid (RA) signaling from proximal to distal. One target of the RA signal is the homeodomain transcription factors Meis1 and Meis2, which become localized to the proximal region of the limb and fin bud. In any case, expression of 5′ HoxD and HoxA genes is involved in determining the three main segments of the limb, and these mechanisms will be discussed in greater detail below.

Dorsoventral polarity is established by the dorsal and ventral compartments of the limb bud ectoderm. Dorsal ectoderm expresses Wnt7a and signals the underlying mesenchyme by activating the LIM transcription factor Lmx1. In the ventral compartment of the ectoderm, the transcription factor En1 prevents the expression of Wnt7a.

Another classical signaling center is involved in patterning the anterior-posterior axis; the so-called zone of polarizing activity or ZPA. This group of mesenchymal cells is located at the posterior edge of the limb bud and was discovered because transplanting these cells to the anterior edge of the wing bud in chickens could result in mirror image duplications of digits. It turned out later that the molecular basis for the polarizing activity of the ZPA was the production of a signaling molecule called Sonic hedgehog or Shh (Riddle, Johnson et al. 1993). Shh has the same activities as those of ZPA cells and Shh knockout results in the loss of all digits, save one, which seems to assume the fate of digit I. The current model posits that Shh influences both the number of digits and their identity. Digits IV and V derive from cells that express Shh, while digit III derives, in part, from cells that express Shh and, in part, from cells that receive Shh signals. Digit II develops from cells that receive Shh signals, and digit I is independent of Shh signaling (McGlinn and Tabin 2006).

The morphology of a Shh -/- mutant showed that there was also a Shh-independent mechanism for anterior-posterior polarity, because the stylopod retained its characteristic anterior-posterior asymmetry, even in the absence of Shh activity (Chiang, Litingtung et al. 2001). This finding pointed to mechanisms that are necessary for posteriorly restricted Shh expression. A key player in this task is the transcription factor dHand, which acts upstream of Shh.

Initially, *dHand* is expressed all through the limb-forming field and later becomes restricted to the posterior part where it induces canonical 5′ HoxD genes: *HoxD13, HoxD12, HoxD11, HoxD10,* and *HoxD9*. During this phase of development, these HoxD genes are expressed in a classical nested pattern of posterior predominance in which *HoxD13* has the most posteriorly restricted expression domain nested within the *HoxD12* domain, which itself is nested within the *HoxD11* domain, and so on. Then the area with the highest level of HoxD gene expression is induced to express Shh.

The initial posterior polarization of *dHand* in the forelimb area is probably under the control of *HoxB8* in the lateral mesoderm, which links the limb bud anterior-posterior polarity to the anterior-posterior polarity of the body axis. During this early establishment of Shh expression, RA also plays an ill-defined role, but one that is often experimentally used to assess whether a system has a ZPA-like activity (i.e., whether Shh expression can be induced by RA). The mechanistic link between limb and body axis polarity is a hint at the gene regulatory network that underlies the integration of the vertebrate body plan.

As briefly mentioned above, these *three patterning systems* are interdependent and integrated with each other such that these systems *form an integrated network* rather than three independent signaling systems. For example, Shh expression in the ZPA is maintained by Fgf signaling from the AER and, in turn, Shh activity is necessary to maintain Fgf4 activity in the AER. This explains why a *Shh-/-* mutant results not only in defects in anterior-posterior polarity, but also in a loss of many of the distal structures. In addition, Wnt7a signaling from the dorsal ectoderm contributes to maintaining the ZPA. It is tempting to speculate that the activity and interdependency of the ZPA, AER, and dorsal ectoderm potentially form a limb/fin bud ChIN, which raises the question to what extent this network is conserved.

Most of the molecular work on fin development is from studies of zebrafish pectoral fin development. Shh is also expressed posteriorly in the fin bud and a loss of its activity, as in the spontaneous *SonicYou* mutation, results in a loss of fins. The AER also exists in the zebrafish fin bud and expresses Fgf10. There are slight differences in the relative timing of gene expression and, in some species, certain members of the Fgf gene family are replaced by others; however, the overall functional structure of the network seems to be conserved.

The ZPA even exists in shark and skate pectoral fin development (Dahn, Davis et al. 2007), and in medial fin buds, which suggests that the basic fin/limb core network is associated with individuated fins among all gnathostomes (Freitas, Zhang et al. 2006). This also suggests that the developmental control network of paired fins arose first in medial fins (see above).

Conservation of the basic gene regulatory network among medial and paired fins, as well as limbs, is remarkable given the morphological disparity that these structures represent. It seems as though all vertebrate appendages (fins and limbs) are serially homologous and that they individuated by acquiring regulatory genes specific to their position. This is best understood with regard to the identity of pectoral and pelvic appendages that are consistently associated with the activity of limb identity genes, *Tbx5* and *Tbx4*, respectively. It is currently not known what individuates the paired appendages relative to the medial fins.

Because the zebrafish pectoral fin and the tetrapod forelimb do not share any homologous skeletal elements (see above), but do share the basic fin/limb core gene regulatory network, it is clear that the identity, if any, of the characteristic skeletal elements is independent of the fundamental similarity of fins and limbs as paired appendages. In order to come closer to the potential developmental mechanisms for the fin-limb transition, we need to take a closer look at developmental mechanisms downstream of the core signaling network—in particular, those involved in the development of the autopod.

The best-investigated class of genes involved in the development of limb segments are the 5' *HoxD* and *HoxA* genes (figure 10.10). In particular, *HoxD* genes have two well-distinguished expression patterns, an early and a late phase of gene expression.[3] Early expression was described above and represented a classical nested pattern with the more 5' genes expressed more posterior than the more 3' genes. This pattern eventually becomes reversed in the autopod under the influence of the Shh signal, which leads to the late phase expression with *HoxD13* expressed along the entire extent of the future autopod and *HoxD12* and *HoxD11* with more posteriorly restricted expression domains. This expression pattern is restricted to the autopod and has been identified as indicative of autopodial identity.

The other pattern of gene expression that can be linked to autopodial development is the expression of the two 5' HoxA genes *HoxA13* and *HoxA11*. When the autopod develops, these two genes assume mutually exclusive expression domains. *HoxA11* remains limited to the zeugopodial segment of the limb bud and distally encounters *HoxA13* expression with a sharp expression boundary. *HoxA13* expression marks the territory of the future

---

[3] This concept originated in a paper by Nelson et al., 1996 who originally described three phases of Hox gene expression, but only two are clearly discernable—namely, their phases II and III.

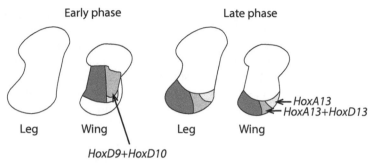

FIGURE 10.10: Hox gene expression domains in chicken limb buds according to Nelson et al., 1996. Shades of gray represent the number of Hox genes expressed, with darker gray indicating the expression of more Hox genes. The difference is that during the early phase, 5' Hox genes (like *HoxD13*) have the most limited expression, while during the late phase, the 5' Hox genes have the most extensive expression and the 3' Hox genes have the most limited expression. The last phase of gene expression is associated with autopod development and is likely a novelty associated with the fin-limb transition.

autopod. The exclusivity of the *HoxA11/HoxA13* expression territories is actively maintained, with *HoxA13* repressing *HoxA11*.

This can be shown in *HoxA13* KO mutations in which *HoxA11* expression extends distally into the autopodial region (Post and Innis 1999). The difference in *HoxA11* and *HoxA13* expression has functional meaning in that *HoxA13* overexpression in the zeugopodium results in a loss of the long bone character of the zeugopodial elements (Yokouchi, Nakazato et al. 1995).

Most informative are double KO mutations of the A and D Hox genes. A double KO of *HoxA13* and *HoxD13* results in a complete loss of acropodial elements, which shows that these two genes are jointly necessary for autopodial development (Fromental-Ramain, Warot et al. 1996). Similarly, a double KO of *HoxA11* and *HoxD11* results in a severe underdevelopment of the zeugopodial elements, and relatively normal autopod development (Davis, Witte et al. 1995), even though *HoxD11* also plays a role in digit development. These experiments showed that, in the mouse, the autopod and the zeugopod are developmental modules.

Evolutionarily, in the frog hind limb the proximal tarsals are transformed into long bones and this transformaton was associated with a distal expression shift of *HoxA11* (Blanco, Misof et al. 1998). Thus, it is plausible that the zeugo- and the autopod are developmentally individualized parts of the limb and are true characters in the sense proposed here. The question is whether the developmentally individualized autopod is an evolutionary novelty or a plesiomorphic feature of the paired appendage.[4] We will consider this

---

[4] The fact that they are also serially homologous is a problem that will be discussed later in this chapter.

question below with a short overview of developmental scenarios for the fin-limb transition.

Developmental Models of the Fin-Limb Transition

The Metapterygial Axis

As noted above, there is a broad consensus that the limb skeleton was derived from the metapterygium of basal gnathostomes. This notion is supported by the arrangement of skeletal elements in stem tetrapod fins, the development of a metapterygial fin skeleton in forms like *Polyodon,* and the pattern of skeletogenesis in the tetrapod limb (see above). These similarities lend support to homologies between basal sarcopterygian skeletal elements and that of the tetrapod limb.

These homology hypotheses are solely based on spatial relationships among elements and how they develop during embryogenesis. Strictly speaking, they are corresponding elements that derive from a pattern formation process that is likely homologous among gnathostomes, but in themselves do not show that particular elements are individualized to be homologs. For example, it is not clear whether the first mesomere in a shark, the Australian lungfish, or the paddlefish pectoral fin is developmentally individualized like it is in a tetrapod forelimb (i.e., the humerus). On morphological grounds the first mesomere is recognized as similar to the stem tetrapod humerus in *Panderichthys*, and is found in similar shapes in *Tiktaalis* and any stem tetrapod higher up toward the crown Tetrapoda.

Hence, it is more likely that the stylo- and zeugopod together are homologous to the metapterygium, but that the individualities of the humerus and femur and that of tetrapod zeugopodial elements evolved in the stem lineage of tetrapods rather than being already present as individualized structures in basal gnathostomes. There is currently no developmental evidence to address this question, although this is a question that would be interesting to pursue.

How the metapterygial axis (mpt-axis) continues into the autopod is more controversial (summarized in Larsson 2007; Wagner and Larsson 2007). For example, Gegenbaur first suggested that the mpt-axis extended through the intermedium to digit I (Gegenbaur 1878) and, thereby, homologized the digits with post-axial radials. This scheme is implied by Gegenbaur's archipterygium theory, which assumed that a structure with both pre-and post-axial radials was primitive. However, Gegenbaur reversed his idea later by suggesting that it went to digit V. In contrast, Watson (Watson 1913) and Jarvik (Jarvik 1980) identified digit IV as the distal end of the mpt-axis. In this model, digits I to III are identified with pre-axial radials. This level of variability in interpretation suggests that the evidence used in this controversy is, at best, ambiguous.

In spite of the controversies summarized above, embryological data show that the first digit to form is digit IV in direct continuation of the posterior axis.[5] This was first noted by Steiner (Steiner 1934) in a study of crocodile limb development and has been confirmed in all modern studies of tetrapod limb development (e.g., Burke and Alberch 1985; Müller and Alberch 1990; and others), with the exception of urodele limb development briefly discussed above and in chapter 11. However, identifying the sequence of distal elements in line with the metapterygial axis is problematic for the following reason.

Following the ulnare/fibulare is distal carpal 4 and then metapodial 4 and digit IV. Thus, this would suggest that distal carpal 4 is the fourth mesomere, that metacarpal 4 is the fifth, and so on (figure 10.5). In the three most basal stem tetrapods known, *Acanthostega, Ichthyostega,* and *Tulerpeton,* the posterior digits directly articulate with the ulnare or the fibulare, respectively. Hence, if distal carpal 4 is the fourth mesomere, it originated phylogenetically after the fifth mesomere, or had to have arisen from a digit-like element after digits arose. This and the general difficulty of finding a coherent scheme to relate metapterygial elements and the developmental pattern of the metapterygium, as shown in sharks, skates, and paddlefish, suggest that autopod development is a radical departure from the plesiomorphic developmental pattern of the metapterygium.

Given these facts, the autopod might be a novel part characteristic of limbs. This conclusion is also supported by the fact that all morphologically archaic stem tetrapods, all the way up to *Tiktaalis,* have a branching skeletal structure in the distal part of the fin, whereas the autopod and the digits never have branching arrangements (with the exception of some pathological situations). Models of "neomorphic autopodium" will be discussed below.

The Neomorphic Autopodium

As summarized above, in the context of the discussion on the metapterygial axis, morphological evidence suggests that the autopod is a radical deviation from the developmental mode of the metapterygium and can be a novel developmental module and, hence, a novel character. This view received major support when it was shown that *HoxD* gene expression in the zebrafish pectoral fin seemed to have only an early, collinear expression pattern, but lacked the typical late autopodial expression with its inverse colinearity (Sordino, Hoeven et al. 1995). Later it was discovered that the late phase expression of 5′ *HoxD* genes was controlled by a special cis-regulatory element, the global control region (GCR), a second element called Prox, as well as many more recently discovered large regions of non-coding DNA (figure 10.11) (Sordino, Hoeven et al. 1995).

---

[5] This is true for most "typical" cases of digit development, but see chapter 11 for deviations from this pattern.

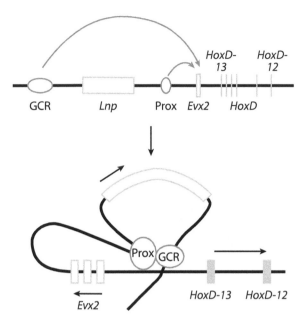

FIGURE 10.11: Model of late phase HoxD expression regulation in the autopod according to Spitz et al. (2003). The two closest regulatory elements are Prox and the GCR that cooperate in regulating the 5′ HoxD genes and prefer *HoxD13* over more distal Hox genes. This model is a simplification, as there are many more cis-regulatory elements contributing to the late phase of gene expression.

The GCR is a 40 kb block of high sequence conservation that is also found in zebrafish and pufferfish that drives reporter gene expression in the central nervous system and in the autopod (Spitz, Gonzalez et al. 2003). The puffer fish GCR, however, did not drive reporter gene expression in a mouse transgenic model, at least not during the stages of development that were examined. This further cemented the idea that late phase *HoxD* gene expression is a derived tetrapod characteristic and may be a genetic mechanism involved in the origin of a neomorphic autopodium.

Of course, the weakness of the genetic evidence for a neomorphic autopodium is that it is all derived from teleost species: zebrafish and pufferfish. As mentioned above, teleosts have a highly derived and reduced endoskeletal paired fin skeleton with no trace of a metapterygium (Grandel and Schulte-Merker 1998; Grandel 2003; Mabee and Noordsy 2004). Gene expression studies of phylogenetically better placed species showed that there seemed to be at least a late phase of *HoxD* gene expression in paddlefish (Davis, Dahn et al. 2007), shark (Freitas, Zhang et al. 2007), and even zebrafish (Ahn and Ho 2008). The lungfish data showed a late expression pattern associated with the development of the radials, both the anterior as well as the posterior radials

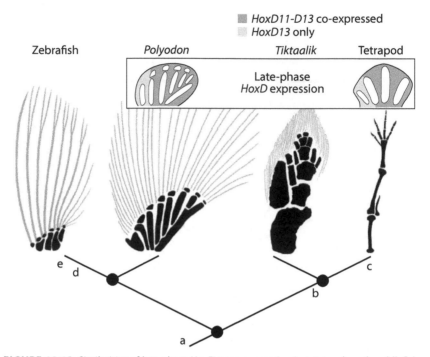

FIGURE 10.12: Similarities of late phase HoxD gene expression in tetrapods and paddlefish, *Polyodon*. In both cases, *HoxD11* and *HoxD13* are co-expressed in the posterior compartment of the limb bud, but only *HoxD13* is expressed in the anterior parts of the bud. Reprinted by permission from Macmillan Publishers Ltd: *Nature*, Davis et al., 2007, copyright 2007.

(Johanson, Joss et al. 2007), although no early *HoxD* expression was reported, probably for technical reasons.

The paddlefish evidence is particularly interesting (Davis, Dahn et al. 2007). At stage 40, the pectoral fin buds show classical early colinear *HoxD* gene expression (figure 10.12). *HoxD13* is expressed at the posterior margin of the fin bud nested within a slightly more extensive *HoxD12* expression domain, which itself is nested in a *HoxD11* expression domain that extends even farther anteriorly. At stage 46, *HoxD* gene expression is mostly found in the distal margin of the mesenchymal fin bud with some hint at an inverse collinear pattern for which *HoxD13* expression extends all the way to the anterior margin and *HoxD12* stops short of extending that far. *HoxD11* expression is not as clear (figure 3a in Davis, Dahn et al. 2007), but even in tetrapods, *HoxD11* has a transient phase during which it extends as far as *HoxD13* (e.g., in alligator; Vargas, Kohlsdorf et al. 2008).

Later, at 10 days post-spawning, the expression pattern extends farther proximally into the inter-radial mesenchyme and shows a clear inverse co-linearity with *HoxD13* found in six inter-radial domains and *HoxD12* and

HoxD11 in only four posterior inter-radials. Hence, it seems that inverse co-linearity may likely be older than the tetrapods. In addition, the autopodial field could perhaps be older than the morphologically recognized autopod.

The evidence from shark and zebrafish is less impressive, although it still suggests a second phase of *HoxD* gene expression, even in the highly derived zebrafish fin. During cat shark pectoral fin stages 29 and 30, there is an early colinear phase of *HoxD* expression, while in stage 32, *HoxD13* and *HoxD12* are expressed in a very thin marginal zone at the distal end of the fin bud (Freitas, Zhang et al. 2007). *HoxD13* expression extends more anteriorly than that of *HoxD12*, but *HoxD13* expression does not reach the anterior margin, as in tetrapods and paddlefish.

In the zebrafish, there is a weak late phase of *HoxD* expression that is more distally restricted than the earlier phase, but a clear inverse colinearity is not detectable (Ahn and Ho 2008). Hence, fin bud *HoxD* expression is not limited to early collinear expression patterns, but may include expression patterns that are reminiscent of autopodial *HoxD* expression, at least in the paddlefish, and to a lesser degree in shark and zebrafish. Before we discuss the implications of these findings, we need to look at the other genetic autopodial marker gene, *HoxA13*, and the exclusion of *HoxA11* from the autopodial field.

In the paddlefish, both *HoxA11* and *HoxA13* are expressed in the pectoral fin bud, with *HoxA13* more distally restricted but nested within the *HoxA11* expression domain (Metscher, Takahashi et al. 2005; Davis, Dahn et al. 2007). The important point here is that there is no evidence for a late phase *HoxA13/HoxA11* expression pattern for which *HoxA13* excludes *HoxA11* from the distal fin bud domain, as is the case in limb bud development. The situation in zebrafish is similar to that in paddlefish, albeit slightly more complicated because of the existence of paralog HoxA clusters. For example, *HoxA13a* is expressed in a distal fin bud domain, but the corresponding *HoxA11a* is not expressed at a similar stage. The b-paralogs, however, are expressed in a pattern at 60 hours post-fertilization that is similar to that in paddlefish (figure 5 in Ahn and Ho 2008). In shark, *HoxA13* and *HoxA11* expression also overlaps (Sakamoto, Onimaru et al. 2009). Unfortunately, there is no available information regarding *HoxA* gene expression in lungfish. However, it is at least clear that, ancestrally in Gnathostomes, *HoxA11* and *HoxA13* had a distal, nested (overlapping) expression pattern in the pectoral fin bud, and that the exclusion of *HoxA11* from the distal region is likely a derived state.

The molecular evidence regarding the phylogenetic age of the autopodial field is mixed. *HoxD* genes expression suggests an ancient origin for inverse colinearity, which in tetrapods is typical for autopod development. However, the *HoxA13/HoxA11* expression pattern is clearly different in fin bud from that in the limb bud. In terms of *HoxD* and *HoxA* gene co-expression, the situation is also ambiguous. At stage 46, both *HoxD13* and *HoxA13* are co-expressed in the distal rim of the fin bud. However, after that,

HoxA13 expression can no longer be detected and is not coincidental with the *HoxD13* expression as it is in the limb. Hence, unlike the situation in limb buds, in paddlefish pectoral fin buds, *HoxD13* and other D-genes play roles independent of that of *HoxA13* late in development.

Where does this leave us with respect to the question of whether the autopod is evolutionarily derived or is an ancestral relative of the tetrapod situation? It is certainly possible that a distal developmental domain could exist that is homologous to the autopodial field in tetrapods, even though the typical morphological autopod features are absent in fins. After all, we acknowledge that homologs are individuals that are united by individuality and continuity of descent rather than defined by some phenotypic features (Ghiselin 2005; Wagner 2007). It is possible that the distal radials in the paddlefish fin and the tiny distal radials in the teleost fins are distant homologs to the autopod. On the other hand, the molecular evidence is ambiguous, with autopodial *HoxD* gene expression being ancestral, whereas *HoxA* expression is derived, at least in sarcopterygians if not for the tetrapods.

The new evidence regarding the ancient origin of the autopodium was also discussed by Woltering and Duboule (Woltering and Duboule 2010) who concluded that no fin gene expression pattern mimicked the typical autopodial situation in mouse and chicken. The main points are that, in tetrapods, the autopodium is characterized by late *HoxD* expression and this is separated from the proximal expression domain by a gap. This zone without *HoxD* expression is the prospective mesopodium. No such pattern has been described in any fin bud. Also, the distal expression in the fin does resemble a distalization of *HoxD* expression in the zeugopod of mice. Thus, the distal focus of expression seen in fin buds is not indicative of an autopodial identity. Woltering and Duboule also emphasized the lack of local exclusivity between *HoxA11* and *HoxA13* expression, which is characteristic of the zeugo-autopodial difference.

Considering the gene regulatory network of autopodial gene expression, Woltering and Duboule further pointed out that the autopodial expression is regulated by a different set of cis-regulatory elements than that in the proximal parts of the limb bud (Woltering and Duboule 2010). There is currently no direct evidence regarding the gene regulatory mechanisms of fin *HoxD* expression, although there is indirect evidence to suggest that the autopdial regulatory loop is not active in fins.

The autopodial CRE, called CsB, regulates not only the five 5' HoxD genes, but also a gene called *Lunapark* (*Lnp*) that is located between the CsB and the HoxD cluster. As a consequence, *Lnp* is also expressed in the developing autopod. *Lnp* also occurs in the zebrafish genome in about the same relative location, but is not expressed in the fin bud. The fugu CsB also does not drive reporter expression in the mouse limb bud and, thus, may not have a fin bud function (Spitz, Gonzalez et al. 2003), although the zebrafish CsB does drive some weak mesopodial and metapodial expression (Schneider, Aneas et al. 2011).

Woltering and Duboule concluded that the sum total of the molecular evidence supported a neo-autopodium (i.e., a derived developmental module) and, thus, a true Type I novelty. Woltering and Duboule argued that the duality of *HoxD* regulation allowed for the origin of a "no-HoxD land" that separated the autopodial and zeugopodial domain, as they said. They argued that without dual regulation of *HoxD* genes, there was no way to create a mesopodium. Hence, the novelty of the fin-limb transition is actually two-fold: a neo-acropodium and a neo-mesopodium.

What then is the developmental evidence that the tetrapod autopodium is a developmental module? As mentioned above in the section on limb development, the difference between the autopod and the zeugopod is most convincingly demonstrated by the fact that both can be nearly obliterated by different mutations that leave the other limb segments largely unaffected. Specifically, *HoxA13/HoxD13* KO results in a nearly complete loss of the autopod (Fromental-Ramain, Warot et al. 1996), while *HoxA11/HoxD11* KO results in nearly losing the zeugopod but without greatly affecting the autopod (Davis, Witte et al. 1995). Similar experiments have not been attempted with paddlefish, which is not surprising given that the paddlefish is not a transgenic model organism. We do not yet have any evidence that shows that the distal fin bud is a developmental module and, thus, individualized in a manner similar to that of autopod, at least in mice.

Here is a scenario that accommodates the facts summarized above. During limb bud development, inverse *HoxD* expression is induced by Shh signaling from the ZPA. Thus, it is part of the secondary anterior-posterior patterning mechanism. In all fins examined to date, Shh plays a similar role and it might be that the inverse collinear *HoxD* expression pattern is an ancient part of Shh signaling in vertebrate appendages, rather than a mechanism that was invented as part of the fin-to-limb transition. It certainly is used in autopod development, but may not have evolved as part of the origin of digits or the autopod.

What, then, about *HoxA* expression? In fins it is clear that there is a proximodistal polarity, probably mediated by the nested expression pattern of *HoxA13*, *HoxA11*, and *HoxA10* for which there is colinearity between proximodistal expression and the 3'–5' positions of the genes on the chromosome. However, developmental polarity is not the same as having a fin bud that contains distinct proximal and distal developmental modules. Taking clues from the differentiation of cell types, one has to suspect that different modules each have subsets of genes that are mutually exclusive between the two modules. No such pair is known from proximal and distal parts of the fin bud, although *HoxA13/HoxA11* exclusivity is the expected pattern.

One would expect, though, that *HoxA13* and *HoxA11* are not the only genes with this pattern but are a part of a small set of genes, which remain to be discovered. If this is the case, then the lack of *HoxA13/HoxA11* exclusivity suggests that the autopodial developmental module was derived, as is the mutually exclusive expression pattern. Testing this idea would require identifying other members of the hypothesized autopodial and zeugopodial core gene regulatory networks.

Concluding Reflection on the Nature of Character Identity

When surveying the information regarding the origin and evolution of fins and limbs presented in this chapter, an intriguing pattern of serial homology, identity, and innovation emerges that *contradicts the notion of hierarchical homology*. The fossil record suggests that paired appendages with endoskeletal support arose first at the pectoral level and then were duplicated to arrive at the serially homologous pectoral and pelvic fins. These were, and are, greatly individualized even in primarily aquatic vertebrates ("fish" in the vernacular). The fin-to-limb transition, at least in one school of thought, added a novel structure to the paired appendages: the autopodium. The autopodium is a novelty that is simultaneously grafted upon two already quite different and probably individualized pairs of fins, which already had their fin/limb identity genes, *Tbx4/5*, which also individualized them at the genetic level. The acquisition of autopodial morphology is different between the forelimb and the hind limb, in that the hind limb tends to show the more derived character states than does the forelimb.

To summarize, during the fin-limb transition, two sets of body parts (the pectoral and pelvic fins), which are both serially homologous as well as individualized, acquired a novel part: the autopod. The autopodium itself is serially homologous between forelimbs and hind limbs, but also individualized as soon as it was detectable in the fossil record, with the foot more advanced than the hand. This suggests a cross-cutting pattern of character identities (figure 10.13A), rather than a hierarchical pattern, a fact that we already encountered with feathers (character swarms; chapter 9). The most parsimonious interpretation of this pattern of cross-cutting character identities seems to be that each part, the forelimbs and hind limbs as well as the autopod, has a genetic identity that may be linked to *Tbx5, Tbx4,* and *HoxA13*, respectively.

The special identities of the forelimb and hind limb autopods ("hand," manus, and "foot," pes) can be understood as a conjunction of limb and autopodial identity mechanisms—for example, Manus = FL & Autopod = {*HoxA13, HoxD13, Tbx5*} and Pes = HL & Autopod = {*HoxA13, HoxD13, Tbx4*} (figure 10.13B). Hence, as soon as the autopodial identity evolved (but see above for contrasting views), it evolved in the context of a preexisting forelimb and hind limb identities, which they immediately acquired.

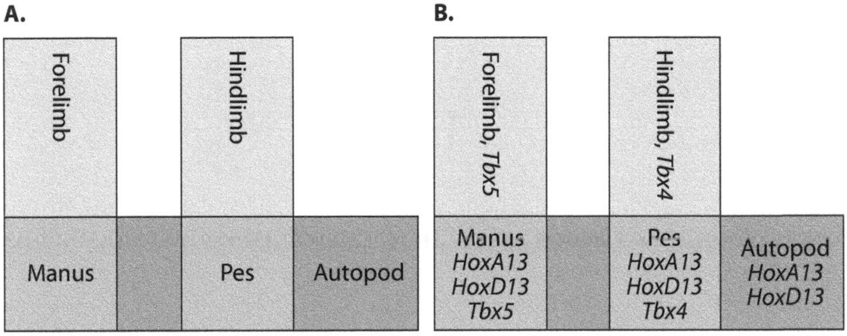

A.

B.

FIGURE 10.13: Cross-cutting character identities. A) The autopod evolved after forelimb and hind limb (fin) and, thus, is an innovation that is found in two locations. The difference between hand (manus) and foot (pes) is due to the interaction between the pre-existing individuality of forelimbs and hind limbs and the novel autopodial identity. B) Model for how the individuality of manus and pes may be determined by the overlap of forelimb and hind limb identities and the autopodial identity.

Hence, the origin of the autopod might also be coincidental with the origin of a *manus* and *pes* identity due to the pre-existing pectoral and pelvic fin identities.

More generally, this scenario suggests that the identity of an individual character can, in some cases, be the conjoint result of multiple character identities acting in that same part of the body, rather than being an atomic property. These ideas are reminiscent of the palimpsest model of developmental modularity for which earlier developmental modules get "over-written" by later reorganizations of the phenotype (Hallgrímsson, Jamniczky et al. 2009).

11

Digits and Digit Identity

Digits are the most distal appendages of the tetrapod limb. They are supported by a series of small long bones, are one of the defining features of the tetrapod limb, and are the locus of many adaptive modifications. For some authors, such as Neil Shubin at the University of Chicago, the origin of the tetrapod limb is synonymous with the origin of digits, while others like Denis Duboule from the University of Geneva, Switzerland think that the origin of the mesopodial joint is more important. Digits have been traced back to the radials of sarcopterygian fins, which raises the interesting question of what the difference is between radials and digits and how this relates to the origin of the autopodium discussed in chapter 10.

In this chapter I want to focus on the evolutionary and developmental biology of digits and examine what digits can teach us about character identity and character origination. The idea that digits have individuality and that it is meaningful to distinguish between the thumb, the index finger, and all the other digits seems intuitive. Yet this is surprisingly controversial once one asks for specific evidence to support this. With it arise questions regarding how digits evolved, how they were gained, lost, and then possibly regained, and whether or not they had changed place in the limb. These questions are particularly difficult due to a lack of clarity regarding the developmental and genetic nature of digits and digit identity. Hence, digit development and evolution is a rich field for studying the developmental evolution of character identity and character origination.

The Origin of Digits

As summarized in chapter 10, the paired appendages of basal gnathostome lineages are remarkably conserved in their proximal elements, the upper and lower arm or leg. The proximal elements of sarcopterygian paired fins are also sufficiently conserved to suggest homologies to the proximal (stylo- and zeugopodial) elements of the tetrapod limb. Distal to these readily recognizable skeletal elements, evolutionary variation is set free and very little, if any, continuity of individual elements between fins and limbs is recognizable. Nevertheless, there are some common patterns that suggest a connection between digits and distal fin elements.

At the most general level, digits are recognized by their small long bones that are arranged in two dimensions. In the proximal-distal direction they form "rays," meaning a head-to-tail row of small long bones, starting with the metacarpals/metatarsals (hereafter called metapodials) and continuing to a variable number of phalanges. Then these digital rays are arranged along the anterior-posterior dimension of the limb to form the hand and foot. Sarcopterygian radials, on the other hand, are also small long bones, but can be arranged in quite a variety of ways from the pennaceous shape of the Australian lungfish fins (see figure 10.1C) to various arrangements seen in the stem tetrapods (see figure 10.7).

However, there are a number of fossils from near the fin-limb transition in which the radials are arranged in an anterior to posterior array similar to digits. This is most obvious in *Sauripterus* and *Tiktaalis*, the latter being the closest known fossil taxon to stem tetrapods with recognizable hands and feet, such as in *Acanthostega* fossils (Daeschler, Shubin et al. 2006). Thus, it is tempting to identify these anterior-posterior arranged small long bones (the radials) with digits, or at least speculate that digits, as are known from extant tetrapods, somehow originated as a modification of these radials.

One of the few qualitative, recognizable differences between the radials of say, *Tiktaalis* and digits in most tetrapods is that digits rarely if ever bifurcate as a wild type condition. In contrast, many of the radials form a bifurcating series of elements rather than as separate rays. Digits have the developmental potential to bifurcate, as is seen in experimental or pathological conditions. Bifurcation is also seen in the highly modified digits of some aquatic tetrapods, as for example in ichthyosaurs, but not in aquatic mammals. However, bifurcating digits are rarely, if ever seen as a species character and, thus, *suppressing the bifurcation potential in digit rays may have been an important developmental innovation that led to digits.*

Digits Come and Go: Is There a Pentadactyl Ground Plan?

Among extant tetrapods, digit number is highly stereotypical. Five digits is the norm and most variations are toward a loss of digits either through rudimentation (degeneration) or specialization (adaptation to specialized functions, like in the hoofed animals; for the distinction between degeneration and specialization, see Steiner and Anders 1946). Even though the pentadactyl limb seems to be typical for the tetrapod limbs, it is not the ancestral state for the oldest known stem tetrapod fossil with clearly recognizable digits.

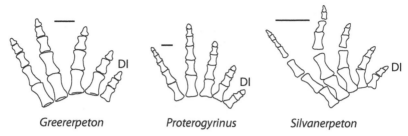

Greererpeton   *Proterogyrinus*   *Silvanerpeton*

FIGURE 11.1: Digits of *Greererpeton* and other five-digit stem-tetrapods. Note that digit I is already characterized by only two phalanges, which is a character maintained among most derived amniotes (modified from Clack 2002).

The earliest fossil with a recognizable autopod is *Acanthostega gunnari* from the upper Devonian of East Greenland. In both the forelimb and the hind limb there are eight digits (figure 10.8). The next fossil up the tetrapod stem lineage is *Ichthyostega stensioei*, also an upper Devonian fossil from East Greenland. Unfortunately, the hand of *Ichthyostega* is not known, although the foot shows seven digits arranged in two groups, with three smaller anterior digits and four larger posterior digits. In addition, in one specimen an anterior spur directly articulating with the tibia is seen. Next is *Tulerpeton curtum* from the upper Devonian of Central Russia with six digits in each hand and foot; in the hand, five anterior digits are large and the posterior digit is slender and directly articulates with the ulna.

From thereon up the phylogenetic tree, none of the stem-tetrapods exhibit more than five digits. An example is the genus *Greererpeton* from the Carboniferous period from Great Britain and North America with a phalangeal formula of 2-3-3-4-3 in the hand and 2-2-3-4-2+ in the foot (figure 11.1). It is notable that the first digit in the oldest pentadactyl fossil already has two phalanges, a character that is conserved throughout most tetrapods (Alexander Vargas, unpublished[1]).

Exactly when the pentadactyl state was reached is controversial, although current consensus is that *Tulerpeton* is a stem-tetrapod (Ruta, Coates et al. 2003). Thus, the pentadactyl limb seems to be an ancestral character of crown group tetrapods. It seems that the tendency to form pentadactyl limbs is a derived condition rather than simply the ancestral state of the autopodium. *Hence, the pentadactyl "type" of the crown tetrapod limb is an acquired variational tendency that evolved after autopodial field and digits were established.* Thus, it is worth asking whether other evidence also points in this direction and what the nature of the pentadactyl type is. To get a more accurate picture,

---

[1] Many of the paleontological facts I learned from a manuscript written by Alexander O. Vargas, although to my knowledge, this paper was never published.

we have to contemplate a number of phenomena related to digit number variation. We will return to the issue of the nature of the pentadactyl "type" later in this chapter.

Developmental and Morphological Heterogeneity of the Tetrapod Hand

Although the ancestral polydactyly of the tetrapod hand/foot had vanished before the most recent common ancestor of crown group tetrapods, nevertheless the three main groups of recent tetrapods, amniotes, anurans, and urodeles,[2] are not uniform in their autopodial development and morphology. Here I will briefly summarize the evidence that anuran and urodele autopods are distinct from each other and from amniotes. Variational and developmental differences exist, although autopodia are clearly homologous among all tetrapods.

In addition to the five "true" digits of crown group tetrapods, an additional digit-like character exists in a variety of tetrapods. These have been called pre-pollex and pre-hallux in the hand and foot, respectively, and have been described in anurans (Fabrezi 2001; Tokita and Iwai 2010, figure 11.2), some temnospondyls, the fossil sister group of living amphibians (Gregory, Minner et al. 1923), occasionally in urodeles (Vorobyeva and Hinchliffe 1996; Hinchliffe and Vorobyeva 1999; figure 5 in Wagner and Chiu 2001; specimen collected and prepared by Jürgen Rienesel), and some mammals, for example opossums, elephants (figure 11.3A; Weissengruber, Egger et al. 2006), moles (figure 11.3B), and pandas (Holmgren 1933; Mitgutsch, Richardson et al. 2011).

In mammals, however, these "pre-digits" seem to be sesamoid bones (i.e., ossifications of tendons), rather than replacement bone as is the case in amphibians. Hence, the mammalian pre-hallux and pre-pollex have been classified as an entirely different phenomenon from the amphibian pre-digits (Gillis and Hopkins 1922; Fabrezi 2001). However, the situation is far from clear, mostly because of the paucity of developmental data for these structures. A notable exception is a recent study on the pre-pollex of the Iberian mole, *Talpa occidentalis* (Mitgutsch, Richardson et al. 2011).

This study showed that the pre-hallux arose out of a typical chondrogenic condensation as judged by the expression of the mesenchymal condensation marker gene *Sox9*. There was also a strong expression of *Msx2*, a gene involved in determining anterior digit identity. However, no anterior extension of the AER was found, as would be expected if the pre-hallux developed like

---

[2] The other major tetrapod group, gymnophthalimds, are limbless animals.

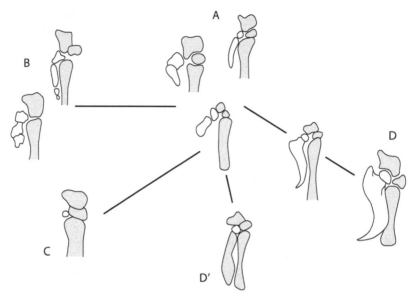

FIGURE 11.2: Variation of the pre-pollex in anurans. The pre-pollex is white and the carpal and first metacarpal are gray. The pre-pollex, as well as the pre-hallux in the hind limb, is a regular part of the anuran autopod and is present in a large fraction of anuran species (after Fabrezi 2001).

other digits. No histological observations were reported and, thus, it is unclear whether the pre-pollex was in fact a sesamoid or not.

A histological study of the European mole, *Talpa europaea*, identified the pre-digit as a sesamoid bone without a chondrified precursor (Prochel 2006). Alternatively, this structure can be seen as an elongated mesocarpal element. Similarly unclear is the situation with the famous panda's thumb. In the absence of developmental observations, it is impossible to determine whether the panda's thumb is a hypertrophic sesamoid (originating in a ligament), a modified periarticular ossicle, or a neomorphic member of the carpus.

Pre-digits are occasional variants found in urodeles, such as *Salamandrella* (Vorobyeva and Hinchliffe 1996) and *Triturus* (Wagner and Chiu 2001). In the latter case, a pre-hallux was found in the normally five-digit hind limb of *Triturus*, which comprises three consecutive small long bones just as normal digit I and hardly distinguishable from other digits in this limb. There is no doubt that this digit is a pre-hallux, as it is articulated with an element Y in direct continuation of the tibia. In other words, in urodeles, the phenotype of the pre-hallux can be quite digit-like, even though out of place in terms of its position and being supernumerary. Nevertheless, to my knowledge, these structures in urodeles are only variants rather than species characters.

In anurans, however, the presence of a pre-digit is an integrated feature of the autopodium. With very few exceptions, at least a rudimentary skeletal

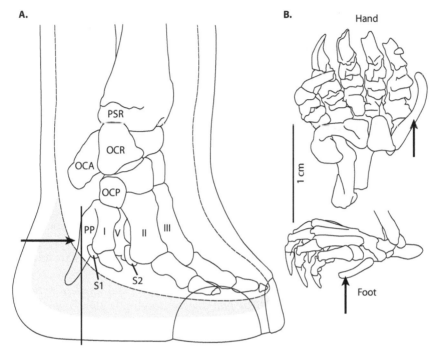

FIGURE 11.3: Presumptive pre-digits in mammals. A) Hand of the African elephant with a pre-pollex (PP). Reprinted by permission of Weissengruber et al., 2006. Copyright © 2006 John Wiley and Sons. Medial aspect of the distal part of the left forelimb in an African elephant. Broken line: outline of soft tissues of the locomotor apparatus (ligaments, muscles, tendons); gray: position of the foot cushion. PSR: processus styloideus radii; OCA: os carpi accessorium; OCR: os carpi radiale; OCP: os carpale primum; PP: prepollex; I: first metacarpal bone; II: second metacarpal bone; III: third metacarpal bone; V: fifth metacarpal bone; S1: promimal sesamoid bone of the 1st digit; S2: medial proximal sesamoid bone of the 2nd digit. B) Hand and foot of the Iberian mole (*Talpa occidentalis*) with prominent pre-digits (arrow). Reprinted by permission of Mit-gutsch et al., 2011, figure 2c. © 2011 The Royal Society.

element at the position of the pre-digits is found in most anurans (Fabrezi 2001). Pre-digits regularly occur in both the hand and the foot and assume various morphologies, from a single nodular element, to a proximal element and a variably shaped elongated distal spur, to a series of up to three or four "phalanges" (figure 11.2).

The question of whether the pre-digits are just an extension of the digit row of the "typical digits" and, thus, whether they are homologous to other digits has been decided negatively for two reasons. One is the timing of pre-digit development relative to the development of the other digits. Typical anuran digits develop in a posterior-to-anterior sequence with the most anterior digit, digit I, developing last. In some hylids and micro-hylids, however, pre-digital cartilages develop before the first typical digit (Fabrezi 2001;

Tokita and Iwai 2010). The development of pre-digits is not integrated into the general developmental sequence of the digital arch.

Similarly, the reduction of digits in anurans starts with digit I (Alberch and Gale 1985). However, pre-digits can still be present when digit I is reduced or lost (Fabrezi 2001). Both developmentally and variationally, pre-digits belong to a different developmental module than do "typical" digits. Pre-digits are not simply a case of polydactyly with an extension of the row of typical digits. Nevertheless, these pre-digits are a digit-like component of the developmental type of the anuran autopod. Hence, anuran autopods, in spite of their superficial similarities to amniote autopods, represent a separate stabilization of an autopodial ground plan.

The most radical deviation from the "typical" pathway of digit development is found in urodeles. In most tetrapods, including anurans and amniotes, digit development is highly stereotypical with the first digit to develop in direct continuation of the posterior axis (i.e., the ulna and fibula, respectively). Then only one digit is added posteriorly, and all the others are added anteriorly.

Urodele digit development is different in many ways (for a recent review, see Fröbisch and Shubin 2011). The first digit to develop is digit II, counted from anterior, then digit I is added anterior and all the others posterior. The order of digit development is the inverse of that in anurans and amniotes in which the predominant developmental polarity is from posterior to anterior. In addition, there are other signs of an anterior prevalence in urodeles as opposed to the posterior dominance in anurans and amniotes. For example, the tibia and radius in urodeles develop sooner than the posterior zeugopodial elements, the fibula and the ulna. Another sign is the early development of the *basale commune*, which is interpreted as the fusion of the distal carpale/tarsale one and two.

It remains unclear what the evolutionary reason is for this mode of limb development. Originally, this mode of development was used as evidence to support the notion that urodeles were the sister group of all other tetrapods and may have evolved limbs independently from the rest of the tetrapods (Holmgren 1933; Jarvik 1942). Phylogenetic evidence from paleontology to DNA sequences, however, makes this idea untenable (Meyer and Wilson 1990; Meyer and Dolven 1992; Zardoya and Meyer 2001). There is strong evidence that the extant amphibians are a clade and, thus, the urodeles cannot be sister to other tetrapods or form a clade with lungfish, as proposed by Holmgren. It has to be mentioned that digit development in urodeles is, to some degree, variable and shows some gradation between the amniote-like and the salamander mode of development (Fröbisch and Shubin 2011).

Urodele limb development is also different in terms of gene expression and function. For example, distal outgrowth in most tetrapods is sustained by Fgf signaling from the AER and a positive feedback between the AER and the ZPA. This feedback loop is mediated by the posterior expression of Fgf4 in the posterior AER. No Fgf4 expression is found during salamander limb development and Fgf8 and Fgf10 expression is not limited to the epidermis (Han, An et al. 2001; Christensen, Weinstein et al. 2002). Thus, Christensen and collaborators suggested that distal outgrowth of the salamander limb bud was likely under a different kind of control than was amniote limb bud growth.

During amniote limb development, the expression patterns of Hox genes are nearly as stereotypical as the embryological events of limb skeletal development. For example, *HoxD* genes are first expressed in a nested pattern consistent with posterior dominance (i.e., more 5′ *HoxD* genes are expressed to the posterior than 3′ *HoxD* genes). Then there is a gap in *HoxD* expression where the mesopodium is found. In the autopodium, there is an inverse *HoxD* gene expression pattern, with *HoxD13* most extensively expressed and *HoxD12* and *HoxD11* slightly less so, but expressed more to the anterior than in the zeugopod. The proximal and mesopodial expression of *HoxD11* in urodeles is similar to that in amniotes. The mesopodium also does not express *HoxD11* as in amniotes, but in the autopod, *HoxD11* expression does not expand to the anterior and is much less strongly expressed than in amniotes (Torok, Gardiner et al. 1998).

In amniotes and anurans, the mutually exclusive expression of *HoxA13* (in the autopod) and *HoxA11* (in the zeugopod) is a hallmark of autopodial individualization (see chapter 10). In salamanders it was found that initially *HoxA11* was expressed in the zeugopod (Wagner, Khan et al. 1999) and *HoxA13* in the autopod (Gardiner, Blumberg et al. 1995). However, when digits 3 and 4 appeared, then there was strong expression of *HoxA11* in the distal tips of the developing digits. Autopodial *HoxA11* expression is unexpected and, thus, adds to the uniqueness of salamander limb development. Urodele limb and digit development is highly unusual by affecting some of the most conserved aspects of limb development, like *Fgf* and *Hox* gene expression in addition to the embryological features described above. Explaining this deviant limb and digit development is one of the most challenging problems in developmental evolution, with no available well-supported hypothesis.

Two theories have been advanced to explain the uniqueness of urodele digit development. Both suggest that the urodele model of digit development is derived. The evidence for this is cladistic, but the parsimony advantage of this hypothesis is not very strong compared to that for two independent derivations of the "typical" mode of development found in anurans and amniotes. New paleontological evidence engenders further uncertainty, even with respect to this conclusion, because there are two fossil forms with evidence regarding digit development. One is *Gerobatrachus* and the other is *Apateon*, a branchiosaurid fossil with large numbers of fossilized larvae (see summary in Fröbisch and Shubin 2011).

In the latter fossils there is clear evidence for an anterior dominance in terms of digit ossification. In *Gerobatrachus* there is fossil evidence for the presence of a basale commune. The problem is that both branchiosaurids and *Gerobatrachus* are phylogenetically placed as sister taxa of Lissamphibia and, thus, are ancestral of the most recent common ancestor of urodeles and anurans. If this phylogenetic hypothesis is correct, then it implies that either anterior dominance in digit development and the basale commune evolved independently in urodeles and these fossil taxa or that anuran limb development is an independent derivation of an amniote-like developmental mode. Hence, as of now, the question of which form of limb development is ancestral and which is derived or convergent is completely open.

One theory proposed that the deviant mode of digit development was due to a heterochronic acceleration of the development of digits one and two (Wake and Shubin 1994), which are used as "balancers" in pond larvae of urodeles. Indeed, the larval digits of pond species are unusually long and those of creek species are shorter and show a less pronounced anterior bias in digit development. The other theory assumed that in the stem lineage of urodeles there was a period of digit loss and that the mode of digit development in urodeles was due to a re-evolution of digits posterior to the remaining two digits (Wagner, Khan et al. 1999). Digits I and II in urodeles would, at least in terms of their developmental position, correspond to digit positions 3 and 4 in amniotes and anurans.

As summarized above, there are some developmental signs that the regulation of digit development in urodeles is different at the genetic level than that of anurans and amniotes (e.g., low *HoxD11* expression in the autopod; Torok, Gardiner et al. 1998) and expression of *HoxA11* in digits, a gene that is usually excluded from the autopod (Wagner, Khan et al. 1999). No fossil evidence, however, supports this idea and at this time neither theory has garnered enough evidence to outcompete the other. More work is required to sort this problem out.

From the evidence summarized above, however, it is clear that, at least in terms of digit development, tetrapod limbs are heterogeneous. Differences in the mode of digit development between urodeles and anurans have variational consequences. The order of digit development largely dictates the sequence of digit loss. Anurans tend to lose digits 1 and 2 first, while urodeles lose digits 5 and 4 first. This was shown both in natural variation and in experimentally induced phenotypes, which showed that the different patterns were not due to natural selection; rather, they were due to internal developmental constraints (Alberch and Gale 1985; Stopper and Wagner 2007). During digit reduction, urodele autopodia access a different set of character states than anuran limbs and, thus, are different character modalities (see chapter 2 for the notion of character modality).

The urodele limb develops differently from that of anurans and amniotes, and anurans have integrated pre-digits into their autopodial ground plan.

Anuran, amniote, and urodele autopods are, thus, different character modalities. The evolutionary history and the causes for these differences are largely obscure.

Digit Loss and Re-evolution in Amniotes

Ever since the pentadactyl ground plan was established in the amniote lineage, the main mode of digit number evolution was digit loss (Greer 1991). In mammals, digit loss was mostly associated with adaptive modifications of the limb, which resulted in the specialization and the relative over-development of a few digits and the reduction of the rest. The paradigm of this mode of evolution is hoofed animals with the horse and horse relatives retaining only one functional digit, while two are retained in cloven hoofed animals. Limb and digit reduction has been most extensively studied in squamates, reptiles for which complete limb loss is characteristic of highly successful clades like the snakes.

Digit and limb loss is correlated with body elongation and, thus, a switch to either a burrowing or grass- or sand-swimming mode of locomotion. It is estimated that in squamates, limb reduction (i.e., partial or complete loss of limb skeleton) occurred 62 times in 52 lineages (Greer 1991). In skinks alone, it occurred 31 times in 25 lineages (Greer 1991). Loss of limbs and limb elements is part of a syndrome of elongated body shape and smaller limbs (Wiens and Slingluff 2001). A snake-like body form has evolved independently at least 26 times in squamates (Wiens, Brandley et al. 2006). This is a fruitful and complex area of ongoing research in terms of its ecological and adaptive significance, although it will not be discussed here. What concerns us here is what the developmental patterns and causes of digit reduction are and what they potentially tell us about the pentadactyl nature of the amniote autopodium and of digit identity.

The suspicion that complex traits, once lost, could re-appear has been around for quite some time. Examples include the second molar tooth in *Lynx* (Kurtén 1963; Werdelin 1987), wings in some phasmid insect species (Whiting, Bradler et al. 2003), and limbs and fully functional eyes in snakes (Laurent 1983; Coates, Ruta et al. 2000; Tchernov, Rieppel et al. 2000). Until recently, however, the difficulty was that inferences about the re-evolution of traits requires strong phylogenetic evidence and appropriate statistical tests. Only during the last decade have phylogenies gained sufficient robustness to re-visit this question.

The other problem is that the results of ancestral character reconstruction methods depend on assumptions regarding the relative likelihood of character transformations. However, there is no independent evidence other than comparative data to parameterize these character evolution models. Finally, it was found that estimating transition rates from phylogenetic data was very difficult and required large amounts of data and, thus, evidence for unequal transition rates was hard to come by (Ree and Donoghue 1999).

However, it turned out that it is possible to test for the likelihood of the hypothesis that certain transitions in character states were irreversible (i.e., Dollo's law), even with moderate amounts of data. This approach was first used in a study of the South American gymnophthalmid genus *Bachia* (Kohlsdorf and Wagner 2006). This study identified three cases for which there was statistical evidence for digit re-evolution. Since then, statistical methods for detecting violations of Dollo's law have improved (Goldberg and Igič 2008) and many other examples of re-evolution of digits and other morphological characters have been found (Brandley, Huelsenbeck et al. 2008; Kohlsdorf, Lynch et al. 2010; Lynch and Wagner 2010; Siler and Brown 2011; Wiens 2011).

In a recent study on tooth re-evolution in the anurans, John Wiens at Stony Brook University, New York pointed out that the number of statistically supported cases of re-evolution is most likely an underestimate of the true number of instances because the amounts and kinds of data required for detecting re-evolution are quite restrictive (Wiens 2011). Intriguingly, there are other morphological signs that cases of inferred re-evolution may, in fact, be cases of re-evolution. It was first pointed out for *Bachia* that the digits that had been inferred to have been re-evolved differed morphologically from ancestral digits. The most common condition is a homogenization of the phalangeal formula. For example, *Bachia panoplia* and *scolecoides* have a phalangeal formula of 2-2-2-2 compared to the ancestral situation of 2-3-4-5-3 (Kohlsdorf and Wagner 2006). Another example is the forelimb of *Bipes*, with 3-3-3-3-3, which also has been inferred to have been likely re-evolved (Brandley, Huelsenbeck et al. 2008).

However, most importantly for the current context is that in no case of digit re-evolution did more than five digits re-evolve. This is remarkable because if re-evolution of digits is just a case of adding digits to the ancestral, in this case oligodactyl phenotype, then why stop at five digits? Is there a cryptic developmental "type" maintained even in species that do not have five digits as an adult? If so, what could the mechanistic explanation be? Below, a possible scenario is discussed based on observations of chicken development, although there is no direct evidence available at this time for this model to apply to squamates.

The Pentadactyl Autopodium (PDA) Type

As we saw above, in amniotes the pentadactyl condition was not only ancestral for the crown group tetrapods, but was also a "ground state" that was rarely exceeded. When needed, additional digit-like skeletal elements were added, but they are likely sesamoid bones (i.e., ossified tendons and membranes) or mesopodial elements, but are not part of the array of typical digits. However, this ground state was not the ancestral state of the tetrapod limb, as seen from the fossils of early tetrapods like *Acanthostega* and *Ichthyostega* (figure 10.8). Thus, the question arises as to what it is about the pentadactyl state that makes it "typical" or the "ground state" for amniotes?

There are two possibly related facts that suggest that the PDA is, in fact, a developmental type rather than just an arbitrary trace of history.[3] The first was that the loss of anatomical digits was not necessarily associated with the loss of digit condensations, and may not ever have occurred. This was first suggested in a paper by Frietson Galis at the University of Leiden in 2001 (Galis, Alphen et al. 2001). Her argument was that, once initiated, changes in digit development had local effects, while mutations that affected the patterning of digit condensations before digits individualized were more likely to have had severe, widespread deleterious pleiotropic effects. This argument is similar to what we encountered earlier in the discussion regarding the conservation of the phylotypic stage (see chapter 1).

This argument is that the developmental processes that lead up to individualized body parts are more vulnerable to severe deleterious effects than those processes that take place after individualization of body parts and within these modular parts. In fact, Galis and collaborators did make an explicit connection with the phylotypic stage of vertebrates by noting that digit condensations occurred during the phylotypic stage in amniotes (but not in amphibians). It is unclear whether the patterning of the autopodial plate is, indeed, tied up in the mechanisms that are active during the phylotypic stage. However, there is one intriguing fact that supports the central claim of Galis and collaborators.

The hands of chicken and ostriches have three digits, but during early development they do have five digit anlagen (figure 11.4). For the chicken this was shown using three different methods that demonstrated that the most anterior digit anlage is just a pre-chondrogenic condensation that then fuses with the second digit and never expresses collagen II, a cartilage component (Larsson and Wagner 2002), Alcian Blue–positive glycoproteins, or incorporation of SO_4^{2-} as would be expected for chondroitin sulfate production (Hinchliffe 1977; Hinchliffe and Griffiths 1983). Nevertheless, there is capillary regression (Kundrát, Seichert et al. 2001; Kundrát, Seichert et al. 2002), mesenchymal cells with peanut agglutinin affinity, typical of condensing pre-cartilaginous cells (Larsson and Wagner 2002), and *Sox9* expression (Welten, Verbeek et al. 2005), as would be expected for condensing pre-chondrogenic cells.

---

[3] Of course a variational type is also a trace of history. The distinction made here is that types are homeostatic historical traces (i.e., have a tendency of being retained in history), unlike the state of a synonymous nucleotide, which is changed by mutation and drift, but has no inherent tendency to return to its ancestral state.

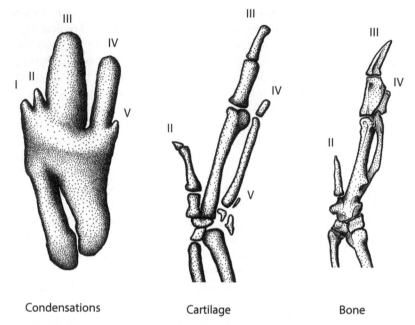

Condensations   Cartilage   Bone

FIGURE 11.4: Digits and digit anlagen in chicken wing development. There are five chondrogenic condensations that lead to four chondrified digits or digit rudiments of which three ossify to form the definite digits of the chicken wing (from Larsson and Wagner, 2002). This shows that digit anlagen could have been maintained for millions of years for digits that are were expressed in the adult skeleton, consistent with the model of Galis, Alphen et al. (2001).

Furthermore, retinoic acid induces an additional anterior digit in the wing, but not in the leg (Nikbakht and McLachlan 1999). This was remarkable, not only for the question of digit identity discussed below, but also for the fact that this condensation had not produced a functional digit for more than 100 million years! The persistence of this condensation through phylogeny is hard to explain by anything other than the type of developmental constraints that Galis and collaborators had proposed.

Unfortunately, no other case of digit loss has been investigated using methods that would detect pre-cartilaginous condensations. In Galis, Alphen et al. (2001) pictures from late nineteenth and early twentieth century embryological investigations are cited and reproduced, although these are notoriously unreliable for two reasons.

(1) At that time, no multiple independent methods were available to test for the presence of these condensations and, thus, this evidence is solely based on cell density and cell shape variation in histological sections. It is not possible to completely exclude fixation artifacts in these preparations.

(2) Authors in the late nineteenth century were influenced by Haeckel's recapitulation theory, which caused some of them to see more embryonic

condensation, for example in the mesopodium, than what can actually be confirmed using modern techniques (Hinchliffe 1977; Hinchliffe 2007). Hence, it will be important to investigate the limb development of species with digit loss and determine whether the findings for the chicken wing, which shows a pentadactyl ground state, are more general, if not universal, in amniotes as predicted by Galis' theory (Galis, Alphen et al. 2001).

The second piece of evidence for the reality of the PDA-type derives from recent evidence regarding the re-evolution of digits in squamates discussed above (Kohlsdorf and Wagner 2006; Brandley, Huelsenbeck et al. 2008; Kohlsdorf, Lynch et al. 2010; Siler and Brown 2011). In all those cases for which re-evolution of digits has been suggested by statistical evidence, the derived form never has more than five digits. In contrast, pathological poly-dactyl conditions are frequent, and experimental *Gli3-/-* knockout results in digit homogenization and far more than five digits (eight are reported by Welscher, Zuniga et al. 2002). Hence, polydactyly is not difficult to achieve mechanistically, but is for whatever reason phylogenetically exceedingly rare.

The two pieces of evidence mentioned above, re-evolution of up to five digits and retention of five digit anlagen after digit loss, may be the same phenomenon. But as long as we do not know the early embryonic stages of digit development in lineages with digit loss from which digit re-evolution occurred, these two observations have to be considered as putative independent pieces of evidence.

Overall, these observations suggest that the PDA is a type or a ground state. This means that there is a variational tendency that was retained during evolution because of developmental constraints of the type proposed by Frietson Galis and colleagues. These constraints are probably shared among all amniotes that at least form a limb with at least one digit (i.e., of course not retained in forms that do not form any limbs or limb buds, like snakes). The PDA is an experimentally accessible candidate for a developmental type[4] that represents more than just the shadow of an ancestral character state.

Developmental Genetics of Digit Identity

Digits are serially repeated structures arranged along the anterior-posterior dimension of the limb. Hence, the developmental biology of digit identity is tied up in the anterior-posterior polarity of limb bud development.

The most investigated factor in anterior-posterior limb pattern is a signaling center located at the posterior margin of the limb bud, called the Zone of Polarizing Activity or ZPA, which was briefly introduced in previous chapters. The ZPA was discovered by experiments in which transplanting the posterior mesenchyme of the wing bud to the anterior edge of another wing bud

---

[4] In our terminology used here, it is a character modality (see chapter 2).

resulted in mirror image duplication of skeletal elements. It was later found that the polarizing activity was due to secretion of a signaling molecule called Sonic Hedgehog, or Shh. Shh plays two roles in limb development (Drossopoulou, Lewis et al. 2000).

Early in development, Shh endows mesenchymal cells with the competence to form digits. Thus, by determining the amount of mesenchymal cells available for digit development, this determines the number of digits that can be formed. Digit number variation has been found to be associated with variation in the amount of Shh signaling. This has been shown experimentally by, for example, using the alkaloid cyclopamine that, in *Ambystoma*, results in different degrees of digit loss (Stopper and Wagner 2007). In *Hemiergis*, an Australian genus of skinks, it was shown that natural variation in digit number among species is associated with differences in Shh expression (Shapiro, Hanken et al. 2003).

During the second phase of Shh signaling, a gradient of Shh activity affects digit identity (Drossopoulou, Lewis et al. 2000). It was originally thought that Shh would be a paradigm of the classical morphogen gradient model in which different signal strength levels would be interpreted by cells in terms of different digit identities. The story, however, seems to be more complicated in that the extracellular level of Shh signals is only relevant for digits II and III, whereas the identity of digits IV and V is determined by a previous history of Shh expression in the cells themselves, although the exact mechanism how Shh expression affects cell state is unclear. Digit I seems to be refractory to Shh signaling and is the only digit identity that can be realized in the absence of Shh.

By influencing both the number and the identity of digits, the dual role of Shh in digit development results in paradoxical knockout (KO) phenotypes (Welscher, Zuniga et al. 2002). A completely homozygous KO genotype of the *Shh* gene results in digit loss, with only one digit that resembles digit I remaining in the hind limb. However, KO of the effector transcription factor of Shh signaling, Gli3, results in polydactyl hands and feet, but these digits all have digit I identity (i.e., have two phalanges). These results are paradoxical because, in terms of digit number, signal loss has an effect that is opposite the loss of the effector transcription factor gene that is regulated by Shh. As is usual for signaling pathways, the explanation of these paradoxical results follows from the somewhat roundabout manner in which the Shh signal is transduced (figure 11.5).

In the absence of Shh signaling, the effector transcription factor of the Shh signaling pathway, Gli3, is proteolytically cleaved, which generates a short isoform that acts as a transcriptional repressor, Gli3-R. Hence, Shh target genes are repressed in the absence of a Shh signal. When Shh binds to its membrane receptor, Gli3 protein cleavage is prevented. Hence, in the presence of Shh, Gli3 remains as its long isoform that is a transcriptional

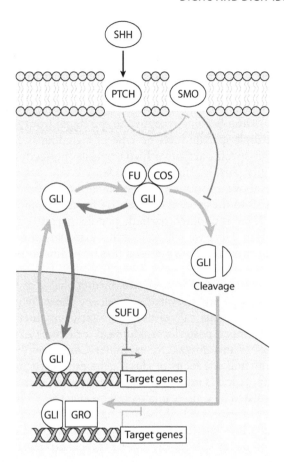

FIGURE 11.5: The Shh signaling pathway (see text for explanation). (After Paul J. Scotting, David A. Walker, and Giorgio Perilongo, *Nature Reviews Cancer* 5 [June 2005], 481–488.)

activator, Gli3-A. When Gli3 is knocked out, it cannot be cleaved to form the repressor Gli3-R and Shh target genes are not repressed, which results in an overproduction of digit-forming mesenchyme cells. However, because Shh cannot modulate Gli3 activity along the anterior-posterior axis, the resulting digits are all of the same kind: digit I identity.

Shh acts long before digits actually form. In fact, most Shh expression is gone when the first signs of cartilage formation are visible. During digit morphogenesis, the information for digit identity is presented by and signaled from the interdigital mesenchyme (Dahn and Fallon 2000). The mesenchymal cells between the digits are secondary signaling centers that affect digit identity. The rule seems to be that the interdigital mesenchyme posterior to the prospective digit signals to the digit rudiment anterior to it. The most likely signaling molecule at this stage of development is BMP4 (Dahn and Fallon 2000).

The target cells of the interdigital signal are in the phalange-forming region (PFR). These PFR cells are derived from the mesenchymal cells under the apical epidermal ridge (AER) where these cells proliferate. When they are in the sub-AER region, they commit to a cartilage-forming fate by expressing *Sox9 and BMPr-1b* and become receptive to the signal from the interdigital mesenchyme. A digit-specific level or type of signaling is translated into pSMAD1/5/8 digit-specific activity, which correlates perfectly with morphological digit identity (Suzuki, Hasso et al. 2008).

The narrative above connects the first signal acting on the autopodial plate from the AER to the process of phalange formation and, thus, to digit morphogenesis. There is another, partially parallel narrative that deals with *Hox* gene regulation. Here we will follow this explanation of digit identity and see what the connections are between these two models of digit identity determination.

During limb development there are two distinct phases of *HoxD* gene expression (see chapter 10 and the cited references). The early phase of *HoxD* expression follows the usual rule of nested posterior to anterior expression domains in which more posterior anatomical structures express more 5' *HoxD* genes (*HoxD13* and *HoxD12* for example). Consequently, the posterior margin of the limb bud is a focus of *HoxD* gene expression, and it is at this focus that *Shh* expression is initiated. The early, colinear expression of *HoxD* in the limb is regulated by a cis-regulatory element at the 3' end of the *HoxD* cluster. When Shh is expressed, Shh signaling feeds back on *HoxD* expression, which results in a late, inverse colinear *HoxD* expression pattern. In this late expression pattern, the 5'-most *HoxD* gene, *HoxD13*, is expressed all over the autopodial plate (i.e., has now the most anterior expression), while *HoxD12* and *HoxD11* are more posterior restricted. Digit I predominantly expresses *HoxD13* and *HoxA13*, while *HoxD12* and *HoxD11* are expressed in digits II and the posterior digits.

The late pattern of expression is governed by cis-regulatory elements, most notably the CsB of the so-called global control region (GCR) and the element Prox (Spitz, Gonzalez et al. 2003). More recent studies with chromatin conformation capture, however, have revealed that many more regions 5' of the *HoxD* cluster participate in *HoxD* gene expression (Montavon, Soshnikova et al. 2011).

Early experiments with the chicken hind limb showed that anterior inappropriate expression of *HoxD11* resulted in a DI → DII transformation (Morgan, Izpisúa-Belmonte et al. 1992). Hence, there is a consensus that late *Hox* gene expression plays a causal role in digit identity. But where is the molecular link between Shh signaling and *HoxD* expression? One link is probably related to the fact that digit identity develops long after the Shh signaling center, the ZPA, has ceased to exist. Nevertheless, Shh target genes remain active.

The explanation is that *HoxD12* and other *HoxD* proteins can physically interact with Gli3 and convert Gli3R into an activator complex: Gli3:*HoxD12* (Chen, Knezevic et al. 2004). In this way, the response to a Shh signal, which turns on *HoxD* genes, becomes fixed by modifying the Gli3 effector transcription factor for Shh signaling.

Overall, there is quite good evidence that there is a genetically distinct digit I identity. Digit I resistance to and independence of Shh signaling and the specific absence of *HoxD12* and *HoxD11* gene expression all contribute to a genetic explanation for digit I identity (Montavon, Garrec et al. 2008). The current evidence, however, is not consistent with the notion of character identity proposed in previous chapters in this book (i.e., the idea that character identity is due to a character-specific gene regulatory network that mediates between external signals and the execution of a character-specific developmental program). This raises the question of whether there are digit-specific, or digit identity genes, because the gene expression markers of digit I identity discussed this far are either shared with all other digits (*HoxD13* and *HoxA13*) or are negative markers (absence of *HoxD12* and *HoxD11*).

Functional genomic techniques, like RNA sequencing, are now available to search for potential digit identity genes in a systematic way, and some candidates have been found. Early in development, prospective digit I–forming cells have been found to express *Mkp3* and *Sef* (Uejima, Amano et al. 2010). *Mkp3* is also known as dual specificity phosphatase 6, which de-phosorylates both serine/threonine and tyrosine residues on MAP kinases and inactivates them. Sef is also a known interleukin-17 receptor, a regulator of FGF signaling. In later stages of chick limb development, my lab found two additional positive markers, *Zic3* and *Lhx9* (Wang, Young et al. 2011). *Zic3* is also expressed in mouse digit I (Cotney, Leng et al. 2012).

Zic3 is an intriguing candidate for a digit I identity gene because its protein is known to physically interact with Gli3 during brain development and, thus, might be a candidate modulator of Shh signals. It was found that Zic proteins bound to a DNA sequence that was similar to Gli binding sites and physically interacted with Gli proteins (Chen, Knezevic et al. 2004). The effect of the Zic-Gli interaction on reporter gene activity can be either activating or repressing, depending on the cell type. However, no experimental analysis of *Zic3* function in limb development has been performed.

For posterior digits, positive markers of digit identity have been identified, *Tbx2* and *Tbx3* in particular (Suzuki, Takeuchi et al. 2004). In the chicken hind limb it has been shown that *Tbx3* expression was necessary and sufficient for digit III identity and co-expression of *Tbx2* and *Tbx3* was necessary and sufficient for digit IV identity.

Given the pivotal role of Shh signaling for digit identity development, it is interesting to reflect upon the evolutionary mechanisms involved in the evolution of digit identity. A posterior signaling center that expresses Shh is not a derived condition for tetrapods. To the contrary, it is found in all paired vertebrate appendages, from zebrafish paired fins to shark and even in medial fins (see references in chapter 10). Hence, it is clear that the genetic changes that give rise to digit identity evolved downstream of a plesiomorphic anterior-posterior signaling system. In this case, the signaling system is older than the character-determining system, probably because Shh signaling is part of the identity network of the limb/paired appendage.

Digit Identity: Real or Imaginary?

Given the different morphogenetic and variational tendencies in the autopod represented by the three major recent tetrapod groups (which in fact have limbs), urodeles, anurans, and amniotes (see above), it seems prudent to limit the discussion of digit identity to amniotes, as the situation in amphibians might be different.

There are three easy ways to identify individual digits, at least in many cases, based on relative size, relative position, and phalangeal number. Most of the time, the thumb is the smallest digit and the one that ancestrally also had the smallest number of phalanges, namely two. The ancestral amniote condition with respect to phalange number is 2-3-4-5-3 in the forelimb and 2-3-4-5-4 in the hind limb. Hence, each one of the four anterior digits is unique in terms of phalange number and, thus, it seems that they are also clearly individualized at the genetic/developmental level.

Of these three traits, two are relational characters, relative position and size, and are problematic as signs of character individuality. The last, phalangeal number, is intrinsic to the digit and could be a sign of digit individuality. The problem, though, is that phalageal number seems to be a quite labile character, at least experimentally, and in some lineages it is associated with limb reduction. However, on a phylogenetic scale, phalangeal number is not as labile as it seems.

For example, the first digit almost always has two phalanges with very few exceptions (e.g., *Bipes* with three, but this limb could be re-evolved; see above). Hence, rather than being a sign of intrinsic digit identity or individuality, phalangeal number could just reflect a gradient in the rate or duration of digit growth. Whether it exists or not, digit identity is hard to pin down.

In some groups of amniotes, however, there are additional stable differences among digits. These are qualitative in nature, which suggest a deeper form of digit individuality. For example, in squamate reptiles the first digit is characterized by a reduced distal mesopodial I, which is small and fused to the base of metapodial I (Fabrezi, Abdala et al. 2007; Young, Caputo et al. 2009; Leal, Tarazona et al. 2010). In the hand of theropod dinosaurs, the three remaining digits are highly differentiated in size, shape, and many detailed anatomical characters (Dececchi and Larsson 2009).

In contrast to the ancestral situation for amniotes in general, the ancestral mammalian phalangeal formula is 2-3-3-3-3 and has been maintained in all generalized mammalian limbs, such as in humans. In this case, there is a lack of morphological differentiation among the posterior digits. Nevertheless, there are qualitative differences in the human hand between digit I and the posterior elements for which metacarpal I has a proximal epiphysis and the posterior metacarpals have a distal epiphysis. In primates, metacarpal I also has a unique proximal joint that allows for opposition.

Overall, the morphological variation among amniote species shows a mixed picture. Some species exhibit limited differences between digits, which can be easily explained as small variations in growth rate and growth period (e.g., the ancestral amniote situation), whereas in other cases qualitative differences have evolved, which are suggestive of true developmental individuality of digits like the hands of theropod dinosaurs, including birds. From these data, one could conclude that digit identity might be a dynamic property, which means that digit identity was not fixed in the ancestral tetrapod or amniote ancestor and, thus, is not universally applicable across all tetrapods, or even all amniotes.

Hence, digit II might exist in one clade, say the forelimb of theropods, but there might be species, like humans, in which digit II is not individualized from digits III to V and, thus, does not exist in this species as an individualized homolog. *Identity of a specific digit might not be a universal property of all tetrapods or even all amniotes, but might be phylogenetically dynamic, present in some but not necessarily in all lineages.*

This proposal to consider digit identity as a phylogenetically dynamic trait is supported by the analogous case of tooth types in mammals. Mammals have a set of clearly defined tooth types: incisors, canines, premolars, and molars. These distinctions are limited to mammals and are not applicable to reptiles. However, even within mammals, tooth types can become lost, as was the case for toothed whales. Hence, developmental and variational individuality (and hence character identity and homology) among serially homologous body parts can be gained and lost and may be limited to certain clades or paraphyletic assemblages (e.g., all mammals except for whales and other lineages that lost tooth identities).

Digit Identities in Mammals and Birds

With this perspective on limited digit individuality, we can return to mammals for which the ancestral phalangeal formula is 2-3-3-3-3. At face value, this could suggest that there are two digit identities: digit I and all others. Anatomically, this classification is supported in primates by the special anatomical characters associated with digit I (mentioned above), which allow for opposability. The individuality of digit I is also supported by gene expression data, which indicate that the first amniote digit has a unique digit identity—absent or low *HoxD12* and *HoxD11* expression, and expression of *Zic3* and others (see above).

The individuality of digit I can also be detected on the variational level. Philip Reno and collaborators have shown that, in the anthropoid hand, there are two variational modules: one comprises the first digit and the second comprises the posterior digits II to V (Reno, McCollum et al. 2008). This pattern corresponds to the differentiation in terms of *HoxD* gene expression and may reflect the existence of two types of digits: digit I and posterior digits.

A more complicated situation is the avian and theropod hand. Birds and theropod dinosaurs have three definite digits that morphologically correspond to digits I, II, and III (see below regarding additional complications). Unlike the situation in mammals and primates in particular for which there is good evidence for only two digit identities, the three digits in the theropod and avian hand are quite distinct in many characters (Wagner and Gauthier 1999; Dececchi and Larsson 2009). This makes it unlikely that they represent only two digit identities.

In particular, digits II and III differ not only in size and phalangeal number, but also in the character and orientation of joints and the size and shape of the metacarpals and other traits. A detailed comparison of the digit transcriptomes at two stages of development in the chicken also showed unique differences between digits II and III, which most importantly did not correspond to similar gene expression differences in the toes (Wang, Young et al. 2011). It turns out that the second and third digits in the wing do not clearly correspond to any of the posterior digits in the foot. More precisely, the signal is mixed, but not as strong and clear as for digit I.

A closer inspection of the transcriptome data revealed that the differences among the posterior wing digits in terms of gene expression were much stronger than those between posterior toes. There was even a gene expressed in digit III, *Socs3*, which was not found in the hind limb or any other digits that have been analyzed. One interpretation of these data is that the identity of the posterior bird fingers is derived and independent of the identities of the posterior toes. Hence, there might not be any homology between the second bird finger and the second or third bird toe, because the identities of the second and third bird fingers are phylogenetically derived, while the posterior toes may be independently individualized.

There is a good chance that hand digits in theropods are unique, derived identities. The proper catalog of homologs for the avian hand would be: digit I, which is clearly plesiomorphic; avian-FL-D-II; and avian-FL-D-III (i.e., forelimb digit II and III identities that are limited to the avian clade, even though digits existed ancestrally in positions 2 and 3). Whether these already had developmental identities ancestrally is not known at this point.

An analysis of alligator digit transcriptomes and those of other reptiles would greatly help clarify whether this interpretation is correct.

Musical Chairs: Digits Changing Places

No other issue has highlighted the problem of the nature of digit identity more than the long-standing controversy regarding avian wing digit identity. Birds have, at most, three ossified digits in their wing. Thus, the natural question is to which of the five digits in the pentadactyl ground state of amniote hand do these three avian digits correspond? This seemingly straightforward question has remained unresolved since the discovery of the bird-dinosaur relationship by Gegenbaur in 1863 (Gegenbaur 1863). The two competing models are that the three avian digits correspond to digits I, II, and III or digits II, III, and IV (reviewed in Vargas and Fallon 2005; Wagner 2005b; Young, Bever et al. 2011).

During the nineteenth century, this problem was discussed in the context of similar problems in lineages other than birds, most notably in the Italian three-toed skink, *Chalcides chalcides* (Fürbringer 1870; Steiner and Anders 1946). Duirng the last 20 to 30 years, a considerable amount has been published that pertains to this issue and it would require an entire chapter to summarize this research in detail. Because previous reviews detailed the progress made since the 1990s (Vargas and Fallon 2005b; Wagner 2005; Bever, Gauthier et al. 2011; Young, Bever et al. 2011), as well as a history of the controversy (Hansen 2010), it will not be necessary to summarize this literature in detail here. Instead, I will provide a sketch of the problem, consider the evidence, and discuss the unsolved problems pertaining to the question of the nature of digit identity.

At its core, the problem of avian digit identity is a conflict between different types of evidence used to assess character identity. Even though the first evidence put forward in favor of the II, III, IV model was anatomical (Owen 1836), the bulk of evidence for the II, III, IV hypothesis is embryological. The chicken wing has four chondrifying digit anlagen and their positions relative to mesocarpal elements as well as to the primary axis of digit development suggested that the three digits of the avian wing developed like digits II, III and IV and, thus, were identified as such (Hinchliffe 1985; Burke and Feduccia 1997; Hinchliffe 2007).

That the three digits of the avian wing, in fact, developed in embryological positions 2, 3, and 4 was shown beyond reasonable doubt, as an anterior digit rudiment was identified using four different histochemical methods that were mentioned above in the discussion on the pentadactyl ground state of the amniote autopod (Kundrát, Seichert et al. 2001; Feduccia and Nowicki 2002; Kundrát, Seichert et al. 2002; Larsson and Wagner 2002; Welten, Verbeek et al. 2005; Larsson, Heppleston et al. 2010). Hence, as discussed above, it was shown that during development, the chicken wing went through a stage of development that represented the pentadactyl ground state of the amniote hand (Larsson, Heppleston et al. 2010).

So far, the embryological data are clear and consistent. The trouble, however, arises from an equally (rock-) solid source of evidence regarding digit identity: the fossil record. Birds belong to the clade of theropod dinosaurs. Basal lineages of theropods already had a partially reduced set of digits in the hand, although there were still metacarpal rudiments of all five digits. This is the case in *Herrerasaurus* as an example in which two reduced metacarpal rudiments are found posterior to the three definite digits and for that reason are identified as digits I, II, and III (for a summary of this evidence, see Wagner and Gauthier 1999). Thus, among amniotes, the theropod hand suffers from a rare form of digit reduction, namely a reduction of the posterior digits, in contrast to the more usual pattern of losing digit I first, then V, and finally II. This also made data interpretation difficult.

A number of hypotheses have been proposed to resolve this conflict. One challenged the phylogenetic inference that birds derived from dinosaurs (Feduccia 1996; Feduccia 2001), while others proposed various developmental models (Shubin 1994; Chatterjee 1998; Garner and Thomas 1998; Wagner and Gauthier 1999; Galis, Kundrát et al. 2005), or questioned the identification of the digits in fossil dinosaurs (Galis, Kundrát et al. 2003). All of these theories have been discussed and evaluated extensively in the published literature (Wagner 2005; Young, Bever et al. 2011). Here, it may suffice that the recent evidence tends to favor the so-called frame shift hypothesis (FSH), which is the idea that both the embryological and paleontological evidence are correct.

What is not correct is the assumption that there is a rigid causal connection between embryological origin/position and digit identity. Specifically, the FSH proposes that, in the stem lineage of birds, after the loss of at least one of the posterior digits, the three digits changed place such that, in birds, digit I develops from embryological position 2, digit II from position 3, and digit III from position 4 (figure 11.6). One can say that digits played musical chairs, if one is receptive to such metaphors.

Of course, the FSH only makes sense if one adopts a specific concept of character identity—namely, one in which character identity is connected to the activation of a gene regulatory network, as proposed here. Needless to say, at that time, this idea was far from the mainstream, and even today is not firmly established. This conceptual assumption, that character identity is a gene regulatory state rather than an embryological or spatial descriptive fact, is probably one reason for the vehement opposition to the FSH (e.g., Feduccia 1999; Galis, Kundrát et al. 2003; Galis, Kundrát et al. 2005; Thulborn 2006).

For the FSH to possibly be true, one has to assume that there are gene expression patterns and mechanisms of gene regulation that are associated

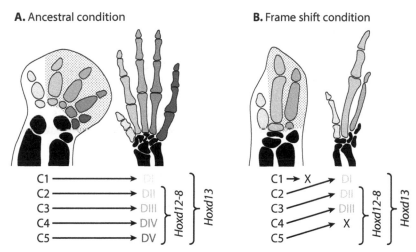

FIGURE 11.6: Frame Shift Hypothesis of avian digit homology. Ancestrally, the five digit anlagen correspond to the five digit identities. In particular, digit I identity is determined by the absence of *HoxD12* and *HoxD11* expression in the anteriormost digit anlage. In birds, three digits develop from digit anlagen 2, 3, and 4, but the digit developing from condensation C2 has the gene expression pattern of digit I (after Vargas and Wagner 2008).

with the identity of each digit. The FSH then predicts that the relocation of digit identities, as seen in morphological evidence, is subscribed by corresponding changes in the expression of regulatory genes, such as transcription factor genes.

The first molecular evidence that digit identities might have changed place in the stem lineage of birds came from an *in situ* hybridization study conducted by Alexander Vargas, now a professor at the Universidad de Chile, and John Fallon at the University of Wisconsin (Vargas and Fallon 2005a). They compared *HoxD13* and *HoxD12* expression between the wing and foot of the chicken. Their rationale was that digit identity in the foot of birds was not in question and, thus, any similarities in gene expression between fingers and toes would provide evidence regarding the identity of wing digits.

What they found was that digit I was characterized by *HoxD13* expression and absence of *HoxD12* both in the foot and in the wing. They also showed that mutations that changed wing and foot morphology, such as *Silkie*, also followed this rule. In the *Silkie* foot, an additional anterior digit developed that had the morphology of digit II and the second toe had the morphology of digit I. Yet the second digit with digit I morphology had low-level *HoxD12* expression, which meant that this signal was not just a sign of the most anterior digit (= MAD; according to Woltering and Duboule 2010). Hence, the *HoxD* signature of digit I identity was not just a sign of MADness (contra, Woltering and Duboule 2010).

Since the 2005 Vargas-Fallon paper, follow-up papers have substantiated the hypothesis that *HoxD* expression in the chicken wing is actually derived (Vargas, Kohlsdorf et al. 2008), and digit I typical expression patterns are also found in an experimentally induced digit identity shift (Vargas and Wagner 2009). The initial findings by Vargas and Fallon were also confirmed with full digit transcriptomes—namely, by showing that wing digit I and foot digit I are most similar in gene expression to the exclusion of other digits (Wang, Young et al. 2011). The transcriptome data also added new positive markers of digit I identity, which were genes that were uniquely expressed in digit I in both the wing and the foot—for example, *Zic3*. It was also recently shown that the third wing digit did not follow the mode of development typical of digit IV in other amniotes, consistent with its identity as a digit III (Tamura, Nomura et al. 2011; Towers, Signolet et al. 2011).

While the *in situ* hybridization results for *Hox* genes and the transcriptome data clearly favor homology of the most anterior wing digit as digit I, the transcriptome data added a new twist to the story as discussed in the previous section—namely, the second and third digits in the hand might be unique derived identities. This scenario implies that posterior digits had not yet acquired distinct identities in the archosaur ancestor, a thesis that is testable by transcriptome work on the alligator or any other crocodilian.

While most work was done on avian digit identity, the question of digit identity of partially reduced limbs is not limited to birds. The second most notorious case is that of the three-toed Italian skink, *Chalcides chalcides*, and related species (Fürbringer 1870; Steiner 1934; Steiner and Anders 1946). Even though early in the history of this problem birds and skinks were discussed together as exemplars of the same problem, during much of the twentieth century the literature on *Chalcides* was published in German and French and, thus, escaped the attention of the broader biological community.[5]

The case of digit identity in the Italian skink is similar to the avian case in many of its morphological and developmental aspects, but is different in others. The similarities include morphological evidence suggesting that the three remaining ossified digits are digits I, II, and III (Steiner and Anders 1946; Renous-Lecuru 1973), whereas the embryological evidence suggests that they are digits II, III, and IV (Raynaud and Clergue-Gazeau 1986; Raynaud, Clergue-Gazeau et al. 1986; Raynaud, Clergue-Gazeau et al. 1987; Raynaud, Perret et al. 1989; Raynaud 1990). The difference is that the digit reduction affects both forelimbs and hind limbs and that digit loss occurred more recently, perhaps as recently as 6 million years ago (Carranza, Arnold et al. 2008).

---

[5] The present author only became aware of this example after reading a German handbook on European reptiles. Credit is due Böhme, W., Ed. (1981): *Handbuch der Reptilien und Amphibien Europas* (Wiesbaden, Germany, Akademische Verlagsgesellschaft).

What is the morphological evidence for identifying the fist digit of *Ch. chalcides* as digit I? As usual, the first digit is smaller and has a stout metacarpal and two phalanges. These are characters of digit I, although not very reliable because size, shape, and phalangeal number can vary. The most important qualitative character is the absence of a distal carpal I, as mentioned above. This means that the proximal articulation point of the first digit is more proximal than that of the posterior digits. During development and during a short time window, which is hard to catch when collecting embryos, a small nodular cartilage appears which then fuses with metacarpal I. This has been found to be the case for the five-digit hand of the gecko (Leal, Tarazona et al. 2010) and the first of the three digits of the three-toed Italian skink (Young, Caputo et al. 2009). Hence, the first digit of the three digits in *Chalcides chalcides* has the morphology characteristic of squamate digit I. The problem, though, is that the embryological evidence suggests that this digit is the second anterior to the primary axis (i.e., develops in position 2 of the pentadactyl ground state).

The evidence for the embryological position of the digits in *Ch. chalcides* is, so far, somewhat indirect, because collecting this species' embryos is not easy. Essentially, the argument is: posterior to the three ossified digits one finds a rudimentary metapodial in both the hand and the foot that is maintained because it is the insertion point for an important tendon. Considering the competing hypotheses, I, II, III or II, III, IV, this metapodial should either be position 5 or position 4. Position 4 would be derived from the primary axis and, thus, should develop first. If it is position 5, then it should develop after the digit anterior to it, which would then be in position 4. Embryological evidence clearly shows that the rudimentary metapodial develops late and, thus, likely derives from position 5. If this inference is correct, then the three definite digits in the skink develop in positions 2, 3, and 4, just as in the avian hand.

To further test the identification of the first digit as digit I in *Ch. chalcides*, the expression of *HoxD11* was examined by *in situ* hybridization. It was found that, in fact, this digit had low *HoxD11* expression as expected for a digit I. The sum total of evidence suggests that the first digit in the Italian skink is, in fact, a digit I and develops from position 2, as with the first digit in the avian hand (Young, Caputo et al. 2009). The inference, of course, is that during digit reduction in this and the avian lineage, digit I had changed place by re-deploying the digit I gene regulation network in embryonic position 2.

After all these details on bird and skink digits, one may ask what the general scientific value is for knowing this much detail. For the current context, the most important implication is that, rather than position and embryonic derivation, character (digit) identity is tied to activating a set of transcription factor genes and that this perspective can solve a long-standing paradox in comparative anatomy and evolutionary biology. Any conceptual idea in the sciences is only as good as the subsequent research that it encourages. For anchoring character identity to gene regulation network activity, the avian and skink digit identity problem has so far been its most productive test case. Character identity reflects a gene regulatory state rather than a positional or descriptive fact.

A Fingerpost on the Nature of Character Identity

The summary provided above on the developmental and evolutionary biology of digits and autopodia highlighted a number of observations that cast a new light on the nature of hands and feet, as well as the nature of digits. I want to summarize the salient points here.

- Tetrapods are a clade and, thus, digits and autopodia are likely to be homologous. But *each of the three principal groups of tetrapods with limbs* (i.e., urodeles, anurans, and amniotes) to the exclusion of gymnophions, which are limbless, *represent a different character modality of the autopod.*
  - Amniotes have a pentadactyl ground state and a posterior to anterior developmental polarity of digits.
  - Anurans are similar to amniotes but have integrated pre-digits (pre-hallux and pre-pollex) into the ground plan of the autopod.
  - Urodeles have an anterior to posterior developmental and associated variational polarity.
- Even within the amniote limb there exist further modalities that are distinguished by their variational tendencies. This has been documented with respect to phalangeal number variations.
  - The best case is mammals that have an ancestral 2-3-3-3-3 phalangeal formula in contrast to the 2-3-4-5-4/3 amniote phalangeal formula. Mammals also differ from other amniotes with much lower variability in phalangeal numbers compared to reptiles (Richardson and Chipman 2003). Within mammals, the exception are the cetaceans, which show that the constraint underlying the character modality is paraphyletic.

How the three principal character modalities of digits and autopod arose is unclear. It could have been the result of divergent stabilization of the autopod ground state in different lineages, or the result of lineage-specific modifications to an ancestral autopodial type, as suggested by the model to explain the unique mode of digit development in urodeles (Shubin and Wake 1997; Wagner, Khan et al. 1999). The mammalian phalangeal character modality was achieved in a unique way by the progressive shortening and eventual loss of the pre-terminal phalange (Vargas, in prep.), which may have resulted in a unique derived mode of digit development leading to a unique, derived set of developmental constraints.

- The pentadactyl state of the anmiote limb modality is both the ancestral phenotype of the most recent common ancestor of amniotes and also a "developmental type." By "type" I mean a set of character states that is actively maintained through developmental or functional constraints. The evidence for this inference derives from two observations:
  - Re-evolution of digits has never been observed to lead to polydactyl hands/feet. In each documented case of re-evolution of digits, the derived (re-evolved) state has either five or fewer digits.
  - Loss of digits does not necessarily affect the developmental ground state of the autopod, which goes through a pentadactyl stage during digit condensation regardless of the number of definite digits (Galis, Alphen et al. 2001).

This conclusion is, perhaps, the most radical deviation from the neo-Darwinian consensus, which sees shared patterns only as traces of ancestral character states. More research on the modes of digit development in different tetrapod lineages is necessary to determine whether this conclusion is robust.

- Digits are serial homologs and their individuality (i.e., digit identity) is not a universal tetrapod trait. Little is known regarding the individuality of amphibian digits, but the comparison of mammalian and bird digits suggests the existence of clade-specific digit identities.
  - In mammals, both morphology with the ancestral 2-3-3-3-3 phalangeal formula and other characters and gene expression evidence suggest the existence of as few as two identities: digit I and posterior digits.
  - In the bird wing, and perhaps in the theropod hand in general, there seem to exist three well-individualized digit identities: D-I; av-D-II; and av-D-III. Of these, only D-I is comparable with the digits of other amniotes, and the other two digits are not even comparable with the digits in the bird hind limb (Wang, Young et al. 2011). It is likely that these are unique forelimb-specific avian digit identities.

The loss and gain of digit identities is, thus, an important mode of developmental evolutionary change that needs additional attention in future research.

- Digit I identity seems to be widespread and phylogenetically stable, at least shared among amniotes if not all tetrapods.

- During development, digit I displays a clear, evolutionarily widely distributed gene expression signature in terms of *Hox* and non-*Hox* genes.
- At the morphological level, there are several signs for the uniqueness of digit I:
    - In squamates, digit I is distinguishable from the rest of the digits by the fusion of distal carpal/tarsal 1 to metacarpal/tarsal I, regardless of digit I position (Young, Caputo et al. 2009; Leal, Tarazona et al. 2010).
    - In primates, digit I is distinguishable by a large set of qualitative anatomical characters and its variational quasi-independence with respect to the posterior digits (Reno, McCollum et al. 2008).
    - The *HoxD* gene expression signature of digit I is shared between chicken (Nelson, Morgan et al. 1996; Vargas and Fallon 2005), alligator (Vargas, Kohlsdorf et al. 2008), mouse (Montavon, Garrec et al. 2008), gecko (Tamura, Nomura et al. 2011), and skink (Young, Caputo et al. 2009).
    - Digit I–specific expression of Zic3, a Gli3 interacting protein, is shared between chicken and mouse and, thus, is likely an amniote character (Wang, Young et al. 2011; Cotney, Leng et al. 2012).

The differentiation and individuality of digit I is, thus, likely to be an ancient character of and thus a fundamental building block of the tetrapod limb.

Overall, these observations suggest that the "tetrapod limb" is a collection of different character modalities and digit identities with limited taxonomic distribution and, thus, likely the result of a dynamic evolution of character identities.

12

Flowers

All previous chapters on specific systems placed a heavy emphasis on vertebrates. This was mostly due to the author's greater familiarity with vertebrate developmental evolution and did not do justice to the outstanding work done in invertebrate systems, in particular insects and other arthropods. The literature on arthropod devo-evo, however, is vast, and would be better synthesized by someone intimately familiar with this field. The current chapter on flower evolution in this book is an exception to the vertebrate bias.

The motivation is that research on character identity and character origination is much more advanced in plant biology than in any of the zoological model systems of which I am aware. When reading this literature, it seems to me that there are two reasons for the advanced state of this field. One may be the excellent collaboration between evolutionary and developmental biologists in seed plant biology. The other may be that the early discovery of a well-supported model of flower organ identity, the ABC model of Meyerowitz and colleagues (Coen and Meyerowitz 1991), gave a boost to research on the origin of flowers and flower organs. Research on the origin of flowers was already well advanced in evolutionary plant biology when the ABC model entered the stage and, together with the developmental genetics of flowers, created a field intensively focused on the evolution of organ identity.

In contrast, much of the zoological literature focuses on other aspects of developmental evolution, such as the controversy regarding the relative importance of cis-regulatory and trans-regulatory evolutionary changes, the causes of organ loss (e.g., the loss of pelvic structures in freshwater stickleback populations), or the modifications of certain quantitative or meristic characters like the numbers and kinds of bristles in different parts of the insect larva or imago. Overall, the problem of organ identity origination has received much less attention in zoological devo-evo than in botany (with the exception of recent work on sex combs in Drosophila by Artyom Kopp's lab at the University of California at Davis; discussed in chapter 6).

Here I will summarize some of the deep insights garnered on the nature of organ identity, organ integration, and the origin of evolutionary novelties from research on the developmental evolution of flowers.

What Is a Flower?

If there is any typological concept that is alive and well, then it is the concept of an angiosperm flower.[1] As we will see in the following sections, there are good reasons why the concept of "flower type" is a meaningful biological term. Flowers are developmentally tightly integrated and the extensive variation among extant angiosperms can be understood within the confines of a "flower *Bauplan*" or "flower developmental type."

Generically, a flower is described as comprising four kinds of organs that form at least four whorls around a determinate shoot axis (i.e., a shoot axis that stops growing at some point; figure 12.1). From the outside to the inside these whorls are: sepals, petals, stamens, and carpels. Typically, sepals form a whorl of green leaves to protect the flower bud before it opens. Petals typically are a whorl of showy modified leaves specialized for attracting pollinators. Stamens are the male reproductive organs, and the carpels surround the female gametophyte. More technically, the flower is a bisexual, compressed, determinate shoot axis with a perianth which consists of the sterile flower organs, typically sepals and petals. In cases for which the sepal and petals are not well differentiated, the structures are called tepals. Less obvious is the occurrence of an endosperm that contributes to the fruit upon fertilization.

The model of the "typical" flower outlined above (figure 12.1) of course does not cover any and all angiosperm reproductive organs that one would casually call "flowers." The "typical flower" described above is a good model for what one could call the flower of core-eudicots, the largest and most advanced group of flowering plants. Beyond that model there are variations of different kinds. On one hand there are different flower phenotypes related to the evolutionary transition from ancestral angiosperm reproductive organs to the core-eudicot flowers, as for example a spirally arranged perianth, rather than whorls, and the undifferentiated perianths of magnoliids. There are also the highly derived reproductive organs of grasses, which are probably adaptations to wind pollination. There are also considerable variations in phenotypes and, perhaps, the identities of perianth organs, as for example the petaloid phenotypes of sepals in the Ranunculids and many monocots (e.g., tulips).

---

[1] Of course the term "angiosperm flower" is intrinsically redundant, as flowers exist only on angiosperms.

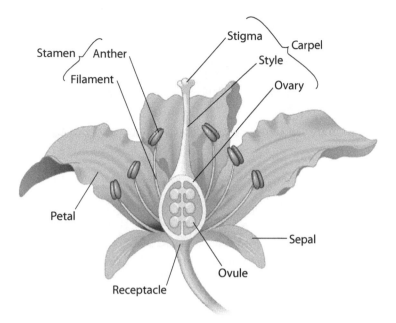

FIGURE 12.1: Basic anatomy of a typical flower comprising four whorls and four types of flower organs. From the outside to the inside, these are sepal, petal, stamen, and carpel.

There are a few cases of truly bizarre flowers that require special attention. One is the flower of a small group of south Pacific trees and shrubs, called *Eupomatia*. This group is unique in that the stamens form the outer most ring of the flower (figure 12.2), followed by apparent "peri"-anth organs that appear to be petaloid, and finally the carpels (Endress 1993). This is the only case for which female and male characters are separated by apparent petals and in which the male organs form the outermost whorl. In order to rescue the flower model, the petaloid organs have been interpreted to be stamoids (i.e., sterile stamens). This makes sense within the typological flower concept, but we need to consider the possibility that it is a homeotic transformation now that the tools for testing this interpretation are available (Soltis, Soltis et al. 2006). In fact, it has been found that a homeotic transformation or translocation of organ identities was the best explanation for this phenotype (Alvarez-Buylla, Ambrose et al. 2010).

Another bizarre flower has been described for *Lacandonia schismatica*. *Lacandonia* is the only species in its genus and is found in the Lacandonian jungle of Chiapas, Mexico. It is a saprophytic monocot with a flower in which the stamens form the center that is surrounded by carpels (Márquez-Guzmán, Engelman et al. 1989). This situation is even worse than the case of *Eupomatia* because there is no scenario that can reconcile this arrangement

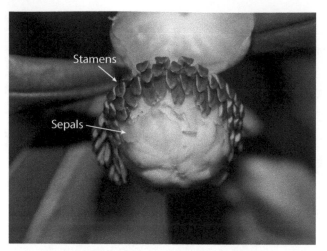

FIGURE 12.2: The flower of *Eupomatia bennettii*, a flower that defies the logic of the typical flower type as exemplified in figure 12.1. In this case, the stamens are located in the outermost whorl and outside of sepal-like green blades. (Based on: http://www.flickr.com/photos /plantnerd/6321898025/.) Reproduced by permission from Lui Weber.

with the typological model of an angiosperm flower. These and other examples alert us to the fact that in plants, homeotic transformations or heterotopy is a possible mode of phenotypic variation among species.

To appreciate the uniqueness of flowers and to define the task of explaining the origin of flowers, it is useful to compare them to the reproductive organs of gymnosperms. Gymnosperms, conifers, ginkgos, gnetales, and cycades, are the sister group of angiosperms. Unfortunately, there is no gymnosperm group more closely related to angiosperms than these (Soltis, Smith et al. 2011). For this reason we have no neontological evidence for the sequence of events that led from the ancestral seed plant to angiosperms (Specht and Bartlett 2009).

Gymnosperm reproductive organs are unisexual and indeterminate shoot axes, colloquially called cones—in particular, the female organs. For example, a spruce has female cones on its upper branches and male "cones" on its lower branches. These details vary among gymnosperms. However, the general scheme of unisexual, extended, and indeterminate reproductive organs is generally applicable to all gymnosperms.

Assuming that the gymnosperm condition is qualitatively representative of the ancestral seed plant situation,[2] then the evolution of flowers had at least

[2] This inference is not supported by cladistic arguments based on recent taxa because of the fact that gymnsperms and angiosperms are each monophyletic. However, both arguments about morphological/developmental complexity as well as fossil forms support the notion that the ancestral seed plants had gymnosperm-like reproductive organs.

four dimensions (Baum and Hileman 2006; Specht and Bartlett 2009): 1) the origin of bisexual shoot axes; 2) the determinative shoot growth and axis compression; 3) the origin of the perianth; and 4) perianth differentiation into sepals and petals. This list will guide the treatment of flower evolution in the following sections. However, before we can dive into the models for flower and flower organ origination, we need to review the comparative biology of flower variation and the basic developmental biology of flowers.

Angiosperm Phylogeny and Flower Character Evolution

As in many other areas of comparative biology, the advent of molecular methods led to large-scale revisions of the phylogenetic tree. Although the gymnosperms were traditionally considered paraphyletic with the gnetales and angiosperms forming a clade, the current consensus is that extant gymnosperms and angiosperms each form a clade, and the gnetales actually are closely related to the conifers (Soltis, Smith et al. 2011). In a way, this is unfortunate because there is now no means to reconstruct the steps that led from a gymnosperm-grade reproductive organ to a flower based on living representatives.

The large-scale relationships within the angiosperms have also been revolutionized (figure 12.3; Doyle and Endress 2000). Current consensus suggests that there are three consecutive basal lineages at the root of the flowering plants, the so-called ANA plants: *Amborella*, Nymphales, and the Austrobaileyales. The *Amborella* lineage is represented by only a single extant species, *Amborella trichopoda*, which is a small understory shrub on New Caledonia in the southern Pacific Ocean east of Australia. The Nymphales are the well-known water lilies, and the Austrobaileyales are a group of about 100 species of woody plants from southern Asia and the southern Pacific, with the star anise being the most broadly known representative.

The clade above the node that unites the Austrobaileyales and the rest of the flowering plants is called the Eu-angiosperms or Mes-angiospermae and comprises three major clades. First is a clade that includes the Chloranthaceae, the peppers, the Magnoliids, and the laurels. Then there are the monocots and a lineage comprising *Ceratophyllum* and finally the Eu-dicots.

Using this recent phylogenetic information, Endress and Doyle attempted to reconstruct the character transformation events that led from the ancestor of flowering plants to the "typical" core-eudicot flower (Endress and Doyle 2009). Many character transformations remain ambiguous, but a few are quite clear and shed some light on the early evolutionary events that resulted in the consolidation of the flower type.

The ancestral flower was probably bisexual, although the most basal lineage, *Amborella*, is unisexual. However, it is important to note here that

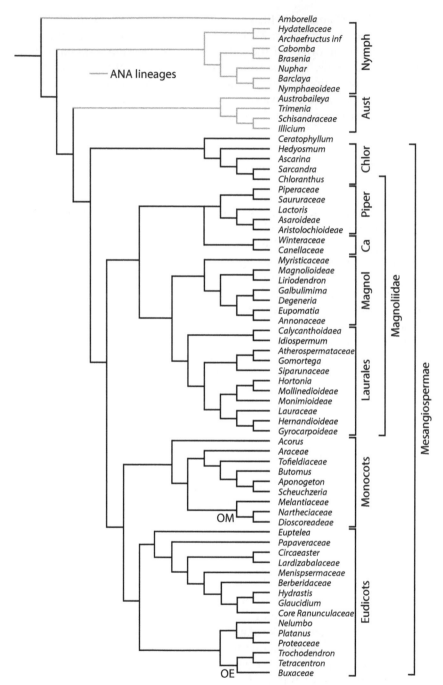

FIGURE 12.3: Angiosperm phylogeny after Doyle and Endress (2000). The basal lineages are called ANA lineages for Amborella, Nymphales, and Austrobaileyales. The rest of the clade are the Mesangiospermae.

Amborella is functionally unisexual but has sterile stamens in the female flower and, thus, is anatomically bisexual. The functional unisexuality of *Amborella* is, thus, likely derived. Hence, one of the key steps during the origin of flowers, the transition from unisexual to bisexual organs, occurred before the most recent common ancestor of extant angiosperms. Models for the origin of this character rely on interpretations of genetic data and the fossil record, to the extent available (*vede sotto*).

The flower of the most recent common ancestor of angiosperms clearly had a perianth with more than two whorls or rings. However, the arrangement of perianth organs is ambiguous because *Amborella* has a spiral arrangement, the Nymphales have whorls, and the Austrobaileyales also have spirals. Surprisingly, the data strongly suggest threefold symmetry of the perianth, so that we know more about the symmetry of the ancestral flower than about its phyllotaxy. The differentiation of the perianth organs is complicated and will be discussed in the section on petal origins and evolution.

As is the case for perianth organs, the stamens were ancestrally trimerous and came in more than two whorls. The number of whorls shifted to two stamen whorls at the base of mesangiospermae (i.e., after the most basal ANA lineages).

In summary, the ancestral flower was bisexual and had more than two whorls of perianth organs and stamens, both in trimerous symmetry and ascidiate (tubular) carpels. Evolution within the crown angiosperms was, thus, mostly focused on perianth differentiation. Later we will consider the origin of petals as one example of a novelty that arose within the extant angiosperms.

Genetics of Canonical Flower Development

Flower development is, like everything in biology, quite complex. Nevertheless, the model of organ identity determination that underlies most evolutionary studies of flower development is relatively simple. I will start here with the classical model of flower organ development and later will add modifications to the model to accommodate more data. The summary provided below follows the reviews by: Coen and Meyerowitz 1991; Theissen 2001; Krizek and Fletcher 2005; Causier, Schwarz-Sommer et al. 2010; Sablowski 2010.

The basic outline for the classical model of flower organ identity is one of combinatorial gene expression, called the ABC model. This model summarizes the effects of homeotic mutations on flower organ phenotypes. It posits that there are three gene functions, called A, B, and C, which in combination determine the four flower organ identities of the eu-dicot flower of *Arabidopsis thaliana*: sepal, petal, stamen, and carpel. A-function defines sepal identity. A+B-function determines petal identity. B+C-function leads to stamen identity. C-function alone leads to carpel identity, as well as flower meristem

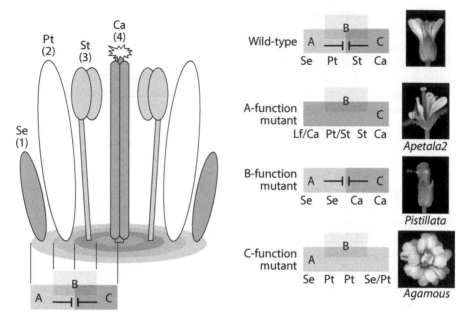

FIGURE 12.4: The ABC model of flower organ identity and the typical homeotic mutations that support it. The four flower organs, sepals, petals, stamen, and carpels, are thought to be determined by the combinatorial expression of three classes of transcription factor genes, called A-, B-, and C-class genes. Sepals are thought to be determined by the expression of A-class genes, petals by the co-expression of A- and B-class genes, and stamens by the expression of B- and C-class genes. Finally, carpels are determined by C-class gene expression. In addition, it is assumed that A- and C-class gene expression is locally exclusive. These principles can explain the mutant phenotypes listed on the right side of the figure. Reprinted from Causier et al., 2010, pp. 73–79, with permission from Elsevier.

identity and determinate axis growth (figure 12.4). For example, the *Arabidopsis* mutation of the *Apetala 2* gene (*AP2*), an A-class gene, results in a stamen into carpel as well as a petal into sepal transformations. Similarly, mutations of *AP3* and *Pistilata* (*PI*) genes, belonging to B-class genes, cause petal into sepal and stamen into carpal transformations. Finally, mutations of the C-class gene *Agamous* (*AG*) results in a stamen into petal as well as a carpel into sepal homeosis. To explain the latter phenotype, we must assume that A- and C-class genes inhibit each other, and that the loss of C-class activity results in an extension of the domain of A-class gene expression.

A small handful of genes were identified that perform the A-, B-, and C-functions initially in *Arabidopsis* and the snapdragon, *Antirrhinum*. Their names are summarized in the table below. The botanical community committed to using the priority rule for gene naming, unlike the situation in the zoological Hox gene literature. A relatively rational, easy to memorize nomenclature was adopted early on for Hox genes (Gehring 1994). In any case,

Table 12.1. Gene symbols for corresponding flower development genes in different species.

Gene class	Arabidopsis thaliana	Antirrhinum majus	Zea mays (maize)	Oryza sativa (rice)
Floral meristem genes	LFY	FLO		
	AP1	SQUA		
	CAL			
	UFO	FIM		
A-class genes	AP1	SQUA	AP1/SQUA-L	
	AP2	LIP1, LIP2		
B-class genes	AP3	DEF	SILKY	SUPERWOMAN
	PI	GLO		
C-class genes	SHP	PLE	ZAG1, ZMM2	MADS3
	AG	FAR		DL

the names of the flower genes are hard to remember because there are many different names for homologous genes across closely related species.

As illustrated in figure 12.4, the ABC model beautifully explains the variety of floral homeotic phenotypes and provides a guide for understanding flower development and evolution. On the mechanistic side, however, it soon became clear that this model described necessary but not sufficient factors for floral organ development. If, for example, *AG* would be sufficient to determine carpel development, then this gene's expression in vegetative leafs should transform them into carpels, at least partially (see Causier, Schwarz-Sommer et al. 2010 for references). But this is not what occurs. Similarly, the ectopic co-expression of *AP3* and *PI* in vegetative leafs also did not change leaf identity.

Hence, it was clear that the role of the ABC genes depended on the floral context. Only in the floral context are these genes able to affect floral organ identity. The first hint regarding the identities of these additional genes derived from an experiment in which the snapdragon B-class genes *DEF* and *GLO* were found to bind to DNA much more strongly in the presence of *SQUA* (*SQUAMOSA*). This was the first indication of what is now called the E-class function, which led to the ABCE model.

In *Arabidopsis*, the E-class genes are represented by three genes called *SEPALLATA*: *SEP1*, *SEP2*, and *SEP3*. In a triple loss of function mutation, the inner floral organs and the petals are replaced with sepals. These genes are expressed after floral meristem identity induction, but before expression of the ABC genes. Further functional analysis confirmed that these genes defined the floral context within which the ABC genes could define floral organ identity. Ectopic expression of *AP1*, *AP3*, and *PI* in conjunction with *SEP* in vegetative leafs results in petaloid transformations, thus proving sufficiency of this gene combination for determining petal identity (within the leaf context).

Interestingly, the expression of *AP1*, *AP3*, and *PI* also results in this phenotype as well as *SEP3* in conjunction with *AP3* and *PI*, which suggests that *SEP* and *AP1* can substitute for each other. Hence, *AP1* has both A and E functions. We will return to the question of A-function below.

The discovery of E-function genes and, together with ABC genes, their ability to transform vegetative leafs into flower organs also has implications for theories of serial homology. More than 200 years ago, during the age of speculative morphology, the German poet and naturalist Johann Wolfgang von Goethe proposed that all floral organs were transformed leafs (i.e., are serially homologous to vegetative leafs, as we would say today). The evidence for how ABC genes do their work (see below) further substantiates this view.

Floral organ identity genes exert their influence by modulating the leaf developmental program. Hence, the structure of developmental control reflects the history of plant organ evolution. If petals and carpels develop by modifying leaf developmental programs, then vegetative leafs have to be the evolutionary precursors of petals and carpels. Before we consider more evidence for this conclusion, we first need to consider how the ABCE functions work on the molecular level.

After the discovery of the ABC function genes, it was also found that their proteins bound to DNA as dimers, either as homodimers or heterodimers. Binding studies with E-function proteins, SEP3 in particular, showed that both C- and B-function proteins formed tetrameric complexes and that these complexes bound to two transcription factor binding sites, which resulted in a DNA loop (figure 12.5). For example, it was found that AP3 and PI DNA binding was not cooperative in the absence of SEP3, but was cooperative if expressed with SEP3. Obligatory formation of higher order protein complexes explains the necessity of each of the participating genes in determining floral organ identity and development. In short, the combinatorial floral organ code according to the ABCE model reads:

$$A + E \rightarrow \text{sepals}$$
$$A + B + E \rightarrow \text{petals}$$
$$B + C + E \rightarrow \text{stamens}$$
$$C + E \rightarrow \text{carpels}$$

SEP proteins do not only function in the floral context, but also seem to be bridge proteins for the formation of MADS box protein complexes in other developmental and physiological contexts. Thus, they are similar to the function of Fox-proteins in vertebrates.

There is one more modification to the classical ABC model that needs to be mentioned, which was proposed recently (Causier, Schwarz-Sommer et al. 2010). This has to do with the notion of homeotic A-function (i.e., that

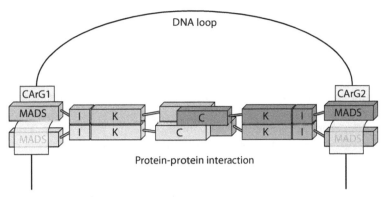

FIGURE 12.5: MADS box transcription factors form a tetrad comprising two dimers, each of which binds to neighboring MADS domain binding sites. These dimers then form a protein-protein interaction through their C-terminal domains. This tetrad is the active transcription factor complex that regulates target genes. The cooperativity and interdependency caused by these obligatory protein-protein interactions is critical for forming distinct organ identities and integrating the floral organs into a flower.

A-function genes *AP1* and *AP2* are sepal identity genes). The problem arose because the typical A-function homeotic phenotype in *Arabidopsis*, carpel, stamen, stamen, carpel, has never been observed in the snapdragon *Antirrhinum*. In fact, even in the classical ABC model, A-function genes are assumed to have two functions—namely, sepal organ identity and restricting the expression of the C-function gene *AG*. These and other difficulties are accommodated in what is called the (A)BCE model.

Briefly, the (A)BCE model removes the A-function genes from the class of organ identity genes and, instead, assigns them a role for determining floral meristem identity. One reason is A-class mutations sometimes, but not exclusively, result in first and second whorl homeotic transformations. Very often they result in bract-like phenotypes, and even additional flowers. A-class genes also have functions other than for floral development, while B- and C-function genes require the floral context to have an effect. A-class genes are expressed earlier than are B- and C-class genes and, together with *SEP* genes, establish floral meristem identity. In this way they establish sepals as the ground state of floral organ identity as well as the context for B- and C-functions.

This is a rough outline of what is known about the genetic mechanisms of flower organ development. We next want to see how this model of flower development enlightened the understanding of floral evolution and origination. We will first look into how well this model explains flower morphology variation within angiosperms and then move to the more difficult questions of flower and perianth origination.

The Developmental Genetic Architecture
of the Flower *Bauplan*

How do ABC organ identity genes bring about the development of their respective organs? How do flower developmental genes contribute to establishing an integrated organ, the flower? Here we want to reflect on these questions by first considering how the gene regulatory network downstream of the ABC genes looks and how it establishes specific phenotypes. Then we want to reflect on the nature of flower organ integration.

The first target gene of a floral identity gene identified was *SHATTER-PROOF* (*SHP*). *SHP* was found in a differential expression screen as a gene that is expressed in the carpel but is absent in *AG* mutation genotypes. *SHP* is a MADS box gene and, thus, is also a transcription factor (regulatory) gene, which plays a role in fruit dehiscence. Genome-wide screens for target genes of floral organ identity genes have revealed a number of interesting patterns (see Sablowski 2010). In terms of gene functions, target genes of B- and C-function genes differ with respect to the stage of development. The early responding genes are enriched for regulatory genes, like the MADS box gene *SHP*. Late responding genes are enriched for genes involved in cell differentiation and metabolism. This pattern also suggests that organ identity genes are involved in controlling all aspects of organ development.

Figure 12.6 gives a partial summary of known target genes of *AG*, the *Arabidopsis* C-function gene. This list includes transcription factors like SHP, SPL, and ATH1, as well as enzymes involved in hormone production (DAD1 is an enzyme involved in the biosynthesis of jasmonic acid) as well as general functions like organ growth (NUB and JAG). Follow-up analysis of these target genes revealed that some of the regulatory genes under *AG* control also control certain restricted parts of the phenotype of the carpel.

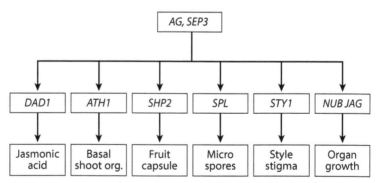

FIGURE 12.6: Target genes of the C-class gene *AG/SEP3*. Each target gene controls a quasi-modular part of carpel development so that carpel identity is realized through the co-expression of a number of developmental modules (for details, see Sablowski 2010).

For example, *STY1* controls the development of the distal carpel, the style, and stigma. It does this in a modular fashion, as style and stigma-like shapes could be induced on vegetative leafs by *STY1*. By comparison, ATH1 controls the development of basal shoot organs not only in the flower, but also in the vegetative parts of the plant. The latter also reinforces the point that floral organ identity genes act upon and modify the ancestral leaf development program, as mentioned above.

Overall, floral organ identity genes control the activities of modules that are responsible for different aspects of final organ phenotype, like the distal parts, the proximal parts, overall growth, and so on. Organ identity comprises and integrates a mosaic of features. The floral organ identity network collects all the necessary parts in one place in which they play their biological roles, but it is not a unitary developmental program. Outside of the carpel context, style and microspore development can experimentally be dissociated, whereas in the wild type they arise together because of their common control through the organ identity gene *AG*.

Molecular identification of genomic regions that bind organ identity proteins also suggests that the expression of organ identity genes is maintained by autoregulatory loops. This means that the predominant transcription factor protein binding targets are regulatory elements of the organ identity genes themselves. This is a structural motif that we already found in cell type identity networks (see chapter 8, Graf and Enver 2009, and Holmberg and Perlmann 2012), as well as hypothesized as a general feature of character identity networks (chapter 3 and Wagner 2007).

Even more importantly, the (A)BCE model of flower development aids in understanding the integration of the flower *Bauplan*. Recall that flowers evolved from the integration of ancestrally separate male and female reproductive structures and integrated vegetative leafs into a coherent structure with, for the most part, fairly stereotypical configurations of flower organs. Exceptions to the eu-angiosperm flower *Bauplan*, as summarized above, are extremely rare. Hence, the question arises: what causes the retention of this configuration?

It could be functional necessity—in other words, stabilizing natural selection for the morphological arrangement. This idea is certainly plausible, given the role of insect pollination in many angiosperm lifecycles. Insect pollination may work best when male and female reproductive organs are at the same site where insects harvest nectar and pollen. Nevertheless, unisexual flowers exist in angiosperms, including the most basal lineage, *Amborella*. However, in these cases, the flowers are functionally unisexual but are anatomically bisexual (i.e., male organs develop but are sterile).

The alternative hypothesis is that the integration of the angiosperm flower resulted from the developmental genetic architecture and, thus, represents what Ron Amundson calls a developmental type. For the flower *Bauplan*, this was first suggested by Melzer, Wang, and Theissen at the University of Jena in Germany (Melzer, Wang et al. 2010). The key observation is that flower organ identity genes, in particular B- and C-class genes, require the floral context to exert their role in flower organ development (see above).

Mechanistically, the floral organ context is provided by the expression of E-class genes, which determine floral meristem identity, and their proteins physically interact with the B- and C-class transcription factors (see above). These physical interactions also provide the mechanistic explanation for why floral organ development is limited to the floral context. The physical interaction between B- and C-class transcription factors and E-class transcription factors also contributes to the functional cooperativity between B- and C-class floral organ genes and the floral identity E-genes. Because B- and C-class floral organ genes require E-class genes for their function, and E-class genes determine floral identity, floral organ development is causally linked to flower development.

The morphological, phenotypic integration of the flower (i.e., its nature as a *Bauplan* or developmental type) is a consequence of the molecular interactions of floral organ transcription factors (B- and C-class) and floral meristem transcription factors (E-class). It is worth noting that this tight integration is not necessary from a mechanistic point of view. It could have been that E-genes simply had the role of activating the B- and C-class genes, but the B- and C-class transcription factors could regulate their target genes independently of the E-class transcription factors. If that would have been the case, then flower organs' integration into a flower *Bauplan* would not have been the consequence, as any other factor that activated some of the B- and C-class genes would induce flower organ development outside of the floral context.

The integration of flower organs into a flower type results specifically from the obligatory cooperativity among B- and C-transcription factors and the E-class transcription factors mediated through the formation of tetrameric transcription factor complexes. The tetrameric transcription factor complexes are, at least in this case, the very locus and root of the macroscopic integration of the developmental type, in this case the flower. This is a fundamental insight that raises hope for all of biology in that we may be close to a deep understanding of the vexing patterns and concepts of comparative anatomy, with names like body plan, developmental type, and homology, which have eluded experimental analysis and mechanistic understanding for centuries.

Flower Variation and Novel Flower Organ Identities

Once the basic *Bauplan* of the angiosperm flower was established, further evolutionary modifications could be categorized in two dimensions. For one there were phenotype modifications of the different flower organs, basically character state changes that respect the basic configuration of different flower organs and organ identities. This was by far the most common route of evolutionary modification.

However, there were also, not too infrequently, changes that could, at least on their face be understood as homeotic variations (i.e., variations in flower organ identity). The most frequent type of change had to do with the character of the two perianth organs: the sepals and the petals. For example, in tulips, lilies, orchids, and the buttercups, to name a few, the sepals assume a petal-like phenotype. The converse case includes, for example, the sorrels (e.g., *Rumex acetosa*) in which the petals assume a sepal-like appearance. Given the misexpression phenotypes caused by B-class flower organ identity genes, it is tempting to speculate that these apparent homeotic transformations might have been caused by changes in the spatial expression of these organ identity genes.

For example, the misexpression of B-class genes in the outer whorl in *Arabidopsis* results in petaloid sepals (Krizek and Meyerowitz 1996). Hence, one would expect that the petaloid sepals of tulips, lilies, and others would have been due to the outward shift in the expression domains of the B-class genes *DEF/GLO*, while in the case of sorrels one would expect an inward shift of the expression domains of these genes. This hypothesis has been called a shifting boundary or sliding boundary model (Bowman 1997; Albert, Gustafsson et al. 1998; Theissen, Becker et al. 2000; Kramer, Stilio et al. 2003). In a number of cases, this model has been supported by gene expression data.

In the case of the two whorls of sepal-like perianth organs of the sorrel, Ainsworth and collaborators (Ainsworth, Crossley et al. 1995) found that B-class genes were only expressed in the stamens, but not in the second whorl that normally would be petals. It is tempting to speculate that the sepaloid phenotype of the second whorl was caused during evolution by an inward shift of the outer expression domain boundary of B-class genes. Similarly, for the petaloid sepals of tulips and lilies (Kanno, Saeki et al. 2003), orchids (Tsai, Kuoh et al. 2004; Xu, Teo et al. 2006), and buttercups (Kramer, Stilio et al. 2003), it has been shown that they express B-class genes in accordance with the classical ABC model and the shifting boundary model.

In contrast, the highly modified perianth of grasses (e.g., rice, *Oryza sativa*) displays a fairly standard ABC type expression pattern. In grasses, there is one putative scale that has been hypothesized to be homologous to petals—the so-called lodicule that is surrounded by the palea and the lemma. As expected, the palea and lemma express only A-class genes while the lodicules express both A- and B-class genes (Whipple, Ciceri et al. 2004; Whipple, Zanis et al. 2007). Hence, there is reason to think that organ identities have been preserved in grasses and that the lodicule is, in fact, a petal, albeit a highly modified one.

These data show that the distinction between character identity and character state differences also is reflected in the genetic basis of development.

Character state changes can be very extensive, like the difference between a usual petal and the lodicule of a grass. Yet the character identity is maintained, as judged by the expression patterns of MADS box genes. However, some phenotypic differences can be ascribed to changes in character identity, as with the sepal and petals of lilies and other plants (see above).

Even though the sliding border model of flower variation provides a satisfying explanation of petaloid sepals and sepaloid petals, the exact phenotype of these flowers points to an interesting wrinkle in the concept of organ identity. Take, for example, the case of orchids. Both perianth whorls produce petaloid organs and, in fact, both express a combination of genes that, within the *Arabidopsis* model system, is necessary for petal identity (i.e., co-expression of A- and B-class genes; Tsai, Kuoh et al. 2004; Xu, Teo et al. 2006). Nevertheless, one is hard pressed to think of these two whorls as simple instances of reiterated petals because the outer and the inner tepals differ dramatically in their morphologies. We can only conclude that the outer and inner tepals, as sepals and petals are called when they are similar, still retain a high degree of developmental autonomy so that they can evolve radically different shapes.

Petaloid sepals that are very similar to the petals are actually quite rare, with *Clermontia* possibly being one of the few exceptions, and to my untrained eyes, the tepals of tulips, as one finds them in flower shops. This seems paradoxical because the A- and B-class genes are supposedly organ identity genes and their co-expression in the first whorl would predict a reiteration of complete petal identity in the outer whorl. A more detailed analysis of buttercup tepal development suggests a solution to this apparent paradox.

To my knowledge, this question was first addressed by Elena Kramer and her colleagues at Harvard University (Kramer, Stilio et al. 2003). They investigated the evolution and expression of B-class genes in nine species of Ranunculaceae, the buttercups and their relatives. This family of plants is rich in species with petaloid sepals, but are still strongly differentiated from the petals on the same flower. In addition, there is often a second type of petal that, although sterile, resembles stamens in many ways (Tamura, 1965). In any case, Ranunculaceae have two whorls of petaloid organs, but each of them is different in many other morphological traits, such as venation, the presence of nectarines, and shape and size. "[. . .], it appears that there exist two distinct petal identity programs functioning in many genera of this family" (Kramer, Stilio et al. 2003, p 8).

Kramer et al. found that the first whorl organs expressed B-class genes, as expected from their petaloid phenotype. Hence, it is unlikely that the petaloid phenotype of the sepals is a simple convergence of sepals (i.e., is not a character state of sepals), as they express petal identity genes. For example, in the columbine (*Aquilegia alpina*), the first three whorls express *Pistillata* (*PI*), an ortholog of one of the two *Arabidopsis* B-class genes. Expression in the second and third whorl would be expected by the ABC model, but *PI* expression in the first whorl accounts for the petaloid character of the "sepals" according to the shifting border model.

Contrary to the shifting border model, however, the expression of paralogs of the AP3 gene family shows differential expression between first and second whorl petals. In particular, *AP3-3* paralogs are consistently associated with the second whorl petals (Rasmussen, Kramer et al., 2009; cited after Sharma, Guo et al. 2011). These data strongly support the notion that in the buttercup family, there may not be conventional sepal-petal identities in their first two whorls, but instead they have two distinct petal identities. Each petal identity is more related to petals than to sepals, but they are distinct from each other through the evolution of paralog B-class genes with differential expression among the whorls. This model has been tested with functional experiments that showed that *AP3-3* was, in fact, necessary for second whorl petal development (Sharma, Guo et al. 2011).

The work on buttercup family perianth development shows that homeotic transformation of sepals played a role both in flower evolution and the origin of novel character identities (i.e., the first whorl petaloid organs). The two petaloid organ types of buttercup flowers are, thus, a novel flower organ character identity found only in the Ranunculacaeae.

The Origin of the Bisexual Flower Developmental Type

Most likely, the first step during the evolution of the angiosperm flower was the origin of a bisexual reproductive organ (i.e., the placement of both male and female reproductive organs on the same shoot axis). In gynmosperms, male and female organs are found on different individuals or different shoots, even when they occur on the same individual like in conifers. In terms of organ identity genes, orthologs of ABC type genes are not found in pterophytes, ferns, or mosses, but clear homologs of B- and C-class genes are found in many of the major gymnosperm groups, like conifers, gnetophytes, *Cycas,* and *Ginkgo* (reviewed in Melzer, Wang et al. 2010).

B- and C-class genes also play a role in reproductive organ development in gymnosperms, as their counterparts do in angiosperms, except that they play these roles on different shoot axes. C-class gene expression is associated with female cones, and C- and B-class genes together are associated with male cones. Some gymnosperm B- and C-class genes can even substitute for their angiosperm orthologs, which supports the view that their function was conserved since the most recent common ancestor of seed plants (Sundstrom and Engstrom 2002; Winter, Weiser et al. 2002). The molecular innovations must have been gene regulatory changes that resulted in the occurrence of C and B/C gene expressing cells on the same shoot axis (Baum and Hileman 2006).

Various scenarios have been proposed, from the out-of-male scenario for which a B-gene free zone evolved on a male cone, to the out-of-female scenario, which invokes a translocation of B-gene expression onto a female cone (for a comprehensive summary of these scenarios, see Specht and Bartlett 2009). None of these models has decisive support at this point and, in a way, their differences are not important in the present context. The important facts are that male and female organ identity has been determined by homologous genes in the ancestors of angiosperms and that the critical first step during flower origination comprised engineering the differential expressions of B- and C-class genes on the same shoot axis.

Another apomorphic feature of flower development is the function of E-class transcription factors. In angiosperms they participate in the formation of tetrameric protein complexes that are obligatory for the function of B- and C-class transcription factor function (see above). Phylogenetic analysis of the MIKC-type MADS box genes revealed that all main clades of floral MADS box genes already existed before the most recent common ancestor of seed plants (reviewed in Melzer, Wang et al. 2010; figure 12.7). This includes homologs to the E-class genes *SEP1* and *AGL6*, as well as A-class genes (i.e., homologs of *SQUAMOSA*). Intriguingly, no *SEP* or *SQUA* genes have been found in any gymnosperm, even though considerable effort has been expended to find them.

The most parsimonious conclusion is that *SEP* and *SQUA* must have been lost in the stem lineage of gymnosperms or were modified to an extent that sequence similarity was lost. Furthermore, it follows that *SEP* genes, which are indispensable for reproductive flower organ development in angiosperms, must have had a different or a substitutable function in the ancestors of seed plants. Even though *SEP* genes existed in the ancestral seed plants, the E-class function, as defined in the angiosperm context, is likely derived. But how did the angiosperm E-class function evolve and what was the ancestral situation with respect to protein-protein interactions among MADS domain proteins?

A study was made of protein-protein interactions in the MIKC-type MADS box transcription factors of the gymnosperm *Gnetum gnemon* (Wang, Melzer et al. 2010). This showed that in *Gnetum* also, many of the proteins have direct protein-protein interactions (see figure 12.8). The difference between the *Gnetum* and the *Arabidopsis* protein-protein interaction networks is that in *Gnetum*, the interactions are more evenly distributed, including the E-class protein AGL6, while in *Arabidopsis*, the majority is by interactions with SEP1 and SEP3, in particular interactions with AG-proteins (C-class).

In *Arabidopsis*, both SEP1 and SEP3 have six confirmed interaction partners, while for the AGL6 E-class proteins in *Gnetum*, each have at most three protein-protein interaction partners. Thus, a likely scenario is that in the ancestors of crown group seed plants, AGL6- and SEP1-like proteins were functionally largely equivalent, which allowed SEP1-like genes to be lost in gymnosperms. Another difference is that in *Gnetum*, a B-class protein called GGM2 directly interacts with the AG ortholog GGM3, as well as the B-sister proteins and AGL6. In *Arabidopsis*, B-class proteins do not

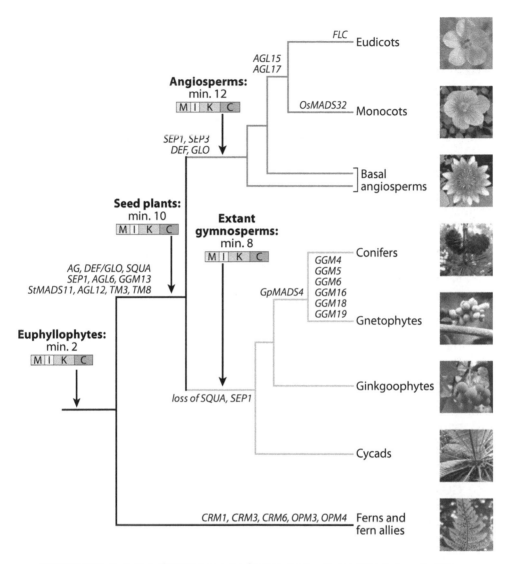

FIGURE 12.7: Evolution of MADS box gene family in tracheophytes. Note that most of the genes involved in floral organ development originated prior to the most recent common ancestor of spermatophytes (i.e., the most recent common ancestor of gymnosperms and angiosperms). Reprinted from Melzer et al., 2010, with permission from Elsevier.

exhibit protein-protein interactions with AG, ABS (B-sister), or AGL6. Thus, it seems that the loss of protein-protein interactions among B-class transcription factors could be a derived feature of angiosperm flower development.

Differences between the protein-protein interaction networks between *Arabidopsis* and *Gnetum* have consequences for the compositions of

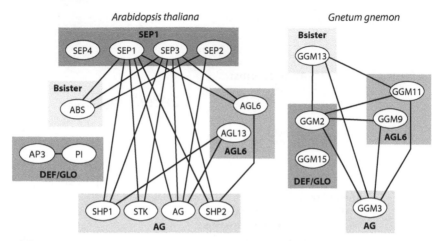

FIGURE 12.8: Protein-protein interaction networks of *Arabidopsis* and *Gnetum* (Wang et al., 2010). Note the preponderance of interactions between SEP (E-class) proteins and the other transcription factor proteins in *Arabidopsis*. In contrast, in *Gnetum* the interactions are more evenly distributed among the different MADS domain protein families with less reliance on SEP proteins.

transcription factor complexes that are active on cis-regulatory elements (figure 12.9). In *Arabidopsis*, the active transcription factor complex consists of two dimers. Each binds to a different binding site and then associates via DNA looping with another dimer bound to another site on the DNA to form a tetrameric complex. For example, carpel (female) organ development is directed by the association between two SEP:AG hetero-dimers: (SEP:AG):(SEP:AG),[3] while stamen (male) development is directed by a tetrad comprising two different heterodimers: (SEP:AG):(AP3:PI).

In contrast, the E-class proteins homologous to AGL6, GGM9 and GGM11, do not participate in the DNA-bound tetrads in *Gnetum*. Female cones form a tetrad of two homodimers: (GGM3:GGM3):(GGM3:GGM3) in which GGM3 is a C-class protein. Male cone development depends on a tetrad of two different homodimers (GGM3:GGM3):(GGM2:GGM2) in which GGM2 is a *Gnetum* B-class homeotic protein. Hence, it seems that the participation of E-class proteins in DNA-bound transcription factor complexes is a derived feature, even though the AGL6 proteins can form tetrads but seem to not participate in the active transcription factor complexes (Wang, Melzer et al. 2010). As mentioned above, in *Gnetum*, the B-class proteins can directly interact with AG and AGL6 proteins, while in *Arabidopsis* they do not.

Given the molecular data for protein-protein interactions in *Arabidopsis* and *Gnetum*, a likely scenario is that tetrameric complex formation among

[3] In this notation, a dimer bound by round brackets binds to DNA as a dimer and then interacts with another DNA-bound dimer.

transcription factors was an ancestral feature, but that in gymnosperm-grade ancestors of angiosperms, tetrad formation was more flexible with many different complexes fulfilling the same or similar functions. In the stem lineage of angiosperms, when the B-class proteins lost their ability to directly interact with other reproductive organ identity proteins, tetrad formation became dependent on SEP proteins (E-class; i.e., the situation we find in *Arabidopsis*). The obligatory dependence of B- and C-class transcription factor function on interactions with E-class proteins then resulted in the developmental integration of reproductive organs in the flower context. In angiosperms, reproductive organs develop only in the floral meristem context, while in gymnosperms this pattern is less strict. For example, ovule development is occasionally found on the vegetative leafs of the ginkgo, but that has never been described in angiosperms, even under experimental conditions.

The flexibility of the gymnosperm system results from the many possible interactions among transcription factor proteins, many of which are probably functionally redundant. By comparison, the rigidity of the angiosperm flower *Bauplan* results from the obligatory interaction among a number of transcription factors, all of which need to be co-expressed for reproductive organ development. This is unlikely to occur spontaneously outside of the flower context, as it would require multiple simultaneous changes.

To my knowledge, this is the first example for which we have a *mechanistically plausible explanation for the origin of a developmental type, the angiosperm flower*. The integration of the developmental type results from the obligatory formation of a multimeric transcription factor complex, which necessitates the co-expression of multiple transcription factor genes with little or no functional redundancy. This situation arose from an ancestral situation in which numerous protein-protein interactions were possible and with many means for functional compensation. The derived canalized state was due to a loss of possible physical protein-protein interactions among transcription factors, which made the activation of target genes dependent on a strictly determined set of factors.

The scenario above provides a plausible mechanistic scenario for the origin of the flower *Bauplan*. However, it does not explain why natural selection would favor the loss of functional redundancy and would select for the relatively rigid development in flowers. This seems counterintuitive given that, in general, robustness seems to be a good thing for organisms (Wagner 2005). A possible explanation, which is also interesting with respect to organ identity determination, has to do with the transition between alternative phenotypes. Recall that the key event during the origin of the bisexual flower was the co-expression of male and female organs in one shoot axis. In gymnosperms, this occurs either on different branches on the same individual (conifers) or on different individuals, as in *Gnetum*. It is also known that the difference between male and female organ identity is caused by quantitative differences in the expression levels of E-class transcription factors or some other factors.

A.

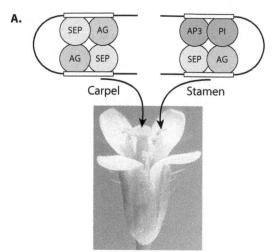

Carpel   Stamen

A. thaliana

B.

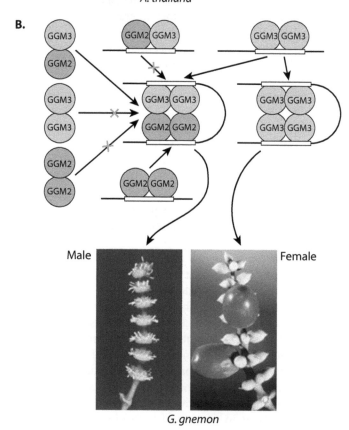

Male   Female

G. gnemon

FIGURE 12.9: Comparison of transcription factor complexes in *Arabidopsis* and *Gnetum*. In *Arabidopsis*, male and female organ formation requires inclusion of E-class transcription factors SEP, while in *Gnetum*, female organs require GGM3 homodimers and male organs require GGM3/GGM2 homodimers. The consequence of this difference is that in *Arabidopsis* and likely in other angiosperms, the formation of male and female organs can occur only in the floral context, while they are independent of a common context in the gymnosperm *Gnetum*. Note that in the *Gnetum*, B and C act alone, while in *Arabidopsis*, SEP has inserted itself as a necessary partner. Redrawn by permission of Wang, Melzer, and Theissen, 2010, *The Plant Journal* 64: 177. © 2010 The Authors. Journal compilation © 2010 Blackwell Publishing Ltd. Reproduced with permission of John Wiley and Sons, Inc.

In gymnosperms, E-class genes function in determining reproductive organ identity, as is the case in angiosperm flowers. In gymnosperms, this role is played by the *AGL6* gene and its relatives. For example, in the Norway spruce, the *AGL6* gene is called *DAL1* and is expressed in cone meristems. The *DAL1* expression level is correlated with male and female identity. High *DAL1* levels are associated with female cone development and lower *DAL1* expression with male cones. The same is true for the *Gnetum* ortholog to *DAL1*, called *GGM11*, with the only difference being that in *Gnetum* male and female cones are found on different individuals.

Hence, it is likely that the ancestral bisexual flower had a gradient of E-class proteins along its axis with high concentrations at the apex and lower concentrations toward the base. In the angiosperm flower this concentration gradient is expressed on a much smaller spatial scale, and it becomes critical to have a sharp transition from male to female organ identity to avoid intermediate phenotypes. Intermediate phenotypes are likely less functional and, thus, decrease fitness compared to an individual with no intermediate phenotypes.

One of the main factors driving the evolution of functional specialization is the occurrence of less functional intermediate phenotypes between two specialized organ identities (Rueffler, Hermisson et al. 2012). Theissen and colleagues suggested that obligatory tetrad formation had the role of increasing the cooperativity of transcription factor function, which resulted in a crisp transition between alternative gene regulatory states along the E-class gene gradient; one for male and the other for female organ development. In this scenario, the evolution of canalized alternative organ phenotypes and the evolution of *Bauplan* integration go hand in hand. Co-expression of alternative organ identities in close proximity leads to both and through the same molecular mechanism to the canalization of organ identities and the integration of the flower *Bauplan*.

Perianth Evolution and the Origin of Petals

In nearly all angiosperms, the reproductive flower organs are surrounded by whorls of sterile blade-like organs called the perianth. The perianth occurs in the most basal angiosperm lineage and, thus, is probably a shared derived character of all angiosperms. In the "typical" flower we have an inner whorl of usually shiny and colored petals and typically green sepals. The evolutionary history of "true" petals is complicated and controversial. Here I will argue that this uncertainty is likely due to the conflation of character identity and character state in the definition of petals.

Classically, petals are defined by a combination of positional and structural traits. Petals are said to be uniquely defined as perianth organs that surround the stamens, usually have a narrow base and one vascular trace, and are showy with a thin adaxial epidermis of conical cells (Hiepko 1965; Endress and Doyle 2009). These "true" petals are distinguished from petaloid organs, like the large bracts of the dogwood flowers and many more. The classical definition of a petal captures the intuitive meaning of a petal, but it becomes problematical as a tool for identifying evolutionary petal identity because of the amount of time and evolutionary modification that occurred since their origin.

With this long history of petal evolution, it is likely that the petal organ could have undergone vast amounts of phenotypic change that obscure some of the traits that were included in the original definition of the term. We encountered a similar situation in the relationship between the insect hind wing and the dipteran haltere. Certainly the haltere is not a wing in the usual sense of the word, but it is clearly derived from a hind wing, which implies that, technically, it IS a hind wing, albeit not one that is used to create lift. Similarly, the discussion on the evolution of the petal is confusing if we exclude the possibilities that some petals do not look like petals in the way that petals were originally perceived. An example is the lodicule of grasses (Whipple, Zanis et al. 2007) and many others. With this preamble, we can discuss the current comparative evidence regarding the origin and evolution of petals.

If we operationally define petals as differentiated inner perianth organs, then the phylogenetic distribution of petals defined in this way is spotty and suggests several independent petal origins (figure 12.10). Zanis and collaborators identified up to six independent origins for petals (Zanis, Soltis et al. 2003). This reading of the evidence is supported by the generally undifferentiated perianth of basal flowers (e.g., *Amborella*) and fossil flowers (e.g., *Archaefructus*; Sun, Ji et al. 2002). In contrast, the evolution of the B-class gene families suggests a more continuous occurrence of an inner perianth identity than the overt morphological differences.

As mentioned above, B-class genes are determinants of petal and stamen identities and were already involved in male organ (stamen) identity in the gymnosperms. One of the B-class genes in *Arabidopsis* is called *AP3* and is the product of a gene duplication at the base of the core-eudicots for which the two paralogs are called *euAP3* and *TM6*. *Arabidopsis AP3* belongs to the *euAP3* lineage. There is strong mounting evidence that *euAP3*'s role in petal development is conserved among core eudicots (reviewed in Irish 2009). Outside the core eudicots, the pre-ortholog of *euAP3* and *TM6* is called

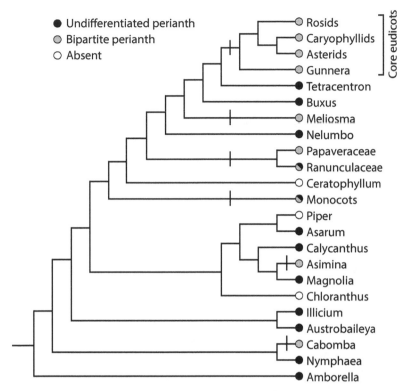

FIGURE 12.10: Taxonomic distribution of overt perianth differentiation. Using morphological criteria suggests that differentiated perianths and, by implication, petals independently evolved up to six times (possible evolutionary originations indicated by cross-marks). However, genetic data suggest that inner perianth identity might be much older than the petaloid phenotype. Inner perianth identity might be present during most of angiosperm phylogeny (see figure 12.12). (After Irish 2009.)

paleoAP3. This gene is involved in stamen development and is expressed and functional in inner perianth development. This has been shown in grasses in which *paleoAP3* is expressed and functions in lodicule development. Similarly, *paleoAP3* is also involved in inner perianth development in other monocots, like *Asparagus officinalis* (Park, Ishikawa et al. 2003). In the poppies (*Papaver somniferum*), which belong to a basal eudicot lineage, there are two recent paralogs of the *paleoAP3* gene, one of which functions in stamen development and one in inner perianth development.

Hence, it is possible that inner perianth organ identity is older than suggested by the phylogenetic mapping of differentiated perianth morphology (a detailed analysis of the comparative data is provided by Hileman and Irish 2009). Of course, the question is not whether there is *AP3* expression in the inner perianth whorl, but whether the petal ChIN is already active in

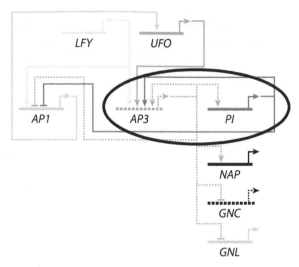

FIGURE 12.11: Gene regulatory network interactions of *AP1*, *AP3*, and *PI* involved in petal identity determination. Note that *AP3* and *PI* form a positive feedback loop (circled) in combination with a feed forward network targeting the downstream genes. (After Hileman and Irish, 2009.)

the basal angiosperm and eudicot flowers. Hence, we need to look for other members of the petal ChIN as found, for example, in *Arabidopsis*.

Functional analysis of *AP3* function in *Arabidopsis* suggests that there is a positive feedback between *AP3* and the other B-class gene *PI* (figure 12.11). This likely is the core network of petal identity, the petal ChIN, which receives inputs from upstream genes like *LFY* and *UFO*, and regulates target genes like *NAP*, *GNC*, and *GNL* (see summary in Hileman and Irish 2009). Hence, the question is how old is the co-expression and auto-co-regulation of *AP3* and *PI* in order to determine the age of the petal identity.

The patterns of *PI* expression and function are similar to those of *AP3*. There are numerous gene duplications that complicate the picture. However, functionally, *PI* semi-orthologs are generally involved in stamen and petal (inner perianth) development in both the core-eudicots and the basal eudicots and monocots (see summary in Irish 2009). Overall, the evidence from *PI* and *AP3* expression is that there is a phylogenetically old developmental domain from which both stamen and inner perianth organs develop. Thus, it is likely that petal identity (although not petal-like phenotype) is a relatively old developmental character identity within the angiosperms and, thus, the inner perianth of basal angiosperms and the petals of core-eudicots are likely the same character (i.e., homologous; Figure 12.12).

Agreement on this question is different from the question of how, when, and how often petal phenotype (a large shiny, colored or white sterile perianth organ) evolved. The latter question involves how the downstream gene regulatory network evolved and, thus, how the petal character state evolved.

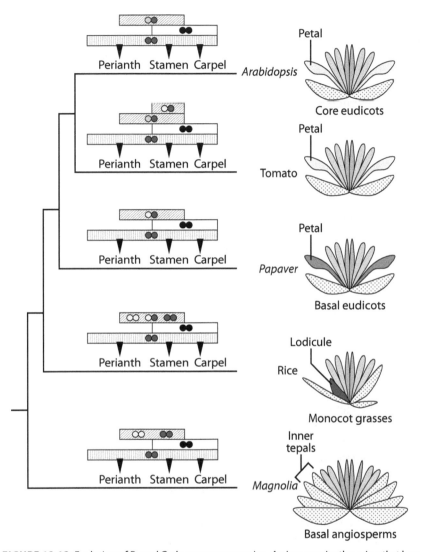

FIGURE 12.12: Evolution of B- and C-class gene expression. An inner perianth region that has more differentially regulated gene expression than the outer perianth is a trait that seems to be as old as the most recent common ancestor of Magnolia and Arabidopsis. Hence, it is possible that the petal, as a developmental evolutionary individuality, is much older than the morphologically overt differentiation of the perianth (figure 12.10). However, the expression of these B-class genes in the inner perianth whorl is not sufficient to conclude that the petal-like gene regulatory network was already established. More functional data are needed to resolve this question. (After Irish 2009, *J. Exp. Bot.* 60:2517-2527.)

The main caveat of my argument above is that I am not aware of any detailed experimental analysis regarding basal eudicots or basal angiosperm perianth development comparable to that for *Arabidopsis*. Furthermore, there is an alternative reading of the data summarized above, namely that of Vivian Irish at Yale University, which review I am referring to in this section. Vivian Irish interprets the conservation of *AP3* and *PI* gene expression in the inner perianth whorl as a case of "deep homology," as suggested by Shubin and collaborators (Irish 2009). Deep homology involves cases for which the gene regulatory networks are older than the characters themselves (Shubin, Tabin et al. 2009).

The answer to the question of whether petals display deep homology or real homology across many taxa outside the core-eudicots depends on whether the inner perianth whorls are developmentally and variationally individualized in the same way as the derived petals are, regardless of how petaloid they look. Probing the target gene sets of *AP3* and *PI* in basal eudicots and determining whether they are distinct from the gene expression profiles and upstream regulation of the outer perianth whorls will clarify this issue. If there are distinct regulatory hierarchies in inner and outer perianth whorls, then the idea that they are already distinct developmental character identities would be supported. If not, then *AP3* and *PI* expression would reflect different positional but not character identities, as suggested by Irish.

Genetics of Organ Identity: Challenges from Gene Duplication

As mentioned earlier in this book, any homology concept has to accommodate the fact that developmental pathways of homologous characters can vary considerably between species without affecting the identity of the characters concerned. The answer proposed here is that not all aspects of development are relevant for character identity; only the gene regulatory network that mediates between the positional information signals and the genes that form the character is relevant (i.e., the character identity network matters for character identity). However, this notion was challenged by Alejandra Jaramillo and Elena Kramer in a seminal paper in 2007 (Jaramillo and Kramer 2007). Thus, it is important to discuss their arguments, which is what I will do in this section.

Jaramillo and Kramer discussed two examples of gene families and their relationship to floral character identity. In one example, they explain the gene trees of the B-class gene *AP3*. *AP3* of *Arabidopsis* and its ortholog *DEF* in *Antirrhinum* are one product of a gene duplication at the base of eudicots. The paralog of *AP3/DEF* is *TM6*, and the outgroup lineage is called *paleoAP3*. *TM6* is not found in the genomes of the model species, but does occur in *Petunia hybrida* and is called *PhTM6*. Petunia also has an ortholog of *AP3*,

PhAP3. *PhTM6* is expressed during stamen and carpel development and functional evidence suggests that *PhTM6* is involved in stamen development, but not in petal development.

Jaramillo and Kramer provided two possible interpretations for these results. One was sub-functionalization where, in the Petunia lineage, one paralog, *TM6*, specialized in stamen development and the other, *AP3*, in petal development. Their other interpretation was that *TM6* retained the original function of *paleoAP3* in male organ development, while *PhAP3* acquired the derived petal function (see also Lamb and Irish 2003). A similar situation, although with a more recent duplication, was discussed for the other B-class gene, *PI/GLO*.

The second example concerns the C-class genes *AG* from *Arabidopsis* and *PLE* from *Antirrhinum*. It turns out that *AG* and *PLE* are actually paralogs. In *Arabidopsis*, the ortholog of *PLE* is *SHATTERPROOF* and the ortholog of *AG* in *Antirrhinum* is *FARINELLI* with various functions in seed development and carpel identity determination.

In both of these examples, character identity of a clearly homologous organ type (e.g., stamen) is controlled by different combinations or paralogs of closely related transcription factor genes. So, let us focus on the homology of stamens among *Arabidopsis*, *Antirrhinum*, and *Petunia* and ask whether the evidence presented by Jaramillo and Kramer constitutes developmental systems drift (DSD).

In all these species, stamen identity is determined by genes that are homologous and with likely continuity of the corresponding biochemical functions related to stamen identity. These genes are all either orthologs or paralogs that arose after the origin of the respective organ identity, which implies continuity of function of *AP* genes *sensu lato*. Because these genes are homologs, there is clear continuity of the function of genes from this family, and the diversification of gene lineages occurred after the origin of the corresponding character identity. Recall that B-class homologs already determine male organ identity in gymnosperms (see above).

In contrast, cases of developmental systems drift are those for which there is no continuity of developmental function between the genes that replace each other in different species. For example, the role of *ftz* in segment development in insects is derived with no prior function for *ftz* or a close paralog, relative to the age of the segments. Developmental systems drift is the replacement or recruitment of genes that are unrelated to the genes that ancestrally were involved in the development of the character.

Clearly, these examples from flower organ development do not qualify as DSD. It is equally clear that gene duplications can give rise to paralogs and that one or the other paralog may continue to perform the function of the pre-duplication character identity gene. It is also clear that duplicated genes from a ChIN can result in the evolution of novel character identities, as in the case of the first whorl petals in buttercups (see above and Kramer, Stilio et al. 2003). However, gene duplication does not disrupt the correspondence between the continuity of character identity at the phenotype level and the continuity of the developmental function of the corresponding character identity network.

Another issue discussed by Jaramillo and Kramer was the fact that the homologs of the B-class genes from *Arabidopsis* and *Antirhinum* in grasses were active in the non-petaloid lodicules of grasses. They also reported that grass orthologs of B-class genes were functionally equivalent in *Arabidopsis* but, in their native context, regulated different target genes corresponding to the different morphological phenotypes of lodicules and petals. Note that lodicules and petals are both second whorl organs and other authors have concluded that they are, in fact, homologs. The morphological differences between lodicules and petals are character state differences.

Consistent with the role of character identity genes, they are agnostic with respect to the downstream target gene and the ultimate phenotype of the character. Remember that homology is the *same organs regardless of form and function* and, thus, the retention of the B-class gene functions in lodicules is exactly what is expected of a gene network that is responsible for character identity rather than character states. The ChIN remains continuously associated with character identity despite major changes in the character state and corresponding differences in the set of target genes.

Summary and Conclusions

The extensive literature on flower development and evolution harbors many important lessons for understanding the origin and nature of character identity. Some of these are similar to those learned from other systems, as for example the importance of distinguishing between character identities and character states or phenotypes. Others have not been learned from the examples discussed in previous chapters. The deepest insight, in my opinion, is the relationship between the evolution of a developmental type, the flower, and its constituent characters and character identities. Many parts of what eventually became the flower already existed before the angiosperm radiation, like male and female reproductive organs and many of their molecular regulators.

What is unique about flowers is how the deployment of the various parts became developmentally integrated into the flower. The developmental integration of the flower is implemented through the dependence on the expression and function of flower meristem genes (the E-class genes) of the organ identity genes (see above). This integration evolved through a derived dependency of the function of the organ identity transcription factors on protein-protein interactions with the flower meristem regulators. *The notion of a developmental type, the angiosperm flower, can be grounded in the obligatory cooperativity among these transcription factor proteins.*

Another important implication of this field of research is the realization that, after gene duplications, additional organ identities can evolve. This was illustrated with the case of the buttercup perianth where Elena Kramer and collaborators showed that the outer petaloid organs were likely a derived second set of developmentally individualized petals.

Most generally, the research on flower development and flower organ evolution exemplifies the claim that the concepts of homology and character identity can be grounded in a deeper understanding of how gene regulatory networks control development and influence evolutionary diversification.

13

Lessons and Challenges

The twelve preceding chapters provided a long argument for the reality of a class of biological entities that have a hard time finding their place in modern biology—that is, a place in a theory of evolution based on genetics and population biology. These entities are things like cell types, homologs, body plans, and so on—that is, entities that Amundson calls developmental types (Amundson 2005). At this point it is impossible to know whether this argument will be successful or not because success in science is historically post hoc. Rather, the question at hand is whether these ideas can productively interact with research programs in labs and field stations.

For that reason, the second part of this book was a confrontation of those ideas with examples that already have empirical data to see whether they are contradicted by known facts about certain well-studied organ systems, like limbs, skin appendages, and flowers. The question also was whether "reading" the existing empirical knowledge from the perspective of a theory of character identity will generate new insights. I believe that it does and I want to summarize some of the lessons that I learned by reviewing the literature on these paradigms of devo-evo research. I also want to reflect on the challenges inherent in this perspective of developmental evolution. First, though, it might be useful to provide a condensed version of the main arguments developed in previous pages.

What Are the Core Claims of This Model of Homology?

There are two points that underlie the view of homology developed here. One is conceptual and the the other is a mechanistic hypothesis. The *conceptual point is that it is essential to distinguish between character identity and character states.* Only then can we truly understand the meaning of the original definition of homology as the "same organ in different species in every variety of shape and function." Character identity, as with the notion of gene identity, is not tied to phenotypic similarity or sequence similarity. A body part can be a wing, even if it does not look like a prototypical wing, as is the case with the elytra of beetles or the halteres of flies. Character identity is a hypothesis about the developmental and variational individuality of a body part and its evolutionary derivation from the same individualized body part in the most recent common ancestor. A homolog is a body part that is free to acquire any shape or function as demanded by natural selection. The lodicule of grasses is a petal not because it looks like a prototypical petal, but because it is derived from the same individualized flower organ in a common ancestor of grasses and other flowering plants.

The mechanistic hypothesis is that character identity has a genetic basis distinct from that of character states. More specifically, the genetic machinery that underlies character identity has the role of individualizing a part of the body during development and enables the execution of a distinct developmental program. It turns out that candidate gene regulatory networks for character identity determination, called Character Identity Networks (ChINs), tend to include positive feedback loops among cooperatively acting transcription factor genes. The conservation and cohesion of character identity are likely caused by obligatory protein-protein interactions among those transcription factors involved in character identity determination. In addition, it seems to be the case that ChINs are more conserved than are other aspects of character development. The continuity of ChINs is tied to the continuity of character identity or homology. Examples and experimental details were discussed in previous chapters.

These two conceptual and mechanistic hypotheses allow us to amend the classical notions of homology.

Special homology: two body parts in two species are homologous if

- They are individualized body parts in both species and
- Correspond to the same individualized body part in the most recent common ancestor of the two species.

Note that this formulation only differs from the classical Darwinian reinterpretation of Owen's homology concept by requiring that homology statements should be restricted to individualized body parts. There is also continuity of character states, although this should be called something different given that it reflects a different biological reality. The term to use for corresponding character states is synapomorphy.

If the idea holds up that homology (i.e., continuity of character identity) is tied to the continuity of ChINs, then the notion of homology can be generalized to situations where historical continuity of the phenotypic character does not apply. There can be dormant ChINs and, thus, there can be continuity of potential character identity despite discontinuity of characters' expression.

> *General homology*: two body parts are homologous if their character identity is based on the continuity of the underlying character identity networks.

Note that this concept does not require historical continuity of the manifest morphological characters, as is suggested by the notions of atavism, cryptic homology, and similar phenomena (Müller 1989; Hall 1994). If the ancestral ChIN survives periods of non-use, re-activation of the ChIN can lead to what is called an atavism, the expression of a character that was present in an ancestral lineage but was lost in the immediate ancestors. The notion of serial homology can also be assimilated, a concept that is incompatible with a strictly historical notion of homology.

> *Serial homology*: two body parts of the same organism are serially homologous if they result from the repeated activation of the same character identity network.

I think it desirable that the manifest fact that the same body part can be repeated within the same body and the fact that the same body part can occur in different bodies should have a common interpretation.

The statements about special, general, and serial homology above read like definitions, but they should not be interpreted as such (Brigandt and Love 2012). In reality they are hypotheses to guide the interpretation of biological diversity and disparity (i.e., hypothetical schemes for analyzing the history of biological diversity). Their scientific value depends on their ability to productively interact with research programs in evolutionary and developmental biology. Some examples of the change in perspective that they can engender are briefly summarized below.

Characters Are Real But Historically Limited

Cell types and anatomical entities, like limbs, fins, petals, and others (many but clearly not all terms found in anatomy books correspond to real entities) are as real as cells, genes, and chromosomes. They are individualized from other such entities through highly cooperative gene regulatory networks that allow differential gene expression across small spatial scales. Individualized characters display historical continuity and are grounded in experimentally analyzable developmental and genetic mechanisms. With our present state of knowledge, this is most clearly seen for cell types, which correspond to self-maintaining gene regulation network states (core networks; see Graf and Enver 2009; Holmberg and Perlmann 2012; but see below). There are some hints that even multicellular organs are associated with highly conserved gene regulation networks (see chapters 3 and 12), but the evidence is not as clear as is the case for cell types (see below).

This strong affirmation of the reality and importance of cell types and anatomical entities in the conceptual inventory of biology comes, however, with a caveat. Characters and character identities have a limited historical lifetime; they originate (i.e., are a novelty at some point during evolutionary history) and may disappear in some lineages, like the mammalian tooth types did in toothed whales. Hence, when one asks the question, "What is the homolog of character X that is found in species A in another species, say B?", there might not be a positive answer. This question could be biologically meaningless, like asking "What is the homolog of the tetrapod hind limb in *Trichoplax*?" (see figure 8.1 for reference).

Clearly, tetrapods and *Trichoplax* share a common evolutionary history because they are both metazoans. However, it is also intuitively obvious that to ask for the limb homolog in *Trichoplax* is meaningless. Yet in less extreme cases, like comparisons between fish fins and tetrapod limbs, it requires some effort to decide whether homology questions are biologically meaningful and these may be undecided given our present state of knowledge; well-informed scientists can disagree on these questions, as in the case of whether there is a part in the fish fins that corresponds to the autopodium in a tetrapod limb (see chapter 10). Characters and character identities have definite historical beginnings and we need to decide first whether there is any basis for proposing a homology hypothesis.

A similar limitation applies to homology statements in development. Body parts form during development and it follows that, during some early stages of development, most if not all the body parts of the adult just do not exist. This seems self-evident, but it is not always clear when a character starts to exist. For example, this is controversial when it comes to digit development and digit identity. Are definite digit identities already present once one can see cellular condensations or the first sign of cartilage formation, or is there a stage at which there are cartilages but no digit identity has yet been determined? There are no universal answers to questions of this sort because they depend on what we think mediates character identity. They also depend on the available empirical knowledge to determine when during development these character identity–determining mechanisms kick in.

Another caveat is that finding some elements of the gene regulation network underlying the identity of a character does not prove that the character is really there. Unless things are clear and obvious, as among closely related organisms, what one needs to show is that the expressed genes are part of a network that conveys developmental individuality to the body part that corresponds to that of the other character we want to identify (Schlosser 2005; Wagner 2007; Woltering and Duboule 2010). This might be a steep requirement, and I certainly do not advocate that we need to meet this criterion for each and every homology hypothesis.

However, we definitely should not allow ourselves to jump to non-intuitive homology hypotheses based on fragmentary evidence. Certainly, gene expression is evidence for or against homology, even though fragmentary, but it cannot become a substitute for critically analyzing what the expression of a particular gene means in terms of the underlying mechanisms of character identity. In other words, knowing gene expression patterns can help us to assess the question of homology, but without information on what the genes are doing and whether they convey developmental individuality, there is no strong case to be made for homology (Woltering and Duboule 2010).

Homology Is Not Hierarchical

A common idea is that homology is hierarchical. This idea is plausible if one considers the fact that any complex organ can contain anatomical entities as parts that, in turn, have character identity and, thus, can be the subject of homology statements (figure 13.1). An example is the fact the "tetrapod limb" has at least three larger units within it, the proximodistal segments (see chapter 10 for details). These examples show that anatomical entities can overlap, but overlapping is not the same as being hierarchical. This became clear duirng the discussion of the origin of tetrapod limbs (chapter 10), which, in one school of thinking, includes the origin of a new developmental module: the autopod. Yet the autopod originated in already existing and individualized "paramorph" body parts, the pectoral and the pelvic fins. Hence, the homology of the autopod cross-cuts the pre-existing individualities of the anterior and posterior paired appendages (see figure 10.11).

We have to emancipate our thinking from the hierarchical concept of how the bodies of organisms are structured. In fact, hierarchy never made sense if one thinks of the body as an integrated system that contains differentiated parts. Integration is primary, differentiation is secondary, and how the body becomes parceled into modular units does not follow a hierarchical logic. Each unit at one level is connected to others through overlapping systems. While it is convenient to think about the tetrapod limb in terms of its skeletal components, in reality the segments of the limb are connected by overlapping

Hierarchy

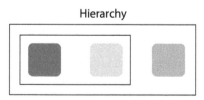

FIGURE 13.1: Example of a nested hierarchy. Hierarchy assumes that identities or units are nested within each other. For example, one can think of the large box as representing a limb and the smaller boxes inside as the parts of the limbs, like certain bones and muscles. This model is a popular way of thinking about multilevel homology in complex organisms. However, this model is misleading, as explained in figure 13.2.

muscular and connective tissue links. At the very least, we should be very clear on what we mean when we say that homology is hierarchical.

Cross-cutting identities become the main feature when we talk about homology and diversification of skin appendages as in the discussion of feathers in chapter 9. There I proposed the term *"character swarm"* for characters that remain in an intermediary state between individuality and serial repetition. Feathers are sufficiently individualized to have evolved functional specializations, but still have a tendency to share derived character states (i.e., are variationally not fully individualized).

There is another sense in which the phrase "homology is hierarchical" is used. The paradigm example is to say that the bird wing is not homologous to the bat wing "as wing" but homologous "as forelimb." While there is some useful truth in this observation, it is also intrinsically confusing. In fact, bird wings and bat wings are homologous, but what is not synapomorphic is their character state as a wing.

This confusion derives from the fact that *sensu lato* one can speak of the homology of characters and the "homology" (shared derived state) of character states. What makes a bird forelimb a wing is a character state of the homolog forelimb (or, to be precise, the anterior paired gnathostome appendage). To subsume the distinction between character identity and character states under the idea of a hierarchy, or conceptually equivalent homology statements, misses the point.

The Quasi-Cartesian Model of Character Identity

The considerations above, based on the discussions of the origin of tetrapod limb identity (chapter 10) and skin appendages (chapter 9), suggest an alternative model of how character identities are arranged in the body: the quasi-Cartesian model (figure 13.2A). This model is based on the realization that anatomical entities come at different levels and are arranged along different axes. For example, they appear arranged along the anterior-posterior body axis, like arthropod segments, and have different identities along this axis, like head, thorax, and abdomen. However, there are other entities, such as appendages, which are arranged along the dorsal-ventral axis and are different along this axis, such as wings versus legs. Wings and legs are also different along the body axis, which leads to cross-cutting identities like the fore wing and the hind wing and the first and second leg in an insect.

This model is called quasi-Cartesian because cross-cutting dimensions are formally similar to what mathematicians call a Cartesian product of two sets. Say we have the set of identities along the body axes and the set of kinds of appendages. Then the Cartesian product of these two sets is all the pairwise combinations of body axis and appendage identities. In reality, the systems of character identities are not fully Cartesian because certain combinations

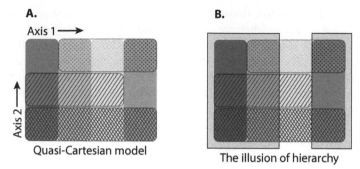

A.

Axis 1 ⟶

Axis 2 ⟶

Quasi-Cartesian model

B.

The illusion of hierarchy

FIGURE 13.2: Quasi-Cartesian model of character identity. A) Character identities are arranged along different axes. For example, axis 1 could be the anterior-posterior body axis and the boxes could represent different segmental identities. The second axis could be the dorsoventral axis of body segments in which different identities are realized, as for example limbs, wings, and others. A singular anatomical unit, say the hindwing of an insect, is then the intersection of body axis identity (say metathorax) and wing identity equals hind wing. Consequently, anatomical identities do not form a nested hierarchy, as suggested by the model in figure 13.1. Rather, they form a system of cross-cutting identities that determine specific character identities at their intersections. This structure is quasi-Cartesian, rather than fully Cartesian, because not all possible cross-cutting identities need to be present, as suggested by the fact that, for example, the stippled horizontal boxes do not overlap with all the vertical boxes from axis 1. B) The illusion of hierarchy. If we screen out some of the boxes from axis 1 and focus on one box and the character identities that cross-cut it, it looks like a nested hierarchy. But this is only illusory as it results from ignoring the distribution of character identities in other parts that belong to axis 1 (refer to part A).

of character identities may not exist, such as wing-like dorsal appendages on the head of the fly. I propose that thinking of character identities and how they are arranged in a body plan in terms of a quasi- or incomplete Cartesian structure is more productive and reflective of the actual structure of multicellular organisms.

The quasi-Cartesian model also explains how the illusion of a hierarchical arrangement of character identities can arise (figure 13.2B). In figure 13.2, for example, if we focus on one character along axis 1, say the body axis, and screen out all the other entities along that axis, then the arrangement of characters within that entity appears hierarchical (figure 13.2B). However, it is also clear that this impression is produced only by ignoring the broader context within which the focal character is placed.

Character Individuality and Gene Regulatory Network Cooperativity

Characters and cell types were originally recognized because of their seemingly qualitative differences. Switching from one identity to another is, to the degree that we understand the mechanisms, subscribed by two self-reinforcing molecular mechanisms and perhaps more. There are positive feedback networks that reinforce and maintain differences in gene regulatory network activity between different cells and tissues. This is best documented in the case of cell type identities (see chapter 8). The other mechanism is functional cooperativity between transcription factor proteins, for which target gene activation or repression depends on the presence of several transcription factors. This is nicely demonstrated in the case of floral organ identity determination (see chapter 12). It is likely that other mechanisms, including miRNAs and long noncoding RNAs, will be part of this machinery that allows the realization and active maintenance of alternative gene regulatory network states.

Open Questions and Difficulties

At various points in this chapter, as well as throughout this book, cell types are held up as an example of how positive feedback among regulatory genes underlies distinct identities. The idea is that each cell identity is caused by a core gene regulatory network. However, there are three challenges that arise based on recent research, as well as from matters of principle. The first challenge is that there is increasing evidence that the gene regulatory network state of a cell is governed not by one core network, but by a mosaic of densely interconnected network modules each of which, in isolation, might look like a core network. One example of this principle has been found in the hematopoietic lineage in which alternative cell types can be understood as different combinations of gene regulatory network modules, each of which is relatively autonomous in its expression and activation (Novershtern, Subramanian et al. 2011).

If this principle holds up to more experimental scrutiny, then the notion of cell identity as a unitary state will have to be revised to allow for a mosaic of character identities. It also might allow for the evolution of novel cell identities by evolutionary recombinations of gene regulatory network modules for generating new combinations of network states. In this case, the unit of historical continuity will shift from the cell to the gene regulatory network module level. These are things that will need to be developed as more information about the causes of cell type identity becomes available.

The other way that cell identity as a paradigm of character identity is limited is that cell identity is an intracellular state and, thus, gene regulatory networks acting within a cell are (perhaps) sufficient to understand cell identity. For cells, everything from the outside is a signal and not part of cell identity itself.[1] For

---

[1] This is likely a simplification, as there is evidence that cell fate decisions are possible only in sufficiently large populations of cells and, thus, cell-cell communication, probably by paracrine mechanisms, might be important duirng the establishment and maintenance of cell type identity. This might be a feature of cells that exist in solid tissues rather than being relevant for blood cells.

multicellular organs, however, cell-cell signaling must be an integral part of the character identity network. Hence, the separation between positional information and a character identity network is not as clear as in the case of cells. In fact, in some paradigms, conserved gene regulatory networks associated with character identity are mostly known as networks of signaling centers.

For example, the AER and the ZPA as essential and conserved parts of vertebrate appendage development are integrated and interdependent signaling centers and are also integrated with dorsoventral signaling. Another example is the segment polarity network, which is quite rigidly associated with insect segmentation. How to understand the limits of positional information and character identity determination in the case of multicellular organs is an unsolved question.

Another challenge is understanding the relationship between cell identity and organ identity. Multiple organs contain the same cell identities, like forelimbs and hind limbs that are both made of chondrocytes, osteocytes, muscle cells, fibroblasts, and skin cells. The cells that make up a limb can also change when skeletal elements either ossify or remain cartilages, even though the character identity does not change. How are cell and organ identity related or independent?

One possible solution is to think of character identities in terms of the quasi-Cartesian model discussed above (figure 13.2A). Cell types and organs could be thought of as different axes of character identities, and specific cell types in certain organs are combinations of cell and organ identities. For example, there are differences between different functional cell types, like fibroblasts and myocytes, but there are also differences in fibroblast identity between limbs, say, because muscle anatomy is determined by the identity of the fibroblasts (Christ and Brand-Saberi 2002).

At this point, there still is much room for speculation regarding these issues. What would be even better are data that could channel our reflections on these issues into empirically meaningful directions.

Population, Tree, and Homology Thinking

The conceptual outline of evolutionary biology has been characterized by different conceptual styles of thinking about the natural world. Most influential was Ernst Mayr's characterization of the New Synthesis Biology as reflected by the phrase "population thinking" (Mayr 1959). This meant the idea that the unit of evolutionary change is the population and that evolutionary change is best understood in terms of the dynamics of genetic variation within populations (Sober 1980). More recently and with the advent of rigorous comparative methods, a second style of thinking has been proclaimed: "tree thinking" (O'Hara 1997). Its claim is that biological diversity is best understood in the context of the network of phylogenetic relationships with all the associated methods based on phylogenetic evidence.

Now, with a renewed focus on the developmental side of evolutionary change, a new type of "thinking" has been proposed—namely, "homology thinking" (Ereshefsky 2007; Ereshefsky 2012). Homology thinking is a form of historical explanation that draws upon information about the phylogenetic origins and developmental underpinnings of body parts that evolve in lineages. It is a complement to functionalist explanations that seek to explain organismal diversity from the point of view of functional need. Ereshefsky (2012) argues that homology explanations are richer and more nuanced than are functional explanations, because they cover more and deeper biological details about the objects of study.

Whatever the ultimate utility of this concept may be, it is interesting to consider the possibility that an approach to evolutionary biology based on a mechanistic understanding of homology and, thus, character identity, may allow us to capture dimensions of organismal biology that have been obscured by other styles of thinking because they have screened out the organism and its structure from consideration.

REFERENCES

Abouheif, E. (1997). "Developmental genetics and homology: a hierarchical approach." *Trends Ecol. Evol.* **12**(10): 405–408.

Abouheif, E. (1999). "Establishing homology criteria for regulatory gene networks: prospects and challenges," in Homology, eds. G. R. Bock and G. Cardew. Chichester, England, J Wiley & Sons: 207–221.

Aghajanova, L., J. A. Horcajadas, et al. (2010). "The bone marrow–derived human mesenchymal stem cell: potential progenitor of the endometrial stromal fibroblast." *Biol. Reprod.* **82**(6): 1076–1087.

Ahn, D. and R. K. Ho (2008). "Tri-phasic expression of posterior Hox genes during development of pectoral fins in zebrafish: implications for the evolution of vertebrate paired appendages." *Dev. Biol.* **322**(1): 220–233.

Ainsworth, C., S. Crossley, et al. (1995). "Male and female flowers of the dioecious plant sorrel show different patterns of MADS box gene expression." *Plant Cell* 7(10): 1583–1598.

Alba, M. M. and R. Guigo (2004). "Comparative analysis of amino acid repeats in rodents and humans." *Genome Res.* **14**(4): 549–554.

Alberch, P. (1983). "Development and evolution: embryos, genes, and evolution." *Science* **221**(4607): 257–258.

Alberch, P. (1983). "Morphological variation in the neotropical salamander genus *Bolitoglossa.*" *Evolution* **37**: 906–919.

Alberch, P. and J. Alberch (1981). "Heterochronic mechanisms of morphological diversification and evolutionary change in the neotropical salamander Bolitoglossa occidentalis (Amphibia: Plethodontidae)." *J. Morphol.* **167**: 249–264.

Alberch, P. and E. A. Gale (1983). "Size dependence during the development of the amphibian foot. Colchicine-induced digital loss and reduction." *J. Embryol. Exp. Morph.* **76**: 177–197.

Alberch, P. and E. A. Gale (1985). "A developmental analysis of an evolutionary trend: Digital reduction in Amphibians." *Evolution* **39**: 8–23.

Albert, V. A., M. H. G. Gustafsson, et al. (1998). "Ontogenetic systematics, molecular developmental genetics, and the angiosperm petal," in Molecular Systematics of Plants II, eds. D. Soltis, P. Soltis and J. J. Doyle. New York, Klewer Academic Publishers: 349–374.

Alibardi, L., L. Dalla Valle, et al. (2009). "Evolution of hard proteins in the sauropsid integument in relation to the cornification of skin derivatives in amniotes." *J. Anat.* **214**(4): 560–586.

Alibardi, L. and M. B. Thompson (2000). "Scale morphogenesis and ultrastructure of dermis during embryonic development in the alligator (Alligator mississipiensis, Crocodilia, Reptilia)." *Acta Zoologica* **81**: 325–338.

Allan, D. W., D. Park, et al. (2005). "Regulators acting in combinatorial codes also act independently in single differntiating neurons." *Neuron* **45**: 689–700.

Allen, C. E., P. Beldade, et al. (2008). "Differences in the selection response of serially repeated color pattern characters: Standing variation, development, and evolution." *BMC Evolutionary Biology* **8**: **94–107**.

Alvarez-Buylla, E. R., B. A. Ambrose, et al. (2010). "B-function expression in the flower center underlies the homeotic phenotype of Lacandonia schismatica (Triuridaceae)." *Plant Cell* **22**(11): 3543–3559.

Amemiya, C. T., T. P. Powers, et al. (2010). "Complete HOX cluster characterization of the coelacanth provides further evidence for slow evolution of its genome." *Proc. Natl. Acad. Sci., USA* **107**(8): 3622–3627.

Amores, A., A. Force, et al. (1998). "Zebrafish hox clusters and vertebrate evolution." *Science* **282**: 1711–1714.

Amundson, R. (2005). The Changing Role of the Embryo in Evolutionary Thought: Roots of Evo-Devo. Cambridge, Cambridge University Press.

Amundson, R. (1989). "The trials and tribulations of selectionist explanations," in Issues in Evolutionary Epistemology, eds. K. Hahlweg and C. A. Hooker. New York, State University of New York Press: 413–432.

Amundson, R. (2007). "Richard Owen and animal form" in On the Nature of Limbs. A Discourse: Richard Owen, ed. B. K. Hall. Chicago and London, University of Chicago Press: XV–LI.

Anan, K., N. Yoshida, et al. (2007). "Morphological change caused by loss of the taxon-specific polyalanine tract in Hoxd-13." *Mol. Biol. Evol.* **24**(1): 281–287.

Ancel, L. W. and W. Fontana (2000). "Plasticity, evolvability and modularity in RNA." *J. Exp. Zool. B. Mol. Dev. Evol.* **288**: 242–283.

Andrews, J., D. Garcia-Estefania, et al. (2000). "OVO transcription factors function antagonistically in the Drosophila female germline." *Development* **127**(4): 881–892.

Angelini, D. R. and T. C. Kaufman (2004). "Functional analyses in the hemipteran Oncopeltus fasciatus reveal conserved and derived aspects of appendage patterning in insects." *Dev. Biol.* **271**(2): 306–321.

Arai, K. M., R. Takahashi, et al. (1986). "The primary structure of feather keratins from duck (Anas platythynchos) and pigeon (Columbia livia)." *Biochim. Biophys. Acta.* **873**: 6–12.

Arendt, D. (2003). "Evolution of eyes and photoreceptor cell-types." *Int. J. Dev. Biol.* **47**: 563–571.

Arendt, D. (2005). "Genes and homology in nervous system evolution: Comparing gene functions, expression patterns, and cell-type molecular fingerprints." *Theor. Biosci.* **124**: 185–197.

Arendt, D. (2008). "The evolution of cell-types in animals: Emerging principles from molecular studies." *Nat. Rev. Genet.* **9**: 868–882.

Arendt, D., H. Hausen, et al. (2009). "The 'division of labour' model of eye evolution." *Philos. Trans. Royal Soc. Lond. B. Biol. Sci.* **364**(1531): 2809–2817.

Arendt, D., K. Tessmar-Raible, et al. (2004). "Cilliary photoreceptors with a vertebrate-type opsin in an invertebrate brain." *Science* **306**: 869–871.

Arnoult, L., K. F. Y. Su, et al. (2013). "Emergence and diversification of a Drosophila pigmentation pattern through evolution of a gene regulatory module." *Science* **339**: 1423–1426.

Assis, L. C. S. and I. Brigandt (2009). "Homology: homeostatic property cluster kinds in systematics and evolution." *Evol. Biol.* **36**: 248–255.

Ax, P. (1984). Das phylogenetische System. Stuttgart, Gustav Fischer Verlag

Balfour, F. M. (1881). "On the development of the skeleton of the paired fins of Elasmobranchii considered in relation to its bearings on the nature of the limbs of the vertebrates." *Proc. Zool. Soc. London* **49**(3): 656–670.

Balinsky, B. I. (1974). "Supernumerary limb induction in the Anura." *J. Exp. Zool.* **188**(2): 195–201.

Barbareschi, M., L. Pecciarini, et al. (2001). "p63, a p53 homologue, is a selective nuclear marker of myoepithelial cells of the human breast." *Am. J. Surg. Pathol.* **25**(8): 1054–1060.

Bateson, W. (1894). Materials for the Study of Variation Treated with Especial Regard to Discontinuity in the Origin of Species. London, Macmillan.

Baum, D. A. and L. C. Hileman (2006). "A developmental genetic model for the origin of the flower," in Flowering and its Manipulation, ed. C. Ainsworth. Sheffield, UK, Blackwell: 3–27.

Bebenek, I. G., R. D. Gates, et al. (2004). "Sine oculis in basal Metazoa." *Dev. Genes Evol.* **214**: 342–351.

Bechtel, W. (2010). "The downs and ups of mechanistic research: circadian rhythm research as an exemplar." *Erkenntinis* **73**: 313–328.

Beldade, P., K. Koops, et al. (2002). "Developmental constraints versus flexibility in morphological evolution." *Nature* **416**: 844–847.

Bellairs, A. d'A. and C. Gans (1983). "A reinterpretation of the amphisbaenian orbitosphenoid." *Nature* **302**: 243–244.

Bennett, K. (2009). "The complete development of the deep-sea cidaroid urchin *Cidaris blakei* (Agassiz, 1878) with an emphasis on the hyaline layer." Masters Thesis, University of Oregon.

Bertalanffy, L. v. (1936). "Wesen und Geschichte des Homologiebegriffs." *Unsere Welt* **28**: 161–168.

Bever, G. S., J. A. Gauthier, et al. (2011). "Finding the frame shift: digit loss, developmental variability, and the origin of the avian hand." *Evol. Dev.* **13**: 269–279.

Bird, A. P. (1993). "Functions for DNA methylation in vertebrates." *C.S.H. Symp. Quant. Biol.* **58**: 281–285.

Blackburn, D. G. (1991). "Evolutionary origins of the mammary gland." *Mammal Rev.* **21**(2): 81–96.

Blanco, M. J. and P. Alberch (1992). "Caenogenesis, development variability, and evolution in the carpus and tarsus of the marbled newt *Triturus marmoratus*." *Evolution* **46**: 677–687.

Blanco, M. J., B. Y. Misof, et al. (1998). "Heterochronic differences of Hoxa-11 expression in Xenopus fore- and hind limb development: evidence for a lower limb identity of the anuran ankle bones." *Dev. Genes Evol.* **208**: 175–187.

Blumer, M. J. F. (1995). "The ciliary photoreceptor in the teleplanic veliger larvae of Smaragdia sp. and Strombus sp. (Mollusca, Gastropoda)." *Zoomorphology* **115**: 73–81.

Bock, W. J. (1974). "Philosophical foundations of classical evolutionary classification." *Syst. Zool.* **22**: 375–392.

Bock, W. J. (1977). "Foundations and methods of evolutionary classification," in Major Patterns in Vertebrate Evolution, eds. M. K. Hecht, P. C. Goody and B. M. Hecht. New York, Plenum Press: 851–895.

Böhme, W., Ed. (1981). Handbuch der Reptilien und Amphibien Europas. Wiesbaden, Germany, Akademische Verlagsgesellschaft.

Boisvert, C. A., E. Mark-Kurik, et al. (2008). "The pectoral fin of Panderichthys and the origin of digits." *Nature* **456**(7222): 636–638.

Bolker, J. A. and R. A. Raff (1996). "Developmental genetics and traditional homology." *BioEssays* **18**: 489–494.

Bongard, J. (2002). "Evolving modular genetic regulatory networks." 2002 Congress on Evolutionary Computation. Honolulu, HI, IEEE: 1872–1877.

Boras, K. and P. A. Hamel (2002). "Alx4 Binding to LEF-1 Regulates N-CAM Promoter Activity." *J. Biol. Chem.* **277**(2): 1120–1127.

Bostwick, K. S. and R. O. Prum (2003). "High-speed video analysis of wing-snapping in two manakin clades (Pipridae: Aves)." *J. Exp. Biol.* **206**(Pt 20): 3693–3706.

Botchkarev, V. A. and R. Paus (2003). "Molecular biology of hair morphogenesis: development and cycling." *J. Exp. Zool. B. Mol. Dev. Evol.* **298**(1): 164–180.

Bourque, G., B. Leong, et al. (2008). "Evolution of the mammalian transcription factor binding repertoire via transposable elements." *Genome Res.* **18**(11): 1752–1762.

Bowman, J. L. (1997). "Evolutionary conservation of angiosperm flower development at the molecular and genetic levels." *J. Biosci.* **22**: 515–527.

Boyd, R. (1991). "Realism, anti-foundationalism, and the enthusiasm of natural kinds." *Philosophical Studies* **61**: 127–148.

Boyer, L. A., T. I. Lee, et al. (2005). "Core transcriptional regulatory circuitry in human embryonic stem cells." *Cell* **122**(6): 947–956.

Bradley, L., D. Wainstock, et al. (1996). "Positive and negative signals modulate formation of the Xenopus cement gland." *Development* **122**(9): 2739–50.

Brakefield, P. M., J. Gates, et al. (1996). "Development, plasticity and evolution of butterfly eyespot patterns." *Nature* **384**: 236–242.

Brandley, M. C., J. P. Huelsenbeck, et al. (2008). "Rates and patterns in the evolution of snake-like body form in squamate reptiles: evidence for repeated re-evolution of lost digits and long-term persistence of intermediate body forms." *Evolution* **62**(8): 2042–2064.

Brandon, R. (1982). "The levels of selection." *PSA* 1982 **1**: 315–333.

Brandon, R. N. (1999). "The units of selection revisited: the modules of selection." *Biol. Phil.* **14**: 167–180.

Brandstätter, R., B. Misof, et al. (1990). "Micro-anatomy of the pectoral fin in blennies (Blenniini, Blennioidea, Telesotei)." *J. Fish Biol.* **37**: 729–743.

Braus, H. (1906). "Die Entwicklung der Form der Extremitäten und des Extremitätenskeletts," in Handbuch der vergleichenden und experimentellen Entwicklungslehre der Wirbeltiere, ed. O. Hertwig. Jena, Gustav Fisher. **3, part 2:** 167–338.

Brayer, K. J., V. J. Lynch, et al. (2011). "Evolution of a derived protein-protein interaction between HoxA11 and Foxo1a in mammals caused by changes in intramolecular regulation." *Proc. Natl. Acad. Sci., USA* **108**(32): E414–420.

Breitling, R. and J. K. Gerber (2000). "Origin of the paired domain." *Dev. Genes Evol.* **210**(12): 644–650.

Bresch, C., G. Müller, et al. (1968). "Genes involved in meiosis and sporulation of a yeast." *Molecular and General Genetics (MGG)* **102**(4): 301–306.

Bretscher, A. (1949). "Die Hinterbeinentwicklung von Xenopus laevis Daud und ihre Beeinflussung durch Colchicin." *Revue Suisse Zool.* **56**: 33–96.

Bridges, C. B. 1935. "Salivary chromosome maps." *J. Hered.* **26**: 60–64.

Bridgham, J. T., E. A. Ortlund, et al. (2009). "An epistatic ratchet constrains the direction of glucocorticoid receptor evolution." *Nature* **461**(7263): 515–519.

Brigandt, I. (2007). "Typology now: homology and developmental constraints explain evolvability." *Biology & Philosophy* **22**: 709–725.

Brigandt, I. and A. C. Love (2010). "Evolutionary novelty and the evo-devo synthesis: field notes." *Evol. Biol.* **37**: 93–99.

Brigandt, I. and A. C. Love (2012). "Conceptualizing evolutionary novelty: moving beyond definitional debates." *J. Exp. Zool. B. Mol. Dev. Evol.* **318**: 417–427.

Brigandt, I. and A. C. Love (2012). "Reductionism in biology," in The Stanford Encyclopedia of Philosophy, ed. E. N. Zalta. Palo Alto, Stanford Encyclopedia of Philosophy.

Britten, R. J. and E. H. Davidson (1971). "Repetitive and non-repetitive DNA sequences and a speculation on the origins of evolutionary novelty." *Quart. Rev. Biol.* **46**(2): 111–138.

Brunetti, C. R., J. E. Selegue, et al. (2001). "The generation and diversification of butterfly eyespot color patterns." *Curr. Biol.* **11**: 1578–1585.

Brush, A. H. (1965). "The structure and pigmentation of the feather tips of the scaled cuckoo (Lepidogrammus cumingi)." *The Auk* **82**(2): 155–160.

Brush, A. H. (1967). "Additional Observations on the structure of unusual feather tips." *The Wilson Bulletin* **79**(3): 322–327.

Brush, A. H. (1996). "On the origin of feathers." *J. Evol. Biol.* **9**: 131–142.

Brush, A. H. (2000). "Evolving a protofeather and feather diversity." *Am. Zool.* **40**: 631–639.

Brush, A. H. and K. Allen (1963). "Astaxanthin in the Cedar Waxwing." *Science* **142** (3588): 47–48.

Brush, A. H. and J. Wyld (1982). "Molecular organization of avian epidermal structures." *Comp. Biochem. Physiol.* **73B**: 313–325.

Bürger, R. (2000). The Mathematical Theory of Selection, Recombination, and Mutation. New York, Wiley.

Burke, A. C. (1989). "Development of the turtle carapace: implications for the evolution of a novel bauplan." *J. Morph.* **199**: 363–378.

Burke, A. C. and P. Alberch (1985). "The development and homology of the chelonian carpus and tarsus." *J. Morph.* **186**: 119–131.

Burke, A. C. and A. Feduccia (1997). "Developmental patterns and the identification of homologies in the avian hand." *Science* **278**: 666–668.

Buss, L. W. (1987). The Evolution of Individuality. New York, Columbia University Press.

Butler, A. B. and W. M. Saidel (2000). "Defining sameness: historical, biological, and generative homology." *Bioessays* **22**: 846–853.

Carranza, S., E. N. Arnold, et al. (2008). "Radiation, multiple dispersal and parallelism in the skinks, Chalcides and Sphenops (Squamata: Scincidae), with comments on Scincus and Scincopus and the age of the Sahara Desert." *Mol. Phylogenet. Evol.* **46**(3): 1071–1094.

Carroll, S. B. (1995). "Homeotic genes and the evolution of arthropods and chordates." *Nature* **376**: 479–485.

Carroll, S. B. (2008). "Evo-Devo and an expanding evolutionary synthesis: a genetic theory of morphological evolution." *Cell* **134**: 25–34.

Carroll, S. B., J. K. Grenier, et al. (2001). From DNA to Diversity. Malden, MA, Blackwell Science.

Causier, B., Z. Schwarz-Sommer, et al. (2010). "Floral organ identity: 20 years of ABCs." *Semin. Cell Dev. Biol.* **21**(1): 73–79.

Cebra-Thomas, J., F. Tan, et al. (2005). "How the turtle forms its shell: a paracrine hypothesis of carapace formation." *J. Exp. Zool. B. Mol. Dev. Evol.* **304**(6): 558–569.

Cerny, A. C., G. Bucher, et al. (2005). "Breakdown of abdominal patterning in the Tribolium Kruppel mutant jaws." *Development* **132**(24): 5353–5363.

Chan, Y. F., M. E. Marks, et al. (2010). "Adaptive evolution of pelvic reduction in sticklebacks by recurrent deletion of a Pitx1 enhancer." *Science* **327**(5963): 302–305.

Charlesworth, B. (1990). "The evolutionary genetics of adaptations," in Evolutionary Innovations, ed. M. H. Nitecki. Chicago, IL, University of Chicago Press: 47–70.

Chatterjee, S. (1998). "Counting the fingers of birds and dinosaurs." *Science* **280**: 355a.

Chen, X., H. Xu, et al. (2008). "Integration of external signaling pathways with the core transcriptional network in embryonic stem cells." *Cell* **133**(6): 1106–1117.

Chen, Y., V. Knezevic, et al. (2004). "Direct interaction with Hoxd proteins reverses Gli3-repressor function to promote digit formation downstream of Shh." *Development* **131**: 2339–2347.

Cheverud, J. M. (2001). "The genetic architecture of pleiotropic relations and differential epistasis," in The Character Concept in Evolutionary Biology, ed. G. P. Wagner. San Diego, Academic Press: 411–433.

Cheverud, J. M., T. H. Ehrich, et al. (2004). "Pleiotropic effects on mandibular morphology II: differential epistasis and genetic variation in morphological integration." *J. Exp. Zool. B. Mol. Dev. Evol* **302B**: 424–435.

Chiang, C., Y. Litingtung, et al. (2001). "Manifestation of the limb prepattern: limb development in the absence of Sonic Hedgehog function." *Dev. Biol.* **236**: 421–435.

Chiu, C.-H., D. Nonaka, et al. (2000). "Evolution of Hoxa-11 in lineages phylogenetically positioned along the fin-limb transition." *Mol. Phylogen. Evol.* **17**: 305–316.

Chow, R. L., C. R. Altmann, et al. (1999). "Pax6 induced ectopic eyes in a vertebrate." *Development* **126**: 4213–4222.

Christ, B. and B. Brand-Saberi (2002). "Limb muscle development." *Int. J. Dev. Biol.* **46**(7): 905–914.

Christensen, R. N., M. Weinstein, et al. (2002). "Expression of fibroblast growth factors 4, 8, and 10 in limbs, flanks, and blastemas of Ambystoma." *Dev. Dyn.* **223**(2): 193–203.

Chudinov, P. K. (1965). "New facts about the fauna of the upper Permian of the USSR." *J. Geol.* **73**: 117–130.

Chuong, C. M. and D. G. Homberger (2003). "Development and evolution of the amniote integument: current landscape and future horizon." *J. Exp. Zool. B. Mol. Dev. Evol.* **298**(1): 1–11.

Chuong, C.-M., R. B. Widelitz, et al. (1996). "Early events during avian skin appendage regeneration: dependence on epithelial-mesenchymal interaction and order of molecular reappearance." *J. Invest. Dermatol.* **107**: 639–646.

Clack, J. A. (2002). Gaining Ground: The Origin and Evolution of Tetrapods. Bloomington, IN, Indiana University Press.

Clark, C. J. and T. J. Feo (2008). "The Anna's hummingbird chirps with its tail: a new mechanism of sonation in birds." *Proc. Biol. Sci.* **275**(1637): 955–962.

Coates, M. I. (2003). "The evolution of paired fins." *Theory Biosci.* **122**: 266–287.

Coates, M. I., J. E. Jeffery, et al. (2002). "Fins to limbs: what the fossils say." *Evolution & Development* **4**(5): 390–401.

Coates, M. I., M. Ruta, et al. (2000). "Early tetrapod evolution." *TREE* **15**: 327–328.

Coen, E. S. and E. M. Meyerowitz (1991). "The war of the whorls: genetic interactions controlling flower development." *Nature* **353**(6339): 31–37.

Coen, E., T. Strachan, et al. (1982). "Dynamics of concerted evolution of ribosomal DNA and histone gene families in the melanogaster species subgroup of Drosophila." *J. Mol. Biol.* **158**(1): 17–35.

Cohn, M. J., J. C. Izpisua-Belmonte, et al. (1995). "Fibroblast growth factors induce additional limb development from the flank of chick embryos." *Cell* **80**(5): 739–746.

Cope, E. D. (1885). "On the evolution of Vertebrata, progressive and retrogressive." *Am. Nat.* **19**: 140–148, 234–247, 341–353.

Cotney, J., J. Leng, et al. (2012). "Chromatin state signatures associated with tissue-specific gene expression and enhancer activity in the embryonic limb." *Genome Res.* **22**(6): 1069–1080.

Cracraft, J. (1978). "Science, philosophy and systematics." *Syst. Zool.* **27**: 213–216.

Craver, C. F. (2005). "Beyond reduction: mechanisms, multifield intergration and the unity of neuroscience." *Studies in the History and Philosophy of Biological and Biomedical Sciences* **36**: 373–395.

Crawford, A. J. and D. B. Wake (1998). "Phylogenetic and evolutionary perspectives on an enigmatic organ: the balancer of larval caudate amphibians." *Zoology* **101**: 107–123.

Crombach, A. and P. Hogeweg (2008). "Evolution of evolvability in gene regulatory networks." *PLoS Comput. Biol.* **4**(7): e1000112.

Crow, K. D., C. T. Amemiya, et al. (2009). "Hypermutability of HoxA13A and functional divergence from its paralog are associated with the origin of a novel developmental feature in zebrafish and related taxa (Cypriniformes)." *Evolution* **63**(6): 1574–1592.

Crow, K. D., P. F. Stadler, et al. (2006). "The "fish specific" Hox cluster duplication is coincidental with the origin of teleosts." *Mol. Biol. Evol.* **23**: 121–136.

Currie, P. J. and P. Chen (2001). "Anatomy of Sinosauropteryx prima from Liaoning, northeastern China." *Can. J. Earth Sci.* **38**(1): 1705–1727.

Czerny, T., G. Halder, et al. (1999). "Twin of eyeless, a second Pax-6 gene of Drosophila, acts upstream of eyeless in the control of eye development." *Mol. Cell.* **3**(3): 297–307.

Daeschler, E. B., N. H. Shubin, et al. (2006). "A Devonian tetrapod-like fish and the evolution of the tetrapod body plan." *Nature* **440**(7085): 757–763.

Dahn, R. D., M. C. Davis, et al. (2007). "Sonic hedgehog function in chondrichthyan fins and the evolution of appendage patterning." *Nature* **445**: 311–314.

Dahn, R. D. and J. F. Fallon (2000). "Interdigital regulation of digit identity and homeotic transformation by modulated BMP signaling." *Science* **289**: 438–441.

Dalla Valle, L., A. Nardi, et al. (2009). "Forty keratin-associated beta-proteins (beta-keratins) form the hard layers of scales, claws, and adhesive pads in the green anole lizard, Anolis carolinensis." *J. Exp. Zool. B. Mol. Dev. Evol.* **314**(1): 11–32.

Damen, W. G. (2007). "Evolutionary conservation and divergence of the segmentation process in arthropods." *Dev. Dyn.* **236**(6): 1379–1391.

Damen, W. G., M. Hausdorf, et al. (1998). "A conserved mode of head segmentation in arthropods revealed by the expression pattern of Hox genes in a spider." *Proc. Natl. Acad. Sci. USA* **95**(18): 10665–10670.

Davidson, E. (2001). Genomic Regulatory Systems. San Diego, Academic Press.

Davidson, E. H. (2006). The Regulatory Genome: Gene Regulatory Networks in Development and Evolution. Amsterdam, Academic Press.

Davidson, E. H. and D. H. Erwin (2006). "Gene regulatory networks and the evolution of animal body plans." *Science* **311**: 796–800.

Davies, H. R. (1889). "Die Entwicklung der Feder und ihre Beziehung zu anderen Integumentbildungen." *Morphologisches Jahrbuch* **15**: 560–645.

Davis, A. P., D. P. Witte, et al. (1995). "Absence of radius and ulna in mice lacking hoxa-11 and hoxd-11." *Nature* **375**: 791–795.

Davis, G. K. and N. H. Patel (2002). "Short, long, and beyond: molecular and embryological approaches to insect segmentation." *Annu. Rev. Entomol.* **47**: 669–699.

Davis, M. C., R. D. Dahn, et al. (2007). "An autopodial-like pattern of Hox expression in fins of a basal actinopterygian fish." *Nature* **447**: 473–476.

Davis, R. L., H. Weintraub, et al. (1987). "Expression of a single transfected cDNA converts fibroblasts to myoblasts." *Cell* **51**(6): 987–1000.

Dawkins, R. (1976). The Selfish Gene. New York, Oxford University Press.

Dawkins, R. (1978). "Replicator selection and the extended phenotype." *Z. Tierpsychol.* **47**: 61–67.

Dawkins, R. (1989). "The evolution of evolvability," in Artificial Life: The Proceedings of an Interdisciplinary Workshop on the Synthesis and Simulation of Living Systems, ed. C. Langton. Santa Fe, NM, Addison Wesley: 202–220.

De Beer, G. R. (1971). Homology, an Unsolved Problem. London, Oxford University Press.

Dececchi, T. A. and H. C. E. Larsson (2009). "Patristic evolutionary rates suggest a punctuated pattern in forelimb evolution before and after the origin of birds." *Paleobiology* **35**(1): 1–12.

DePinna, M. (1991). "Concepts and tests of homology in the cladistic paradigm." *Cladistics* **7**: 367–394.

Deutsch, J. (2005). "Hox and wings." *Bioessays* **27**: 673–675.

Dhainaut-Courtois, N. (1965). "Sur la présence d'un organe photorécepteur dans le cerveau de Nereis pelagica L. (Annélide Polychète)." *Acad. Sci. Paris* **261**: 1085–1088.

Dhouailly, D. (2009). "A new scenario for the evolutionary origin of hair, feather, and avian scales." *J. Anat.* **214**(4): 587–606.

Dhouailly, D. and T. T. Sun (1989). "The mammalian tongue filiform papillae: a theoretical model for primitive hairs," in Trends in Human Hair Growth and Alopecia Research, eds. D. v. Neste, J.-M. LaChapelle and J. L. Anoine. Boston, Kluwer Acad. Publ.: 29–34.

DiFerdinando, A., R. Calabretta, et al. (2001). "Evolving modular architectures for neural networks" in Proceedings of the Sixth Neural Computation and Psychology Workshop: Evolution, Learning, and Development, eds. R. French and J. Sougné. London, Springer Verlag: 253–262.

Donoghue, M. J. (1992). "Homology," in Keywords in Evolutionary Biology, eds. E. F. Keller and E. A. Lloyd. Cambridge, MA, and London, Harvard University Press: 171–179.

Donoghue, M. J., R. H. Ree, et al. (1998). "Phylogeny and the evolution of flower symmetry in the Asteridae." *Trends Plant Sci.* **3**: 311–317.

Donoghue, M. J. and R. H. Ree (2000). "Homoplasy and developmental constraint: a model and an example from plants." *Am. Zool.* **40**: 759–769.

Donoghue, P. C. J. and M. A. Purnell (2005). "Gene duplication, extinction, and vertebrate evolution." *Trends Ecol. Evol.* **20**: 312–319.

Donner, A. L. and R. L. Maas (2004). "Conservation and non-conservation of genetic pathways in eye specification." *Int. J. Dev. Biol.* **48**: 743–753.

Doyle, J. A. and P. K. Endress (2000). "Morphological phylogenetic analysis of basal angiosperms." *Int. J. Plant Sci.* **161** (Supplement): S121–S153.

Draghi, J. and G. P. Wagner (2008). "Evolution of evolvability in a developmental model." *Evolution* **62**(2): 301–315.

Draghi, J. and G. P. Wagner (2009). "The evolutionary dynamics of evolvability in a gene network model." *J. Evol. Biol.* **22**(3): 599–611.

Drossopoulou, G., K. E. Lewis, et al. (2000). "A model for anterioposterior patterning of the vertebrate limb based on sequential long- and short-range Shh signaling and Bmp signaling." *Development* **127**: 1337–1348.

Duellman, W. E. and L. Trueb (1986). Biology of Amphibians. New York, McGraw-Hill, Inc.

Dunker, N., M. H. Wake, et al. (2000). "Embryonic and larval development in the caecilian Ichthyophis kohtaoensis (Amphibia, gymnophiona): a staging table." *J. Morphol.* **243**(1): 3–34.

Eakin, R. M. and J. A. Westfall (1964). "Fine structure of the eye of a chaetognath." *J. Cell Biol.* **21**: 115–132.

Eberhard, W. G. (2001). "Multiple origins of a major novelty: moveable abdominal lobes in male sepsid flies (Diptera: [S]epsidae), and the question of developmental constraints." *Evolution and Development* **3**: 206–222.

Eckhart, L., L. D. Valle, et al. (2008). "Identification of reptilian genes encoding hair keratin-like proteins suggests a new scenario for the evolutionary origin of hair." *Proc. Natl. Acad. Sci., USA* **105**(47): 18419–18423.

Eldredge, N. (1979). "Alternative approaches to evolutionary theory." *Bull. Carnegie Mus. Nat. Hist.* **13**: 7–19.

Elias, H. and S. Bortner (1957). "On the phylogeny of hair." *Amer. Mus. Novitates* **1820**: 1–15.

Emera, D., C. Casola, et al. (2011). "Convergent evolution of endometrial prolactin expression in primates, mice, and elephants through the independent recruitment of transposable elements." *Mol. Biol. Evol.* **29**(1): 239–247.

Emera, D., R. Romero, et al. (2011). "The evolution of menstruation: a new model for genetic assimilation: explaining molecular origins of maternal responses to fetal invasiveness." *Bioessays* **34**(1): 26–35.

Emera, D. and G. P. Wagner (2012). "Transposable element recruitments in the mammalian placenta: impacts and mechanisms." *Brief Funct. Genomics* **11**(4): 267–276.

Endress, P. K. (1993). "Eupomatiaceae," in The Families and Genera of Vascular Plants, eds. K. Kubitzki, J. G. Rohwer and V. Bittrich. Berlin, Springer. **2**: 296–298.

Endress, P. K. (1999). "Symmetry in flowers: diversity and evolution." *Int. J. Plant Sci.* **160**: S3–S23.

Endress, P. K. and J. A. Doyle (2009). "Reconstructing the ancestral angiosperm flower and its initial specializations." *Am. J. Bot.* **96**(1): 22–66.

Ereshefsky, M. (2007). "Psychological categories as homologies: lessons from ethology." *Biol. Phil.* **22**(5): 659–674.

Ereshefsky, M. (2012). "Homology thinking." *Biol. Phil.* **27**(3): 382–400.

Eriksson, B. J., N. N. Tait, et al. (2009). "The involvement of engrailed and wingless during segmentation in the onychophoran *Euperipatoides kanangrensis* (Peripatopsidae: Onychophora) (Reid 1996)." *Dev. Genes Evol.* **219**(5): 249–264.

Evans, T. M. and J. M. Marcus (2006). "A simulation study of the genetic regulatory hierarchy for butterfly eyespot focus determination." *Evol. Dev.* **8**(3): 273–283.

Fabrezi, M. (2001). "A survey of prepollex and prehallux variation in anuran limbs." *Zool. J. Linn. Soc.* **131**: 227–248.

Fabrezi, M., V. Abdala, et al. (2007). "Developmental basis of limb homology in lizards." *Anat. Rec. (Hoboken)* **290**(7): 900–912.

Falconer, D. S. and T. F. C. Mackay (1996). Introduction to Quantitative Genetics. Essex, England, Longman.

Falk, R. (1986). "What is a gene?" *History and Philosophy of the Life Sciences* **17**: 133–173.

Farzana, L. and S. J. Brown (2008). "Hedgehog signaling pathway function conserved in Tribolium segmentation." *Dev. Genes Evol.* **218**(3–4): 181–192.

Feduccia, A. (1996). The Origin and Evolution of Birds. New Haven, Yale University Press.

Feduccia, A. (1999). "1,2,3 = 2,3,4: accommodating the cladogram." *Proc. Natl. Acad. Sci., USA* **96**: 4740–4742.

Feduccia, A. (2001). "Digit homology of birds and dinosaurs: accommodating the cladogram." *TREE* **16**: 285–286.

Feduccia, A. and J. Nowicki (2002). "The hand of birds revealed by early ostrich embryos." *Naturwissenschaften* **89**: 391–393.

Ferrier, D. K., C. Minguillón, et al. (2000). "The amphioxus Hox cluster: deuterostome posterior flexibility and Hox14." *Evol. Dev.* **2**: 284–293.

Feschotte, C. (2008). "Transposable elements and the evolution of regulatory networks." *Nature Rev. Genetics* **9**: 397–405.

Fisher, R. A. (1930). The Genetical Theory of Natural Selection. Oxford, Clarendon Press.

Fitch, W. M. (1970). "Distinguishing homologous from analogous proteins." *Syst. Zool.* **19**(2): 99–113.

Fondon, III, J. W., and H. R. Garner (2004). "Molecular origins of rapid and continuous morphological evolution." *Proc. Natl. Acad. Sci., USA* **101**(52): 18058–18063.

Fontana, W. (2002). "Modelling 'evo-devo' with RNA." *Bioessays* **24**(12): 1164–1177.

Force, A., M. Lynch, et al. (1999). "Preservation of duplicate genes by complementary, degenerative mutations." *Genetics* **151**: 1531–1545.

Foucher, I., M. L. Montesinos, et al. (2003). "Joint regulation of the MAP1B promoter by HNF3beta/Foxa2 and Engrailed is the result of a highly conserved mechanism for direct interaction of homeoproteins and Fox transcription factors." *Development* **130**(9): 1867–1876.

Franke, W. W., E. Schmid, et al. (1980). "Intermediate-sized filaments of the prekeratin type in myoepithelial cells." *J. Cell Biol.* **84**(3): 633–654.

Fraser, G. J. and M. M. Smith (2011). "Evolution of developmental pattern for vertebrate dentitions: an oro-pharyngeal specific mechanism." *J. Exp. Zool. B. Mol. Dev. Evol.* **316B** (2): 99–112.

Fraser, H. B. (2006). "Coevolution, modularity and human disease." *Curr. Opin. Genet. Dev.* **16**(6): 637–644.

Freitas, R., G. Zhang, et al. (2006). "Evidence that mechanisms of fin development evolved in the midline of early vertebrates." *Nature* **442**(7106): 1033–1037.

Freitas, R., G. J. Zhang, et al. (2007). "Biphasic Hoxd gene expression in shark paired fins reveals an ancient origin of distal limb domain." *PLoS ONE* **2**(8): e754.

Friedrich, M. (2006). "Ancient mechanisms of visual sense organ development based on comparison of the gene networks controlling larval eye, ocellus, and compound eye specification in Drosophila." *Arthropod. Struct. Dev.* **35**(4): 357–378.

Friedrich, M. (2008). "Opsins and cell fate in the Drosophila Bolwig organ: tricky lessons in homology inference." *Bioessays* **30**(10): 980–993.

Fröbisch, N. B. and N. H. Shubin (2011). "Salamander limb development: integrating genes, morphology, and fossils." *Dev. Dyn.* **240**(5): 1087–1099.

Fromental-Ramain, C., X. Warot, et al. (1996). "Hoxa-13 and Hoxd-13 play a crucial role in the patterning of the limb autopod." *Development* **122**: 2997–3011.

Frost, D. R., T. Grant, et al. (2006). "The Amphibian Tree of Life." *Bull. Am. Mus. Nat. Hist.* **297**: 1–370.

Fuchs, E. (2007). "Scratching the surface of skin development." *Nature* **445**(7130): 834–842.

Fuchs, E. and V. Horsley (2008). "More than one way to skin." *Genes Dev.* **22**(8): 976–985.

Fürbringer, M. (1870). Die Knochen und Muskeln der Extremitaten bei den schlangenahnlichen Sauriern: vergleichend anatomische Abhandlung. Leipzig, Verlag von Wilhelm Engelmann.

Futuyma, D. J. (1998). Evolutionary Biology. Sunderland, MA, Sinauer Associates.

Fuxreiter, M., P. Tompa, et al. (2007). "Local structural disorder imparts plasticity on linear motifs." *Bioinformatics* **23**(8): 950–956.

Gabriel, W. N. and B. Goldstein (2007). "Segmental expression of Pax3/7 and engrailed homologs in tardigrade development." *Dev. Genes Evol.* **217**: 421–433.

Galant, R. and S. B. Carroll (2002). "Evolution of a transcriptional repression domain in an insect Hox protein." *Nature* **415**: 910–913.

Galis, F., M. Kundrát, et al. (2003). "An old controversy solved: bird embryos have five fingers." *TREE* **18**: 7–9.

Galis, F., M. Kundrát, et al. (2005). "Hox genes, digit identities and the theropod/bird transition." *J. Exp. Zool. Part. B. Mol. Dev. Evol.* **304B**:198–205.

Galis, F. and J. A. Metz (2001). "Testing the vulnerability of the phylotypic stage: on modularity and evolutionary conservation." *J. Exp. Zool. B. Mol. Dev. Evol.* **291**: 195–204.

Galis, F., J. J. M. v. Alphen, et al. (2001). "Why five fingers? Evolutionary constraints on digit numbers." *TREE* **16**: 637–646.

Gans, C. and R. G. Northcutt (1983). "Neural crest and the origin of vertebrates: a new head." *Science* **220**: 268–274.

Gao, F. and E. H. Davidson (2008). "Transfer of a large gene regulatory apparatus to a new developmental address in echinoid evolution." *Proc. Natl. Acad. Sci., USA* **105**(16): 6091–6096.

Garcia-Pacheco, J. M., C. Oliver, et al. (2001). "Human decidual stromal cells express CD34 and STRO-1 and are related to bone marrow stromal precursors." *Mol. Hum. Reprod.* **7**(12): 1151–1157.

Gardiner, D. M., B. Blumberg, et al. (1995). "Regulation of HoxA expression in developing and regenerating axolotl limbs." *Development* **121**: 1731–1741.

Garner, J. P. and A. L. R. Thomas (1998). "Counting the fingers of birds." *Science* **280**: 355.

Gauthier, J., A. G. Kluge, et al. (1988). "Amniote phylogeny and the importance of fossils." *Cladistics* **4**: 105–209.

Geeta, R. (2003). "Structure trees and speies trees: what they say about morphological development and evolution." *Evol. Devel.* **5**: 609–621.

Gegenbaur, C. (1863). "Vergleichend-anatomische Bemerkungen über das Fusskelet der Vögel." *Arch. Anat. Phys. u. wiss. Med.* **1863**: 450–472.

Gegenbaur, C. (1876). "Zur Morphologie der Gliedmassen der Wirbeltiere." *Morphologisches Jahrbuch* **2**: 396–420.

Gegenbaur, C. (1878). Elements of Comparative Anatomy. London, Macmillan.

Gehring, W. J. (1994). "A history of the homeobox," in Guidebook to the Homeobox Genes, ed. D. Duboule. Oxford, Oxford University Press: 1–10.

Gehring, W. J. (1998). Master Control Genes in Development and Evolution: The Homeobox Story. New Haven, CT, Yale University Press.

Gehring, W. J. and K. Ikeo (1999). "Pax6 mastering eye morphogenesis and eye evolution." Trends Genet. 15: 371–377.

Geison, G. L. (1969). "Darwin and heredity: the evolution of his hypothesis of pangenesis." J. Hist. Med. Allied Sci. 26(4): 375–411.

Gellersen, B. and J. Brosens (2003). "Cyclic AMP and progesterone receptor cross-talk in endometrium: a decidualizing affair." J. Endocrinol. 178: 357–372.

Gellersen, B., R. Kempf, et al. (1994). "Nonpituitary human prolactin gene transcription is independent of Pit-1 and differentially controlled in lymphocytes and in endometrial stroma." Mol. Endocrinol. 8(3): 356–373.

Ghiselin, M. T. (1966). "An application of the theory of definitions to systematic principles." Syst. Zool. 15: 127–130.

Ghiselin, M. T. (1969). The Triumph of the Darwinian Method. Berkeley, CA, University of California Press.

Ghiselin, M. T. (1974). "A radical solution to the species problem." Syst. Zool. 23: 536–544.

Ghiselin, M. T. (1997). Metaphysics and the Origin of Species. New York, State University of New York Press.

Ghiselin, M. T. (2003). "Carl Gegenbaur versus Anton Dohrn." Theory in Biosciences 122: 142–147.

Ghiselin, M. T. (2005). "Homology as a relation of correspondence between parts of individuals." Theory in Biosciences 124: 91–103.

Gibson, G. and D. S. Hogness (1996). "Effect of polymorphism in the Drosophila regulatory gene Ultrabithorax on homeotic stability." Science 271: 200–203.

Gidley, J. W. (1906). "Evidence bearing on tooth-cusp development." Washington Academy of Sciences Proceedings 8: 91–110.

Gilbert, S. F. (2010). Developmental Biology. Sunderland, MA, Sinauer Associates.

Gilbert, S. F. and J. A. Bolker (2001). "Homologies of process and modular elements of embryonic construction," in The Character Concept in Evolutionary Biology, ed. G. P. Wagner. San Diego, CA, Academic Press: 435–454.

Gilbert, S. F. and D. Epel (2009). Ecological Developmental Biology: Integrating Epigenetics, Medicine, and Evolution. Sunderland, MA, Sinauer Associates.

Gillis, C. D. and P. W. Hopkins (1922). "The phylogenetic significance of the prehallux and prepollex: a theory." Proc. Royal Soc. Queensland 33: 30–38.

Gillis, J. A., R. D. Dahn, et al. (2009). "Shared developmental mechanisms pattern the vertebrate gill arch and paired fin skeletons." Proc. Natl. Acad. Sci., USA 106(14): 5720–5724.

Goldberg, E. E. and B. Igic (2008). "On phylogenetic tests of irreversible evolution." Evolution 62: 2727–2741.

Goldschmidt, R. B. (1938). Physiological Genetics. New York, McGraw-Hill.

Gompel, N., B. Prud'homme, et al. (2005). "Chance caught on the wing: cis-regulatory evolution and the origin of pigment patterns in Drosophila." Nature 433: 481–487.

Goodman, F. R., S. Mundlos, et al. (1997). "Synpolydactyly phenotypes correlate with size of expansions in HOXD13 polyalanine tract." Proc. Natl. Acad. Sci., USA 94(14): 7458–7463.

Goodwin, B., "Beyond the Darwinian paradigm: understanding biological forms," in Evolution: The First Four Billion Years, eds. Michael Ruse and Joseph Travis. Cambridge: Harvard University Press, 2009: 299–312.

Goodwin, B. C. and L. E. H. Trainor (1983). "The ontogeny and phylogeny of the pentadactyl limb," in Development and Evolution, eds. B. C. Goodwin, N. Holder and C. C. Wylie. London, Cambridge University Press: 75–98.

Gould, S. J. (1977). Ontogeny and Phylogeny. Cambridge, MA, The Belknap Press of Harvard University Press.

Gould, S. J. (2002). Structure of Evolutionary Theory. Cambridge. MA, Belknap Press.

Gould, S. J. and R. Lewontin (1978). "The Spandrels of San Marco and the Panglossian Paradigm: A Critique of the Adaptationist Programme." Proc. R. Soc. London 205: 581–598.

Graf, T. and T. Enver (2009). "Forcing cells to change lineages." Nature 462: 587–594.

Grainger, R. M. (1992). "Embryonic lens induction: shedding light on vertebrate tissue determination." Trends Genet. 8(10): 349–355.

Grandel, H. (2003). "Approaches to a comparison of fin and limb structure and development." Theory in Biosciences 122: 288–301.

Grandel, H. and S. Schulte-Merker (1998). "The development of the paired fins in the zebrafish (Danio rerio)." Mech. Devel. 79: 99–120.

Grbic, M., L. M. Nagy, et al. (1996). "Polyembryonic development: insect pattern formation in a cellularized environment." Development 122(3): 795–804.

Greenwold, M. J. and R. H. Sawyer (2010). "Genomic organization and molecular phylogenies of the beta (β) keratin multigene family in the chicken (Gallus gallus) and zebra finch (Taeniopygia guttata): implications for feather evolution." BMC Evol. Biol. 10: 148.

Greer, A. E. (1991). "Limb reduction in Squamates: identification of the lineages and discussion of the trends." J. Herpetology 25: 166–173.

Gregory, W. K., R. Minner, et al. (1923). "The carpus of Eryops and the structure of the primitive chiropterygium." Bull. Am. Mus. Nat. Hist. 48: 279–288.

Grell, K. G. and G. Benwitz (1971). "Die Ultrastruktur von Trichoplax adhaerens F. E. Schulze." Cytobiologie 4: 216–240.

Grenier, J. K. and S. B. Carroll (2000). "Functional evolution of the Ultrabithorax protein." Proc. Natl. Acad. Sci., USA 97: 704–709.

Griffiths, P. E. (2007). "The phenomena of homology." Biol. Phil. 22: 643–658.

Griffiths, P. E. (1997). What Emotions Really Are. The Problem of Psychological Categories. Chicago, IL, University of Chicago Press.

Haig, D. (1993). "Genetic conflicts in human pregnancy." Quart. Rev. Biol. 68(4): 495–532.

Halder, G., P. Callaerts, et al. (1995). "Induction of ectopic eyes by targeted expression of the eyeless gene in Drosophila." Science 267(5205): 1788–1792.

Hall, B. K. (1984). "Developmental processes underlying heterochrony as an evolutionary mechanism." Can. J. Zool. 62: 1–7.

Hall, B. K. (1994). "Homology and Embryonic Development," in Evolutionary Biology, eds. M. K. Hecht, R. J. MacIntyre, and M. T. Clegg. New York, Plenum Press. 28: 1–30.

Hall, B. K. (1998). Evolutionary Developmental Biology. London, Chapman and Hall.

Hall, B. K. (2003). "Descent with modification: the unity underlying homology and homoplasy as seen through an analysis of development and evolution." Biol. Rev. Camb. Philos. Soc. 78: 409–433.

Hall, B. K. and T. Miyake (1992). "The membranous skeleton: the role of cell condensations in vertebrate skeletogenesis." *Anatomy and Embryology* **186**: 107–124.

Hall, B. K. and T. Miyake (1995). "Divide, accumulate, differentiate: cell condensation in skeletal development revisited." *Int. J. Dev. Biol.* **39**: 881–893.

Hall, B. K., R. Pearson, et al., eds. (2003). Environment, Development and Evolution. The Vienna Series in Theoretical Biology. Cambridge, MIT Press.

Hallgrímsson, B., H. Jamniczky, et al. (2009). "Deciphering the palimpsest: studying the relationship between morphological integration and phenotypic covariation." *Evol. Biol.* **36**(4): 355–379.

Halmos, P. R. (1972). Naive Set Theory. Princeton, Nostrand Co.

Han, M. J., J. Y. An, et al. (2001). "Expression patterns of Fgf-8 during development and limb regeneration of the axolotl." *Dev. Dyn.* **220**(1): 40–48.

Hansen, K. L. (2010). "A history of digit identification in the manus of theropods (including Aves)." *Geological Society, London, Special Publications* **343**: 265–275.

Hansen, T. F. (2003). "Is modularity necessary for evolvability? Remarks on the relationship between pleiotropy and evolvability." *BioSystems* **69**: 83–94.

Hansen, T. F. (2006). "The evolution of genetic architecture." *Annu. Rev. Ecol. Systematics* **37**: 123–157.

Hansen, T. F. (2013). "Why epistasis is important for selection and adaptation." *Evolution* **67**: 3501–3511.

Hansen, T. F., J. M. Alvarez-Castro, et al. (2006). "Evolution of genetic architecture under directional selection." *Evolution* **60**(8): 1523–1536.

Harris, M. P., J. F. Fallon, et al. (2002). "A Shh-Bmp2 developmental module in feather development and evolution." *J. Exp. Zool. B. Mol. Dev. Evol.* **294**: 160–176.

Harris, M. P., S. Williamson, et al. (2005). "Molecular evidence for an activator-inhibitor mechanism in development of embryonic feather branching." *Proc. Natl. Acad. Sci., USA* **102**(33): 11734–11739.

Hartwell, L. H., J. J. Hopfield, et al. (1999). "From molecular to modular cell biology." *Nature* **402**(6761 Suppl): C47–52.

Hazkani-Covo, E., D. Wool, et al. (2005). "In search of the vertebrate phylotypic stage: a molecular examination of the developmental hourglass model and von Baer's third law." *J. Exp. Zool. B. Mol. Dev. Evol.* **304**(2): 150–158.

Heffer, A., J. W. Shultz, et al. (2010). "Surprising flexibility in a conserved Hox transcription factor over 550 million years of evolution." *Proc. Natl. Acad. Sci., USA* **107**(42): 18040–18045.

Heimberg, A. M., L. F. Sempere, et al. (2008). "MicroRNAs and the advent of vertebrate morphological complexity." *Proc. Natl. Acad. Sci., USA* **105**(8): 2946–2950.

Hendrikse, J. L., T. E. Parsons, et al. (2007). "Evolvability as the proper focus of evolutionary developmental biology." *Evol. Dev.* **9**(4): 393–401.

Hennig, W. (1966). Phylogenetic Systematics. Urbana, IL, University of Illinois Press.

Henry, J. J. and R. M. Grainger (1990). "Early tissue interactions leading to embryonic lens formation in Xenopus laevis." *Dev. Biol.* **141**(1): 149–163.

Hermisson, J., T. F. Hansen, et al. (2003). "Epistasis in polygenic traits and the evolution of genetic architecture under stabilizing selection." *Am Nat* **161**: 708–734.

Hermisson, J. and G. P. Wagner (2004). "The population genetic theory of hidden variation and genetic robustness." *Genetics* **168**: 2271–2284.

Hertel, J., M. Lendemeyer, et al. (2006). "The expansion of the metazoan microRNA repertoire." *BMC Genomics* **7**: 25.

Hiepko, P. (1965). "Vergleichend-morphologische und entwicklungsgeschichtliche Untersuchungen üb das Perianth bei den Polycarpicae." *Bot. Jb f System* **84**: 359–426.

Hileman, L. C. and V. F. Irish (2009). "More is better: the uses of developmental genetic data to reconstruct perianth evolution." *Am. J. Bot.* **96**(1): 83–95.

Hinchliffe, J. R. (1977). "The chondrogenic pattern in chick limb morphogenesis: a problem of development and evolution," in Vertebrate Limb and Somite Morphogenesis, eds. D. A. Ede, J. R. Hinchliffe and M. Balls. Cambridge, UK., Cambridge University Press: 293–309.

Hinchliffe, J. R. (1985). "'One, two, three' or 'two, three, four': an embryologist's view of the homologies of the digits and carpus of modern birds," in The Beginnings of Birds: Proceedings of the International Archaeopteryx Conference, Eichstätt., eds. M. K. Hecht, J. H. Ostrom, G. Viohl and P. Wellnhofer. Willibaldsburg, Eichstätt: Freunde des Jura-Museums Eichstätt: 141–147.

Hinchliffe, J. R. and P. J. Griffiths (1983). "The prechondrogenic patterns in tetrapod limb development and their phylogenetic significance," in Development and Evolution, eds. B. C. Goodwin, N. Holder and C. C. Wylie. Cambridge, Cambridge Univ. Press: 99–121.

Hinchliffe, J. R. and D. R. Johnson (1980). The Development of the Vertebrate Limb. New York, Oxford University Press.

Hinchliffe, J. R. and E. I. Vorobyeva (1999). "Developmental basis of limb homology in urodeles: heterochronic evidence from the primitive hynobiid family," in Homology, eds. G. R. Bock and G. Cardew. Chichester, England, J. Wiley.

Hinchliffe, R. (2007). "Bird wing digits & their homologies: reassessment of developmental evidence for a 2,3,4 identity." *ORYCTOS 7*: 7–12.

Hinman, V. F. and E. H. Davidson (2007). "Evolutionary plasticity of developmental gene regulatory network architecture." *Proc. Natl. Acad. Sci., USA* **104**(49): 19404–19409.

Hinman, V. F., A. T. Nguyen, et al. (2003). "Developmental gene regulatory network architecture across 500 million years of echinoderm evolution." *Proc. Natl. Acad. Sci., USA* **100**(23): 13356–13361.

Holland, L. Z., M. Kene, et al. (1997). "Sequence and embryonic expression of the amphioxus engrailed gene (AmphiEn): the metameric pattern of transcription resembles that of its segment-polarity homolog in Drosophila." *Development* **124**(9): 1723–1732.

Holmberg, J. and T. Perlmann (2012). "Maintaining differentiated cellular identity." *Nat. Rev. Genet.* **13**(6): 429–439.

Holmgren, N. (1933). "On the origin of the tetrapod limb." *Acta Zoologica* **14**: 187–248.

Holtfreter, J. (1935). "Über das Verhalten von Anurenektoderm in Urodelenkeim." *Roux' Arch* **133**: 427–494.

Holtfreter, J. (1936). "Regionale Induktionen in xenoplastisch zusammengesetzten Explantaten." *Roux' Arch* **134**: 466–550.

Houle, D., B. Morikawa, et al. (1996). "Comparing mutational variabilities." *Genetics* **143**: 1467–1483.

Hu, S, A. Mamedova, and R. S. Hegde. (2008). DNA-binding and regulation mechanisms of the SIX family of retinal determination proteins. *Biochemistry* **47**: 3586–3594.

Hughes, A. J. and D. M. Lambert (1984). "Functionalism, structuralism, and ways of seeing." *J. Theor. Biol.* **111**: 787–800.

Hull, D. L. (1978). "A matter of individuality." *Philosophy of Science* **45**: 335–360.

Hull, D. L. (1980). "Individuality and selection." *Annu. Rev. Ecol. Syst.* **11**: 311–332.

Hull, D. L. (2002). Varieties of Reductionism: Derivation and Gene Selection. Promises and Limits of Reductionism in the Biomedical Sciences. M. H. V. v. Regenmortel and D. L. Hull. Chichester, John Wiley & Sons.

Ihde, A. J. (1964). The Development of Modern Chemistry. New York, Dover Publ., Inc.

Irie, N. and S. Kuratani (2011). "Comparative transcriptome analysis reveals vertebrate phylotypic period during organogenesis." *Nat. Commun.* **2**: 248.

Irie, N. and A. Sehara-Fujisawa (2007). "The vertebrate phylotypic stage and an early bilaterian-related stage in mouse embryogenesis defined by genomic information." *BMC Biol.* **5**: 1.

Irish, V. F. (2009). "Evolution of petal identity." *J Exp Bot* **60**(9): 2517–2527.

Irish, V. F. and A. Litt (2005). "Flower development and evolution: gene duplication, diversification and redeployment." *Curr. Opin. Genet. Devel.* **15**: 454–460.

Jaramillo, M. A. and E. M. Kramer (2007). "The role of developmental genetics in understanding homology and morphological evolution in plants." *Int. J. Plant Sci.* **168**(1): 61–72.

Jarvik, E. (1942). "On the structure of the snout of crossopterygians and lower gnathostomes in general." *Zool. Bidr. Upps.* 21: 235–675.

Jarvik, E. (1980). Basic Structure and Evolution of Vertebrates. New York and London, Academic Press.

Jernvall, J. and H. S. Jung (2000). "Genotype, phenotype, and developmental biology of molar tooth characters." *Am. J. Phys. Anthropol. Suppl.* **31**: 171–190.

Jernvall, J. and I. Thesleff (2000). "Reiterative signaling and patterning during mammalian tooth morphogenesis." *Mech. Devel.* **92**(1): 19–29.

Ji, Q., Z. X. Luo, et al. (1999). "A Chinese triconodont mammal and mosaic evolution of the mammalian skeleton (vol. 398, pg. 326, 1999)." *Nature* **402**(6764): 898–898.

Joachimsthal, G. (1900). "Verdoppelung des linken Zeigefingers und Dreigliederung des rechten Daumens." *Berliner Klinische Wochenschrift* **37**: 835–838.

Johanson, Z., J. Joss, et al. (2007). "Fish fingers: digit homology in sarcopterygian fish fins." *J. Exp. Zool. B. Mol. Dev. Evol.* **308B**: 757–768.

Johanson, Z. and M. M. Smith (2005). "Origin and evolution of gnathostome dentitions: a question of teeth and pharyngeal denticles in placoderms." *Biol. Rev. Camb. Philos. Soc.* **80**(2): 303–345.

Joss, J. and T. Longhurst (2001). "Lungfish paired fins," in Major Events in Early Vertebrate Evolution: Paleontology, ed. P. E. Ahlberg. London and New York, Taylor & Francis: 370–376.

Kaelin, J. (1945). "Zur Morphogenese des Panzers bei den Schildkröten." *Acta Anat.* **1**: 144–176.

Kalinka, A. T., K. M. Varga, D. T. Gerrard, S. Preibisch, D. L. Corcoran, J. Jarrells, U. Ohler, C. M. Bergman, and P. Tomancak (2010). "Gene expression divergence recapitulates the developmental hourglass model." *Nature* **468**(7325): 811–814.

Kamm, K., B. Schierwater, et al. (2006). "Axial patterning and diversification in the cnidaria predate the Hox system." *Curr. Biol.* **16**(9): 920–926.

Kanno, A., H. Saeki, et al. (2003). "Heterotopic expression of class B floral homeotic genes supports a modified ABC model for tulip (Tulipa gesneriana)." *Plant Mol. Biol.* **52**(4): 831–841.

Kapranov, P., A. T. Willingham, et al. (2007). "Genome-wide transcription and the implications for genomic organization." *Nat. Rev. Genet.* **8**(6): 413–423.

Kardon, G., T. A. Heanue, et al. (2004). "The Pax/Six/Eya/Dach network in development and evolution," in Modularity in Development and Evolution, eds. G. Schlosser and G. P. Wagner. Chicago and London, University of Chicago Press: 59–80.

Karlin, S. and C. Burge (1996). "Trinucleotide repeats and long homopeptides in genes and proteins associated with nervous system disease and development." *Proc. Natl. Acad. Sci., USA* **93**(4): 1560–1565.

Kashtan, N. and U. Alon (2005). "Spontaneous evolution of modularity and network motifs." *Proc. Natl. Acad. Sci., USA* **102**(39): 13773–13778.

Kemp, A. (1986). "The biology of the Australian lungfish, Neoceratodus forsteri (Krefft 1870)." *J. Morph. Suppl.* **1**: 181–198.

Kenney-Hunt, J. P., B. Wang, et al. (2008). "Pleiotropic patterns of quantitative trait loci for 70 skeletal traits." *Genetics* **178**: 2275–2288.

Kenyon, K. L., D. J. Li, et al. (2005). "Fly SIX-type homeodomain proteins Sine oculis and Optix partner with different cofactors during eye development." *Dev. Dyn.* **234**(3): 497–504.

Kenyon, K. L., D. Yang-Zhou, et al. (2005). "Partner specificity is essential for proper function of the SIX-type homeodomain proteins Sine oculis and Optix during fly eye development." *Dev. Biol.* **286**(1): 158–168.

Keys, D. N., D. L. Lewis, et al. (1999). "Recruitment of a hedgehog regulatory circuit in butterfly eyespot evolution." *Science* **283**: 532–534.

Kiontke, K., A. Barriere, et al. (2007). "Trends, stasis, and drift in the evolution of nematode vulva development." *Curr. Biol.* **17**(22): 1925–1937.

Kirk, D. L. (1998). Volvox. Cambridge, New York, Cambridge University Press.

Kjaer, K. W., J. Hedeboe, et al. (2002). "*HOXD13* polyalanine tract expansion in classical synpolydactyly type Vordingborg." *Am. J. Med. Genetics* **110**(2): 116–121.

Knezevic, V., R. DeSanto, et al. (1997). "Hoxd-12 differentially affects preaxial and post-axial chondrogenic branches in the limb and regulates Sonic hedgehog in a positive feedback loop." *Development* **124**: 4523–4536.

Koestler, A. (1971). The Case of the Midwife Toad. London, Hutchinson.

Kohlsdorf, T., V. J. Lynch, et al. (2010). "Data and data interpretation in the study of limb evolution: a reply to Galis et al. on the reevolution of digits in the lizard genus Bachia." *Evolution* **64**: 2477–2485.

Kohlsdorf, T. and G. P. Wagner (2006). "Evidence for the reversibility of digit loss: a phylogenetic study of limb evolution in Bachia (Gymnophthalmidae: Squamata)." *Evolution* **60**(9): 1896–1912.

Kontges, G. and A. Lumsden (1996). "Rhombencephalic neural crest segmentation is preserved throughout craniofacial ontogeny." *Development* **122**(10): 3229–3242.

Koopman, K. F. (1993). "Chiroptera," in Mammalian Species of the World, eds. D. E. Wilson and D. M. Reeder. Washington, D. C., Smithsonian Press: 137–241.

Kopp, A. (2011). "Drosophila sex combs as a model of evolutionary innovations." *Evol. Dev.* **13**(6): 504–522.

Koyanagi, M., K. Kubokawa, et al. (2005). "Cephalochordate melanopsin: evolutionary linkage between invertebrate visual cells and vertebrate photosensitive retinal ganglion cells." *Curr. Biol.* **15**(11): 1065–1069.

Kramer, E. M., V. S. D. Stilio, et al. (2003). "Complex patterns of gene duplication in the Apetala3 and Pistillata lineages of the Ranunculaceae." *Int. J. Plant Sci.* **164**: 1–11.

Kriebel, M., F. Muller, et al. (2007). "Xeya3 regulates survival and proliferation of neural progenitor cells within the anterior neural plate of Xenopus embryos." *Dev. Dyn.* **236**(6): 1526–1534.

Krizek, B. A. and J. C. Fletcher (2005). "Molecular mechanisms of flower development: an armchair guide." *Nat. Rev. Genet.* **6**(9): 688–698.

Krizek, B. A. and E. M. Meyerowitz (1996). "Mapping the protein regions responsible for the functional specificities of the Arabidopsis MADS domain organ-identity proteins." *Proc. Natl. Acad. Sci.. USA* **93**(9): 4063–4070.

Kunarso, G., N. Y. Chia, et al. (2010). "Transposable elements have rewired the core regulatory network of human embryonic stem cells." *Nat. Genet.* **42**(7): 631–634.

Kundrát, M., V. Seichert, et al. (2001). "Developmental remnants of the first avian metacarpus." *J. Morph.* 248: 252A.

Kundrát, M., V. Seichert, et al. (2002). "Pentadactyl pattern of the avian wing autopodium and pyramid reduction hypothesis." *J. Exp. Zool. B. Mol. Dev. Evol.* **294**: 152–159.

Kuraku, S., R. Usuda, et al. (2005). "Comprehensive survey of carapacial ridge-specific genes in turtle implies co-option of some regulatory genes in carapace evolution." *Evol. Dev.* **7**(1): 3–17.

Kuratani, S., S. Kuraku, et al. (2011). "Evolutionary developmental perspective for the origin of turtles: the folding theory for the shell based on the developmental nature of the carapacial ridge." *Evol. Dev.* **13**(1): 1–14.

Kurtén, B. (1963). "Return of a lost structure in the evolution of the felid dentition." *Soc. Sci. Fenn. Comm. Biol.* 4: 1–12.

Kuwada Y. (1911). "Meiosis in the pollen mother cells of Zea Mays L." *Bot. Mag.* **25**:163.

Laiosa, C. V., M. Stadtfeld, et al. (2006). "Reprogramming of committed T cell progenitors to macrophages and dendritic cells by C/EBP alpha and PU.1 transcription factors." *Immunity* **25**(5): 731–744.

Lamb, R. S. and V. F. Irish (2003). "Functional divergence within the APETALA3/PISTILLATA floral homeotic gene lineages." *Proc. Natl. Acad. Sci., USA* **100**(11): 6558–6563.

Land, M. F. and R. D. Fernald (1992). "The evolution of eyes." *Annu. Rev. Neurosci.* **15**: 1–29.

Lang, R. A. (2004). "Pathways regulating lens induction in the mouse." *Int. J. Dev. Biol.* **48**(8–9): 783–791.

Lankester, R. (1870). "On the use of the term homology." *Ann. and Mag. of Natural History, Zoology, Botany and Geology* **6**(31): 34–43.

LaPorte, J. (2004). Natural Kinds and Conceptual Change. Cambridge, UK, Cambridge University Press.

Larson, A. and W. W. Dimmick (1993). "Phylogenetic relationships of the salamander families: an analysis of congruence among morphological and molecular characters." *Herpetological Monographs* **7**: 77–93.

Larsson, H. C. E. (2007). "MODES of developmental evolution: origin and definition of the autopodium," in Major Transitions in Vertebrate Evolution, eds. J. S. Anderson and H.-D. Sues. Bloomington, Indiana University Press: 150–281.

Larsson, H. C. E., A. C. Heppleston, et al. (2010). "Pentadactyl ground state of the manus of Alligator mississippiensis and insight into the evolution of digit reduction in Archosauria." *J. Exp. Zool. B. Mol. Dev. Evol.* **314B**: 571–579.

Larsson, H. C. E. and G. P. Wagner (2002). "The pentadactyl ground state of the avian wing." *J. Exp. Zool. B. Mol. Dev. Evol.* **294**(2): 146–151.

Laubichler, M. D. and J. Maienschein (2003). "Ontogeny, anatomy, and the problem of homology: Carl Gegenbaur and the American tradition of cell lineage studies." *Theory Biosci.* **122**: 194–203.

Laurent, R. F. (1983). "Irreversibility: a comment." *Syst. Zool.* **32**(75).

Leal, F., O. A. Tarazona, et al. (2010). "Limb development in the gekkonid lizard Gonatodes albogularis: a reconsideration of homology in the lizard carpus and tarsus." *J. Morphol.* **271**(11): 1328–1341.

Lee, S., B. Lee, et al. (2008). "A regulatory network to segregate the identity of neuronal subtypes." *Dev. Cell* **14**(6): 877–889.

Lee, S. K. and S. L. Pfaff (2001). "Transcriptional networks regulating neuronal identity in the developing spinal cord." *Nat. Neurosci.* **4 Suppl**: 1183–1191.

Lenser, T., G. Theissen, et al. (2009). "Developmental robustness by obligate interaction of class B floral homeotic genes and proteins." *PLoS Comput. Biol.* **5**(1): e1000264.

Leroi, A. M. (2000). "The scale independence of evolution." *Evol. Devel.* **2**: 67–77.

Li, C., X. C. Wu, et al. (2008). "An ancestral turtle from the Late Triassic of southwestern China." *Nature* **456**(7221): 497–501.

Li, H., R. Helling, et al. (1996). "Emergence of preferred structures in a simple model of protein folding." *Science* **273**(5275): 666–669.

Li, H., Y. Sun, et al. (2009). "Exploring pathways from gene co-expression to network dynamics." *Methods Mol. Biol.* **541**: 249–267.

Li, Y., F. Wang, et al. (2006). "MicroRNA-9a ensures the precise specification of sensory organ precursors in Drosophila." *Genes Dev.* **20**(20): 2793–2805.

Lieberkind, I. (1937). Vergleichende Studien über die Morphologie und Histogenese der larvalen Haftorgane bei den Amphibien. Copenhagen, C. A. Reitzels.

Lilversage, R. A. (1978). "Oscar E. Schotté." *Am. Zool.* **18**(4): 825–827.

Löhr, U., M. Yussa, et al. (2001). "Drosophila fushi tarazu: a gene on the border of homeotic function." *Curr. Biol.* **11**(18): 1403–1412.

Löhr, U., and L. Pick (2005). Cofactor-interaction motifs and the cooption of a homeotic Hox protein into the segmentation pathway of Drosophila melanogaster. *Curr. Biol.* **15**: 643–649.

Lombardi, J. (1998). Comparative Vertebrate Reproduction. Boston, Kluwer Academic Publishers.

Loosli, F., S. Winkler, et al. (1999). "Six3 overexpression initiates the formation of ectopic retina." *Genes Dev.* **13**(6): 649–654.

Loredo, G. A., A. Brukman, et al. (2001). "Development of an evolutionarily novel structure: fibroblast growth factor expression in the carapacial ridge of turtle embryos." *J. Exp. Zool. B. Mol. & Dev. Evol.* **291**(3): 274–81.

Lorenz, K. (1981). The Foundations of Ethology. New York, Springer-Verlag.

Love, A. C. (2003). "Evolutionary morphology, innovation and the synthesis of evolutionary and developmental biology." *Biol. Phil.* **18**: 309–345.

Love, A. C. (2009). "Typology reconfigured: from the metaphysics of essentialism to the epistemology of representation." *Acta Biotheoretica* **57**: 51–57.

Love, A. C. and R. A. Raff (2006). "Larval ectoderm, organizational homology, and the origins of evolutionary novelty." *J. Exp. Zool. B. Mol. Dev. Evol.* **306**(1): 18–34.

Lowe, C. B., G. Bejerano, et al. (2007). "Thousands of human mobile element fragments undergo strong purifying selection near developmental genes." *Proc. Natl. Acad. Sci., USA* **104**(19): 8005–8010.

Lucas, A. M. and P. R. Settenheim (1972). Avian Anatomy: Integument I. Washington DC, US Department of Agriculture.

Lucas, A. M. and P. R. Stettenheim (1972). Avian Anatomy: Integument II. Washington, DC, US Department of Agriculture.

Lynch, M. (2007). "The frailty of adaptive hypotheses for the origins of organismal complexity." *Proc. Natl. Acad. Sci. USA.* **104**: 8597–8604.

Lynch, M. (2007). The Origins of Genome Architecture. Sunderland, MA, Sinauer Associates.

Lynch, M. and J. S. Conery (2000). "The evolutionary fate and consequences of duplicate genes." *Science* **290**: 1151–1155.

Lynch, M. and A. Force (2000). "Gene duplication and the origin of interspecific incompatibility." *Am. Nat.* **156**: 590–605.

Lynch, M. and B. Walsh (1998). Genetics and Analysis of Quantitative Traits. Sunderland, MA, Sinauer Associates.

Lynch, V. J., R. D. Leclerc, et al. (2011). "Transposon-mediated rewiring of gene regulatory networks contributed to the evolution of pregnancy in mammals." *Nat. Genet.* **43**(11): 1154–1159.

Lynch, V. J., G. May, et al. (2011). "Regulatory evolution through divergence of a phosphoswitch in the transcription factor CEBPB." *Nature* **480**(7377): 383–386.

Lynch, V. J., J. J. Roth, et al. (2004). "Adaptive evolution of HoxA-11 and HoxA-13 at the origin of the uterus in mammals." *Proc. Royal Soc. B: Biol. Sci.* **271**(1554): 2201–2207.

Lynch, V. J., A. Tanzer, et al. (2008). "Adaptive changes in the transcription factor HoxA-11 are essential for the evolution of pregnancy in mammals." *Proc. Natl. Acad. Sci., USA* **105**(39): 14928–14933.

Lynch, V. J. and G. P. Wagner (2008). "Resurrecting the role of transcription factor change in developmental evolution." *Evolution* **62**: 2131–2154.

Lynch, V. J. and G. P. Wagner (2010). "Did egg-laying boas break Dollo's law? Phylogenetic evidence for reversal to oviparity in sand boas (Eryx: Boidae)." *Evolution* **64**(1): 207–216.

Lynch, V. J. and G. P. Wagner (2011). "Revisiting a classic example of transcription factor functional equivalence: are Eyeless and Pax6 functionally equivalent or divergent?" *J. Exp. Zool. B. Mol. Dev. Evol.* **316B**(2): 93–98.

Mabee, P. M. (2000). "Developmental data and phylogenetic systematics: evolution of the vertebrate limb." *Amer. Zool.* **40**: 789–800.

Mabee, P. M. and M. Noordsy (2004). "Development of paired fins in the paddlefish, Polyodon spathula." *J. Morphology* **261**: 334–344.

Maderson, P. F. (2003). "Mammalian skin evolution: a reevaluation." *Exp. Dermatol.* **12**(3): 233–236.

Maderson, P. F. A. (1971). "On how an archosaurian scale might have given rise to an avian feather." *Am. Nat.* **106**: 424–428.

Maderson, P. F. A. (1972). "When? Why? and How?: some speculations on the evolution of the vertebrate integument." *Am. Zool.* **12**: 159–171.

Maderson, P. F. A. and L. Alibardi (2000). "The development of the sauropsid integument: a contribution to the problem of the origin and evolution of feathers." *Am. Zool.* **40**: 513–529.

Mangan, S. and U. Alon (2003). "Structure and function of the feed-forward loop network motif." *Proc. Natl. Acad. Sci., USA* **100**: 11980–11985.

Mar Albà, M., M. F. Santibáñez-Koref, et al. (1999). "Amino acid reiterations in yeast are overrepresented in particular classes of proteins and show evidence of a slippage-like mutational process." *J. Molec. Evol.* **49**(6): 789–797.

Mariani, F. V. and G. R. Martin (2003). "Deciphering skeletal patterning: clues from the limb." *Nature* **423**(6937): 319–325.

Marino-Ramirez, L., K. C. Lewis, et al. (2005). "Transposable elements donate lineage-specific regulatory sequences to host genomes." *Cytogenet. Genome Res.* **110**(1–4): 333–341.

Márquez-Guzmán, J., M. Engelman, et al. (1989). "Anatomia reproductiva de *Lacandonia schismatica* (Lacadoniaceae)." *Ann. Missouri Bot. Gard.* **76**: 124–127.

Masel, J. and A. Bergman (2003). "The evolution of the evolvability properties of the yeast prion [PSI+]." *Evolution* **57**(7): 1498–1512.

Maynard-Smith, J., R. Burian, et al. (1985). "Developmental constraints and evolution." *Quart. Rev. Biol.* **60**: 265–287.

Maynard-Smith, J. and K. Sondhi (1960). "The genetics of a pattern." *Genetics* **45**: 1039–1050.

Mayr, E. (1942). Systematics and the Origin of Species. New York, Columbia University Press.

Mayr, E. (1959). "Typological and population thinking," in Evolution and Anthropology: A Centennial Appraisal, eds. B. J. Meggers. Washington, The Anthroplogical Society of Washington: 409–412.

Mayr, E. (1960). "The emergence of evolutionary novelties," in Evolution after Darwin, ed. S. Tax. Cambridge, MA, Harvard University Press. **1**: 349–380.

Mayr, E. (1982). The Growth of Biological Thought. Cambridge, London, The Belknap Press.

Mayr, E. (1983). "How to carry out the adaptationist program?" *Am. Nat.* **121**: 324–334.

Mayr, E. (1987). "The ontological status of species: scientific progress and philosophical terminology." *Biol. Phil.* **2**: 145–166.

McBride, R. C., C. B. Ogbunugafor, et al. (2008). "Robustness promotes evolvability of thermotolerance in an RNA virus." *BMC Evol. Biol.* **11**: 231.

McGinnis, N., M. A. Kuziora, et al. (1990). "Human Hox-4.2 and Drosophila deformed encode similar regulatory specificities in Drosophila embryos and larvae." *Cell* **63**: 969–976.

McGlinn, E. and C. J. Tabin (2006). "Mechanistic insights into how Shh patterns the vertebrate limb." *Curr. Opin. Genet. Devel.* **16**: 426–432.

McGregor, A. P., V. Orgogozo, et al. (2007). "Morphological evolution through multiple cis-regulatory mutations at a single gene." *Nature* **448**(7153): 587–590.

McKitrick, M. (1994). "On homology and the ontological relationships of parts." *Syst. Biol.* **43**: 1–10.

Meiklejohn, C. D. and D. L. Hartl (2002). "A single model of canalization." *TREE* **17** (468–473).

Melzer, R., Y. Q. Wang, et al. (2010). "The naked and the dead: the ABCs of gymnosperm reproduction and the origin of the angiosperm flower." *Semin. Cell Dev. Biol.* **21**(1): 118–128.

Merabet, S. and B. Hudry (2011). "On the border of the homeotic function: re-evaluating the controversial role of cofactor-recruiting motifs: the role of cofactor-recruiting motifs in conferring Hox evolutionary flexibility may critically depend on the protein environment." *Bioessays* **33**(7): 499–507.

Mercader, N. (2007). "Early steps of paired fin development in zebrafish compared with tetrapod limb development." *Dev. Growth Differ.* **49**(6): 421–437.

Metscher, B. D. and P. E. Ahlberg (1999). "Zebrafish in context: uses of a laboratory model in comparative studies." *Dev. Biol.* **210**: 1–4.

Metscher, B. D., K. Takahashi, et al. (2005). "Expression of Hoxa-11 and Hoxa-13 in the pectoral fin of a basal rayfinned fish, Polyodon spathula: implications for the origin of tetrapod limbs." *Evol. Dev.* **7**: 186–195.

Meyer, A. and S. I. Dolven (1992). "Molecules, fossils and the origin of tetrapods." *J. Mol. Evol.* **35**: 102–113.

Meyer, A. and A. C. Wilson (1990). "Origin of tetrapods inferred from their mitochondrial DNA affiliation to lungfish." *J. Mol. Evol.* **31**: 359–364.

Michaud, J. L., F. Lapointe, et al. (1997). "The dorsoventral polarity of the presumptive limb is determined by signals produced by the somites and by the lateral somatopleure." *Development* **124**(8): 1453–1463.

Michod, R. E. (1999). Darwinian Dynamics: Evolutionary Transitions in Fitness and Individuality. Princeton, NJ, Princeton University Press.

Michod, R. E. (2006). "The group covariance effect and fitness trade-offs during evolutionary transitions in individuality." *Proc. Natl. Acad. Sci., USA* **103**(24): 9113–9117.

Mikkola, M. L. (2007). "Genetic basis of skin appendage development." *Semin. Cell Dev. Biol.* **18**(2): 225–236.

Mikkola, M. L. and I. Thesleff (2003). "Ectodysplasin signaling in development." *Cytokine Growth Factor Rev.* **14**(3–4): 211–224.

Milton, C. C., B. Huynh, et al. (2003). "Quantitative trait symmetry independent of Hsp90 buffering: distinct modes of genetic canalization and developmental stability." *Proc. Natl. Acad. Sci., USA* **100**(23): 13396–13401.

Milton, C. C., C. M. Ulane, et al. (2006). "Control of canalization and evolvability by Hsp90." *PLoS ONE* **1**: e75.

Minelli, A. (2002). "Homology, limbs, and genitalia." *Evol. Dev.* **4**(2): 127–132.

Minelli, A. and G. Fusco (2005). "Conserved versus innovative features in animal body organization." *J. Exp. Zool. B. Mol. Dev. Evol.* **304**(6): 520–525.

Mitgutsch, C., M. K. Richardson, R. Jiménez, J. E. Martin, P. Kondrashov, M. A. de Bakker and M. R. Sánchez-Villagra (2011). "Circumventing the polydactyly 'constraint': the mole's 'thumb.'" *Biol. Lett.* **8**(1): 74–77.

Mivart, S. G. (1879). "Notes on the fins of elasmobranchs, with considerations on the nature and homologies of vertebrate limbs." *Trans. Zool. Soc. London* **10**: 439–484.

Montavon, T., J.-F. Le Garrec, et al. (2008). "Modeling HOX gene regulation in digits: reverse collinearity and the molecular origin of thumbness." *Genes Dev.* **22**: 346–359.

Montavon, T., N. Soshnikova, et al. (2011). "A regulatory archipelago controls Hox gene transcription in digits." *Cell* **147**(5): 1132–1145.

Monteiro, A. (2008). "Alternative models for the evolution of eyespots and of serial homology on lepidopteran wings." *Bioessays* **30**(4): 358–366.

Monteiro, A., G. Glaser, et al. (2006). "Comparative insights into questions of lepidopteran wing pattern homology." *BMC Dev. Biol.* **6**: 52.

Monteiro, A., J. Prijs, et al. (2003). "Mutants highlight the modular control of butterfly eyespot patterns." *Evol. & Dev.* **5**: 180–187.

Montville, R., R. Froissart, et al. (2005). "Evolution of mutational robustness in an RNA virus." *PLoS Biol.* **3**(11): e381.

Morgan, B. A., J.-C. Izpisúa-Belmonte, et al. (1992). "Targeted misexpression of Hox-4.6 in the avian limb bud causes apparent homeotic transformations." *Nature* **358**: 236–239.

Morse, E. (1872). "On the carpus and tarsus of birds." Ann. Lyc. Nat. Hist., New York.

Mountjoy, D. J. and R. T. Robertson (1988). "Why are waxwings "waxy"? Delayed plumage maturation in the cedar waxwing." *The Auk* **105**(1): 61–69.

Müller, G. B. (1989). "Ancestral patterns in bird limb development: a new look at Hampé's experiment." *J. Evol. Biol.* **2**: 31–47.

Müller, G. B. (1990). "Developmental mechanisms at the origin on morphological novelty: a side-effect hypothesis," in Evolutionary Innovations, ed. M. H. Nitecki. Chicago, The University of Chicago Press: 99–130.

Müller, G. B. (2003). "Embryonic motility: environmental influences and evolutionary innovation." Evol. Dev. 5(1): 56–60.

Müller, G. B. (2003). "Homology: the evolution of morphological organization," in Origination of Organismal Form, eds. G. B. Müller and S. A. Newman. Cambridge, MA, MIT Press: 51–69.

Müller, G. B. (2010). "Epigenetic innovation," in Evolution—the Extended Synthesis, eds. M. Pigliucci and G. B. Müller. Boston, MA, MIT Press: 307–332.

Müller, G. B. and P. Alberch (1990). "Ontogeny of the limb skeleton in Alligator mississippiensis: developmental invariance and change in the evolution of Archosaur limbs." J. Morph. 203: 151–164.

Müller, G. B. and S. A. Newman (1999). "Generation, integration, autonomy: three steps in the evolution of homology" in Homology, eds. G. R. Bock and G. Cardew. New York, John Wiley & Sons: 65–79.

Müller, G. B. and J. Streicher (1989). "Ontogeny of the syndesmosis tibiofibularis and the evolution of the bird hindlimb: a caenogenetic feature triggers phenotypic novelty." Anatomy and Embryology 179: 327–339.

Müller, G. B., J. Streicher, et al. (1996). "Homeotic duplication of the pelvic body segment in regenerating tadpole tails induced by retinoic acid." Dev. Genes Evol. 206: 344–348.

Müller, G. B. and G. P. Wagner (1991). "Novelty in evolution: restructuring the concept." Ann. Rev. Ecol. Syst. 22: 229–256.

Muñoz-Fernández, R., F. J. Blanco, et al. (2006). "Follicular dendritic cells are related to bone marrow stromal cell progenitors and to myofibroblasts." J. Immunol. 177(1): 280–289.

Murtha, M. T., J. F. Lechman, et al. (1991). "Detection of homeobox genes in development and evolution." Proc. Natl. Acad. Sci., USA 88: 10711–10715.

Nagashima, H., S. Kuraku, et al. (2007). "On the carapacial ridge in turtle embryos: its developmental origin, function and the chelonian body plan." Development 134(12): 2219–2226.

Nagashima, H., F. Sugahara, M. Takechi, R. Ericsson, Y. Kowashima-Ohya, Y. Narita, and S. Kuratani (2009). "Evolution of the turtle body plan by the folding and creation of new muscle connections." Science 325(5937): 193–196.

Nathan, D. F., M. H. Vos, et al. (1997). "In vivo functions of the Saccharomyces cerevisiae Hsp90 chaperone." Proc. Natl. Acad. Sci., USA 94: 12949–12956.

Neduva, V. and R. B. Russell (2005). "Linear motifs: evolutionary interaction switches." FEBS Letters 579(15): 3342–3345.

Nelson, C. E., B. A. Morgan, et al. (1996). "Analysis of Hox gene expression in the chick limb bud." Development 122: 1449–1466.

Nelson, G. J. (1978). "Ontogeny, phylogeny, paleontology and the biogenetic law." Syst. Zool. 27: 324–345.

Nelson, J. S. (1994). Fishes of the World. New York, John Wiley & Sons, Inc.

Nemeschkal, H. L. (1999). "Morphometric correlation patterns of adult birds (Fringillidae: Passeriformes and Columbiformes) mirror the expression of developmental control genes." Evolution 53: 899–918.

Newman, S. A. and G. B. Müller (2001). "Epigenetic mechanisms of character origination," in The Character Concept in Evolutionary Biology, ed. G. P. Wagner. San Diego, Academic Press: 559–580.

Newman, S. A. and G. B. Müller (2005). "Origination and innovation in the vertebrate limb skeleton: an epigenetic perspective." *J. Exp. Zool. B. Mol. Dev. Evol.* **304**(6): 593–609.

Niemann, C., A. B. Unden, et al. (2003). "Indian hedgehog and beta-catenin signaling: role in the sebaceous lineage of normal and neoplastic mammalian epidermis." *Proc. Natl. Acad. Sci., USA* **100 Suppl 1**: 11873–11880.

Nijhout, H. F. (1991). The Development and Evolution of Butterfly Wing Patterns. Washington and London, Smithsonian Institution Press.

Nikbakht, N. and J. C. McLachlan (1999). "Restoring avian wing digits." *Proc. Royal Soc. Lond. G* **266**: 1101–1104.

Nimwegen, E. V., J. P. Crutchfield, et al. (1999). "Neutral evolution of mutational robustness." *Proc. Natl. Acad. Sci., USA* **96**: 9716–9720.

Niwa, H., Y. Sekita, et al. (2008). "Platypus Pou5f1 reveals the first steps in the evolution of trophectoderm differentiation and pluripotency in mammals." *Evol. Dev.* **10**(6): 671–682.

Niwa, H., Y. Toyooka, et al. (2005). "Interaction between Oct3/4 and Cdx2 determines trophectoderm differentiation." *Cell* **123**(5): 917–929.

Norell, M. A. and X. Xu (2005). "Feathered dinosaurs." *Ann. Rev. Earth Plant. Sci.* **33**: 277–299.

Novershtern, N., A. Subramanian, et al. (2011). "Densely interconnected transcriptional circuits control cell states in human hematopoiesis." *Cell* **144**(2): 296–309.

Nowak, M. A. (2006). Evolutionary Dynamics: Exploring the Equations of Life. Cambridge, MA, Belknap Press of Harvard University Press.

Nyhart, L. K. (1995). Biology Takes Form. Animal Morphology and the German Universities, 1800–1900. Chicago, IL, University of Chicago Press.

Nyhart, L. K. (2002). "Learning from history: morphology's challenges in Germany ca. 1900." *J. Morph.* **252**: 2–14.

Oakley, T. H. (2003). "The eye as a replicating and diverging, modular developmental unit." *Trends Ecol. Evol.* **18**: 623–627.

Oftendal, O. T. (2002). "The mammary gland and its origin during synapsid evolution." *J. Mammary Gland Biol. & Neoplasia* **7**(3): 225–252.

O'Hara, R. J. (1997). "Population thinking and tree thinking in systematics." *Zoologica Scripta* **26**(4): 323–329.

Ohno, S. (1970). Evolution by Gene Duplication. New York, Springer Verlag.

Oliver, C., M. J. Montes, et al. (1999). "Human decidual stromal cells express alpha-smooth muscle actin and show ultrastructural similarities with myofibroblasts." *Hum. Reprod.* **14**(6): 1599–1605.

Oliver, J. C., J. M. Beaulieu, et al. (2013). "Nymphalid eyespot serial homologs originate as a few individualized modules."; submitted.

Oliver, K. R. and W. K. Greene (2009). "Transposable elements: powerful facilitators of evolution." *BioEssays* **31**: 703–714.

Olivera-Martinez, I., J. Thelu, et al. (2004). "Molecular mechanisms controlling dorsal dermis generation from the somitic dermomyotome." *Int. J. Dev. Biol.* **48**(2–3): 93–101.

Oliveri, P., Q. Tu, et al. (2008). "Global regulatory logic for specification of an embryonic cell lineage." *Proc. Natl. Acad. Sci., USA* **105**(16): 5955–5962.

Olson, E. C. and R. L. Miller (1958). Morphological Integration. Chicago, IL, Chicago University Press.

Olson, S. L. (1970). "Specializations of some carotenoid-bearing feathers." The Condor 72(4): 424–430.

Orzack, S. H. and E. Sober, eds. (2001). Adaptation and Optimality. Cambridge Studies in Philosophy and Biology. Cambridge, UK, Cambridge University Press.

Osborn, H. F. (1888). "The evolution of mammalian molars to and from the tritubercular type." Am. Nat. 22: 1067–1079.

Owen, R. (1836). "Aves." Todd's Cyclopedia of Anatomy and Physiology. Todd. 1: 265–358.

Owen, R. (1848). On the Archetype and Homologies of the Vertebrate Skeleton. London, Voorst.

Owen, R. (1849). On the Nature of Limbs. A Discourse. London, John van Voorst.

Pappu, K. S., E. J. Ostrin, et al. (2005). "Dual regulation and redundant function of two eye-specific enhancers of the Drosophila retinal determination gene dachshund." Development 132(12): 2895–2905.

Park, J. H., Y. Ishikawa, et al. (2003). "Expression of AODEF, a B-functional MADS-box gene, in stamens and inner tepals of the dioecious species Asparagus officinalis L." Plant. Mol. Biol. 51(6): 867–875.

Park, M., C. Lewis, et al. (1998). "Differential rescue of visceral and cardiac defects in Drosophila by vertebrate tinman-related genes." Proc. Natl. Acad. Sci., USA 95(16): 9366–9371.

Patel, N. H., E. F. Ball, et al. (1992). "Changing role of even-skipped during the evolution of insect pattern formation." Nature 357: 339–342.

Patterson, C. (1982). "Morphological characters and homology," in Problems of Phylogenetic Reconstruction, eds. K. A. Joysey and A. E. Friday. London and New York, Academic Press: 21–74.

Paulus, H. F. (1989). "Das Homologisieren in der Feinstrukturforschung: das Bolwig-Organ der höheren Dipteren und sein Homologisierung mit Stemmata und Ommatidien eines ursprünglichen Facettenauges der Mandibulata." Zool. Beitr. N. F. 32: 437–478.

Pavlicev, M., J. M. Cheverud, et al. (2011). "Evolution of adaptive phenotypic variation patterns by direct selection for evolvability." Proc. Biol. Sci. 278(1713): 1903–1912.

Pavlicev, M., J. P. Kenney-Hunt, et al. (2008). "Genetic variation in pleiotropy: differential epistasis as a source of variation in the allometric relationship between long bone lengths and body weight." Evolution 62: 199–213.

Pavlicev, M., E. A. Norgard, et al. (2011). "Evolution of pleiotropy: epistatic interaction pattern supports a mechanistic model underlying variation in genotype-phenotype map." J. Exp. Zool. B. Mol. Dev. Evol. 316(5): 371–385.

Pertea, M. and S. L. Salzberg (2010). "Between a chicken and a grape: estimating the number of human genes." Genome Biol. 11(5): 206.

Peterson, K. J., M. R. Dietrich, et al. (2009). "MicroRNAs and metazoan macroevolution: insights into canalization, complexity, and the Cambrian explosion." Bioessays 31(7): 736–747.

Pigliucci, M. (2008). "Is evolvability evolvable?" Nat. Rev. Genet. 9(1): 75–82.

Pinkus, F. (1922). "Neue Befunde zur Entstehung des Haarkleides der Säugetiere." Die Naturwissenschaften 10(23): 521–525.

Plachetzki, D. C., C. R. Fong, et al. (2010). "The evolution of phototransduction from an ancestral cyclic nucleotide gated pathway." Proc. Biol. Sci. 277(1690): 1963–1969.

Platnick, N. I. and H. D. Cameron (1977). "Cladistic methods in textual, linguistic and phylogenetic analysis." *Syst. Zool.* **26**: 380–385.

Post, L. C. and J. W. Innis (1999). "Altered Hox expression and increased cell death distinguish Hyperdactyl from Hoxa13 null mice." *Int. J. Dev. Biol.* **43**: 287–294.

Powers, T. P. and C. T. Amemiya (2004). "Evidence for a Hox14 paralog group in vertebrates." *Curr. Biol.* **14**(5): R183–184.

Prince, V. and F. B. Pickett (2002). "Splitting pairs: the diverging fates of duplicated genes." *Nature Reviews Genetics* **3**: 827–837.

Prochel, J. (2006). "Early skeletal development in Talpa europaea, the common European mole." *Zoolog. Sci.* **23**(5): 427–434.

Prohaska, S. J., P. F. Stadler, et al. (2006). "Evolutionary genomics of Hox gene clusters" in HOX Gene Expression, ed. S. Papageorgiou. New York, Landes Bioscience.

Prud'homme, B., R. de Rosa, et al. (2003). "Arthropod-like expression patterns of engrailed and wingless in the annelid Platynereis dumerilii suggest a role in segment formation." *Curr. Biol.* **13**(21): 1876–1881.

Prud'homme, B., N. Gompel, et al. (2006). "Repeated morphological evolution through cis-regulatory changes in a pleiotropic gene." *Nature* **440**: 1050–1053.

Prud'homme, B., N. Gompel, et al. (2007). "Emerging principles of regulatory evolution." *Proc. Natl. Acad. Sci., USA* **104** (suppl 1): 8605–8612.

Prud'homme, B., C. Minervino, et al. (2011). "Body plan innovation in treehoppers through the evolution of an extra wing-like appendage." *Nature* **473**(7345): 83–86.

Prum, R. O. (1990). "Phylogenetic analysis of the evolution of display behaviour in the Neotropical manakins (Aves: Pipridae)." *Ethology* **84**: 202–231.

Prum, R. O. (1999). "Development and evolutionary origin of feathers." *J. Exp. Zool. A. Mol. Dev. Evol.* **285**: 291–306.

Prum, R. O. and A. H. Brush (2002). "The evolutionary origin and diversification of feathers." *Quart. Rev. Biol.* **77**: 261–295.

Prum, R. O. and J. Dyck (2003). "A hierarchical model of plumage: morphology, development, and evolution." *J. Exp. Zool. B. Mol. Dev. Evol.* **298**(1): 73–90.

Pueyo, J. I., R. Lanfear, et al. (2008). "Ancestral Notch-mediated segmentation revealed in the cockroach Periplaneta americana." *Proc. Natl. Acad. Sci., USA* **105**(43): 16614–16619.

Punzo, C., M. Seimiya, et al. (2002). "Differential interactions of eyeless and twin of eyeless with the sine oculis enhancer." *Development* **129**(3): 625–634.

Queiroz, K. D. (1985). "The ontogenetic method for determining character polarity and its relevance to phylogenetic systematics." *Syst. Zool.* **34**: 280–299.

Queitsch, C., T. A. Sangster, et al. (2002). "Hsp90 as a capacitor of phenotypic variation." *Nature* **417**: 618–624.

Raff, R. (1996). The Shape of Life. Chicago, IL, Chicago University Press.

Raff, R. A. and T. C. Kaufman (1983). Embryos, Genes, and Evolution. New York, Macmillan Publishing Co., Inc.

Raincrow, J. D., K. Dewar, et al. (2011). "Hox clusters of the bichir (Actinopterygii, Polypterus senegalus) highlight unique patterns of sequence evolution in gnathostome phylogeny." *J. Exp. Zool. B. Mol. Dev. Evol.* **316**(6): 451–464.

Ranganayakulu, G., D. A. Elliott, et al. (1998). "Divergent roles for NK-2 class homeobox genes in cardiogenesis in flies and mice." *Development* **125**(16): 3037–3048.

Rao, M. S. and M. Jacobson (2005). Developmental Neurobiology. New York, Kluwer Academic.

Rasmussen, D. A., E. M. Kramer, et al. (2009). "One size fits all? Molecular evidence for a commonly inherited petal identity program in Ranundulares." *Am. J. Bot.* **96**(1): 96–109.

Raynaud, A. (1990). "Developmental mechanisms involved in the embryonic reduction of limbs in reptiles." *Int. J. Dev. Biol.* 34: 233–243.

Raynaud, A. and M. Clergue-Gazeau (1986). "Identification des doigts réduits ou manquants dans les pattes des embryons de lézard vert (Lacerta viridis) tarités par la cytosine-arabinofuranoside. Camparaison avec les dréductions digitales naturalles des espèces de reptiles sependtiformes." *Arch. Biol. (Bruxelles)* **97**: 279–299.

Raynaud, A., M. Clergue-Gazeau, et al. (1986). "Remarques preliminaires sur la structure de la patte du Seps tridactyle (Chalcides chalcides, L.)." *Bull. Soc. Hist. Nat., Toulouse* **122**: 109–111.

Raynaud, A., M. Clergue-Gazeau, et al. (1987). "Nouvelles observations, fondees sur les caracteres du metatarsien lateral et sur la structure du tarse, relative a la formule digitale du Seps tridactyle (Chalcides chalcides, L.)." *Bull. Soc. Hist. Nat., Toulouse* **123**: 127–132.

Raynaud, A., J.-L. Perret, et al. (1989). "Modalités de la réduction digitale chez quelques Scincidés africains (Reptiles)." *Rev. Suisse Zool.* **96**: 779–802.

Rebay, I., S. J. Silver, et al. (2005). "New vision from eyes absent: transcription factors as enzymes." *Trends Genet.* **21**(3): 163–171.

Rebollo, R., M. T. Romanish, et al. (2012). "Transposable elements: an abundant and natural source of regulatory sequences for host genes." *Annu. Rev. Genet.* **46**: 21–42.

Ree, R. H. and M. J. Donoghue (1999). "Inferring rates of change in flower symmetry in asterid angiosperms." *Syst. Biol.* **48**: 633–341.

Reed, R. D. and M. S. Serfas (2004). "Butterfly wing pattern evolution is associated with changes in a Notch/Distal-less temporal pattern formation process." *Curr. Biol.* **14**(13): 1159–1166.

Reeve, H. K. and P. W. Sherman (1993). "Adaptations and the goals of evolutionary research." *Quart. Rev. Biol.* **68**: 1–32.

Remane, A. (1952). Die Grundlagen des Natürlichen Systems, der vergleichenden Anatomie und der Phylogenetik. Leipzig, Akademische Verlagsgesellschaft Geest & Portig.

Rendel, J. M. (1967). Canalization and Gene Control. New York, Logos Press, Academic Press.

Reno, P. L., M. A. McCollum, et al. (2008). "Patterns of correlation and covariation of anthropoid distal forelimb segments correspond to Hoxd expression territories." *J. Exp. Zool. B. Mol. Dev. Evol.* **310B**(3): 240–258.

Renous-Lecuru, S. (1973). "Morphologie comparee du carpe chez les Lepidosauriens actuels (Rhynchocephales, Lacertilens, Amphisbeniens)." *Gegenbaurs morph. Jahrb., Leipzig* **119**: 727–766.

Rhee, H., L. Polak, et al. (2006). "Lhx2 maintains stem cell character in hair follicles." *Science* **312**(5782): 1946–1949.

Rheinberger, H.-J. (2000). "Genetic concepts: fragments from the perspective of molecular biology," in The Concept of the Gene in Development and Evolution, eds. P. J. Beurton, R. Falk and H.-J. Rheinberger. Cambridge, Cambridge University Press: 219–239.

Richardson, M. K. and A. D. Chipman (2003). "Developmental constraints in a comparative framework: a test case using variations in phalanx number during amniote evolution." *J. Exp. Zool. B. Mol. Dev. Evol.* **296**(1): 8–22.

Richardson, M. K., J. Hanken, et al. (1997). "There is no highly conserved embryonic stage in the vertebrates: implications for current theories of evolution and development." *Anat. Embryol. (Berlin)* **196**(2): 91–106.

Riddle, R. D., R. L. Johnson, et al. (1993). "Sonic hedgehog mediates the polarizing activity of the ZPA." *Cell* **75**: 1401–1416.

Riedl, R. (1977). "A systems-analytical approach to macro-evolutionary phenomena." *Quart. Rev. Biol.* **52**: 351–370.

Riedl, R. (1978). Order in Living Organisms: A Systems Analysis of Evolution. New York, Wiley.

Rienesl, J. and G. P. Wagner (1992). "Constancy and change of basipodial variation patterns: a comparative study of crested and marbled newts—*Triturus cristatus, Triturus marmoratus*—and their natural hybrids." *J. Evol. Biol.* **5**: 307–324.

Rieppel, O. (1996). "Testing homology by congruence: the pectoral girdle of turtles." *Proc. R. Soc. Lond. B* **263**: 1395–1398.

Rieppel, O. (2001). "Turtles as hopeful monsters." *Bioessays* **23**(11): 987–991.

Rieppel, O. and R. R. Reisz (1999). "The origin and early evolution of turtles." *Annu. Rev. Ecol. System.* **30**: 1–22.

Roff, D. A. (1997). Evolutionary Quantitative Genetics. New York, Chapman and Hall.

Romer, A. S. (1956). Osteology of Reptiles. Chicago, IL, University of Chicago Press.

Ronshaugen, M., N. McGinnis, et al. (2002). "Hox protein mutation and macroevolution of the insect body plan." *Nature* **415**: 914–917.

Rorick, M. M. and G. P. Wagner (2011). "Protein structural modularity and robustness are associated with evolvability." *Genome Biol. Evol.* **3**: 456–475.

Rossetti, V., B. E. Schirrmeister, et al. (2010). "The evolutionary path to terminal differentiation and division of labor in cyanobacteria." *J. Theor. Biol.* **262**(1): 23–34.

Roth, V. L. (1984). "On homology." *Biol. J. Linn. Sec.* **22**: 13–29.

Roth, V. L. (1988). "The biological basis of homology," in Ontogeny and Systematics, ed. C. J. Humphries. New York, Columbia University Press: 1–26.

Rotmann, E. (1935). "Reiz und Reizbeantwortung in der Amphibienentwicklung." *Verh. d. Zool. Ges.* **1935**: 76–83.

Ruddle, F. H., J. L. Bartels, et al. (1994). "Evolution of Hox genes." *Annu. Rev. Genet.* **28**: 423–442.

Rueffler, C., J. Hermisson, et al. (2012). "Evolution of functional specialization and division of labor." *Proc. Natl. Acad. Sci., USA* **109**(6): E326–335.

Rugg, E. L. and I. M. Leigh (2004). "The keratins and their disorders." *Am. J. Med. Genet. C. Semin. Med. Genet.* **131C**(1): 4–11.

Rupke, N. A. (1994). Richard Owen: Victorian Naturalist. New Haven, CT, Yale University Press.

Ruta, M., M. I. Coates, et al. (2003). "Early tetrapod relationships revisited." *Biol. Rev.* **78**: 251–345.

Rutherford, S. L. (2000). "From genotype to phenotype: buffering mechanisms and the storage of information." *BioEssays* **22**: 1095–1105.

Rutherford, S. L. (2003). "Between genotype and phenotype: protein chaperones and evolvability." *Nat. Rev. Genet.* **4**(4): 263–274.

Rutherford, S. L. and S. Lindquist (1998). "Hsp90 as a capacitor for morphological evolution." *Nature* **396**: 336–342.

Ryan, J. F., M. E. Mazza, et al. (2007). "Pre-bilaterian origins of the Hox cluster and the Hox code: evidence from the sea anemone, Nematostella vectensis." *PLoS One* **2**(1): e153.

Sablowski, R. (2010). "Genes and functions controlled by floral organ identity genes." *Semin. Cell Dev. Biol.* **21**(1): 94–99.

Sakamoto, K., K. Onimaru, et al. (2009). "Heterochronic shift in Hox-mediated activation of sonic hedgehog leads to morphological changes during fin development." *PLoS One* **4**(4): e5121.

Salazar-Ciudad, I. (2007). "On the origins of morphological variation, canalization, robustness, and evolvability." *Integrat. Compar. Biol.* **47**(3): 390–400.

Salvini-Plawen, L. v. and E. Mayr (1977). "On the evolution of photoreceptors and eyes." *Evol. Biol.* **10**: 207–263.

Salvini-Plawen, L. v. and H. Splechtna (1979). "Zur Homologie der Keimblätter." *Z. f. Zool. Systematik u. Evolutionsforschung* **17**: 10–30.

Sander, K. (1983). "The evolution of patterning mechanisms: gleanings from insect embryogenesis and spermatogenesis," in Development and Evolution, eds. B. C. Goodwin, N. Holder and C. C. Wylie. Cambridge, Cambridge University Press: 137–159.

Satoh, N. (2003). "The ascidian tadpole larva: comparative molecular development and genomics." *Nat. Rev. Genet.* **4**(4): 285–295.

Sattler, R. (1984). "Homology—a continuing challenge." *Systematic Botany* **9**(4): 382–394.

Savard, J., H. Marques-Souza, et al. (2006). "A segmentation gene in tribolium produces a polycistronic mRNA that codes for multiple conserved peptides." *Cell* **126**(3): 559–569.

Sawyer, R. H. (1972). "Avian scale development I: Histogenesis and morphogenesis of the epidermis and dermis during formation of the scale ridge." *J. Exp. Zool.* **181**: 365–384.

Sawyer, R. H., T. C. Glenn, et al. (2000). "The expression of beta (b) keratins in the epidermal appendages of reptiles and birds." *Am. Zool.* **40**: 530–539.

Sawyer, R. H., L. Rogers, et al. (2005). "Evolutionary origin of the feather epidermis." *Dev. Dyn.* **232**(2): 256–267.

Sawyer, R. H., B. A. Salvatore, et al. (2003). "Origin of feathers: feather S-keratins are expressed in discrete epidermal cell populations of embryonic scutate scales." *J. Exp. Zool. B. Mol. Dev. Evol.* **295**(1): 12–24.

Schank, J. C. and W. C. Wimsatt (1986). "Generative entrenchment and evolution." *PSA* **2**: 33–60.

Scheiner, S. M. (1993). "Genetics and the evolution of phenotypic plasticity." *Annu. Rev. Ecol. Syst.* **24**: 35–68.

Schlichting, C. D. and M. Pigliucci (1998). Phenotypic Evolution: A Reaction Norm Perspective. Sunderland, MA, Sinauer Associates.

Schlosser, G. (2005). "Evolutionary origins of vertebrate placodes: insights from developmental studies and from comparisons with other deuterostomes." *J. Exp. Zool. B. Mol. Dev. Evol.* **304B**: 347–399.

Schmalhausen, I. I. (1986). Factors of Evolution. The Theory of Stabilizing Selection. Chicago and London, University of Chicago Press.

Schneider, I., I. Aneas, et al. (2011). "Appendage expression driven by the Hoxd Global Control Region is an ancient gnathostome feature." *Proc. Natl. Acad. Sci., USA* **108**(31): 12782–12786.

Scholes-III, E. (2008). "Evolution of the courtship phenotype in the bird of paradise genus *Parotia* (Aves: Paradisaeidae): homology, phylogeny, and modularity." *Biol. J. Linn. Soc.* **94**: 491–504.

Schotté, O. E. and M. V. Edds (1940). "Xenoplastic induction of Rana pipiens adhesive discs on balancer site of Amblystoma punctatum." *J. Exp. Zool.* **84**: 199–221.

Schughart, K., C. Kappen, et al. (1987). "Mammalian homeobox-containing genes: genomic organization, structure, expression and evolution." *Brit. J. Cancer* **58**: 9–13.

Schuster, P., W. Fontana, et al. (1994). "From sequences to shapes and back: a case study in RNA secondary structure." *Proc. Roy. Soc. (London) B* **255**: 279–284.

Schuster, P. and J. Swetina (1988). "Stationary mutant distributions and evolutionary optimization." *Bull. Math. Biol.* **50**: 635–660.

Schweitzer, M. H., J. A. Watt, et al. (1999). "Beta-keratin specific immunological reactivity in feather-like structures of the cretaceous alvarezsaurid, Shuvuuia deserti." *J. Exp. Zool.* **285**(2): 146–157.

Scotting, P. J., D. A. Walker, et al. (2005). "Childhood solid tumours: a developmental disorder." *Nat. Rev. Cancer* **5**(6): 481–488.

Sears, K. E., A. K. Bormet, et al. (2011). "Developmental basis of mammalian digit reduction: a case study in pigs." *Evol. Dev.* **13**(6): 533–541.

Seaver, E. C., D. A. Paulson, et al. (2001). "The spatial and temporal expression of Ch-en, the engrailed gene in the polychaete Chaetopterus, does not support a role in body axis segmentation." *Dev. Biol.* **236**(1): 195–209.

Seidel, F. (1960). "Körpergrundgestalt und Keimstruktur. Eine Erörterung über die Grundlagen der vergleichenden und experimentellen Embryologie und deren Gültigkeit bei phylogenetischen Überlegungen." *Zool. Anz.* **164**: 245–305.

Seimiya, M. and W. J. Gehring (2000). "The Drosophila homeobox gene optix is capable of inducing ectopic eyes by an eyeless-independent mechanism." *Development* **127**(9): 1879–1886.

Sempere, L. F., C. N. Cole, et al. (2006). "The phylogenetic distribution of metazoan microRNAs: insights into evolutionary complexity and constraint." *J. Exp. Zool. B. Mol. Dev. Evol.* **306B**: 575–588.

Sengel, P. (1976). Morphogenesis of Skin. Cambridge, Cambridge University Press.

Serb, J. M. and T. H. Oakley (2005). "Hierarchical phylogenetics as a quantitative analytical framework for evolutionary developmental biology." *Bioessays* **27**(11): 1158–1166.

Shapiro, M. D., J. Hanken, et al. (2003). "Developmental basis of evolutionary digit loss in the Australian lizard Hemiergis." *J. Exp. Zool. B. Mol. Dev. Evol.* 297B: 48–56.

Shapiro, M. D., M. E. Marks, et al. (2004). "Genetic and developmental basis of evolutionary pelvic reduction in threespine sticklebacks." *Nature* **428**(6984): 717–723.

Sharma, B., C. Guo, et al. (2011). "Petal-specific subfunctionalization of an APETALA3 paralog in the Ranunculales and its implications for petal evolution." *New Phytol.* **191**(3): 870–883.

Shearman, R. M. and A. C. Burke (2009). "The lateral somitic frontier in ontogeny and phylogeny." *J. Exp. Zool. B. Mol. Dev. Evol.* **312**(6): 603–612.

Shirasaki, R. and S. L. Pfaff (2002). "Transcriptional codes and the control of neuronal identity." *Ann. Rev. Neurosci.* **25**: 251–281.

Shubin, N. H. (1994). "History, ontogeny, and evolution of the archetype," in Homology: The Hierarchical Basis of Comparative Biology, ed. B. K. Hall. San Diego, Academic Press, Inc.: 249–271.

Shubin, N. H. (1994). "The phylogeny of development and the origin of homology," in Interpreting the Hierarchy of Nature, eds. L. Grande and O. Rieppel. San Diego, Academic Press.

Shubin, N. H. and P. Alberch (1986). "A morphogenetic approach to the origin and basic organization of the tetrapod limb." *Evol. Biol.* **20**: 319–387.

Shubin, N., C. Tabin, et al. (1997). "Fossils, genes and the evolution of animal limbs." *Nature* **388**: 639–648.

Shubin, N., C. Tabin, et al. (2009). "Deep homology and the origins of evolutionary novelty." *Nature* **457**(7231): 818–823.

Shubin, N. and D. Wake (1996). "Phylogeny, variation and morphological integration." *Am. Zool.* **36**: 51–60.

Shubin, N. and D. Wake (1997). "Variation and stability in urodele limb development." *J. Morphol.* 232(3): 323.

Siegal, M. L. and A. Bergman (2003). "Waddington's canalization revisited: developmental stability and evolution." *Proc. Natl. Acad. Sci., USA* **99**: 10528–10532.

Siler, C. D. and R. M. Brown (2011). "Evidence for repeated acquisition and loss of complex body-form characters in an insular clade of Southeast Asian semi-fossorial skinks." *Evolution* 65(9): 2641–2663.

Silva, J. C., S. A. Shabalina, et al. (2003). "Conserved fragments of transposable elements in intergenic regions: evidence for widespread recruitment of MIR- and L2-derived sequences within the mouse and human genomes." *Genet. Res.* **82**(1): 1–18.

Simpson, G. G. (1961). Principles of Animal Taxonomy. New York, Columbia University Press.

Sive, H. and L. Bradley (1996). "A sticky problem: the Xenopus cement gland as a paradigm for anteroposterior patterning." *Dev. Dyn.* **205**(3): 265–80.

Slack, J. M. W. (2003). "Phylotype and Zootype," in Keywords and Concepts in Evolutionary Developmental Biology, eds. B. K. Hall and W. M. Olson. Cambridge, MA, Harvard University Press: 309–317.

Slack, J. M. W., P. W. H. Holland, et al. (1993). "The zootype and the phylotypic stage." *Nature* **361**: 490–492.

Smith, A. B. (1992). "Echinoderm phylogeny: morphology and molecules approach accord." *Trends Ecol. Evol.* **7**: 224–229.

Smith, A. G., J. K. Heath, et al. (1988). "Inhibition of pluripotential embryonic stem cell differentiation by purified polypeptides." *Nature* **336**(6200): 688–690.

Sober, E. (1980). "Evolution, population thinking and essentialism." *Philosophy of Science* **47**(3): 350–383.

Sober, E. (1987). The Nature of Selection. Cambridge, MA, MIT Press.

Soltis, D. E., S. A. Smith, et al. (2011). "Angiosperm phylogeny: 17 genes, 640 taxa." *Am. J. Bot.* **98**(4): 704–730.

Soltis, P. S., D. E. Soltis, et al. (2006). "Expression of floral regulators in basal angiosperms and the origin and evolution of ABC-function." *Adv. Bot. Res.* **44**: 483–506.

Sommer, R. J. and P. W. Sternberg (1994). "Changes of induction and competence during the evolution of vulva development in nematodes." *Science* **265**(5168): 114–118.

Song, M. R., Y. Sun, et al. (2009). "Islet-to-LMO stoichiometries control the function of transcription complexes that specify motor neuron and V2a interneuron identity." *Development* **136**(17): 2923–2932.

Sordino, P., F. v. d. Hoeven, et al. (1995). "Hox gene expression in teleost fins and the origin of vertebrate digits." *Nature* **375**: 678–681.

Specht, C. D. and M. E. Bartlett (2009). "Flower evolution: the origin and subsequent diversification of the angiosperm flower." *Ann. Rev. Ecol. System.* **40**: 217–243.

Spemann, H. (1915). "Zur Geschichte und Kritik des Begriffs der Homologie," in Allgemeine Biologie, eds. C. Chun and W. Johannsen. Leipzig, Germany, Teubner. **3**: 63–86.

Spemann, H. and O. Schotté (1932). "Über xenoplastische Transplantation als Mittel zur Analyse der embryonalen Induktion." *Die Naturwissenschaften* **20**(25): 31–35.

Spitz, F., F. Gonzalez, et al. (2003). "A global control region defines a chromosomal regulatory landscape containing the HoxD cluster." *Cell* **113**: 405–417.

Stadler, B. M. R., P. F. Stadler, et al. (2001). "The topology of the possible: formal spaces underlying patterns of evolutionary change." *J. Theor. Biol.* **213**: 241–274.

Stauber, M., H. Jackle, et al. (1999). "The anterior determinant bicoid of Drosophila is a derived Hox class 3 gene." *Proc. Natl. Acad. Sci., USA* **96**(7): 3786–3789.

Stearns, S. C. (1986). "Natural selection and fitness, adaptation and constraint," in Patterns and Processes in the History of Life, eds. D. M. Raup and D. Jablonski. Berlin, Heidelberg, New York, Springer Verlag: 23–44.

Stearns, S. C. (1989). "The evolutionary significance of phenotypic plasticity." *BioScience* **39**(7): 436–445.

Stearns, S. C. (1992). The Evolution of Life Histories. Oxford, New York, Tokyo, Oxford University Press.

Stearns, S. C. and T. J. Kawecki (1994). "Fitness sensitivity and the canalization of life history traits." *Evolution* **48**: 1438–1450.

Stedman, S. and A. Stedman (1989). "Notes on waxy appendages on cedar waxwings at an Ohio and a Florida banding station." *The North American Bird Bander* **14**(3): 75–77.

Steiner, H. (1934). "Über die embryonale Hand- und Fuss-Skelett-anlage bei den Crocodiliern, sowie über ihre Beziehung zur Vogel-Flügelanlage und zur ursprünglichen Tetrapoden-Extremität." *Rev. Suisse de Zool.* **41**: 383–396.

Steiner, H. and G. Anders (1946). "Zur Frage der Entstehung von Rudimenten. Die Reduktion der Gliedmassen von Chalcides tridactylus Laur." *Rev. Suisse Zool.* **53**: 537–546.

Stenn, K. S., Y. Zheng, et al. (2008). "Phylogeny of the hair follicle: the sebogenic hypothesis." *J. Invest. Dermatol.* **128**(6): 1576–1578.

Sterelny, K. (2000). "Development, evolution and adaptation." *Phil. Science* **67**: S369–S387.

Stern, D. L. (1998). "A role of Ultrabithorax in morphological differences between Drosophila species." *Nature* **396**: 463–466.

Stern, D. L. (2011). Evolution, Development & the Predictable Genome. Greenwood Village, CO, Roberts and Company Publishers.

Stern, D. L. and V. Orgogozo (2008). "The loci of evolution: how predictable is genetic evolution?" *Evolution* **62**(9): 2155–2177.

Stevens, P. F. (2000). "On characters and character states: do overlapping and non-overlapping variation, morphology and molecules all yield data of the same value?" in Homology and Systematics, eds. R. Scotland and R. T. Pennington. Boca Raton, London, New York, CRC Press: 81–105.

Stopper, G. F. and G. P. Wagner (2007). "Inhibition of sonic hedgehog signaling leads to posterior digit loss in Ambystoma mexicanum: parallels to natural digit reduction in urodeles." *Devel. Dynamics* **236**: 321–331.

Stotz, K. and P. E. Griffith (2004). "Genes: philosophical analyses put to the test." *History and Philosophy of the Life Sciences* **26**: 5–28.

Strand, D. J. and J. F. McDonald (1985). "Copia is transcriptionally responsive to environmental stress." *Nucleic Acids Res.* **13**(12): 4401–4410.

Streicher, J. and G. B. Müller (1992). "Natural and experimental reduction of the avian fibula: developmental thresholds and evolutionary constraint." *J. Morph.* **214**: 269–285.

Stümpke, H. (1967). The Snouters: Form and Life of the Rhinogrades. New York, American Museum of Natural History Press.

Sucena, E., I. Delon, et al. (2003). "Regulatory evolution of shavenbaby/ovo underlies multiple cases of morphological parallelism." *Nature* **424**(6951): 935–938.

Sucena, E. and D. L. Stern (2000). "Divergence of larval morphology between Drosophila sechellia and its sibling species caused by cis-regulatory evolution of ovo/shavenbaby." *Proc. Natl. Acad. Sci., USA* **97**(9): 4530–4534.

Sun, G., Q. Ji, et al. (2002). "Archaefructaceae, a new basal angiosperm family." *Science* **296**(5569): 899–904.

Sundstrom, J. and P. Engstrom (2002). "Conifer reproductive development involves B-type MADS-box genes with distinct and different activities in male organ primordia." *Plant J.* **31**(2): 161–169.

Suzuki, T., S. M. Hasso, et al. (2008). "Unique SMAD1/5/8 activity at the phalanx-forming region determines digit identity." *Proc. Natl. Acad. Sci., USA* **105**(11): 4185–4190.

Suzuki, T., J. Takeuchi, et al. (2004). "Tbx genes specify posterior digit identity through Shh and BMP signaling." *Dev. Cell* **6**(1): 43–53.

Syed, T. and B. Schierwater (2002). "Trichoplax adhaerens: discovered as a missing link, forgotten as a hydrozoan, re-discovered as a key to metazoan evolution." *Vie Milieu* **52**(4): 177–187.

Tamura, M. (1965). "Morphology, ecology and phylogeny of the Ranunculaceae." *Sci. Rep. Osaka Univ.* **14**: 53–71.

Tamura, K., N. Nomura, et al. (2011). "Embryological evidence identifies wing digits in birds as digits 1, 2, and 3." *Science* **331**(6018): 753–757.

Tanaka, K., O. Barmina, et al. (2011). "Evolution of sex-specific traits through changes in HOX-dependent doublesex expression." *PLoS Biol.* **9**(8): e1001131.

Tanaka, M., A. Munsterberg, et al. (2002). "Fin development in cartilaginous fish and the origin of vertebrate limbs." *Nature* **416**: 527–531.

Tanaka, M., K. Tamura, et al. (1997). "Induction of additional limb at the dorsal-ventral boundary of a chick embryo." *Dev. Biol.* **182**(1): 191–203.

Taylor, J. S., I. Braasch, et al. (2003). "Genome duplication, a trait shared by 22,000 species of ray-finned fish." *Genome Res.* **13**: 382–390.

Taylor, J. S. and J. Raes (2004). "Duplication and divergence: the evolution of new genes and old ideas." *Annu. Rev. Genet.* **38**: 615–643.

Templeton, A. (1989). "The meaning of species and speciation: a genetic perspective" in Speciation and its Consequences, eds. D. Otte and J. A. Endler. Sunderland, MA, Sinauer Associates: 3–27.

Tchernov, E., O. Rieppel, et al. (2000). "A fossil snake with limbs." *Science* **287**(5460): 2010–2012.

Thamm, K. and E. C. Seaver (2008). "Notch signaling during larval and juvenile development in the polychaete annelid Capitella sp. I." *Dev Biol.* **320**(1): 304–318.

Thanos, D. and T. Maniatis (1995). "Virus induction of human IFN beta gene expression requires the assembly of an enhanceosome." *Cell* **83**(7): 1091–1100.

Thatcher, J. K. (1877). "Median and paired fins, a contribution to the history of vertebrate limbs." *Trans. Conn. Acad.* **3**: 281–310.

Theissen, G. (2001). "Development of floral organ identity: stories from the MADS house." *Curr. Opin. Plant Biol.* **4**(1): 75–85.

Theissen, G., A. Becker, et al. (2000). "A short history of MADS-box genes in plants." *Plant Mol. Biol.* **42**(1): 115–149.

Thompson, D. A. and F. W. Stahl (1999). "Genetic control of recombination partner preference in yeast meiosis. Isolation and characterization of mutants elevated for meiotic unequal sister-chromatid recombination." *Genetics* **153**(2): 621–641.

Thomson, K. (2009). *The Young Charles Darwin.* New Haven, CT, Yale University Press.

Thulborn, R. A. (2006). "Theropod dinosaurs, progenesis and birds: homology of digits in the manus." *N. Jb. Geol. Paläont. Abh.* **242**: 205–241.

Tischler G. (1915). "Chromosomenzahl, Form und Individualität in Planzenreiche." *Progr. Rei Bot.* **5**:164

Tokita, M. and N. Iwai (2010). "Development of the pseudothumb in frogs." *Biol. Lett.* **6**(4): 517–520.

Tomoyasu, Y., S. R. Wheeler, et al. (2005). "Ultrabithorax is required for membranous wing identity in the beetle Tribolium castaneum." *Nature* **375**: 58–61.

Toni, M., L. D. Valle, et al. (2007). "Hard (beta-) keratins in the epidermis of reptiles: composition, sequence, and molecular organization." *J. Proteome Res.* **6**(9): 3377–3392.

Torok, M. A., D. M. Gardiner, et al. (1998). "Expression of HoxD genes in developing and regenerating axolotl limbs." *Dev. Biol.* **200**: 225–233.

Towers, M., J. Signolet, et al. (2011). "Insights into bird wing evolution and digit specification from polarizing region fate maps." *Nat. Commun.* **2**: 426.

Tsai, W. C., C. S. Kuoh, et al. (2004). "Four DEF-like MADS box genes displayed distinct floral morphogenetic roles in Phalaenopsis orchid." *Plant Cell Physiol.* **45**(7): 831–844.

Uejima, A., T. Amano, et al. (2010). "Anterior shift in gene expression precedes anteriormost digit formation in amniote limbs." *Dev. Growth. Diff.* **52**: 223–234.

Valentine, J. W., A. G. Collins, et al. (1994). "Morphological complexity increase in metazoans." *Paleobiology* **20**: 131–142.

Van de Peer, Y., J. A. Fawcett, et al. (2009). "The flowering world: a tale of duplications." *Trends Plant Sci.* **14**(12): 680–688.

Vanderpoele, K., W. D. Vos, et al. (2004). "Major events in the genome evolution of vertebrates: paranome age and size differ considerably between ray-finned fishes and land vertebrates." *Proc. Natl. Acad. Sci., USA* **101**: 1638–1643.

van Eeden, F. J., S. A. Holley, et al. (1998). "Zebrafish segmentation and pair-rule patterning." *Dev. Genet.* **23**(1): 65–76.

VanValen, L. (1965). "The study of morphological integration." *Evolution* **19**: 237–249.

VanValen, L. (1982). "Homology and causes." *J. Morphol.* **173**: 305–312.

Vargas, A. O. and J. F. Fallon (2005a). "Birds have dinosaur wings: the molecular evidence." *J. Exp. Zool. B. Mol. Dev. Evol.* **304B**: 86–90.

Vargas, A. O. and J. F. Fallon (2005b). "The digits of the wing of birds are 1, 2, and 3. A review." *J. Exp. Zool. B. Mol. Dev. Evol.* **304B**: (206–219).

Vargas, A. O., T. Kohlsdorf, et al. (2008). "The evolution of HoxD-11 expression in the bird wing: insights form Alligator mississippiensis." *PLoS ONE* **3**(10): e3325.

Vargas, A. O. and G. P. Wagner (2009). "Frame-shift of digit identity in bird evolution and Cyclopamine-treated wings." *Evol. Dev.* **11**(2): 163–169.

Vickaryous, M. K. and B. K. Hall (2006). "Human cell-type diversity, evolution, development, and classification with special reference to cells derived from the neural crest." *Biol. Rev. Camb. Philos. Soc.* **81**(3): 425–455.

Vincent, C., M. Bontoux, et al. (2003). "Msx genes are expressed in the carapacial ridge of turtle shell: a study of the European pond turtle, Emys orbicularis." *Dev. Genes Evol* **213**(9): 464–469.

Virta, V. C. and M. S. Cooper (2009). "Ontogeny and phylogeny of the yolk extension in embryonic cypriniform fishes." *J. Exp. Zool. B. Mol. Dev. Evol.* **312B**(3): 196–223.

Virta, V. C. and M. S. Cooper (2011). "Structural components and morphogenetic mechanics of the zebrafish yolk extension, a developmental module." *J. Exp. Zool. B. Mol. Dev. Evol.* **316**(1): 76–92.

Vorobyeva, E. I. and J. R. Hinchliffe (1996). "Developmental pattern and morphology of Salamandrella keyserlingii limbs (Amphibia, Hynobiidae) including some evolutionary aspects." *Russ. J. Herpetol.* **3**: 68–81.

Waddington, C. H. (1942). "Canalization of development and the inheritance of acquired characters." *Nature* **150**: 563–565.

Waddington, C. H. (1957). The Strategy of the Genes. London, George Allen & Unwin.

Wagner, A. (1996). "Does evolutionary plasticity evolve?" *Evolution* **50**: 1008–1023.

Wagner, A. (2005). Robustness and Evolvability in Living Systems. Princeton, NJ, Princeton University Press.

Wagner, A. (2007). "Robustness and evolvability: a paradox solved." *Proc. Roy. Soc. B* **275**: 91–100.

Wagner, A. (2011). The Origin of Evolutionary Innovations. Oxford, Oxford University Press.

Wagner, A. and P. F. Stadler (1999). "Evolved robustness in RNA secondary structure." *J. Exp. Zool. B. Mol. Dev. Evol.* **285**: 119–133.

Wagner, G. P. (1989). "The biological homology concept." *Ann. Rev. Ecol. Syst.* **20**: 51–69.

Wagner, G. P. (1989). "The origin of morphological characters and the biological basis of homology." *Evolution* **43**(6): 1157–1171.

Wagner, G. P. (1994). "Homology and the Mechanisms of Development," in Homology: The Hierarchical Basis of Comparative Biology, ed. B. K. Hall. San Diego, Academic Press.

Wagner, G. P. (1995). "The biological role of homologues: a building block hypothesis." *N. Jb. Geol. Paläont. Abh.* **195**: 279–288.

Wagner, G. P. (1996). "Homologues, natural kinds and the evolution of modularity." *Amer. Zool.* **36**: 36–43.

Wagner, G. P. (2001). "What is the promise of developmental evolution? Part II: a causal explanation of evolutionary innovations may be impossible." *J. Exp. Zool. B. Mol. Dev. Evol.* **291**: 305–309.

Wagner, G. P. (2005). "The developmental evolution of avian digit homology: an update." *Theor. Biosci.* **124**: 165–183.

Wagner, G. P. (2007). "The developmental genetics of homology." *Nat. Rev. Gen.* **8**: 473–479.

Wagner, G. P. (2010). "Evolvability: the missing piece in the Neo-Darwinian Synthesis," in Evolution Since Darwin: The First 150 Years, eds. M. A. Bell, D. J. Futuyma, W. F. Eanes and J. S. Levinton. Sunderland, MA, Sinauer Associates: 197–213.

Wagner, G. P. and L. Altenberg (1996). "Complex adaptations and the evolution of evolvability." *Evolution* **50**(3): 967–976.

Wagner, G. P., G. Booth, et al. (1997). "A population genetic theory of canalization." *Evolution* **51**(2): 329–347.

Wagner, G. P., C. Chiu, et al. (2000). "Developmental evolution as a mechanistic science: the inference from developmental mechanisms to evolutionary processes." *Am. Zool.* **40**: 819–831.

Wagner, G. P. and C. Chiu (2001). "The tetrapod limb: a hypothesis on its origin." *J. Exp. Zool. B. Mol. Dev. Evol.* **291**: 226–240.

Wagner, G. P. and J. Draghi (2010). "Evolution of evolvablity," in Evolution—The Extended Synthesis, eds. M. Pigliucci and G. B. Müller. Boston, MA, MIT Press: 378–399.

Wagner, G. P. and J. A. Gauthier (1999). "1,2,3=2,3,4: A solution to the problem of the homology of the digits in the avian hand." *Proc. Natl. Acad. Sci., USA* **96**: 5111–5116.

Wagner, G. P., J. P. Kenney-Hunt, et al. (2008). "Pleiotropic scaling of gene effects and the 'cost of complexity.'" *Nature* **452**: 470–472.

Wagner, G. P., P. A. Khan, et al. (1999). "Evolution of Hoxa-11 expression in amphibians: is the urodele autopodium an innovation?" *Am. Zool.* **39**: 686–694.

Wagner, G. P. and H. C. E. Larsson (2007). "Fins and limbs in the study of evolutionary novelties," in Fins into Limbs: Evolution, Development, and Transformation, ed. B. K. Hall. Chicago and London, University of Chicago Press: 49–61.

Wagner, G. P. and M. D. Laubichler (2000). "Character identification: the role of the organism." *Theory in Biosci.* **119**: 20–40.

Wagner, G. P. and M. D. Laubichler (2004). "Rupert Riedl and the re-synthesis of evolutionary and developmental biology: body plans and evolvability." *J. Exp. Zool. B. Mol. Dev. Evol.* **302B**: 92–102.

Wagner, G. P. and V. J. Lynch (2005). "Molecular evolution of evolutionary novelties: the vagina and uterus of therian mammals." *J. Exp. Zool. B. Mol. Dev. Evol.* **304B**: 580–592.

Wagner, G. P. and V. J. Lynch (2008). "The gene regulatory logic of transcription factor evolution." *Trends Ecol. Evol.* **23**(7): 377–385.

Wagner, G. P. and V. J. Lynch (2010). "Evolutionary novelties." *Curr. Biol.* **20**(2): R48–52.

Wagner, G. P., J. Mezey, et al. (2005). "Natural selection and the origin of modules," in Modularity: Understanding the Development and Evolution of Natural Complex Systems, eds. W. Callebaut and D. Rasskin-Gutman. Cambridge MA, MIT Press: 33–49.

Wagner, G. P. and B. Y. Misof (1993). "How can a character be developmentally constrained despite variation in developmental pathways?" *J. Evol. Biol.* **6**: 449–455.

Wagner, G. P. and G. B. Müller (2002). "Evolutionary innovations overcome ancestral constraints: a re-examination of character evolution in male sepsid flies (Diptera: Sepsidae)." *Evol. & Dev.* **4**(1): 1–6.

Wagner, G. P., M. Pavlicev, et al. (2007). "The road to modularity." *Nat. Rev. Genetics* **8**: 921–931.

Wake, D. B. (1999). "Homoplasy, homology and the problem of 'sameness' in biology." *Novartis Found Symp* **222**: 24–33; discussion 33–46.

Wake, D. B. (2003). "Homology and homoplasy," in Keywords and Concepts in Evolutionary Developmental Biology, eds. B. K. Hall and W. M. Olson. Cambridge, MA, Harvard University Press: 191–201.

Wake, D. B. and N. Shubin (1994). "Urodele limb development in relation to phylogeny and life history." *J. Morphol.* **220**(3): 407–408.

Wang, Y. Q., R. Melzer, et al. (2010). "Molecular interactions of orthologues of floral homeotic proteins from the gymnosperm Gnetum gnemon provide a clue to the evolutionary origin of 'floral quartets.'" *Plant J.* **64**(2): 177–190.

Wang, Z., R. L. Young, et al. (2011). "Transcriptomic analysis of avian digits reveals conserved and derived digit identities in birds." *Nature* **477**(7366): 583–586.

Wang, T., J. Zeng, et al. (2007). "Species-specific endogenous retroviruses shape the transcriptional network of the human tumor suppressor protein p53." *Proc. Natl. Acad. Sci., USA* **104**(47): 18613–18618.

Wardle, F. C. and H. L. Sive (2003). "What's your position? the Xenopus cement gland as a paradigm of regional specification." *Bioessays* **25**(7): 717–726.

Warren, R. W., L. Nagy, et al. (1994). "Evolution of homeotic gene regulation and function in flies and butterflies." *Nature* **372**: 458–461.

Watson, D. M. (1913). "On the primitive tetrapod limb." *Anat. Anzeiger* **44**: 24–27.

Wedeen, C. J. and D. A. Weisblat (1991). "Segmental expression of an engrailed-class gene during early development and neurogenesis in an annelid." *Development* **113**(3): 805–814.

Weisrock, D. W., L. J. Harmon, et al. (2005). "Resolving deep phylogenetic relationships in salamanders: analyses of mitochondrial and nuclear genomic data." *Syst. Biol.* **54**(5): 758–777.

Weiss, K. M. (1990). "Duplication with variation: Metameric logic in evolution from genes to morphology." *Yearbook of Physical Anthropology* **33**: 1–23.

Weissengruber, G. E., G. F. Egger, et al. (2006). "The structure of the cushions in the feet of African elephants (Loxodonta africana)." *J. Anat.* **209**(6): 781–792.

Welscher, P. T., A. Zuniga, et al. (2002). "Progression of vertebrate limb development through SHH-mediated counteraction of GLI3." *Science* **298**: 827–830.

Welten, M. C. M., F. J. Verbeek, et al. (2005). "Gene expression and digit homology in the chicken embryo wing." *Evol. Dev.* **7**: 18–28.

Werdelin, L. (1987). "Supernumerary teeth in Lynx lynx and the irreversibility of evolution." *J. Zool. Lond.* **211**: 259–266.

West-Eberhard, M. J. (2003). Developmental Plasticity and Evolution. Oxford, Oxford University Press.

Whipple, C. J., P. Ciceri, et al. (2004). "Conservation of B-class floral homeotic gene function between maize and Arabidopsis." *Development* **131**(24): 6083–6091.

Whipple, C. J., M. J. Zanis, et al. (2007). "Conservation of B-class gene expression in the second whorl of a basal grass and outgroups links the origin of lodicules and petals." *Proc. Natl. Acad. Sci., USA* **104**(3): 1081–1086.

Whitear, M. (1986). "The skin of fishes including cyclostomes: epidermis," in Biology of the Integument, eds. J. Bereiter-Hahn, A. G. Matoltsy and K. S. Richards. Heidelberg, Springer. **2**: 8–38.

Whiting, M. F., S. Bradler, et al. (2003). "Loss and recovery of wings in stick insects." *Nature* **421**(6920): 264–267.

Whittle, A. C. and D. W. Golding (1974). "The fine structure of prostomial photoreceptors in Eulalia viridis (Polychaeta; Annelida)." *Cell Tissue Res.* **154**(3): 379–398.

Widelitz, R. B., T. X. Jiang, et al. (2000). "Beta-catenin in epithelial morphogenesis: Conversion of part of avian foot scales into feather buds with beta-catenin." *Dev. Biol.* **219**: 98–114.

Widelitz, R. B., T. X. Jiang, et al. (2003). "Molecular biology of feather morphogenesis: a testable model for evo-devo research." *J. Exp. Zool. B. Mol. Dev. Evol.* **298**(1): 109–122.

Wienholds, E., W. P. Kloosterman, et al. (2005). "MicroRNA expression in zebrafish embryonic development." *Science* **309**(5732): 310–311.

Wiens, J. J. (2011). "Re-evolution of lost mandibular teeth in frogs after more than 200 million years, and re-evaluating Dollo's law." *Evolution* **65**(5): 1283–1296.

Wiens, J., R. Bonett, et al. (2005). "Ontogeny discombobulates phylogeny: paedomorphosis and higher-level salamander relationships." *Syst. Biol.* **54**(1): 91–110.

Wiens, J. J., M. C. Brandley, et al. (2006). "Why does a trait evolve multiple times within a clade? Repeated evolution of snakelike body form in squamate reptiles." *Evolution* **60**(1): 123–141.

Wiens, J. J. and J. L. Slingluff (2001). "How lizards turn into snakes: a phylogenetic analysis of body-form evolution in anguid lizards." *Evolution* **55**(11): 2303–2318.

Wiley, E. O. (1975). "Karl R. Popper, systematics and classification: a reply to Walter Bock and other evolutionary taxonomists." *Syst. Zool.* **24**: 233–243.

Wiley, E. O. (2007). "Homology, identity and transformation," in Mesozoic Fishes, eds. G. Arrtia and H.-P. Schultze. Munchen, Germany, Verlag Dr F Pfeil.

Wiley, E. O. (2008). "Homology, identity and transformation." Mesozoic Fishes 4: Homology and Phylogeny. G. Arrtia, H.-P. Schultze and M. V. H. Wilson. Munchen, Germany, Verlag Dr F Pfeil: 9–21.

Wilke, C., J. L. Wang, et al. (2001). "Evolution of digital organisms at high mutation rates leads to survival of the flattest." *Nature* **412**: 331–333.

Wilkins, A. S. (2002). The Evolution of Developmental Pathways. Sunderland, MA, Sinauer Associates.

Williams, G. C. (1966). Adaptation and Natural Selection. Princeton, Princeton University Press.

Williams, G. C. (1992). Natural Selection: Domains, Levels and Challenges. Oxford, Oxford University Press.

Williams, T. A. (1994). "The nauplius larva of crustaceans: functional diversity and the phylotypic stage." *Amer. Zool.* **34**(4): 562–569.

Wilson, R. A., M. J. Barker, et al. (2007). "When traditional essentialism fails: biological natural kinds." *Philosophical Topics* **35**: 189–215.

Winter, K. U., C. Weiser, et al. (2002). "Evolution of class B floral homeotic proteins: obligate heterodimerization originated from homodimerization." *Mol. Biol. Evol.* **19**(5): 587–596.

Witten, F. E. and A. Huysseune (2007). "Mechanisms of chondrogenesis and osteogenesis in fins," in Fins into Limbs: Evolution, Development and Transformation, ed. B. K. Hall. Chicago and London, University of Chicago Press: 79–92.

Wittkopp, P. J., J. R. True, et al. (2002). "Reciprocal functions of the Drosophila yellow and ebony proteins in the development and evolution of pigment patterns." *Development* **129**(8): 1849–1858.

Wittkopp, P. J., K. Vaccaro, et al. (2002). "Evolution of yellow gene regulation and pigmentation in Drosophila." *Curr. Biol.* **12**(18): 1547–1556.

Wolk, C. P. (1996). "Heterocyst formation." *Annu. Rev. Genet.* **30**: 59–78.

Woltering, J. M. and D. Duboule (2010). "The origin of digits: expression patterns versus regulatory mechanisms." *Dev. Cell* **18**(4): 526–532.

Wray, G. A. and E. Abouheif (1998). "When is homology not homology?" *Curr. Opin. Genet. Devel.* **8**: 675–680.

Wu, K. C., J. Streicher, et al. (2001). "Role of motility in embryonic development I: Embryo movements and amnion contractions in the chick and the influence of illumination." *J. Exp. Zool.* **291**(2): 186–194.

Wu, P., L. Hou, et al. (2004). "Evo-devo of amniote integuments and appendages." *Int. J. Dev. Biol.* **48**(2–3): 249–270.

Xu, X., M. A. Norell, et al. (2004). "Basal tyrannosauroids from China and evidence for protofeathers in tyrannosauroids." *Nature* **431**(7009): 680–684.

Xu, Y., L. L. Teo, et al. (2006). "Floral organ identity genes in the orchid Dendrobium crumenatum." *Plant J.* **46**(1): 54–68.

Ye, C., X. Wu, et al. (2010). "Y-keratins in crocodiles reveal amino acid homology with avian keratins." *Mol. Biol. Rep.* **37**: 1169–1174.

Yokouchi, Y., S. Nakazato, et al. (1995). "Misexpression of Hoxa-13 induces cartilage homeotic transformation and changes cell adhesiveness in chick limb buds." *Genes Dev.* **9**: 2509–2522.

Yonei-Tamura, S., G. Abe, et al. (2008). "Competent stripes for diverse positions of limbs/fins in gnathostome embryos." *Evol. Dev.* **10**(6): 737–745.

Young, E. T., J. S. Sloan, et al. (2000). "Trinucleotide Repeats Are Clustered in Regulatory Genes in Saccharomyces cerevisiae." *Genetics* **154**(3): 1053–1068.

Young, N. M. and B. Hallgrímsson (2005). "Serial homology and the evolution of mammalian limb co-variation structure." *Evolution* **59**: 2691–2704.

Young, N. M., G. P. Wagner, et al. (2010). "Development and the evolvability of human limbs." *Proc. Natl. Acad. Sci., USA* **107**: 3400–3405.

Young, R. L., G. S. Bever, et al. (2011). "Identity of the avian wing digits: problems resolved and unsolved." *Dev. Dyn.* **240**(5): 1042–1053.

Young, R. L., V. Caputo, et al. (2009). "Evolution of digit identity in the three-toed skink Chalcides chalcides: a new case of digit identity frame shift." *Evol. Dev.* **11**(6): 647–658.

Yu, M., P. Wu, et al. (2002). "The morphogenesis of feathers." *Nature* **420**: 308–312.

Yunick, R. P. (1970). "An examination of certain aging and sexing criteria for the cedar waxwing (Bombycilla cedorum)." *Bird-Banding* **41**(4): 291–299.

Zanis, M. J., P. S. Soltis, et al. (2003). "Phylogenetic analysis and perianth evolution in basal angiosperms." *Ann. Missouri Bot. Grad.* **90**(2): 129–150.

Zardoya, R. and A. Meyer (2001). "On the origin of and phylogenetic relationships among living amphibians." *Pro. Natl. Acad. Sci., USA* **98**: 7380–7383.

Zhao, X., M. Sun, et al. (2007). "Mutations in HOXD13 underlie syndactyly type V and a novel brachydactyly-syndactyly syndrome." *Am. J. Human Genetics* **80**(2): 361–371.

Zuber, M. E., G. Gestri, et al. (2003). "Specification of the vertebrate eye by a network of eye field transcription factors." *Development* **130**: 5155–5167.

INDEX